Vegetable Genetics and Genomics

Vegetable Genetics and Genomics

Guest Editors

Xixiang Li
Yunsong Lai

MDPI

Basel • Beijing • Wuhan • Barcelona • Belgrade • Novi Sad • Cluj • Manchester

Guest Editors
Xixiang Li
Institute of Vegetables and Flowers
Chinese Academy of Agricultural Sciences
Beijing
China

Yunsong Lai
College of Horticulture
Sichuan Agricultural University
Chengdu
China

Editorial Office
MDPI AG
Grosspeteranlage 5
4052 Basel, Switzerland

This is a reprint of the Special Issue, published open access by the journal *International Journal of Molecular Sciences* (ISSN 1422-0067), freely accessible at: https://www.mdpi.com/journal/ijms/special_issues/Vegetable_Genetics.

For citation purposes, cite each article independently as indicated on the article page online and as indicated below:

Lastname, A.A.; Lastname, B.B. Article Title. *Journal Name* **Year**, *Volume Number*, Page Range.

ISBN 978-3-7258-5375-5 (Hbk)
ISBN 978-3-7258-5376-2 (PDF)
https://doi.org/10.3390/books978-3-7258-5376-2

Cover image courtesy of Yunsong Lai

Contents

International Journal of *Molecular Sciences*

MDPI

Editorial

The Development and Application of Vegetable Genomics Increase the Efficiency of Exploring New Gene Resources for Vegetables

Xi-Xiang Li [1,*] and Yun-Song Lai [2,*]

1 State Key Laboratory of Vegetable Biobreeding, Institute of Vegetables and Flowers, Chinese Academy of Agricultural Sciences, Beijing 100081, China
2 College of Horticulture, Sichuan Agricultural University, Chengdu 611100, China
* Correspondence: lixixiang@caas.cn (X.-X.L.); laiys@sicau.edu.cn (Y.-S.L.)

Vegetables, as indispensable non-staple foods in people's daily diet, provide a variety of essential vitamins, minerals, and other nutrients, as well as special phytochemicals, which are recognized as functional components for human nutritional balance or medicinal purposes. To effectively improve the quality and production performance of vegetables, global researchers have been working hard to understand the genetic bases and regulation mechanisms of important horticultural traits and biotic/abiotic stress tolerance.

The emergence and application of Next-Generation Sequencing (NGS) at the beginning of the 21st century, and subsequently the development of the third-generation sequencing technology represented by the platforms of Oxford nanopore sequencing (ONS) and PacBio sequencing that generate extra-long reads and therefore make genome/mRNA/ncRNA assembly much easier, greatly promote the advances in genetics and genomics of vegetable crops [1,2]. Cucumber is the first vegetable crop with a released genome whose genomic sequencing was accomplished in 2009, and henceforth, more than 30 vegetable crops have been genome-released [3]. To meet the demand of today's bioinformatic analysis, the released genomes keep updating, e.g., hot pepper [4], tomato [5], eggplant [6–8], cucumber [9], water melon [10], and melon [11], by using single molecule long reads and high-throughput chromosome conformation capture (Hi-C).

In order to follow up on the frontier research of vegetable crops, we launched this Special Issue, "Vegetable Genetics and Genomics", at the end of 2021, which aimed to recruit both original research articles and review papers that target horticultural traits and reveal the underlying genes and genetic mechanisms by using advanced technologies of molecular genetics and multiple omics.

This Special Issue attracted researchers from 25 institutes all over the world who contributed 12 research articles and 1 review article. There are six articles about Solanaceae plants, three articles about Cruciferae plants, two articles about Cucurbitaceae plants, and one article each about garlic leaks and olives. Basically, common vegetables are of much wider research interest. The flower stalks and inflorescence of rapeseed are a very popular vegetable in East Asian countries, and processed olives are taken as a vegetable in Europe countries. Therefore, we included these two articles in this Special Issue.

Comparative transcriptome analysis is widely used to profile the expression change of gene sets and therefore identify the candidate genes controlling the target phenotypes. The wax gourd (*Benincasa hispida*) is an important Asian fruit vegetable crop. Luo et al. identified 68 small auxin-upregulated RNA (SAUR) genes in gourd and found *BhSAUR60* displayed distinct expression patterns and might play a role in fruit development [12]. Potatoes have differential performance in tuberization ability, which can be quantified by MS medium culture. Valencia-Lozano et al. found the regulated genes involved in potato microtuber development by comparing two potatoes with extreme tuberization traits; they constructed a protein–protein interaction network of related transcription factors involving light and photosynthesis, hormones, and hormone-mediated signaling pathways [13]. Olive is a fruit and vegetable crop originated in Mediterranean coast regions, and a Spanish team profiled the transcriptome change at the anthesis stage and six fruit-development stages. Combining the cytological and hormonal data, they discovered the mutual coordination of candidate transcription factors, cell cycle-regulating genes, plant hormone regulators, cell wall biosynthesis, and signaling genes in olive fruit development [14]. This article is a good example of making full use of multi-faceted supportive data.

Integrated transcriptome, metabolome, and other analytical strategies were also used to dissect certain physiological phenomena or biochemical metabolic pathways that may be involved in vegetable product quality. Radish is known for its accumulation of beneficial glucosinolates (GSLs). Li et al. profiled the differential expression of GLS pathway genes between low and high GSL content radishes and proposed the central role of MYB28 in GLS biosynthesis [15]. Zhou et al. compared five differentially colored eggplants (anthocyanin pigmented fruits, green fruits, and white fruits) and indicated the activation of early phenylpropanoid biosynthesis genes (*SmPAL*, *SmC4H*, and *Sm4CL*) was responsible for anthocyanin accumulation, while *SmF3′5′H* was the key factor for the formation of a purple color. Two transcription factors, *SmGL2* and *SmGATA26*, were also associated with anthocyanin accumulation [16]. Villa-Rivera et al. compared the expression changes of carotenoid biosynthesis genes between the fruits of 12 chili pepper accessions during the development and ripening processes. They identified the relevant transcription factor for carotenoid biosynthesis by coexpression analysis [17]. By combining comparative genomics and expression pattern analysis with VIGS, subcellular localization, yeast one-hybrid, and dual-luciferase reporter assays, Liu et al. systematically identified members of the CaMYB-related family, predicted their possible biological functions, and revealed that CaMYB37 is critical for the transcriptional regulation of capsaicin biosynthesis in peppers [18]. Coupling RNA-seq and metabolomes, Xia et al. identified garlic odor compounds, free amino acids, and sugars from wild Chinese chives and compared them with cultivated Chinese chives. They cloned the structural genes of garlic odor compounds in wild Chinese chives and found that a deletion in the FMO gene resulted in an impaired protein sequence and the subsequently changed garlic odor compound [19].

Resistance to biotic and abiotic stresses is a major vegetable breeding target trait. Xiao et al. constructed a ChIP-Seq library to profile the transcriptionally targeted genes of Teosinte branched 1/cycloidea/proliferating cell factor (TCP) during the infection of bacterial wilt in eggplant and showed that SmTCP7a positively regulated bacterial wilt that was caused by *Ralstonia solanacearum* [20]. Yan et al. identified 65 glyoxalase (GLY) genes in the rapeseed genome and clarified its possible regulation role in freezing tolerance in Arabidopsis [21]. Compared with molecular markers used in conventional gene/QTL mapping, genomic sequencing-based genotyping is a more convenient and effective alternative. Liu et al. conducted Genome-Wide Association Studies (GWASs) by re-sequencing 220 cucumber core accessions and found that a total of seven loci were associated with salt tolerance in cucumber seedlings and finally predicted five candidate genes controlling salt tolerance [22].

Cytoplasmic male sterility (CMS) is widely used to produce hybrid seeds in vegetable crops, and great efforts have been made to identify male sterility and restorer genes. Ogura is a well-known CMS source found in Japanese radishes, but now it has been transferred into many other Cruciferae crops. Ren et al. well summarized the explosion and utilization of Ogura CMS sources in Cruciferae crops [23]. This is the only review article included in this Special Issue. One research article identified a new candidate gene for restorer-of-fertility (Rf) in capsicum pepper. The single dominant gene $CaRf_{HZ}$ was mapped in a QTL of 533 kb through bulked segregant analysis (BSA)-sequencing [24]. Male sterility is widely used to produce hybrid F1 seeds, and these studies contribute to exploring and developing new male sterility resources based on conventional ones.

To sum up, the articles compiled in this Special Issue provide valuable and novel knowledge of the genetic and molecular mechanisms behind certain important traits of vegetable crops. Nevertheless, further laboratory experiments should be carried out in the future, which can provide more biological evidence to confirm the proposed gene function. In addition, there are hundreds of vegetables on the market, and vegetable science has an interesting scope of thousands of plant species, which is much more than other agricultural research fields like crop science, pomology, ornamental plants, and so on. A lot of vegetable crops have a lot of interesting agronomic traits that need to be clarified. It is worthwhile to expect that the application of advanced genetic/genomic strategies and technology can accelerate the process. We have the support of the editorial office to launch a new Special Issue focusing on the same research theme, entitled "Vegetable Genetics and Genomics 2.0". We will make a continuous effort on this topic and sincerely invite researchers from all over the world to contribute their latest research results.

Funding: This research was supported by grants from the National Key Research and Development Program of China (2016YFD0100204-02) and Natural Science Foundation of China (31772303 and 32072599).

Conflicts of Interest: The authors declare no conflict of interest.

References

1. Rhoads, A.; Au, K.F. PacBio Sequencing and Its Applications. *Genom. Proteom. Bioinform.* **2015**, *13*, 278–289. [CrossRef] [PubMed]
2. Wang, Y.; Zhao, Y.; Bollas, A.; Wang, Y.; Au, K.F. Nanopore sequencing technology, bioinformatics and applications. *Nat. Biotechnol.* **2021**, *39*, 1348–1365. [CrossRef] [PubMed]
3. Huang, S.; Li, R.; Zhang, Z.; Li, L.; Gu, X.; Fan, W.; Lucas, W.J.; Wang, X.; Xie, B.; Ni, P.; et al. The genome of the cucumber, *Cucumis sativus* L. *Nat. Genet.* **2009**, *41*, 1275–1281. [CrossRef] [PubMed]
4. Hulse-Kemp, A.M.; Maheshwari, S.; Stoffel, K.; Hill, T.A.; Jaffe, D.; Williams, S.R.; Weisenfeld, N.; Ramakrishnan, S.; Kumar, V.; Shah, P.; et al. Reference quality assembly of the 3.5-Gb genome of Capsicum annuum from a single linked-read library. *Hortic. Res.* **2018**, *5*, 4. [CrossRef] [PubMed]
5. Hosmani, P.S.; Flores-Gonzalez, M.; van de Geest, H.; Maumus, F.; Bakker, L.V.; Schijlen, E.; van Haarst, J.; Cordewener, J.; Sanchez-Perez, G.; Peters, S.; et al. An improved de novo assembly and annotation of the tomato reference genome using single-molecule sequencing, HiC proximity ligation and optical maps. *bioRxiv* **2019**. [CrossRef]
6. Barchi, L.; Pietrella, M.; Venturini, L.; Minio, A.; Toppino, L.; Acquadro, A.; Andolfo, G.; Aprea, G.; Avanzato, C.; Bassolino, L.; et al. A chromosome-anchored eggplant genome sequence reveals key events in Solanaceae evolution. *Sci. Rep.* **2019**, *9*, 11769. [CrossRef] [PubMed]
7. Li, D.; Qian, J.; Li, W.; Jiang, Y.; Gan, G.; Li, W.; Chen, R.; Yu, N.; Li, Y.; Wu, Y.; et al. Genome sequence and analysis of the eggplant (*Solanum melongena* L.). *bioRxiv* **2019**. [CrossRef]
8. Wei, Q.; Wang, J.; Wang, W.; Hu, T.; Hu, H.; Bao, C. A high-quality chromosome-level genome assembly reveals genetics for important traits in eggplant. *Hortic. Res.* **2020**, *7*, 153. [CrossRef]
9. Li, Q.; Li, H.; Huang, W.; Xu, Y.; Zhou, Q.; Wang, S.; Ruan, J.; Huang, S.; Zhang, Z. A chromo-some-scale genome assembly of cucumber (*Cucumis sativus* L.). *GigaScience* **2019**, *8*, giz072. [CrossRef]
10. Guo, S.; Zhao, S.; Sun, H.; Wang, X.; Wu, S.; Lin, T.; Ren, Y.; Gao, L.; Deng, Y.; Zhang, J.; et al. Resequencing of 414 cultivated and wild watermelon accessions identifies selection for fruit quality traits. *Nat. Genet.* **2019**, *51*, 1616–1623. [CrossRef]
11. Zhang, H.; Li, X.; Yu, H.; Zhang, Y.; Li, M.; Wang, H.; Wang, D.; Wang, H.; Fu, Q.; Liu, M.; et al. A high-quality melon genome assembly provides insights into genetic basis of fruit trait improvement. *iScience* **2019**, *22*, 16–27. [CrossRef] [PubMed]
12. Luo, C.; Yan, J.; He, C.; Liu, W.; Xie, D.; Jiang, B. Genome-Wide Identification of the *SAUR* Gene Family in Wax Gourd (*Benincasa hispida*) and Functional Characterization of *BhSAUR60* during Fruit Development. *Int. J. Mol. Sci.* **2022**, *23*, 14021. [CrossRef] [PubMed]
13. Valencia-Lozano, E.; Herrera-Isidrón, L.; Flores-López, J.A.; Recoder-Meléndez, O.S.; Barraza, A.; Cabrera-Ponce, J.L. *Solanum tuberosum* Microtuber Development under Darkness Unveiled through RNAseq Transcriptomic Analysis. *Int. J. Mol. Sci.* **2022**, *23*, 13835. [CrossRef] [PubMed]
14. Camarero, M.C.; Briegas, B.; Corbacho, J.; Labrador, J.; Gallardo, M.; Gomez-Jimenez, M.C. Characterization of Transcriptome Dynamics during Early Fruit Development in Olive (*Olea europaea* L.). *Int. J. Mol. Sci.* **2023**, *24*, 961. [CrossRef] [PubMed]
15. Li, X.; Wang, P.; Wang, J.; Wang, H.; Liu, T.; Zhang, X.; Song, J.; Yang, W.; Wu, C.; Yang, H.; et al. A Comparative Transcriptome and Metabolome Combined Analysis Reveals the Key Genes and Their Regulatory Model Responsible for Glucoraphasatin Accumulation in Radish Fleshy Taproots. *Int. J. Mol. Sci.* **2022**, *23*, 2953. [CrossRef] [PubMed]
16. Zhou, X.; Liu, S.; Yang, Y.; Liu, J.; Zhuang, Y. Integrated Metabolome and Transcriptome Analysis Reveals a Regulatory Network of Fruit Peel Pigmentation in Eggplant (*Solanum melongena* L.). *Int. J. Mol. Sci.* **2022**, *23*, 13475. [CrossRef] [PubMed]

17. Villa-Rivera, M.G.; Martínez, O.; Ochoa-Alejo, N. Putative Transcription Factor Genes Associated with Regulation of Carotenoid Biosynthesis in Chili Pepper Fruits Revealed by RNA-Seq Coexpression Analysis. *Int. J. Mol. Sci.* **2022**, *23*, 11774. [CrossRef] [PubMed]
18. Liu, Y.; Zhang, Z.; Fang, K.; Shan, Q.; He, L.; Dai, X.; Zou, X.; Liu, F. Genome-Wide Analysis of the MYB-Related Transcription Factor Family in Pepper and Functional Studies of *CaMYB37* Involvement in Capsaicin Biosynthesis. *Int. J. Mol. Sci.* **2022**, *23*, 11667. [CrossRef] [PubMed]
19. Xia, S.-W.; Hang, L.-F.; Ali, S.; Xu, X.-Y.; Liu, Y.-J.; Yan, Q.-Q.; Luo, Q.-Y.; Li, Y.; Lin, L.-J.; Li, H.-X.; et al. Biosynthesis and Metabolism of Garlic Odor Compounds in Cultivated Chinese Chives (*Allium tuberosum*) and Wild Chinese Chives (*Allium hookeri*). *Int. J. Mol. Sci.* **2022**, *23*, 7013. [CrossRef]
20. Xiao, X.; Lin, W.; Feng, E.; Wu, C.; Ou, X. Genome-Wide Identification of Binding Sites for SmTCP7a Transcription Factors of Eggplant during Bacterial Wilt Resistance by ChIP-Seq. *Int. J. Mol. Sci.* **2022**, *23*, 6844. [CrossRef]
21. Yan, G.; Zhang, M.; Guan, W.; Zhang, F.; Dai, W.; Yuan, L.; Gao, G.; Xu, K.; Chen, B.; Li, L.; et al. Genome-Wide Identification and Functional Characterization of Stress Related Glyoxalase Genes in *Brassica napus* L. *Int. J. Mol. Sci.* **2023**, *24*, 2130. [CrossRef] [PubMed]
22. Liu, D.; Dong, S.; Miao, H.; Liu, X.; Li, C.; Han, J.; Zhang, S.; Gu, X. A Large-Scale Genomic Association Analysis Identifies the Candidate Genes Regulating Salt Tolerance in Cucumber (*Cucumis sativus* L.) Seedlings. *Int. J. Mol. Sci.* **2022**, *23*, 8260. [CrossRef] [PubMed]
23. Ren, W.; Si, J.; Chen, L.; Fang, Z.; Zhuang, M.; Lv, H.; Wang, Y.; Ji, J.; Yu, H.; Zhang, Y. Mechanism and Utilization of Ogura Cytoplasmic Male Sterility in Cruciferae Crops. *Int. J. Mol. Sci.* **2022**, *23*, 9099. [CrossRef] [PubMed]
24. Nie, Z.; Song, Y.; Wang, H.; Chen, J.; Niu, Q.; Zhu, W. Fine Mapping and Gene Analysis of *restorer-of-fertility* Gene $CaRf_{HZ}$ in Pepper (*Capsicum annuum* L.). *Int. J. Mol. Sci.* **2022**, *23*, 7633. [CrossRef]

International Journal of
Molecular Sciences

MDPI

Article

Genome-Wide Identification and Functional Characterization of Stress Related Glyoxalase Genes in *Brassica napus* L.

Guixin Yan, Meili Zhang, Wenjie Guan, Fugui Zhang, Wenjun Dai, Lili Yuan, Guizhen Gao, Kun Xu, Biyun Chen, Lixia Li and Xiaoming Wu *

The Key Laboratory of Biology and Genetic Improvement of Oil Crops, The Ministry of Agriculture and Rural Affairs of the PRC, Oil Crops Research Institute of the Chinese Academy of Agricultural Sciences, Wuhan 430062, China

* Correspondence: wuxm@oilcrops.cn

Abstract: Rapeseed (*Brassica napus* L.) is not only one of the most important oil crops in the world, but it is also an important vegetable crop with a high value nutrients and metabolites. However, rapeseed is often severely damaged by adverse stresses, such as low temperature, pathogen infection and so on. Glyoxalase I (GLYI) and glyoxalase II (GLYII) are two enzymes responsible for the detoxification of a cytotoxic metabolite methylglyoxal (MG) into the nontoxic S-D-lactoylglutathione, which plays crucial roles in stress tolerance in plants. Considering the important roles of glyoxalases, the *GLY* gene families have been analyzed in higher plans, such as rice, soybean and Chinese cabbage; however, little is known about the presence, distribution, localizations and expression of glyoxalase genes in rapeseed, a young allotetraploid. In this study, a total of 35 *BnaGLYI* and 30 *BnaGLYII* genes were identified in the *B. napus* genome and were clustered into six and eight subfamilies, respectively. The classification, chromosomal distribution, gene structure and conserved motif were identified or predicted. BnaGLYI and BnaGLYII proteins were mainly localized in chloroplast and cytoplasm. By using publicly available RNA-seq data and a quantitative real-time PCR analysis (qRT-PCR), the expression profiling of these genes of different tissues was demonstrated in different developmental stages as well as under stresses. The results indicated that their expression profiles varied among different tissues. Some members are highly expressed in specific tissues, *BnaGLYI11* and *BnaGLYI27* expressed in flowers and germinating seed. At the same time, the two genes were significantly up-regulated under heat, cold and freezing stresses. Notably, a number of *BnaGLY* genes showed responses to *Plasmodiophora brassicae* infection. Overexpression of *BnGLYI11 gene* in *Arabidopsis thaliana* seedlings confirmed that this gene conferred freezing tolerance. This study provides insight of the *BnaGLYI* and *BnaGLYII* gene families in allotetraploid *B. napus* and their roles in stress resistance, and important information and gene resources for developing stress resistant vegetable and rapeseed oil.

Keywords: glyoxalase; *Brassica napus*; genome-wide analysis; expression; stress

1. Background

Rapeseed (*Brassica napus* L.) is one of the most important oil crops in the world. It not only provides high quality edible oil for human beings, but also is a new type of vegetable [1]. Low erucic acid and glucosinolate rapeseed has rich vitamins, fiber and minerals in its stems and leaves and tastes good, which shows great market prospect [2,3]. However, rapeseed is frequently subjected to adverse damage by stresses caused by low temperature, pathogen infection and so on, which not only seriously reduce the yield, but also threatens the health of animals and human beings when it enters the food chain [4]. Thus, exploring and utilizing stress-tolerant genes in rapeseed is of great theoretical and practical significance for vegetable and oilseed production.

Previous reports reveal that the dominant methylglyoxal (MG) concentration in plants can be increased under stress conditions compared with normal conditions [5]. The glyoxalase pathway is the dominant MG scavenging pathways of plants, which includes two enzymes, GLYI and GLYII [6]. GLYI catalyzes the MG conversion into S-D-lactoylglutathione in the presence of reduced glutathione (GSH) [7]. Subsequently, GLYII hydrolyzes S-D-lactoylglutathione into non-toxic D-lactic acid and GSH [8]. The depletion of GSH and the activity of GLYI lead to the toxicity of MG [9]. GLYII is the second rate-limiting enzyme in the system [10]. Lack of GLYII leads to the accumulation of lactoylglutathione [11]. Recently, the third enzyme GLYIII, found to convert MG directly into D-lactose, is far less efficient than the GLYI and GLYII enzymes [11].

In plants, the roles of GLY systems in resistance to biotic and abiotic stresses have been getting more attention recently. The role of *GLY* systems in resisting salt, extreme temperature and heavy metal stresses were reported [12–16]. *GLYI* and *GLYII* can be induced by stresses at transcription and translation levels. For example, in *Brassica juecea*, *GLYII* expression was significantly up-regulated under zinc stress [17]. The GLYI protein in rice roots at the seedling stage was significantly induced by low temperature stress [18]. The GLYI protein in tomato roots was significantly increased under heavy metal Al stress [19]. Similarly, GLYI and GLYII activities in onion were also induced by low temperature [20]. The level of transcript, protein and specific activity of GLYI was markedly enhanced under water stress [12]. In addition to abiotic stress responses, *GLY* genes are also regulated by biotic stresses. The expression of *GLYI* was induced in *Brassica* after infection by *Sclerotinia sclerotiorum* [21]. In *B. rapa*, *BraGLYI1*, *BraGLYI6*, *BraGLYI11* and *BraGLYI16* were up-regulated after inoculation of *Plasmodiophora brassicae* in both clubroot-resistant and susceptible lines of *B.rapa*. *BraGLYII13* was more highly induced in the resistant line than the susceptible line at 12 h after inoculation [22]. Overall, the enzyme system is considered as a biomarker of plant stress tolerance [23].

A high concentration of MG can impair plant cells or cell components and can even destroy DNA or cause mutation, resulting in the death of plant cells and tissues [16]. Overexpression or higher activity of glyoxalase enzymes eliminate MG toxicity and confer stress tolerance [23]. For example, overexpression of the *GLYI* gene in tobacco showed better stress tolerance than the normal plant to MG and high salinity [13]. The transgenic mustard overexpressing the *GLYI* gene enabled tolerance to salt, heavy metals and drought stresses [5]. Overexpressing of wheat's *GLYI* gene also enhanced the tobacco tolerance to $ZnCl_2$ [24]. Double *GLY* transgenic tomato plants (*BjGlyI* from *B. juncea* and *PgGlyII* from *Pennisetum glaucum*) showed improved salinity resistance, probably by decreasing the oxidative stress [15]. The transgenic tobacco overexpressing both *GLYI* and *GLYII* genes showed significant resistance to $ZnCl_2$ [25]. The overexpression of *Manihot esculenta GLYI-13* enhanced the growth ability of transgenic yeast under iron stress [26]. Therefore, overexpression of the glyoxalase pathway or individual genes harbors the potential to confer tolerance to multiple stresses.

Both *GLYI* and *GLYII* genes were present as multi-gene families. To date, genome-wide identification of the glyoxalase family has been carried out in *Arabidopsis thaliana*, *Oryza sativa*, *Glycine max*, *Medicago truncatula*, *Brassica rapa*, *B. oleracea*, *Vitis vinifera* and *Manihot esculenta. Crantz* [22,26–33]. There are at least 11 *GLYI* homologous genes in *A. thaliana* and rice, 24 *GLYI* members in the soybean genome and 15 in *B. rapa*. According to *GLYII* genes, five *GLYII* genes are in *A. thaliana*, three in rice, twelve *GLYII* members in the soybean genome, and sixteen *GLYII* in *B. rapa* [22,28]. Moreover, some *GLYI* and *GLYII* family genes have been proved to play an important role in response to drought, heavy metals and other stress responses [34–36]. Rapeseed (*B. napus*) is a young allotetraploid, which has more gene copied than its diploid progenitor *B. rapa* and *B. oleracea* [37]. In this study, the members of *GLYI* and *GLYII* families were identified from the *B. napus* database by bioinformatics analysis, and their genetic structure, evolutionary relationships, conserved motifs, chromosome localization, expression in different tissues and response to stresses were also studied. The study will provide an important basis for clarifying the evolutionary level and functional differentiation of *GLYI* and *GLYII* family genes in polyploidy crops and lay a theoretical foundation for further understanding the response mechanisms of stress in *Brassica*.

2. Results

2.1. Identification of GLYI and GLYII Genes in the B. napus Genome

Glyoxalase proteins (GLYI and GLYII) had been identified in Brassicaceae plant *A. thaliana* and *B. rapa*. Thus, *AtGLYI*/*AtGLYII* and *BraGLYI*/*BraGLYII* sequences were used as queries to blast the genome database of *B. napus*, and the sequences with an E-value under 1.0×10^{-10} were screened out. Then, proteins that included the glyoxalase domain (PF00903) and the metallo-beta-lactamase domain (PF00753) were classified as BnaGLYI and BnaGLYII proteins, respectively. In this study, 35 *BnaGLYI* and 30 *BnaGLYII* genes were identified in *B. napus*. Table 1 summarizes the gene ID, chromosomal distribution, number of introns, CDS and amino acid length, predicted sub-cellular localization, calculated molecular weight and isoelectric points of the putative *BnaGLYI* and *BnaGLYII*.

Table 1. Detail information of *BnaGLYI* genes identified in *Brassica napus* L. genomes.

Gene Symbol	Locus Identifier	Location	Gene Start (bp)	Gene Stop (bp)	Strand	No. of Introns	CDS Length (bp)	PP Length (aa)	MW (kDa)	pI	Localization
BnaGLYI01	BnaA02g19970D	A02	12337272	12338438	–	2	504	168	18,818.68	5.81	Cy [a]; Nu [b]
BnaGLYI02	BnaA03g10440D	A03	4700661	4701804	+	3	582	194	21,785.91	8.43	Cy [a]; Mt [b]
BnaGLYI03	BnaA05g35240D	A05	670044	672049	–	3	414	138	15,252.22	5.46	Cy [a]; Nu [b]
BnaGLYI04	BnaA06g04170D	A06	2547123	2548306	–	3	516	172	19,274.7	7.77	Cy [a]; Ch [b]
BnaGLYI05	BnaA06g04580D	A06	2708014	2710119	+	7	714	238	26,635.17	8.33	Cy [a]; Om [a]; Ch [b,c]
BnaGLYI06	BnaA06g07360D	A06	3915659	3923909	+	26	3648	1216	134,984.62	8.86	Cy [a,b]
BnaGLYI07	BnaA06g10060D	A06	5342612	5343786	+	2	525	175	19,778.79	5.68	Cy [a]; Ch [b]
BnaGLYI08	BnaA07g13890D	A07	12263110	12263994	+	0	555	185	20,887.37	4.78	Cy [a,b]
BnaGLYI09	BnaA07g26290D	A07	19358162	19361477	–	8	1026	342	37,894.44	6.48	Cy [a]; Ch [b,c]
BnaGLYI10	BnaA08g23870D	A08	16843513	16844540	–	2	525	175	19,846.96	5.88	Cy [a]; Ch [b]
BnaGLYI11	BnaA08g25110D	A08	17357805	17359871	–	7	852	284	31,872.41	5.26	Cy [a,b]
BnaGLYI12	BnaA09g49270D	A09	32814199	32815336	+	2	414	138	15,394.22	5.84	Cy [a]; Ch [b]
BnaGLYI13	BnaA09g49870D	A09	33093200	33095241	+	1	1323	441	48,072.12	5.72	Cy [a];Om [a];Cysk [b]
BnaGLYI14	BnaA09g56790D	A09	3889968	3891223	+	2	504	168	18,943.89	5.86	Cy [a]; Ch [b]
BnaGLYI15	BnaA10g04310D	A10	2272334	2273834	–	1	1338	446	48,973.02	5.45	Om [a]; Cy [b]
BnaGLYI16	BnaA10g11070D	A10	9346288	9347612	–	3	588	196	21,959	6.71	Cy [a]; Mt [b]
BnaGLYI17	BnaC02g23290D	C02	20281012	20282133	–	2	504	168	18,762.53	5.66	Cy [a]; Nu [b]
BnaGLYI18	BnaC02g46640D	C02	2394713	2395834	+	2	504	168	18,804.61	5.66	Cy [a]; Nu [b]
BnaGLYI19	BnaC03g13130D	C03	6313042	6314254	+	3	582	194	21,712.86	7.77	Cy [a]; Ch [b]
BnaGLYI20	BnaC03g51010D	C03	35479070	35480133	–	4	453	151	16,992.13	5.25	Cy [a]; Nu [b]
BnaGLYI21	BnaC05g04530D	C03	2231485	2233489	–	1	1338	446	48,857.97	5.65	Cy [a,b]; Om [a]
BnaGLYI22	BnaC05g05340D	C05	2609787	2610529	–	2	417	139	15,596.41	5.94	Cy [a,b]
BnaGLYI23	BnaC05g05770D	C05	2848240	2850316	+	7	708	236	26,438.95	8.74	Cy [a]; Pe [a];Ch [b,c]
BnaGLYI24	BnaC05g08770D	C05	4659651	4665014	+	19	2175	725	80,672.51	7.49	Cy [a,b]; Om [a]
BnaGLYI25	BnaC05g11680D	C05	6800739	6802181	+	2	525	175	19,733.75	5.68	Cy [a]; Ch [b]
BnaGLYI26	BnaC06g28360D	C06	29576211	29578606	–	8	1038	346	38,368.97	6.19	Cy [a];Ch [b,c]
BnaGLYI27	BnaC08g15100D	C08	19658835	19661229	+	7	852	284	31,814.27	5.26	Cy [a,b]
BnaGLYI28	BnaC08g16660D	C08	20576966	20578018	+	2	525	175	19,821.86	5.89	Cy [a]; Ch [b]
BnaGLYI29	BnaC08g38920D	C08	34930459	34931660	–	2	522	174	19,568.62	5.86	Cy [a]; Mt [b]
BnaGLYI30	BnaC08g44820D	C08	37849659	37851160	–	1	1323	441	48,162.21	5.46	Cy [a]; Cysk [b]
BnaGLYI31	BnaC09g53670D	C09	3644531	3644929	+	0	399	133	15,196.72	7.77	Cy [a,b]
BnaGLYI32	BnaCnng38880D	Cnn	37521585	37522696	–	2	414	138	15,502.36	6.2	Cy [a]; Ch [b]
BnaGLYI33	BnaCnng47290D	Cnn	46764251	46765432	–	5	432	144	16,180.32	5.39	Cy [a,b]
BnaGLYI34	BnaCnng47280D	Cnn	46758036	46759221	–	5	324	108	12,264.1	5.01	Cy [a]; Nu [b]
BnaGLYI35	BnaCnng59150D	Cnn	58862580	58863862	–	3	414	138	15,253.16	5.45	Cy [a]; Nu [b]

Abbreviations: CDS: coding DNA sequence, PP: polypeptide length, MW: molecular weight, PI: isoelectric point, bp: base pair, aa: amino acid, kDa: kilodalton, Ch: chloroplast, Cy: cytosol, Mt: mitochondria, Nu: nucleus, Pe: periplasm, Ou: outermembrane, Ec: extracellular matrix. [a] Localization prediction by pSORT (http://www.genscript.com/wolf-psort.html, (accessed on 20 May 2020)). [b] Localization prediction by TargetP 1.1 Server (http://www.cbs.dtu.dk/services/TargetP/, (accessed on 20 May 2020)). [c] Chloroplast localization signal confirmed by ChloroP (http://www.cbs.dtu.dk/services/ChloroP/, (accessed on 20 May 2020)).

The CDS length of *BnaGLYI* members ranged from 324 bp (*BnaGLYI34*) to 3648 bp (*BnaGLYI06*). Accordingly, *BnaGLYI06* encodes for the largest protein with a polypeptide length of 1216 aa and a molecular weight of 134.9 kDa, while *BnaGLYI34* encodes for the smallest protein with a polypeptide length of 108 aa and a molecular weight of 12.2 kDa (Table 1). The isoelectric point (pI) of BnaGLYI proteins vary from 4.78 (BnaGLYI08) to 8.86 (BnaGLYI06). Most of the BnaGLYI members showed acidic pI value, and only four proteins (BnaGLYI02, BnaGLYI05, BnaGLYI06 and BnaGLYI23) showed obvious alkalinity pI value (around 8.0) (Table 1). Moreover, there were also four members (BnaGLYI04, BnaGLYI19, BnaGLYI24 and BnaGLYI31) showing a basic pI value (Table 1). Sub-cellular localization of all identified BnaGLYI proteins were analyzed using three tools (CELLO, Wolf-pSORT and ChloroP). The results predicted that all BnaGLYI, except BnaGLYI15, were localized in cytoplasm. However, the TargetP predicted that most BnaGLYI were localized in the chloroplast, followed by the cytoplasm, nucleus, mitochondrion and cytoskeleton (Table 1).

Similarly, the CDS length of *BnaGLYII* genes also had very wide ranges, from 777 bp (*BnaGLYII09*) to 5745 bp (*BnaGLYII24*). Accordingly, BnaGLYII09 was the smallest protein (259 aa, 28.57 kDa), while *BnaGLYII24* was the largest protein (1915 aa, 212.36 kDa) (Table 2). The pI value of BnaGLYII proteins varied from 5.21 (BnaGLYII29) to 9.04 (BnaGLYII04). Most of the BnaGLYII proteins showed alkalinity with a pI value around or more than 7.0, and only seven BnaGLYII proteins had an acidic pI value, including BnaGLYII09, BnaGLYII12, BnaGLYII13, BnaGLYII15, BnaGLYII17, BnaGLYII29 and BnaGLYII30 (Table 2). Sub-cellular locations of BnaGLYII proteins were different from BnaGLYI proteins. BnaGLYII protein members were localized in different sub-cellular compartments, such as cytoplasm, chloroplast, extracellular matrix, outermembrane, periplasm, nucleus and mitochondrion (Table 2).

Table 2. Basic characteristics of *BnaGLYII* genes identified in *Brassica napus* L. genomes.

Gene Symbol	Locus Identifier	Location	Gene Start (bp)	Gene Stop (bp)	Strand	No. of Introns	CDS Length (bp)	PP Length (aa)	MW (kDa)	pI	Localization
BnaGLYII01	BnaA01g03370D	A01	1597636	1600850	−	10	1056	352	39,338.98	7.93	Cy [a]; Ch [b,c]
BnaGLYII02	BnaA02g25890D	A02	18962115	18965111	+	12	1848	616	68,076.64	6.62	Cy [a,b]
BnaGLYII03	BnaA03g14380D	A03	6611723	6613870	+	7	987	329	36,085.29	8.81	Cy [a]; Ch [b,c]
BnaGLYII04	BnaA03g20230D	A03	9624333	9627034	−	8	1026	342	37,496.84	9.04	Ec [a]; Ch [b,c]
BnaGLYII05	BnaA03g33050D	A03	15994860	15997418	+	10	1074	358	39,988.96	7.17	Cy [a]; Ch [b,c]
BnaGLYII06	BnaA04g25070D	A04	18183840	18186578	−	7	993	331	36,397.62	8.82	Ec [a]; Ch [b,c]
BnaGLYII07	BnaA05g03320D	A05	1854330	1860877	−	20	2868	956	106,401.82	8.51	Ou [a]; Ch [b,c]
BnaGLYII08	BnaA05g11390D	A05	6376663	6378817	−	7	981	327	35,883.92	8.07	Pe [a]; Ch [b,c]
BnaGLYII09	BnaA05g28080D	A05	20057585	20059255	+	6	777	259	28,689.47	6.1	Pe [a]; Cy [b]
BnaGLYII10	BnaA06g00850D	A06	603532	606586	+	8	1167	389	43,530.78	7.74	Cy [a]; Pe[a]; Nu [b]
BnaGLYII11	BnaA06g22690D	A06	15848234	15855367	+	23	3450	1150	126,907.01	9.02	Cy [a]; Ch[b,c]
BnaGLYII12	BnaA08g01070D	A08	806559	808627	+	7	786	262	28,574.49	6.19	Pe [a]; Nu [b]
BnaGLYII13	BnaA09g28480D	A09	21351908	21354940	−	11	1566	522	57,465.81	6.19	Cy [a,b]
BnaGLYII14	BnaA09g50050D	A09	33176404	33178430	+	7	969	323	35,263.31	7.16	Cy [a]; Ch [b,c]
BnaGLYII15	BnaAnng27360D	Ann	31283476	31286304	−	4	2085	695	77,448.79	5.81	Cy [a]; Nu [b]
BnaGLYII16	BnaC01g04650D	C01	2443496	2446457	−	10	1059	353	39,822.41	6.96	Cy [a]; Ch [b,c]
BnaGLYII17	BnaC01g28980D	C01	27137154	27139952	−	4	2076	692	77,157.53	5.81	Cy [a]; Mt [b]
BnaGLYII18	BnaC02g47740D	C02	3827265	3830275	+	12	1848	616	68,049.53	6.43	Cy [a,b]
BnaGLYII19	BnaC03g17440D	C03	8920776	8922877	+	7	975	325	35,687.84	8.85	Cy [a]; Mt [b]; Ch [c]
BnaGLYII20	BnaC03g24200D	C03	13557749	13560223	−	7	981	327	35,840.97	8.85	Ec [a]; Pe [a]; Ch [b,c]
BnaGLYII21	BnaC03g38140D	C03	23395036	23397633	+	10	1077	359	39,993.97	7.14	Cy [a]; Ch [b,c]
BnaGLYII22	BnaC03g50930D	C03	35336691	35341520	−	16	2682	894	98,725.64	8.7	Cy [a]; Ch [b,c]
BnaGLYII23	BnaC03g69760D	C03	59489758	59491865	−	7	870	290	31,928.74	8.57	Cy [a]; Ch [b,c]
BnaGLYII24	BnaC04g02920D	C04	2066379	2078825	−	31	5745	1915	212,357.58	8.24	Ou [a]; Ch [b,c]
BnaGLYII25	BnaC04g48930D	C04	47330538	47333344	−	7	996	332	36,386.59	8.7	Ec [a]; Ch [b,c]
BnaGLYII26	BnaC05g20710D	C05	14195563	14198830	+	11	1566	522	57,524.85	6.47	Cy [a,b]
BnaGLYII27	BnaC06g06420D	C06	6895047	6896947	−	7	909	303	33,196.01	7.21	Pe [a]; Ch [b,c]
BnaGLYII28	BnaC08g44630D	C08	37747241	37749209	−	7	984	328	35,631.65	6.83	Pe [a]; Ch [b,c]
BnaGLYII29	BnaCnng30690D	Cnn	29167588	29174293	+	10	2448	816	93,535.33	5.21	Cy [a,b]
BnaGLYII30	BnaCnng46950D	Cnn	46379687	46381389	−	6	777	259	28,653.35	5.72	Pe [a]; Ch [b,c]

Abbreviations: CDS: coding DNA sequence, PP: polypeptide length, MW: molecular weight, PI: isoelectric point, bp: base pair, aa: amino acid, kDa: kilodalton, Ch: chloroplast, Cy: cytosol, Mt: mitochondria, Nu: nucleus, Pe: periplasm, Ou: outermembrane, Ec: extracellular matrix. [a] Localization prediction by pSORT (http://www.genscript.com/wolf-psort.html, (accessed on 20 May 2020)). [b] Localization prediction by TargetP

1.1 Server (http://www.cbs.dtu.dk/services/TargetP/, (accessed on 20 May 2020)). [c] Chloroplast localization signal confirmed by ChloroP ((http://www.cbs.dtu.dk/services/ChloroP/, accessed on 20 May 2020)).

2.2. Chromosome Localization and Gene Duplication Analysis of BnaGLYI and BnaGLYII

In order to get a more intuitive view of the distribution of *GLY* genes in *B. napus* chromosomes, the information of their starting and ending position in chromosomes were collected (Table 1), and a chromosome distribution map was drawn on the MG2C (Map Gene2 Chromosome v2) website (Figure 1). For the *BnaGLYI* gene family, there were 31 genes that were unevenly distributed on 14 chromosomes, and no genes were found on chromosomes A01, A04, C01, C04 and C07. Four genes (*BnaGLYI32-35*) were located in the unmapped scaffolds in the Cnn_ random genome (Figure 1a). Chromosomes A06, C05 and C08 contained the maximum number of *BnaGLYI* genes, four genes each, while five chromosomes (A02, A03, A05, C06 and C09) had only one *BnaGLYI* gene (Figure 1a). In total, 16 *BnaGLYI* genes were distributed on the An-subgenome and 19 were from the Cn-subgenome. Similarly, 27 *BnaGLYII* genes were unevenly located in 15 different chromosomes (Figure 1b). There were four chromosomes that contained no *BnaGLYII* genes (A07, A10, C07 and C09) (Figure 1b). Chromosome C03 harbored five *BnaGLYII* genes, which contained the most *BnaGLYI* genes. However, most of chromosomes, like A01, A02, A04, A08, C02, C05, C06 and C08, contained only one *BnaGLYII* gene (Figure 1b). The remaining three genes could not be mapped into chromosomes of the rapeseed genome, which were distributed on Ann and Cnn (Figure 1b). Moreover, BnaGLYII proteins were mainly localized in the chloroplast and cytosol.

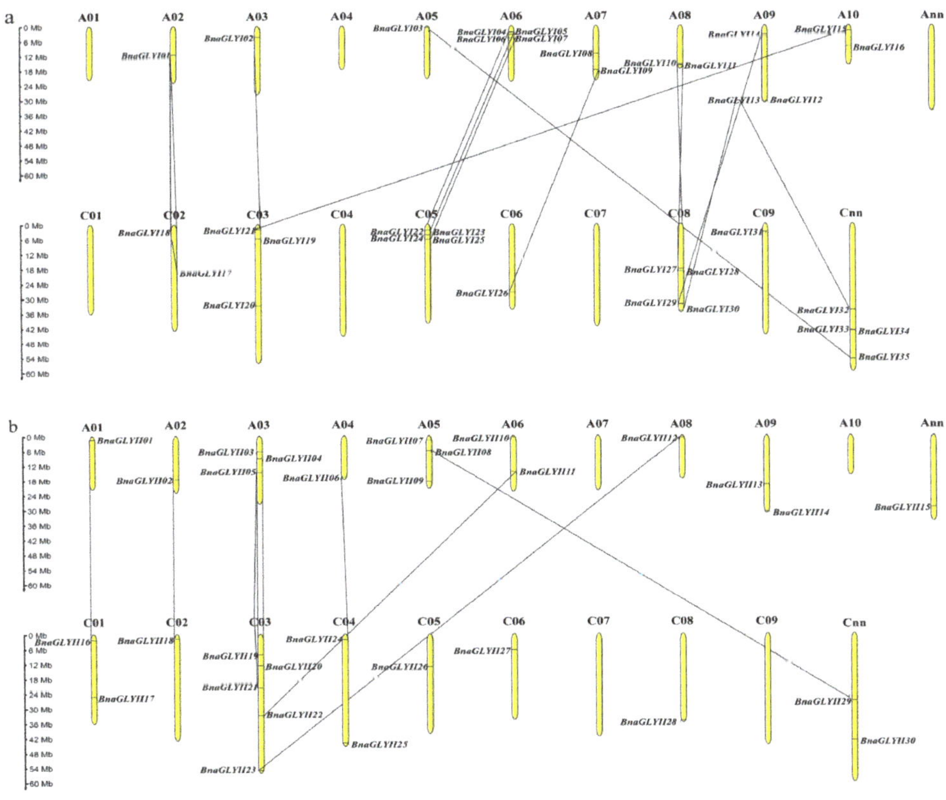

Figure 1. Chromosomal localization of *BnaGLYI* (**a**) and *BnaGLYII* (**b**) genes in the *B. napus* chromosomes. Duplicated genes are connected by black lines.

Gene replication is the main way of gene family expansion. This study also analyzed the gene replication of the *GLY* gene family in *B. napus*. Among BnaGLYI proteins, there were 15 pairs of duplicated genes. Three genes categorized into one group that formed three gene pairs (*BnaGLYI01/17/18*). Other gene pairs all included two genes (*BnaGLYI02/19, BnaGLYI03/35, BnaGLYI04/22, BnaGLYI05/23, BnaGLYI07/25, BnaGLYI09/26, BnaGLYI10/28, BnaGLYI11/27, BnaGLYI12/32, BnaGLYI13/30, BnaGLYI14/29* and *BnaGLYI15/21*). Moreover, there were nine gene pairs with over 95% sequence similarity in BnaGLYII proteins. They were *BnaGLYII01/16, BnaGLYII02/18, BnaGLYII03/19, BnaGLYII04/20, BnaGLYII05/21, BnaGLYII06/24, BnaGLYII08/29, BnaGLYII11/22* and *BnaGLYII12/23*, respectively (Figure 1b). Such a high level of sequence similarity implied the possibility of segmental duplication of those genes. In addition, there were no tandem duplications found in both the BnaGLYI and BnaGLYII proteins.

2.3. Gene Structure and Phylogenetic Analysis of the BnaGLYI and BnaGLYII

To further investigate the gene structure, the CDS and genomic sequences of the *BnaGLY* genes were aligned. One to twenty-six introns varied from *BnaGLYI* genes, and there were no introns in the *BnaGLYI08* and *BnaGLYI31* (Figure 2b, Table 1). *BnaGLYI13, BnaGLYI15, BnaGLYI21* and *BnaGLYI30* only contained one intron. *BnaGLYI06* gene had the largest number of introns (26). Similarly, the *BnaGLYII* genes also contained varied numbers of introns (4–31), e.g., thirty-one introns were identified in *BnaGLYII17*, and four introns were predicted in *BraGLYII24* (Figure 2b, Table 2). In addition, the GLY proteins that clustered in the same group possessed a similar structure (Figure 2).

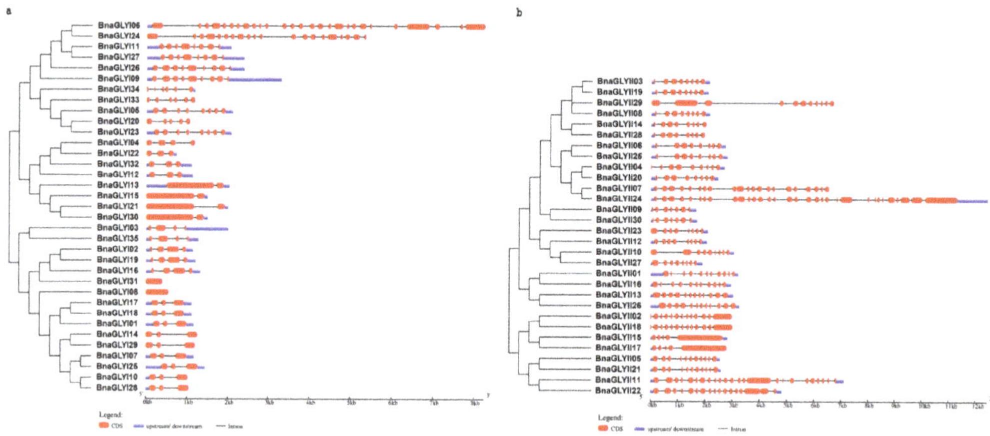

Figure 2. Phylogenetic relationship and gene structure of *BnaGLYI* and *BnaGLYII*. (**a**) BnaGLYI, (**b**) BnaGLYII. An unrooted tree was generated by MEGA 7.0 software (the bootstrap value is 1000) using the full-length amino acid sequences. Exons and untranslated regions (UTR) are indicated by rectangles (red and blue) and introns by lines. CDS and amino acid sequences of *BnaGLYI* and *BnaGLYII* are listed in additional data sheet (Table S5 and S6).

To understand the evolutionary relationships and functions among the predicted BnaGLY proteins, a phylogenetic tree was drawn with all BnaGLY protein sequences by MEGA7.0. The thirty-five BnaGLY proteins were resolved into six distinguishing groups (Group A–F; Figure 3a). The largest clade Group F, containing thirty-two GLYI members, included nineteen GLYI in Brassicaceae, whereas the smallest group (Group B), containing ten members, included six GLYI in Brassicaceae (Figure 3a). Group A included sixteen members of Brassicaceae, and one protein from rice. Group C included twenty-six proteins, which contained fourteen members of Brassicaceae (six BnaGLYI, three BraGLYI, three BolGLYI and two AtGLYI). Group D and E contained 13 and 14 members, respectively. Generally, BnaGLYI members first clustered with proteins from *B. rapa* and *B. oleracea*, followed by AtGLYIs, forming a small Brassicaceae clade, and subsequently further clustered with GLYI members from rice (Figure 3a).

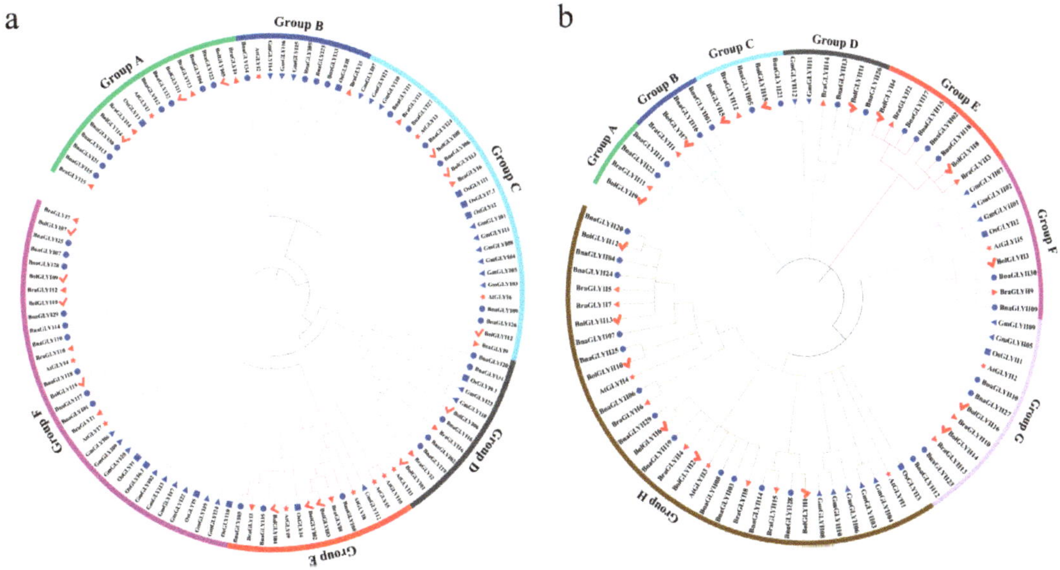

Figure 3. Phylogenetic relationships of BnaGLYI (**a**) and BnaGLYII (**b**) from various plant species. A phylogenetic tree was constructed based on the multiple alignment results, using MEGA 7.0 software with the Neighbor-Joining method. Bootstrap support from 1,000 reiterations is indicated above the branches. The 112 GLYI proteins from *B. napus* (35), *B. rapa* (16), *B. oleracea* (15), *Arabidopsis* (11) *G. max* (24) and *O sativa* (11) were used in phylogenetic analysis of BnaGLYI (**a**), and the 81 GLYII proteins from *B. napus* (30), *B. rapa* (15), *B. oleracea* (16), *Arabidopsis* (5), *G. max* (24) and *O sativa* (3) were used in phylogenetic analysis of BnaGLYII (**b**). Only the first splice variants were considered in the case of multiple members, respectively. "At", "Bra", "Bol", "Gm" and "Os" refer to the proteins in *A. thaliana, B. rapa, B. oleracea, G. max* and *O sativa*. Proteins from *B. napus, B. rapa, B. oleracea, Arabidopsis, G. max* and *O sativa* are marked with solid circles, red triangles, check marks, stars, blue triangles and squares, respectively. The sequences used in the analysis are listed in additional data sheet (Table S7).

Similarly, the BnaGLYII proteins formed eight distinct groups (Figure 3b). Two BnaGLYII proteins were clustered in groups A, B, C, D and F, respectively; four BnaGLYII proteins were classified in group E and G, and twelve BraGLYII proteins were in Group H (Figure 3b).

To explore the selective pressure on *BnaGLY* genes, we calculated the non-synonymous/synonymous mutation ratio (Ka/Ks); $Ka/Ks > 1$ indicates positive selection, $Ka/Ks = 1$ indicates neutral selection, and $Ka/Ks < 1$ indicates purifying selection [38]. The Ka/Ks ratio for 10 *BnaGLYI* genes was >1, ranging from 1.0664 (*BnaGLY11*) to 6.3118 (*BnaGLY102*), which indicated that these genes were experienced higher positive selection pressure during the evolutionary history of *B. napus*. The other genes showed $Ka/Ks < 1$, which experienced higher purifying selection pressure (Table S1). Moreover, most of the *BnaGLYII* genes had a $Ka/Ks < 1$, and only five *BnaGLYII* genes experienced positive selection pressure (Table S2). A comparison of Ka/Ks ratios among the *BnaGLYI* and *BnaGLYII* genes indicated that the average Ka/Ks ratio of *BnaGLYI* was higher than that of *BnaGLYII* genes. These results suggested that the *BnaGLYI* gene family may have suffered robust purifying selective pressure during evolution.

2.4. Conserved Motif and the Enzyme Activity of BnaGLY

Based on the conserved metal binding sites H/QEH/QE of GLYI proteins and multiple alignment results, the expected enzyme activity of the putative BnaGLYI proteins was predicted. Overall, out of a total thirty-five putative BnaGLYI proteins, sixteen have all the four conserved residues and are expected to have functional GLYI enzyme activity (Figure 4 and Table S3). The other 19 BnaGLYI proteins may have variant function. Additionally, GLYII from microorganisms and plants contained the well-conserved metal binding motif (THXHXDH) and active site motif (C/GHT) [31,39], which play an important role in GLYII enzyme activity. Therefore, based on the protein sequence alignment results, the presence of enzymatic activity of the putative BnaGLYII proteins were evaluated (Figure 4, Table S4). Fourteen of thirty putative BnaGLYII proteins possess all the conserved metal binding residues and may have the functional GLYII enzyme activity (Table S4).

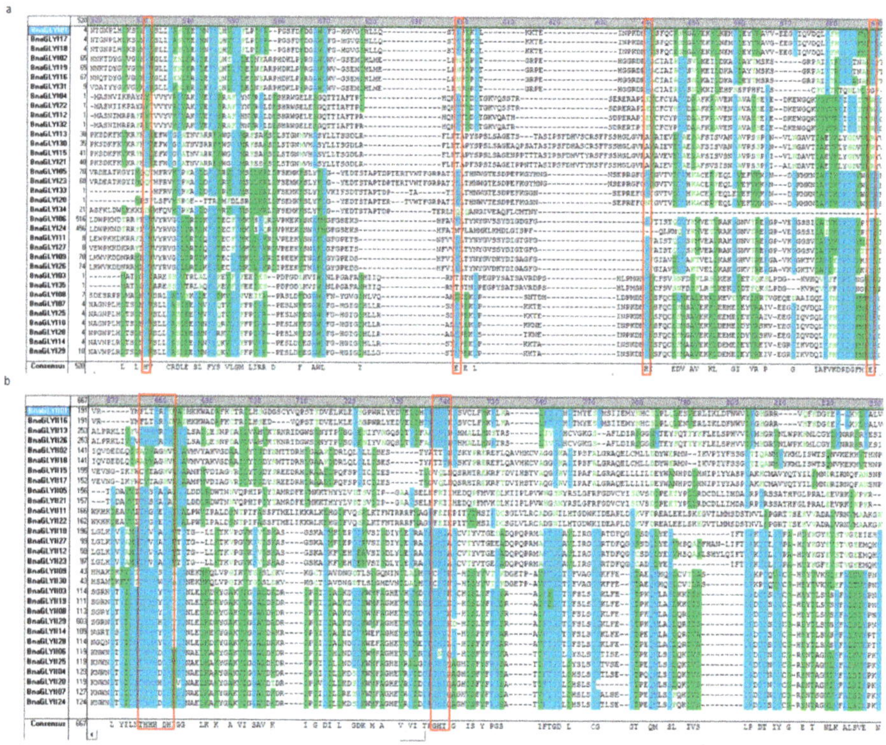

Figure 4. Multiple sequence alignment of GLY domain of all BnaGLYI and BnaGLYII proteins. (**a**) The conserved amino acids (H/QEH/QE) of BnaGLYI were marked by red boxes. (**b**) The red boxes indicate the most conserved metal binding motif (THHHXDH) and active site motif (G/CHT) of BnaGLYII proteins.

To better understand the protein sequence features of BnaGLY, the conserved motifs of each protein were also identified using the MEME Suite 5.1.1 (http://meme-suite.org/tools/meme, (accessed on 28 May 2020)) (Figure S1). We found that most proteins in the same group had similar motifs, and the LOGOs of these protein motifs were obtained by MEME (Figure S2). In BnaGLYI proteins, motif 1 was observed in all the 34 BnaGLYI proteins, except the BnaGLYI20. Motifs 1, 2, 3 and 4 contained the conserved sites of H/QEH/QE. Similarly, motif 1 was detected in most BnaGLYII proteins, except the BnaGLY16 and BnaGLY26. Motifs 1 and 3 contain the metal binding site (THXHXDH) and active site (C/GHT), respectively. Moreover, we also found that most proteins in the same group had similar motifs (Figure S2). However, the conserved domain of the *BnaGLY* protein was divergent (Figure S3a,b). The structural domain is not as conservative as expected.

2.5. Tissue Expression Profiles of BnaGLYI and BnaGLYII

To investigate the tissue-specific expression profiles of the *BnaGLY* genes, we selected 20 types of tissues in database (https://biodb.swu.edu.cn/brassica/, (accessed on 15 June 2020)) to analyze, including 10 tissues at the initial flowering stage and 10 tissues at the different development stages of the podding. The expression levels of the *BnaGLY* genes showed diverse expression patterns (Figure S3). Six *BnaGLYI* members (*BnaGLYI09*, *BnaGLYI11*, *BnaGLYI16*, *BnaGLYI26*, *BnaGLYI27* and *BnaGLYI35*) were highly expressed in most of the tissues. Their average fragments per kilo base of exons per million fragments mapped (FPKM) values were over twenty-five, and the value in a single tissue was at least over three. In contrast, the expression levels of *BnaGLYI22*, *BnaGLYI29*, *BnaGLYI32* and *BnaGLYI34* were not detected in most tissues. They only showed faint expression in one to three tissues, e.g., the expression of *BnaGLYI32* was only detected in silique pericarps (SP) at 24 and 27 days after flowering (DAF). The expression of *BnaGLYI14* was not detected in all the 10 tissues (Figure S4a). In *BnaGLYII* genes, *BnaGLYII28* was only weakly expressed in SP at 27 DAF, while the FPKM values of *BnaGLYII18* and *BnaGLYII29* were less than one. The other *BnaGLYII* genes were constitutive expression. However, some genes were expressed at higher levels in specific tissues, e.g., *BnaGLYII19* was highly expressed in anther (Figure S4b). These results indicated that most *BnaGLY* genes were expressed at different levels in different tissues.

The expression levels of some *BnaGLY* genes were verified in eight different tissues and organs (including root, stem, leaf, flower, flower bud, 25DAF seed, silique and silique wall by qRT-PCR (Figure 5). A total of 19 *BnaGLYI* genes and 33 *BnaGLYII* genes were selected after verifying the specificity for each primer pair. According to the data, the expression level of some *BnaGLYI* and *BnaGLYII* genes was consistent with the results of database; for example, the expression level of most *BnaGLYI* was not detected, such as *BnaGLYI29* and *BnaGLYII28* (Figure S5 and Table S8), whereas most *BnaGLYII* showed constitutive expression. Some genes expressed highly in specific tissues; for example, *BnaGLYI02*, *BnaGLYI09* and *BnaGLYI19* showed relatively high expression levels in the leaf and silique wall, *BnaGLYII12* and *BnaGLYII15* were highly expressed in seeds and *BnaGLYII05* was highly expressed in flowers. However, the expression level of several genes was inconsistent with the GEO data; for example, *BnaGLYI11* and *BnaGLYI27* showed particularly high expression levels in the flowers (Figure 5a), and *BnaGLYII20* and *BnaGLYII26* were strongly expressed in the root (Figure 5b). Moreover, qRT-PCR results indicated that *BnaGLYII07* and *BnaGLYII24* were much more highly expressed in the leaf, not in the stem as shown in Figure 5b.

The identified paralogous pairs of *BnaGLYI* and *BnaGLYII* genes revealed similar expression patterns, such as BnaGLYI02/19. However, some of the paralogous gene pairs showed a high level of expression divergence in different tissues. For example, *BnaGLYII06* showed high level expression in flowers, while its paralogous *BnaGLYII24* showed high level expression in leaf. The diverse expression patterns suggested that these members might play diverse functions.

Moreover, the expression levels of *BnaGLY* in developing seeds at two, four, six and eight weeks after pollination (WAP) were analyzed using the gene expression omnibus (GEO) database (GSE77637) (Figure S6). The expression of *BnaGLYI11* and *BnaGLYI27* genes was high in the four stages, and the FPKM values were all over 2000. *BnaGLYI01*, *BnaGLYI04*, and *BnaGLYI32* showed high expression level at eight WAP. *BnaGLYI10*, *BnaGLYI14*, *BnaGLYI28*, *BnaGLYI29* and *BnaGLYI34* showed low expression levels in all the seed development stages (Figure S6a). In the case of *BnaGLYII*, the expression of *BnaGLYII09*, *BnaGLYII14* and *BnaGLYII28* was up-regulated at eight WAP. The expressions of *BnaGLYII03*,

BnaGLYII06, *BnaGLYII24*, *BnaGLYII13*, *BnaGLYII25*, *BnaGLYII19*, *BnaGLYII20* and *BnaGLYII26* were gradually decreased in the four development stages (Figure S6b).

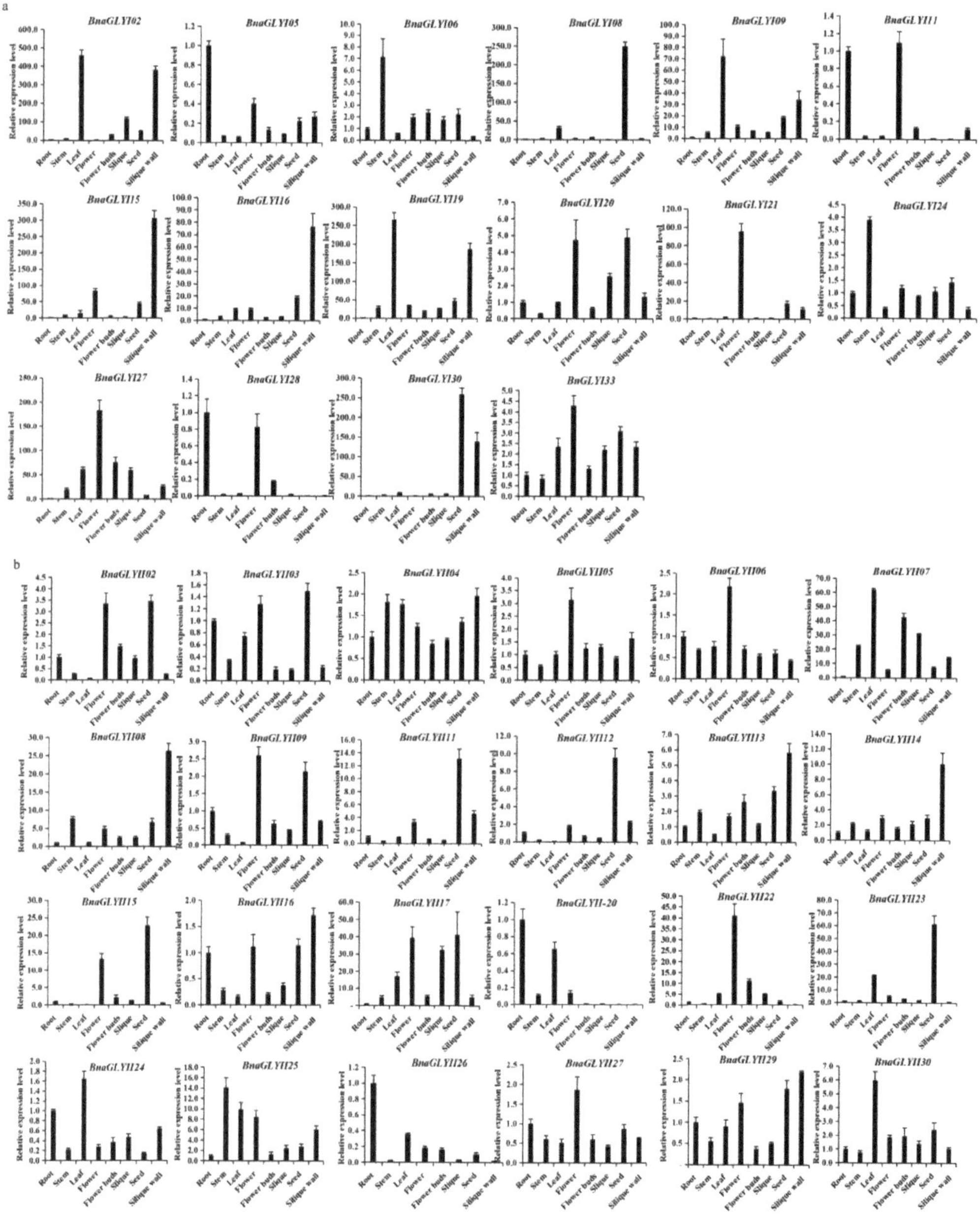

Figure 5. Relative expressions of the *BnaGLYI* (**a**) and *BnaGLYII* (**b**) genes in different tissues of *B. napus* confirmed by qRT-PCR. The normalized relative quantity in the root was set as "1".

2.6. Expression of BnaGLY in B. napus Seed Germination

To dissect the influence of *GLY* pathways in the seed germination of *B. napus*, the expression level of *GLY* genes was analyzed using the data in GEO (GSE13723). The expression level of *BnaGLYI* in winter oilseed rape (WOSR) accessions with different germination rates (high, medium and low; C129, C033 and C032, respectively) at different germination times showed similar expression patterns, e.g., the expression of *BnaGLYI11* and *BnaGLYI27* were significantly up-regulated at 72 h after imbibition (hai) in all the three accessions, and the expressions of *BnaGLYI04, BnaGLYI07, BnaGLYI17, BnaGLYI21* and *BnaGLYI25* obviously decreased at 36 hai and 72 hai. Meanwhile, *BnaGLYI06, BnaGLYI10, BnaGLYI14, BnaGLYI24, BnaGLYI28, BnaGLYI29* and *BnaGLYI34* showed weak expression in all the accessions at the four germination stages (Figure S7a). In contrast, the average expression level of *BnaGLYII* genes was much higher than that of *BnaGLYI*. *BnaGLYII5, BnaGLYII11, BnaGLYII21* and *BnaGLYII22* showed significant up-regulation at 36 hai and 72 hai in all the three accessions, while *BnaGLYII1* was only up-regulated at 72 hai in the three accessions. *BnaGLYII9* showed high expression at early stages of germination; however, it was sharply decreased at 36 hai, and then was significantly up-regulated at 72 hai. *BnaGLYII3* and *BnaGLYII19* were up-regulated at the three germination time points in the three accessions. *BnaGLYII17* and *BnaGLYII25* showed very low expression levels in accession C033 from 0 to 72 hai (Figure S7b). *BnaGLY* showed diverse expression patterns, though some genes were clustered within the same group.

2.7. The Response of the BnaGLYI and BnaGLYII on Pathogen Infection

The RNA-seq data sets available in GEO (GSE81545) were used for analysis of the *BnaGLY* gene expression in response to the fungal pathogen [40]. In the *BnaGLYI* gene, the expression levels of *BnaGLYI07, BnaGLYI24* and *BnaGLYI25* were significantly up-regulated after *S. sclerotiorum* pathogen infection in the resistant cultivar (ZY821) and sensitive cultivar (Westar). *BnaGLYI02, BnaGLYI19, BnaGLYI31* and *BnaGLYI35* were significantly down-regulated after pathogen infection in both cultivars, while the expression of *BnaGLYI08* and *BnaGLYI10* was down–regulated in sensitive cultivar; however, their expression was up–regulated in resistant cultivar. Most interestingly, the *BnaGLYI17* and *BnaGLYI18* genes were up-regulated in the susceptible cultivar; however, they were down-regulated in the resistant cultivar (Figure 6a).

In contrast to the *BnaGLYI* genes, most *BnaGLYII* genes showed significantly decreased expression levels after pathogen infection, e.g., *BnaGLYII06, BnaGLYII08, BnaGLYII14, BnaGLYII23, BnaGLYII25, BnaGLYII28* and *BnaGLYII30* were significantly down-regulated in both cultivars. Only *BnaGLYII02, BnaGLYII03, BnaGLYII18* and *BnaGLYII19* were significantly induced in both cultivars after being infected by pathogen (Figure 6b). Moreover, *BnaGLYII16* was up-regulated in the susceptible cultivars and down-regulated in the resistant cultivars.

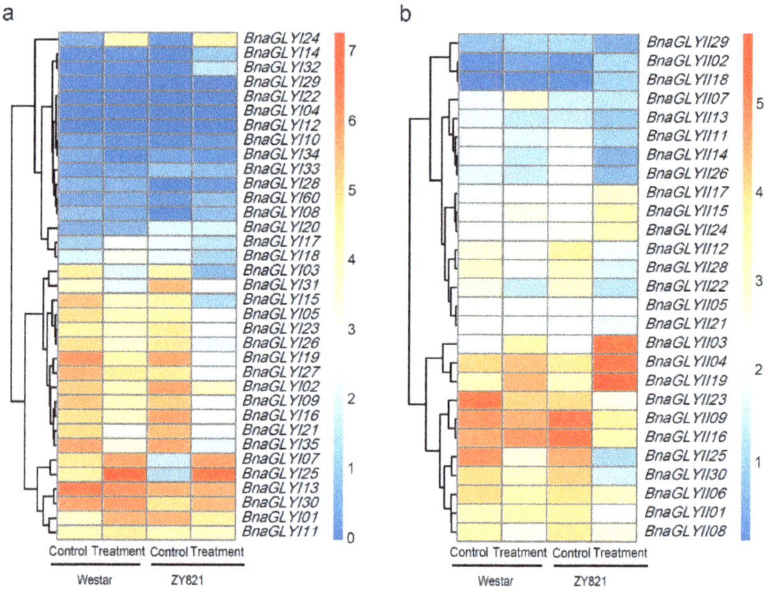

Figure 6. Expression profiles of the *BnaGLY* genes in response to *P. brassicae* infection. Relative expression data of available *BnaGLYI* (**a**) and *BnaGLYII* (**b**) genes under *P. brassicae* infection were obtained from BrassicaEDB database (https://biodb.swu.edu.cn/brassica/, accessed on 15 June 2020). The expression level of the inoculated leaves in susceptible (Westar) and tolerant (ZY821) genotypes were analyzed. Note: CK: Control condition; TR: Treatment under *P. brassicae* infection.

2.8. Expression of BnaGLY in Response to Abiotic Stresses

We also investigated the response of *BnaGLY* genes to drought and heat stresses at expression level (Figure 7, Table S9). The expression levels of 12 *BnaGLYI* genes (*BnaGLYI01, BnaGLYI03, BnaGLYI05, BnaGLYI08, BnaGLYI11, BnaGLYI19, BnaGLYI20, BnaGLYI27, BnaGLYI31, BnaGLYI33* and *BnaGLYI35*) were up-regulated under heat stress, while the expression of 12 *BnaGLYI* genes (*BnaGLYI02, BnaGLYI06, BnaGLYI09, BnaGLYI13, BnaGLYI15, BnaGLYI17, BnaGLYI18, BnaGLYI21, BnaGLYI24, BnaGLYI26, BnaGLYI30* and *BnaGLYI34*) were down-regulated under heat stress. Meanwhile, twenty *BnaGLYI* were highly expressed under heat stress, and five *BnaGLYI* were lowly expressed under drought stress. Ten *BnaGLYI* genes (*BnaGLYI01, BnaGLYI03, BnaGLYI08, BnaGLYI16, BnaGLYI19, BnaGLYI20, BnaGLYI27, BnaGLYI31, BnaGLYI33* and *BnaGLYI35*), whose expression levels were up-regulated under heat stress, were also highly expressed under drought stress, while four *BnaGLYI* genes (*BnaGLYI06, BnaGLYI09, BnaGLYI24* and *BnaGLYI34*) that were down-regulated under heat stress were equally low expressed under drought stress. At the same time, the expression levels of *BnaGLYI11* increased under heat stress but decreased under drought stress, while six *BnaGLYI* genes (*BnaGLYI13, BnaGLYI15, BnaGLYI17, BnaGLYI18, BnaGLYI21* and *BnaGLYI30*) decreased under heat stress but increased under drought stress (Figure 7a).

In the case of *BnaGLYII*, twenty and fourteen *BnaGLYII* genes were induced under heat and drought stresses, respectively, while six and ten *BnaGLYII* genes were lowly expressed under heat and drought stresses, respectively. Among these genes, ten genes (*BnaGLYII01, BnaGLYII06, BnaGLYII09, BnaGLYII10, BnaGLYII14, BnaGLYII16, BnaGLYII20, BnaGLYII23* and *BnaGLYII28*) were up-regulated and two genes (*BnaGLYII04* and *BnaGLYII29*) were down-regulated under both stresses (Figure 7b). At the same time, the expression levels of six *BnaGLYII* genes (*BnaGLYII02, BnaGLYII05, BnaGLYII11, BnaGLYII13, BnaGLYII18* and *BnaGLYII24*) increased under heat stress but decreased under drought stress, while four

BnaGLYII genes (*BnaGLYII08, BnaGLYII19, BnaGLYII26* and *BnaGLYII30)* decreased under heat stress but increased under drought stress (Figure 7b).

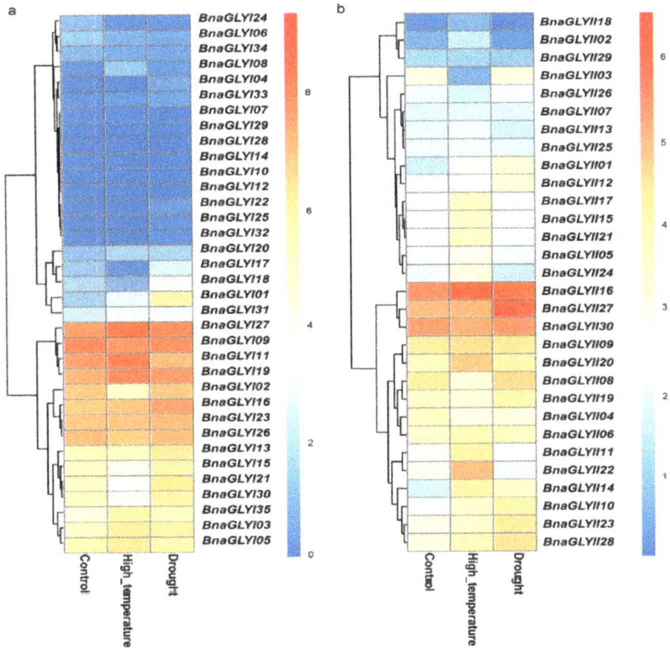

Figure 7. Expression profiles of the *BnaGLY* genes in response to heat and drought stresses. Relative expression data of available *BnaGLYI* (**a**) and *BnaGLYII* (**b**) genes under heat (40 °C, 3 h) and drought (withdrawing water, 3 days) treatment were obtained from NCBI GEO database (GSE156029, https://www.ncbi.nlm.nih.gov/geo/query/acc.cgi?acc=GSE156029, (accessed on 16 December 2022)).

To detect the response of *BnaGLY* genes to low temperature stress, the FPKM value of each gene (Table S10), treated with different cold tolerance under cold accumulations at chilling (CA) and freezing (FA) temperature and cold shocks at the same temperature (CB and FB) condition in two *B. napus* varieties, were determined using the data from GEO (GSE129220: https://www.ncbi.nlm.nih.gov/geo/query/acc.cgi?acc=GSE129220, (accessed on 16 December 2022)). A total of nine, twelve, ten and thirteen *BnaGLYI* genes were up-regulated under CA, FA, CB and FB treatments compared with their controls in *B. napus* variety HX17, respectively. Among these up-regulated genes, there were one (*BnaGLYI11*), two (*BnaGLYI27* and *BnaGLYI11*) and one (*BnaGLYI11*) genes significantly up-regulated under CA, FA and CB treatment (Figure 8a and Table S9). There were nine, eight, four and seven *BnaGLYI* genes up-regulated under CA, FA, CB and FB treatments, and two (*BnaGLYI27* and *BnaGLYI11*), one (*BnaGLYI11*) and three *BnaGLYI* (*BnaGLYI07, BnaGLYI11* and *BnaGLYI27*) genes significantly up-regulated under CA, FA, and FB treatments in *B. napus* variety HX58.

In the case of *BnaGLYII*, 13, 21, 17 and 17 *BnaGLYII* genes were induced in HX17, while 16, 25, 17 and 18 *BnaGLYII* genes were induced under CA, FA, CB and FB treatments HX58, respectively (Figure 8b and Table S10). Among these genes, three (*BnaGLYII11, BnaGLYII22* and *BnaGLYII02*), four (*BnaGLYII02, BnaGLYII11, BnaGLYII14* and *BnaGLYII22*) and four (*BnaGLYII01, BnaGLYII09, BnaGLYII14* and *BnaGLYII30*) were up-regulated significantly under CA, CB and FB in HX17. Three (*BnaGLYII04, BnaGLYII11* and *BnaGLYII22*), eight (*BnaGLYII01, BnaGLYII04, BnaGLYII16, BnaGLYII18, BnaGLYII22, BnaGLYII24, BnaGLYII25* and *BnaGLYII29*), eight (*BnaGLYII02, BnaGLYII10, BnaGLYII11, BnaGLYII14, BnaGLYII17,*

BnaGLYII18, BnaGLYII22 and *BnaGLYII28*) and nine (*BnaGLYII02, BnaGLYII10, BnaGLYII11, BnaGLYII14 BnaGLYII15, BnaGLYII17, BnaGLYII18, BnaGLYII22* and *BnaGLYII28*) *BnaGLYII* were up-regulated significantly under CA, FA, CB and FB in HX58.

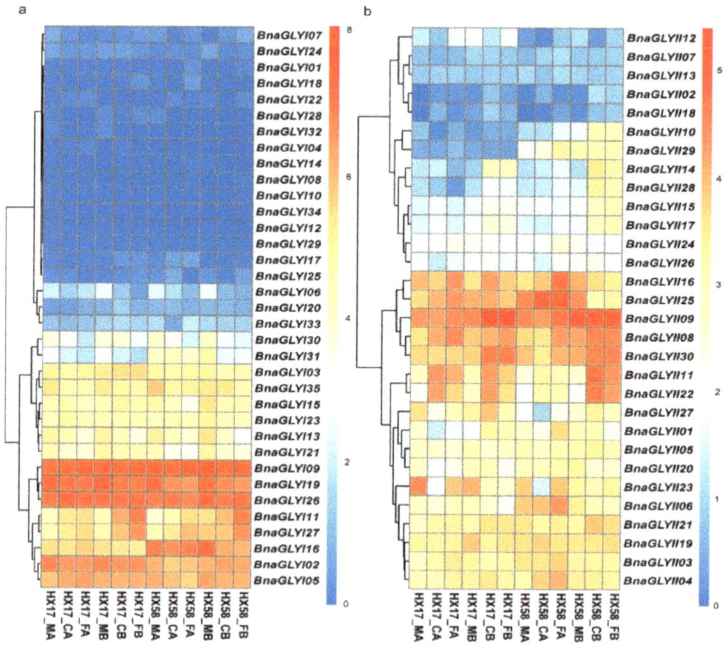

Figure 8. Expression of *BnaGLY* genes under different low-temperature treatments. (**a**) *BnaGLYI* genes; (**b**) *BnaGLYI* genes. The expression levels of each gene ($\log_2$ (FPKM values + 1)) in each sample is indicated by different color rectangles. The transcripts of each gene in leaves treated with chilling and freezing, with or without cold acclimation, were determined using the data from GEO (GSE129220). HX17 and HX58 were two early maturing semi-winter rapeseed varieties. This shows the expression level of *BnGLY* genes in the two *B. napus* varieties with different cold tolerances under cold accumulations at chilling (CA) and freezing (FA) temperature and cold shocks at the same temperature (CB and FB) condition. MA and MB were the control with and without cold accumulations, respectively.

2.9. Overexpression of BnGLYI Confer Enhanced Arabidopsis Seedlings Freezing Tolerance

The above results revealed that *BnaGLYI11* was significantly induced under both chilling and freezing stresses regardless of cold acclimation, which indicated that the two genes play important roles in cold tolerance in *B. napus*. To elucidate the function of *BnaGLY11* gene, we examined the freezing tolerance of *BnaGLYI11*–overexpressing transgenic *Arabidopsis thaliana* seedlings (*BnaGLYI11*-OE). *BnaGLYI11* was indeed overexpressed in these transgenic plants (Figure 9a). After the freezing treatment, the *BnGLYI11*-OE plants showed higher survival rates than the wild-type plants under freezing conditions. The results indicated that the plants over expressing the *BnGLYI11* gene possessed increased capacities to tolerate freezing temperatures (Figure 9b,c).

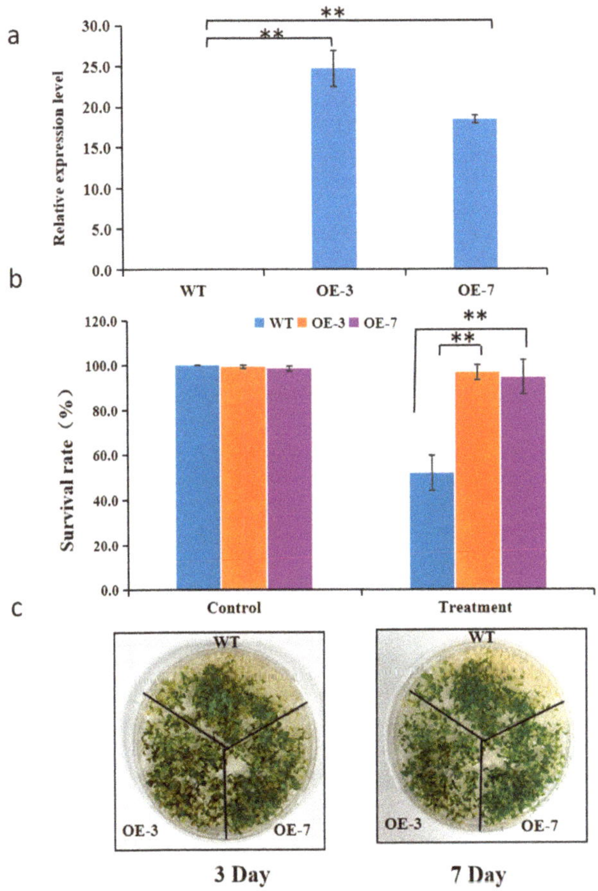

Figure 9. The function of the *BnaGLYI11* gene in response to cold stress. (**a**) The expression level of *BnaGLYI11* in overexpression *A. thaliana* lines OE-3 and OE-7 in 7-day-old seedling. (**b**) Survival rate after −10 °C for cold stress treatment. Two-week-old seedlings grown on MS medium (22 °C with 8 h of light daily) were treated at −10 °C for 210 min after 4 °C for three days. After cold treatment, the seedlings were incubated at 4 °C in the dark for 12 h and then transferred to normal grown conditions. The survival rates of the seedlings were scored visually after 7 days. WT: wild-type (Col-0), OE: *BnaGLYI11* gene overexpression lines. (**c**) phenotype of wild-type (Col-0) and transgenic *BnaGLYI11* gene after −10 °C treatment. Asterisks represent the statistical significance between the means for each treatment. ** $p < 0.001$. Vertical bars in the graph represent mean ± SD.

3. Discussions

Abiotic and biotic stresses severely affect the overall growth of plants and reduce their total productivity. MG is a cytotoxic metabolite, which is produced more under abiotic and biotic stresses [23]. Excessive accumulation of MG results in the disruption of the antioxidant defense system, biomembrane structures and cellular functions, including the peroxidation of lipids, the oxidation of protein and fatty acids and other metabolic dysfunctions [23,41]. The glyoxalase pathway, as the main MG detoxification system, could protect DNA and protein by converting MG into D-lactate and improve plant adaptation to multiple stresses. Many studies have revealed that the glyoxalase-overexpressing transgenic plants restrict MG accumulation and confer stress tolerance [23]. Therefore, genome-wide analysis of the glyoxalase enzymes has been carried out previously in rice,

Arabidopsis, soybean, *B. rapa*, grape and *M. truncatula* based on their conserved metal ion and substrate binding sites [22,27–29,31,42]. However, these families have not been studied in *B. napus*, the important oil and vegetable crop. In this study, we identified 35 *BnaGLYI* and 30 *BnaGLYII* in *B. napus* genome. *B. napus* (AACC) originates from natural crossing of *B. rapa* (AA) and *B. oleracea* (CC) [43]. Sixteen *BraGLYI* and fourteen *BraGLYII* were located in the AA subgenome, while fifteen and thirteen *BolGLYI* and *BolGLYII* were located in the CC subgenome. The number of *BnaGLYI* gene number was much higher than the ancestral species plants *B. rapa* [22] and *B. oleracea* [33] (total 31 *GLYI*). The GLY families display varied sequences, which might lead to diversity in their biochemical functions. Furthermore, phylogenetic relations, gene structures, protein motifs, Ks/Ka and the expression pattern of *BnaGLY* gene families were systematically investigated in the study.

According to the recent GLYI screening criteria, not only metal binding sites but also GSH binding sites, active sites and dimer interfaces should be considered [44]. Meanwhile, fourteen conserved amino acids of GLYIIs were important for substrate binding, metal binding and catalytic activity [45]. Using these criteria, only three *A. thaliana* genes (*AtGLYI2, AtGLYI3* and *AtGLYI6)* and two rice genes (*OsGLYI8* and *OsGLYI11.2*) were functional and contained all the binding sites. In this study, different BnaGLYs subfamily proteins shared the different type of conserved domains, which suggested their functional diversity. Analysis of all BnaGLYI proteins revealed that 16 BnaGLYI possessed conserved metal binding sites (H/QEH/QE) and may have expected enzyme activity of the putative BnaGLYI proteins along with the catalytic domain (PF00903). In addition, fourteen of thirty putative BnaGLYII proteins contained a metal binding motif (THHHXDH) and an active site motif (C/GHT), which were similar to *Escherichia coli*, *Saccharomyces cerevisiae*, *Salmonella typhimurium*, *L. infantum*, *Homo sapiens* and higher plants (*A. thaliana*, *B. juncea*, *O. sativa*) [39].

Previous reports indicated that genes with less introns showed rapid gene activation and timely responses to various stresses [46]. In our study, *BnaGLYI* and *BnaGLYII* genes showed significantly varied gene length and exon-intron structure. However, the genes with less or no introns did not show higher expression levels as reported [46,47], e.g., *BnaGLYI08* and *BnaGLYI31* containing no introns showed very low level of expression, which did not show significant expression responses to the stress conditions. Meanwhile, we found that some genes expressed higher in specific tissues; for example, *BnaGLYI02*, *BnaGLYI09* and *BnaGLYI19* were highly expressed in the leaf, whereas *BnaGLYII12* and *BnaGLYII15* were highly expressed in seeds and *BnaGLYII05* was highly expressed in flowers. These results indicate that most *BnaGLY* genes were expressed at different levels in different tissues. In addition, the expression of several *BnaGLYI* and *BnaGLYII* genes detected using qRT-PCR was different from the published database. This was understandable, as there were different sampling periods and growth conditions.

There is a firm link between GLY enzymes and stress tolerance in plants. The expression level of GLYI and GLYII can be induced under various stress treatments in different plants [34]. The transcripts of *GLYI* in tomatoes were up-regulated under salinity stress and phytohormonal and osmotic stimulation [34]. In pumpkin seedlings, the transcription level of *GLYI* was induced by salinity, heavy metal, white light and MG treatments [48]. Therefore, the expression patterns of the *BnaGLYI* and *BnaGLYII* genes were first analyzed using publicly available expression data in the study. The *BnaGLY* did not show any significant effect on winter oilseed rape germination rates, because the genes showed a similar expression pattern in accessions with different germination rates during germination. In a recent study, the glyoxalase enzyme of rice not only increased tolerance abiotic stresses (salinity, drought and extreme temperatures) but also reduced damage from biotic stresses (*Rhizoctonia solani*) [49]. In grapes, most glyoxalase genes had two periods of high expression after downy mildew inoculation [27]. In our study, the induced genes after infection may have a response on the *S. sclerotiorum* pathogen, e.g., *BnaGLYI07*, *BnaGLYI24*, *BnaGLYI25*, *BnaGLYII02*, *BnaGLYII03*, *BnaGLYII18* and *BnaGLYII19* genes, whose expressions were highly up-regulated in both cultivars. Furthermore, the genes that were only up–regulated in resistant cultivar may have related to the disease resistances of plants, such as *BnaGLYI08* and *BnaGLYI10*. In addition, the genes up-regulated in the susceptible cultivar and down-regulated in the resistant cultivar require further study, such as the *BnaGLYI17* and *BnaGLYI18*. However, whether glyoxalase proteins participate in *S. Sclerotiorum* pathogen response require further investigations.

In this study, we also uncovered that *BnaGLYI* and *BnaGLYII* genes are highly responsive to drought, heat and cold stresses, which implied these genes contribute to multiple stress responses in *B. napus*. For example, *BnaGLYI27* was highly up-regulated under both heat drought stresses, and *BnaGLYII22* was significantly induced under both chilling and freezing stresses regardless of cold acclimation. Most importantly, *BnaGLYI11* was significantly up-regulated under all temperature stresses (heat, chilling and freezing stresses, regardless of cold acclimation). Considering rapeseed oil suffers cold stress during vegetative stage, the rapeseed varieties with low cold tolerance have higher risk of freeze injury in cold winter and spring. Hence, it is vital to understand the cold-induced molecular responses in rapeseed. Therefore, *BnaGLYI11* was overexpressed in *A. thaliana* and the transgenic *A. thaliana* with the *BnaGLYI11* gene confers freezing tolerance. Previously, we found *BnaGLYI* (BnaA06g04580D, *BnaGLYI05*) transgenic yeast cells enhanced their tolerance to extreme temperature stress [30]. Similarly, overexpression of *AtGLYI2*, *AtGLYI3* and *AtGLYI6* in *E. coli* provides multi-stress heat tolerance [44]. BnaGLYI11 shared approximately 96% identity with ATGLYI3, and BnaGLYI05 showed 98% identity with ATGLYI2 (Table S11). The alignment results indicated that BnaGLY proteins shared a high sequence similarity, which may show significant function similarity. However, the freezing-tolerant BnaGLYI11 showed low similarity to BnaGLYI05 (35%), which indicates that the cold tolerance function may be due to the conserved function domains. Further investigations should explore the mechanism of the response of the glyoxalase pathway to stress tolerance in plants to generate more stress-tolerant varieties using molecular approaches.

4. Conclusions

This study provides a systematic knowledge of *BnaGLY* gene family in *B. napus*. A total of 35 *BnaGLYI* and 30 *BnaGLYII* genes were identified. The genes in the same subfamily showed similar structure and motif composition. Phylogenetic comparison and synteny analysis of *BnGLY* genes provide valuable clues for their evolutionary characteristics. Moreover, the gene expression pattern and their responses to seed gemination rate, *S. Sclerotiorum* pathogen heat and drought and temperature stresses were also determined. The transgenic *A. thaliana* with the temperature stresses responding to *BnaGLYI-11* confers cold tolerance. These results provide an important foundation for further understanding the biological functions of *BnaGLY* genes and their utilization in rapeseed breeding.

5. Methods

5.1. Plant Materials

The *B. napus* cultivar Zhongshuang11 used for edible oil and vegetable plants was obtained from the National Mid-term Gene Bank of Oil Crops Research Institute, Chinese Academy of Agriculture Sciences. It was planted in a growth chamber at 20 ± 2 °C with 12 h light and 12 h dark. The roots were sampled from 3-leaf stage seedlings. Fresh flower buds were obtained, and stems, leaves, siliques and seeds were sampled 25 days after flowering (DAF). Three biological replicates of each tested tissue were collected from three different individuals. All the samples were immediately frozen in liquid nitrogen and stored at −80 °C until RNA isolation. Wild-type *Arabidopsis* (Columbia ecotype) and overexpression transgenic lines were planted in a growth room with a constant temperature and light cycle (16 h light/8 h dark, 22 ± 1 °C, humidity 60%).

5.2. Identification of the GLYI and GLYII Genes Family in B. napus

Firstly, AtGLYI and AtGLYII protein sequences were acquired from the Arabidopsis Information Resource-TAIR (http://www.arabidopsis.org, (accessed on 12 May 2020)).

The genes and proteins of *B. napus* were downloaded from the website of the *B. napus* Genome Browser (http://www.genoscope.cns.fr/brassicana pus/, (accessed on 18 May 2020)) databases. Secondly, Arabidopsis GLY protein sequences were used as query to perform the Blast P against *B. napus* genome. Then the protein sequences with the e-value under 1.0×10^{-10} were confirmed again by the Pfam protein family database, according to the particular domains (PF00903) in GLYI proteins and the metallo-beta-lactamase domain (PF00753) in the GLYII proteins. All identified putative GLY proteins were designated as *BnaGLYI01* to *BnaGLYI35* and *BnaGLYII01* to *BnaGLYII30* according to their location in the chromosome. Localization of proteins were predicted using the CELLO v.2.5 website (http://cello.life.nctu.edu.tw/, accessed on 20 May 2020) [50] and the WOLF PSORT website (https://www.genscript.com/wolf-psort.html, accessed on 20 May 2020) [51]. Chloroplast localization was confirmed by ChloroP (http://www.cbs.dtu.dk/services/ChloroP/, accessed on 20 May 2020) [52]. The isoelectric points (pI) and molecular weights (Mw) were calculated using the ExPASy proteomics server database (http://www.expasy.org/tools/, (accessed on 26 May 2020))

5.3. Chromosomal Location, Duplication and Ka (Non-Synonymous Mutation Rate)/Ks (Synonymous Mutation Rate) Analysis

The location information of *GLY* genes on chromosomes were acquired from the website of the *B. napus* Genome Browser. Mapping of these *GLY* genes was performed using the online website MG2C (http://mg2c.iask.in/mg2c_v2.0/, (accessed on 10 August 2020)). Gene duplication was defined according to the criteria described in previous studies [53,54]: the aligned region of two sequences covers over 70% of the longer sequence and the similarity of the aligned region is over 95%. In addition, the Ka and Ks values of the repeating gene pairs were calculated using DnaSp software, and then evolutionary rate and the type of selection pressure on the gene were determined based on the Ka/Ks ratio [38].

5.4. Phylogenetic Analysis, Gene Structure and Conserved Motif

The Cluster X program was used to perform multiple sequence alignments with default parameters [55]. Phylogenetic trees were subsequently constructed by the MEGA 6.0 software using Neighbor-Joining (NJ) [56]. Bootstrap analysis was conducted with 1000 replications. The iTOL website (https://itol.embl.de/, (16 August 2020)) was used to better visualize the phylogenetic tree [57].

The exon/intron of *GLY* genes were inferred through comparison of genomic sequences and CDS sequences in the gene structure display server (http://gsds.cbi.pku.edu.cn/index.php) [58].

The MEME online tool (http://meme-suite.org/, (accessed on 28 May 2020)) [59] was employed to identify conserved motifs in *BnaGLY* genes with the following parameters: distribution of motifs, the optimum width of motif, 6–50; the maximum number of motif, 5; the number of repetitions, any. Only motifs with an e-value $< 1.0 \times 10^{-10}$ were retained for further analysis.

The conserved domains were retrieved, using NCBI Conserved Domain Database (https://www.ncbi.nlm.nih.gov/Structure/cdd/wrpsb.cgi, (accessed on 23 December 2022)) and Visualized NCBI CDD pattern of TB-tools (https://github.com/CJ-Chen/TBtools, (accessed on 27 December 2022)) [60].

5.5. Expression Analysis of BnaGLY Genes at Different Developmental Stages and Under Stress Treatments

The temporal and spatial expression patterns of *BnaGLY* were analyzed using the RNA-seq data online (https://biodb.swu.edu.cn/brassica/, (accessed on 15 June 2020)), including roots, stems, leaves, flowers, seeds and silique tissues and germination, bolting, initial flowering and podding stages. Stem-1 was sampled at the bolting stage. Stem-2, anther, filament, pedicel, inflorescence tip, mature leaf, young leaf, petal and carpel root were sampled at the initial flowering stage. The embryo, inner integument, outer integument, seed, silique pericarp, embryo and seed coat were sampled at podding stage. All the materials were planted in the field.

The publicly available data of the gene expression omnibus (GEO) database (GSE13723) was used to reveal the probable function of *BnaGLY* genes in the germination of winter oilseed rape with different germination rate [61]. The expression of all *BnaGLYI* genes in response to *Sclerotinia* infection was analyzed using the reported RNA-seq data (GSE81545). The inoculated leaves in both susceptible (Westar) and tolerant (ZY821) genotypes of *B. napus* were analyzed [40]. The RNA-seq data of a 35-day-old leaf with treatment of heat (40 °C, 3 h) and drought (withdrawing water, 3 days) (GSE156029) was used to analysis the response of *BnaGLY* genes to heat and drought stresses. The transcriptome data of *B. napus* under cold (4 °C, 12 h) and freezing (−4 °C, 12 h) (GSE129220) temperatures was downloaded from the NCBI database [62].

The data was used to generate a heatmap using the Heat map Illustrator (HemI, http://hemi.biocuckoo.org/down.php, (accessed on 15 October 2020)) package [45].

5.6. Plasmid Construction and Plant Transformation

The coding sequence of *BnaA08g25110D* was amplified by PCR and cloned into PBI121s [63] to generate the over-expression plasmids 35S::BnaGLYI, 35S::BnaGLYI, and 35S::BnaGLYI, respectively. Primers containing restriction enzyme sites for *XbaI* I and *Kpn* I were used for PCR as follows: BnaGLYI11-F: 5′ AGCTTTCGCGAGCTCGGTAC-CACAGAAAACTCTCAAAGCCC 3′, BnaGLYI11-R: 5′ TGCCTGCAGGTCGACTCTA-GAAACACAGAGAAACACGACAC 3′. The constructs used were confirmed by PCR and sequencing.

All the plasmids were transformed into *Agrobacterium tumefaciens* strain GV3101, and the positive clones identified by sequencing were introduced into 5-week-old Arabidopsis plants (Columbia ecotype) using the floral dip method. Positives were selected on MS plates containing 50 mg L^{-1} Kanamycin. The positive transformants was detected using a forward primer designed to the CaMV 35S sequence (35s up-F: 5′ ATTGATGT-GAACATGGTGGAG 3′) and a reverse primer of the gene expression (BnaGLYI11-RT-R: 5′ TTGTCTACCAGGACTGTTTTCC 3′). T1 seeds of PCR-positive transformants were harvested and grown to generation, and T3 generation was used for phenotype identification and gene expression.

5.7. RNA Isolation and Transcript Analysis

The gene expression of the definitive positive plants was analyzed by qRT-PCR. The total RNA from WT and Transgenetic plants was isolated using a polysaccharide and polyphenol total RNA isolation kit (BioTeke, RP3201, Wuxi, China). The cDNA was synthesized using a synthesis kit (EasyScript®First-Strand cDNA Synthesis SuperMix, TransGen Biotech, Beijing, China), and the qRT-PCR and data analysis were carried out as descried by Li et al. [64]. Each sample was quantified in triplicate with three biological replicates. The specific primers (BnaGLYI11-RTF: 5′ AGCTGACCTATAACTACGGCG 3′, BnaGLYI11-RTR: 5′ TTGTCTACCAGGACTGTTTTCC3′) designed are used for the detection.

5.8. Freezing Tolerance Assay

The freezing tolerance assay was performed as previously described [65], with some modifications. Briefly, seeds were grown 14 days on MS medium at 22 °C with 8 h of light daily. Before cold treatment, the 2-week-old seedlings were plated to 4 °C for three days, and then subjected to 210 min at −10 °C for cold stress treatment. After cold treatment, the seedlings were incubated at 4 °C in the dark for 12 h and then transferred to light at 22 °C. The survival rates of the seedlings were scored visually after 7 days.

Supplementary Materials: The following supporting information can be downloaded at: https://www.mdpi.com/article/10.3390/ijms24032130/s1.

Author Contributions: Designed the experiments: G.Y. and X.W. Analyzed the data: G.Y., M.Z., F.Z., W.G., L.Y., W.D., G.G., K.X., B.C. and L.L. Wrote the paper: G.Y. All authors have read and agreed to the published version of the manuscript.

Funding: This research was funded by National Natural Science Foundation of China grant number 32072106, Agricultural Science and Technology Innovation Project grant number CAAS-ZDRW202105, CAAS-OCRI-ZDRW-202201 and CAAS-ASTIP, National Crop Germplasm Resources Center grant number NCGRC2022-016, China Agriculture Research System grant number CARS-12.

Institutional Review Board Statement: Not applicable.

Informed Consent Statement: Not applicable.

Data Availability Statement: Not applicable.

Conflicts of Interest: The authors declare that they have no conflict of interest.

Abbreviations

aa: amino acid; bp: base pair; Br: *Brassica rapa*; BRAD: *Brassica* database; CDS: Coding DNA sequence; Ch: chloroplast; Cy: cytosol; Ec: extracellular matrix; FPKM: Fragments Per Kilo base of exon sper Million fragments mapped; GLYI: glyoxalase I; GLYII: glyoxalase II; GSH: Glutathione (reduced); hai: Hours after inoculation; Kb: Kilo base pair; MG: Methylglyoxal; Mt: mitochondria; MW: molecular weight; NJ: Neighbor-joining; Nu: nucleus; GEO: gene expression omnibus; ROS: Reactive oxygen species; Os: *Oryza sativa*; Ou: outermembrane, Pe: periplasm; *Pg*: *Pennisetum glaucum*; PI: isoelectric point; PP: polypeptide length; ROS: Reactive oxygen species; hai: hour after imbibition

References

1. Cartea, M.E.; Lema, M.; Francisco, M.; Velasco, P.; Sadowski, J.; Kole, C. Basic information on vegetable Brassica crops. In *Genetics, Genomics and Breeding of Vegetable Brassicas*; CRC Press: Boca Raton, FL, USA, 2011; pp. 1–34.
2. Wang, H.Z. New-demand oriented oilseed rape industry developing strategy. *Chin. J. Oil Crop Sci.* **2018**, *40*, 613–617. (In Chinese)
3. Zhang, Z.; Yin, Y.; Liu, F.; Wang, J.J.; Fu, T.D. Current situation and development countermeasures of Chinese rapeseed multifunctional development and utilization. *Chin. J. Oil Crop Sci.* **2018**, *40*, 618–623. (In Chinese)
4. Clemens, S.; Aarts, M.G.; Thomine, S.; Verbruggen, N. Plant science: The key to preventing slow cadmium poisoning. *Trends Plant Sci.* **2013**, *18*, 92–99. [CrossRef] [PubMed]
5. Rajwanshi, R.; Kumar, D.; Yusuf, M.A.; DebRoy, S.; Sarin, N.B. Stress-inducible overexpression of glyoxalase I is preferable to its constitutive overexpression for abiotic stress tolerance in transgenic *Brassica juncea*. *Mol. Breed.* **2016**, *36*, 76. [CrossRef]
6. Deponte, M. Glutathione catalysis and the reaction mechanisms of glutathione-dependent enzymes. *Biochim. Biophys. Acta* **2013**, *1830*, 3217–3266. [CrossRef] [PubMed]
7. Silva, M.S.; Ferreira, A.E.N.; Gomes, R.; Tomas, A.M.; Freire, A.P.; Cordeiro, C. The glyoxalase pathway in protozoan parasites. *Int. J. Med. Microbiol.* **2012**, *302*, 225–229. [CrossRef] [PubMed]
8. Hanssen, N.M.; Stehouwer, C.D.; Schalkwijk, C.G. Methylglyoxal and glyoxalase I in atherosclerosis. *Biochem. Soc. Trans.* **2014**, *42*, 443–449.
9. Edwards, L.G.; Adesida, A.; Thornalley, P.J. Inhibition of human leukaemia 60 cell growth by S-D-lactoylglutathione In Vitro. Mediation by metabolism to N-D-lactoylcysteine and induction of apoptosis. *Leuk. Res.* **1996**, *20*, 17–26. [CrossRef]
10. Marasinghe, G.P.; Sander, I.M.; Bennett, B.; Periyannan, G.; Yang, K.W.; Makaroff, C.A.; Crowder, M.W. Structural studies on a mitochondrial glyoxalase II. *J. Biol. Chem.* **2005**, *280*, 40668–40675. [CrossRef]
11. Ghosh, A.; Kushwaha, H.R.; Hasan, M.R.; Pareek, A.; Sopory, S.K.; Singla-Pareek, S.L. Presence of unique glyoxalase III proteins in plants indicates the existence of shorter route for methylglyoxal detoxification. *Sci. Rep.* **2016**, *6*, 18358. [CrossRef]
12. Veena, R.V.S.; Sopory, S.K. Glyoxalase I from *Brassica juncea*: Molecular cloning, regulation and its over-expression confer tolerance in transgenic tobacco under stress. *Plant J.* **1999**, *17*, 385–395. [PubMed]
13. Singla-Pareek, S.L.; Reddy, M.K.; Sopory, S.K. Genetic engineering of the glyoxalase pathway in tobacco leads to enhanced salinity tolerance. *Proc. Natl. Acad. Sci. USA* **2003**, *100*, 14672–14677. [CrossRef] [PubMed]
14. Wu, C.; Ma, C.; Pan, Y.; Gong, S.; Zhao, C.; Chen, S.; Li, H. Sugar beet M14 glyoxalase I gene can enhance plant tolerance to abiotic stresses. *J. Plant Res.* **2013**, *126*, 415–425. [CrossRef] [PubMed]
15. Viveros, M.F.; Inostroza-Blancheteau, C.; Timmermann, T.; Gonzalez, M.; Arce-Johnson, P. Overexpression of *GlyI* and *GlyII* genes in transgenic tomato (*Solanum lycopersicum* Mill.) plants confers salt tolerance by decreasing oxidative stress. *Mol. Biol. Rep.* **2013**, *40*, 3281–3290. [CrossRef]

16. Hasanuzzaman, M.; Nahar, K.; Hossain, M.S.; Mahmud, J.A.; Rahman, A.; Inafuku, M.; Oku, H.; Fujita, M. Coordinated actions of Glyoxalase and antioxidant defense systems in conferring abiotic stress tolerance in Plants. *Int. J. Mol. Sci.* **2017**, *18*, 200. [CrossRef]
17. Saxena, M.; Bisht, R.; Roy, S.D.; Sopory, S.K.; Bhalla-Sarin, N. Cloning and characterization of a mitochondrial glyoxalase II from *Brassica juncea* that is upregulated by NaCl, Zn, and ABA. *Biochem. Biophys. Res. Commun.* **2005**, *336*, 813–819. [CrossRef]
18. Lee, D.G.; Ahsan, N.; Lee, S.H.; Lee, J.J.; Bahk, J.D.; Kang, K.Y.; Lee, B.H. Chilling stress-induced proteomic changes in rice roots. *J. Plant Physiol.* **2009**, *166*, 1–11. [CrossRef]
19. Zhou, S.; Sauve, R.; Thannhauser, T.W. Proteome changes induced by aluminium stress in tomato roots. *J. Exp. Bot.* **2009**, *60*, 1849–1857. [CrossRef]
20. Hossain, M.A.; Fujita, M. Purification of glyoxalase I from onion bulbs and molecular cloning of its cDNA. *Biosci. Biotechnol. Biochem.* **2009**, *73*, 2007–2013. [CrossRef]
21. Liang, Y.; Srivastava, S.; Rahman, M.H.; Strelkov, S.E.; Kav, N.N. Proteome changes in leaves of Brassica napus L. as a result of Sclerotinia sclerotiorum challenge. *J. Agric. Food Chem.* **2008**, *56*, 1963–1976. [CrossRef]
22. Yan, G.; Xiao, X.; Wang, N.; Zhang, F.; Gao, G.; Xu, K.; Chen, B.; Qiao, J.; Wu, X. Genome-wide analysis and expression profiles of *glyoxalase* gene families in Chinese cabbage (*Brassica rapa* L). *PLoS ONE* **2018**, *13*, e0191159. [CrossRef] [PubMed]
23. Kaur, C.; Singla-Pareek, S.L.; Sopory, S.K. Glyoxalase and methylglyoxal as biomarkers for plant stress tolerance. *Crit. Rev. Plant Sci.* **2014**, *33*, 429–456. [CrossRef]
24. Lin, F.; Xu, J.; Shi, J.; Li, H.; Li, B. Molecular cloning and characterization of a novel glyoxalase I gene *TaGly I* in wheat (*Triticum aestivum* L.). *Mol. Biol. Rep.* **2010**, *37*, 729–735. [CrossRef] [PubMed]
25. Singla-Pareek, S.L.; Yadav, S.K.; Pareek, A.; Reddy, M.K.; Sopory, S.K. Transgenic tobacco overexpressing glyoxalase pathway enzymes grow and set viable seeds in zinc-spiked soils. *Plant Physiol.* **2006**, *140*, 613–623. [CrossRef]
26. Tang, F.L.; Li, R.M.; Zhou, Y.; Wang, S.; Zhou, Q.; Ding, Z.; Yao, Y.; Liu, J.; Wang, Y.; Hu, X.; et al. Genome-wide Identification of Cassava *Glyoxalase I* genes and the potential function of *MeGLYI-13* in iron toxicity tolerance. *Int. J. Mol. Sci.* **2022**, *23*, 5212. [CrossRef]
27. Li, T.; Cheng, X.; Wang, Y.; Yin, X.; Li, Z.; Liu, R.; Liu, G.; Wang, Y.; Xu, Y. Genome-wide analysis of *glyoxalase-like* gene families in grape (*Vitis vinifera* L.) and their expression profiling in response to downy mildew infection. *BMC Genom.* **2019**, *20*, 362. [CrossRef]
28. Mustafiz, A.; Singh, A.K.; Pareek, A.; Sopory, S.K.; Singla-Pareek, S.L. Genome-wide analysis of rice and Arabidopsis identifies two glyoxalase genes that are highly expressed in abiotic stresses. *Funct. Integr. Genom.* **2011**, *11*, 293–305. [CrossRef]
29. Ghosh, A. Genome-wide identification of *glyoxalase* genes in *medicago truncatula* and their expression profiling in response to various developmental and environmental stimuli. *Front. Plant Sci.* **2017**, *8*, 836. [CrossRef]
30. Yan, G.; Lv, X.; Gao, G.; Li, F.; Li, J.; Qiao, J.; Xu, K.; Chen, B.; Wang, L.; Xiao, X.; et al. Identification and characterization of a *glyoxalase I* gene in a rapeseed cultivar with seed thermotolerance. *Front. Plant Sci.* **2016**, *7*, 150. [CrossRef]
31. Ghosh, A.; Islam, T. Genome-wide analysis and expression profiling of glyoxalase gene families in soybean (*Glycine max*) indicate their development and abiotic stress specific response. *BMC Plant Biol.* **2016**, *16*, 87. [CrossRef]
32. Li, Z.G. Role of methylglyoxal and its detoxifcation system in plant thermotolerance. *Acta Physiol. Plant.* **2022**, *44*, 69. [CrossRef]
33. Zhang, M.; Yuan, L.; Zhang, F.; Guan, W.; Yan, G.; Wu, X. Identification and expression analysis of *Glyoxalase* gene family in *Brassica oleracea* L. 2021. Available online: https://kns.cnki.net/kcms/detail/46.1068.S.20210603.1558.012.htm (accessed on 25 November 2022). (In Chinese).
34. Espartero, J.; Sanchez-Aguayo, I.; Pardo, J.M. Molecular characterization of glyoxalase-I from a higher plant; upregulation by stress. *Plant Mol. Biol.* **1995**, *29*, 1223–1233. [CrossRef] [PubMed]
35. Sethi, U.; Basu, A.; Guha-Mukherjee, S. Control of cell proliferation and differentiation by regulating polyamine biosynthesis in cultures of Brassica and its correlation with glyoxalase–I activity. *Plant Sci.* **1988**, *56*, 167–175. [CrossRef]
36. Ramaswamy, O.; Pal, S.; Guha-Mukherjee, S.; Sopory, S.K. Correlation of glyoxalase I activity with cell proliferation in Datura callus culture. *Plant Cell Rep.* **1984**, *3*, 121–124. [CrossRef] [PubMed]
37. Chalhoub, B.; Denoeud, F.; Liu, S.; Parkin, I.A.; Tang, H.; Wang, X.; Chiquet, J.; Belcram, H.; Tong, C.; Samans, B.; et al. Early allopolyploid evolution in the post-Neolithic *Brassica napus* oilseed genome. *Science* **2014**, *345*, 950–953. [CrossRef]
38. Nekrutenko, A.; Makova, K.D.; Li, W.H. The K(A)/K(S) ratio test for assessing the protein-coding potential of genomic regions: An empirical and simulation study. *Genome Res.* **2002**, *12*, 198–202. [CrossRef]
39. Ghosh, A.; Pareek, A.; Sopory, S.K.; Singla-Pareek, S.L. A glutathione responsive rice glyoxalase II, OsGLYII-2, functions in salinity adaptation by maintaining better photosynthesis efficiency and anti-oxidant pool. *Plant J.* **2015**, *80*, 93–105. [CrossRef]
40. Girard, I.J.; Tong, C.; Becker, M.G.; Mao, X.; Huang, J.; de Kievit, T.; Fernando, W.G.D.; Liu, S.; Belmonte, M.F. RNA sequencing of *Brassica napus* reveals cellular redox control of *Sclerotinia* infection. *J. Exp. Bot.* **2017**, *68*, 5079–5091. [CrossRef]
41. Hoque, T.S.; Hossain, M.A.; Mostofa, M.G.; Burritt, D.J.; Fujita, M.; Tran, L.S. Methylglyoxal: An emerging signaling molecule in plant abiotic stress responses and tolerance. *Front. Plant Sci.* **2016**, *7*, 1341. [CrossRef]
42. Ridderstrom, M.; Cameron, A.D.; Jones, T.A.; Mannervik, B. Involvement of an active-site Zn^{2+} ligand in the catalytic mechanism of human glyoxalase I. *J. Biol. Chem.* **1998**, *273*, 21623–21628. [CrossRef]
43. Nagaharu, U. Genome analysis in *Brassica* with special reference to the experimental formation of *B. napus* and peculiar mode of fertilization. *Jpn. J. Bot.* **1935**, *7*, 389–452.

44. Jain, M.; Batth, R.; Kumari, S.; Mustafiz, A. *Arabidopsis thaliana* Contains Both Ni^{2+} and Zn^{2+} dependent Glyoxalase I enzymes and ectopic expression of the latter contributes more towards abiotic stress tolerance in *E. coli*. *PLoS ONE* **2016**, *11*, e0159348. [CrossRef] [PubMed]
45. Limphong, P.; Crowder, M.W.; Bennett, B.; Makaroff, C.A. *Arabidopsis thaliana* GLX2-1 contains a dinuclear metal binding site, but is not a glyoxalase 2. *Biochem. J.* **2009**, *417*, 323–330. [CrossRef]
46. Jeffares, D.C.; Penkett, C.J.; Bahler, J. Rapidly regulated genes are intron poor. *Trends Genet.* **2008**, *24*, 375–378. [CrossRef]
47. Chung, B.Y.; Simons, C.; Firth, A.E.; Brown, C.M.; Hellens, R.P. Effect of 5′-UTR introns on gene expression in *Arabidopsis thaliana*. *BMC Genom.* **2006**, *7*, 120. [CrossRef]
48. Hossain, M.A.; Hossain, M.Z.; Fujita, M. Stress-induced changes of methylglyoxal level and glyoxalase I activity in pumpkin seedlings and cDNA cloning of *glyoxalase I* gene. *Aust. J. Crop Sci.* **2009**, *3*, 53–64.
49. Gupta, B.K.; Sahoo, K.K.; Ghosh, A.; Tripathi, A.K.; Anwar, K.; Das, P.; Singh, A.K.; Pareek, A.; Sopory, S.K.; Singla-Pareek, S.L. Manipulation of glyoxalase pathway confers tolerance to multiple stresses in rice. *Plant Cell Environ.* **2018**, *41*, 1186–1200. [CrossRef]
50. Yu, C.S.; Hwang, J.K. Prediction of protein subcellular localization. *IEEE Comput. Soc.* **2006**, *64*, 643–651. [CrossRef] [PubMed]
51. Horton, P.; Park, K.J.; Obayashi, T.; Fujita, N.; Harada, H.; Adams-Collier, C.J.; Nakai, K. WoLF PSORT: Protein localization predictor. *Nucleic Acids Res.* **2007**, *35*, W585–W587. [CrossRef]
52. Emanuelsson, O.; Nielsen, H.; von Heijne, G. ChloroP, a neural network-based method for predicting chloroplast transit peptides and their cleavage sites. *Protein Sci.* **1999**, *8*, 978–984. [CrossRef]
53. Yang, S.; Zhang, X.; Yue, J.X.; Tian, D.; Chen, J.Q. Recent duplications dominate NBS-encoding gene expansion in two woody species. *Mol. Genet. Genom.* **2008**, *280*, 187–198. [CrossRef] [PubMed]
54. Zhou, T.; Wang, Y.; Chen, J.Q.; Araki, H.; Jing, Z.; Jiang, K.; Shen, J.; Tian, D. Genome-wide identification of NBS genes in japonica rice reveals significant expansion of divergent non-TIR NBS-LRR genes. *Mol. Genet. Genom.* **2004**, *271*, 402–415. [CrossRef] [PubMed]
55. Thompson, J.D.; Gibson, T.J.; Plewniak, F.; Jeanmougin, F.; Higgins, D.G. The ClustalX windows interface: Flexible strategies for multiple sequence alignment aided by quality analysis tools. *Nucleic Acids Res.* **1997**, *25*, 4876–4882. [CrossRef] [PubMed]
56. Tamura, K.; Stecher, G.; Peterson, D.; Filipski, A.; Kumar, S. MEGA6: Molecular evolutionary genetics analysis version 6.0. *Mol. Biol. Evol.* **2013**, *30*, 2725–2729. [CrossRef] [PubMed]
57. Letunic, I.; Bork, P. Interactive tree of life (iTOL) v3: An online tool for the display and annotation of phylogenetic and other trees. *Nucleic Acids Res.* **2016**, *44*, W242–W245. [CrossRef]
58. Guo, A.Y.; Zhu, Q.H.; Chen, X.; Luo, J.C. GSDS: A gene structure display server. *Hereditas* **2007**, *29*, 1023–1026. [CrossRef]
59. Bailey, T.L.; Williams, N.; Misleh, C.; Li, W.W. MEME: Discovering and analyzing DNA and protein sequence motifs. *Nucleic Acids Res.* **2006**, *34*, W369–W373. [CrossRef]
60. Chen, C.; Chen, H.; Zhang, Y.; Thomas, H.R.; Frank, M.H.; He, Y.; Xia, R. TBtools: An iIntegrative tToolkit developed for in-teractive analyses of big biological data. *Mol. Plant* **2020**, *13*, 9. [CrossRef]
61. Boter, M.; Calleja-Cabrera, J.; Carrera-Castano, G.; Wagner, G.; Hatzig, S.V.; Snowdon, R.J.; Legoahec, L.; Bianchetti, G.; Bouchereau, A.; Nesi, N.; et al. An integrative approach to analyze seed germination in *Brassica napus*. *Front. Plant Sci.* **2019**, *10*, 1342. [CrossRef]
62. Xin, H.; Xianchao, N.; Pan, X.; Wei, L.; Min, Y.; Yu, K.; Qin, L.; Hua, W. Comparative transcriptome analyses revealed conserved and novel responses to cold and freezing stress in *Brassica napus* L. *G3 Genes | Genomes | Genet.* **2019**, *9*, 2723–2737. [CrossRef]
63. Liu, F.; Xiong, X.J.; Wu, L.; Fu, D.H.; Hayward, A.; Zeng, X. *Braltp1*, a lipid transfer protein gene involved in epicuticular wax deposition, cell proliferation and flower development in *Brassica napus*. *PLoS ONE* **2014**, *9*, e110272. [CrossRef] [PubMed]
64. Li, J.; Gao, G.; Xu, K.; Chen, B.Y.; Yan, G.X.; Li, F.; Qiao, J.W.; Zhang, T.Y.; Wu, X.M. Genome-wide survey and expression analysis of the putative non-specific lipid transfer proteins in *Brassica rapa* L. *PLoS ONE* **2014**, *9*, e84556. [CrossRef] [PubMed]
65. Ye, K.; Li, H.; Ding, Y.; Shi, Y.; Song, C.; Gong, Z.; Yang, S. Brassinosteroid-insensitive 2 negatively regulates the stability of transcription factor ICE1 in response to cold stress in *Arabidopsis*. *Plant Cell* **2019**, *31*, 2682–2696. [PubMed]

International Journal of
Molecular Sciences

MDPI

Article

Characterization of Transcriptome Dynamics during Early Fruit Development in Olive (*Olea europaea* L.)

Maria C. Camarero [1], Beatriz Briegas [1], Jorge Corbacho [1], Juana Labrador [1], Mercedes Gallardo [2] and Maria C. Gomez-Jimenez [1,*]

1 Laboratory of Plant Physiology, University of Extremadura, Avda de Elvas s/n, 06006 Badajoz, Spain
2 Laboratory of Plant Physiology, University of Vigo, Campus Lagoas-Marcosende s/n, 36310 Vigo, Spain
* Correspondence: mcgomez@unex.es

Abstract: In the olive (*Olea europaea* L.), an economically leading oil crop worldwide, fruit size and yield are determined by the early stages of fruit development. However, few detailed analyses of this stage of fruit development are available. This study offers an extensive characterization of the various processes involved in early olive fruit growth (cell division, cell cycle regulation, and cell expansion). For this, cytological, hormonal, and transcriptional changes characterizing the phases of early fruit development were analyzed in olive fruit of the cv. 'Picual'. First, the surface area and mitotic activity (by flow cytometry) of fruit cells were investigated during early olive fruit development, from 0 to 42 days post-anthesis (DPA). The results demonstrate that the cell division phase extends up to 21 DPA, during which the maximal proportion of 4C cells in olive fruits was reached at 14 DPA, indicating that intensive cell division was activated in olive fruits at that time. Subsequently, fruit cell expansion lasted as long as 3 weeks more before endocarp lignification. Finally, the molecular mechanisms controlling the early fruit development were investigated by analyzing the transcriptome of olive flowers at anthesis (fruit set) as well as olive fruits at 14 DPA (cell division phase) and at 28 DPA (cell expansion phase). Sequential induction of the cell cycle regulating genes is associated with the upregulation of genes involved in cell wall remodeling and ion fluxes, and with a shift in plant hormone metabolism and signaling genes during early olive fruit development. This occurs together with transcriptional activity of subtilisin-like protease proteins together with transcription factors potentially involved in early fruit growth signaling. This gene expression profile, together with hormonal regulators, offers new insights for understanding the processes that regulate cell division and expansion, and ultimately fruit yield and olive size.

Keywords: cell division; cell expansion; flow cytometry; fruit growth; hormone; olive; ploidy; transcriptome

1. Introduction

The olive tree (*Olea europaea* L., Oleacea) produces fruit that rank among the world's leading oil crops [1]. Although many authors have described polyploidy [2,3], the species is diploid (2n = 2X = 46). Its genome size is approximately 1800 Mb [4]. Though it is technically a fleshy drupe from hermaphroditic flowers, the fruit is commonly called a 'stone fruit' because the seed coat is enclosed in a stone-hard, lignified endocarp [5]. The olive fruit is composed of two parts: the flesh, which is of maternal origin, and the embryo, which has different genetic origins that explain certain characteristics of the oil composition in different environments [6].

The size increase of the olive fruit depends on two distinct mechanisms: (1) young fruit growth (cell division and expansion) before endocarp lignification, and subsequently, (2) fruit pulp growth (cell expansion) after endocarp lignification (late fruit growth). As the young olive fruit accumulates water, its cells swell and also rapidly proliferate. This early developmental feature of the fruit becomes patent some 50 days following ovary fertilization, as most of the nutritional resources become mobilized at this early stage while the other parameters tend to slow down [5,6]. After this early stage and until fruit ripening, fruit pulp growth is due solely to cell expansion, and oil starts to accumulate. The oil content and composition are determined by the cultivar with little environmental influence, whereas other compounds (phenols and sterols) depend more on the environment [6]. Because of the increasing oil content during fruit pulp growth, which can reach 30% (fresh weight) at full ripening [6], and due to the high commercial value of the oil, most studies to date have attempted to elucidate the molecular bases of olive fruit ripening [7–20]; however, to date, few studies have shown candidate genes associated with early olive fruit development [21–28]. In particular, the transcriptomic control of changes associated with early fruit development in olive remains unknown.

In fleshy fruit, early development requires a precise orchestration of cell division, cell expansion, and cell differentiation by integrating endogenous signals and various environmental cues [29–34]. Fruit growth begins immediately after the ovary is pollinated, triggering vigorous cell division. This cell division is the primary driver for growth within the ovary. Spatial changes in cell division rate or duration within the ovary thus holds sway over the final fruit size [34]. In tomato (*Solanum lycopersicum*), a model system for fleshy fruit development [34,35], the prime factors controlling ultimate tomato fruit size are cell division and cell expansion [30,31,36]. Moreover, in many species, including tomato, the transition from cell division to expansion phases is accompanied by endoreduplication [31–33,37,38]. Because cell size for a certain cell type is generally proportional to the quantity of nuclear DNA, an effective strategy of cell growth is endoreduplication, which often occurs in differentiated cells that are large or metabolically highly active [39–41]. An understanding of the interaction involving endoreduplication, cell division, and cell expansion processes is vital for predicting the appearance of key morphological traits (fruit size, shape, mass, and texture). In olive, however, the ploidy levels during early fruit development have not yet been explored, and several questions remain regarding not only the rate and duration of cell division during early fruit development but also its relative importance on the final fruit size. According to previous studies analyzing the olive fruit histology by qualitative (cross-section) methods, cell division and cell expansion occur concomitantly in the first phase of fruit growth [42–44]. However, no available studies have investigated the mitotic activity of olive fruits by flow cytometry during early fruit development. Recently, it has been demonstrated that under water stress during the fruit pulp growth (cell expansion) in olive, even though olive fruit growth appears to stop, stress-induced cell division occurs, and the final size of the olive fruit is not severely diminished when the trees are irrigated again during the period of fruit pulp growth [45].

In recent years, notable progress has been made in understanding the molecular mechanisms that coordinate the interaction of transcriptional control and hormonal activity in early fruit development [38,46–54]. The modification of genes that act in hormonal regulation can alter fruit size, given that hormone contents can undergo major shifts during fruit growth [34]. In fact, using mutants or exogenous treatments, many studies have indicated the decisive role of hormones for early fruit-growth regulation [55–59]. In olive, high levels of cytokinins (CKs), gibberellins (GAs), and polyamines (PAs) have been reported in the developing fruits [27,60,61]. Moreover, the application of exogenous brassinosteroid (BR) promoted early fruit growth in olive, whereas blocking BR synthesis with brassinazole (Brz) slowed down the fruit growth rate [62]. Likewise, the data have provided new findings on the role of BRs in modulating the composition and gene expression of sterols and sphingolipids, these being lipophilic membrane components essential for cellular functions during early olive fruit growth [62,63]. This latter study also revealed the upregulation of β-sitosterol biosynthesis by BR at the transcriptional level during early olive fruit growth. In addition, previous data have demonstrated that free endogenous PAs may regulate olive flower anthesis mainly through S-adenosyl methionine decarboxylase (SAMDC) enzyme activity and an expression gene localized in the ovary. The same study has indicated that PAs appear to correlate positively with cell division during early fruit development in olive [27]. Nevertheless, knowledge remains scant on the hormone content and composition of olive fruits during early fruit development.

The purpose of the present study was to gain a more complete picture of gene expression and hormonal control during early fruit development in olive. For this, the cytology, ploidy, and transcriptome dynamics associated with early fruit development in combination with data on hormonal content were analyzed. The cell area and mitotic activity of the developing olive fruits were first investigated by flow-cytometric analysis in in order to characterize the duration of the cell division and expansion phases. Next, transcriptional and hormonal changes in olive fruit were investigated during early fruit development as a means of establishing candidate genes associated with distinct phases of early development. Specific potential genes and their associated hormonal pathways were determined to participate in the phases of cell division and expansion during early olive fruit development.

2. Results and Discussion

2.1. Morphological and Cytological Changes during Early Fruit Development in Olive

The early fruit development in olive was explored by monitoring weight, longitudinal and transverse diameters, and size of cells in an olive cultivar 'Picual' with large sized and elongated fruits. As shown in Figure 1, the olive fruit, after ovule fertilization (concomitant with anthesis), showed a highly reproducible growth progression in terms of increases in weight, longitudinal and transverse diameters, index shape, and growth rate up to 42 DPA, followed by a period of endocarp lignification for fruit under the experimental conditions used (Figure 1A–F). Immediately following fertilization, the developing fruits grew more in length than in width (Figure 1C,D), resulting in an increase in the fruit-shape index (Figure 1E). This trend is characteristic of the elongated fruit shape of the 'Picual' olive, and consequently its fruit size (weight) was augmented 128-fold (±6.0) from anthesis to 42 DPA. Greater cell size was apparent in both the epicarp and mesocarp (pericarp) cells of the fruit during the early stages (0–42 DPA) (Figure 1H,I). In fact, the epicarp- and mesocarp-cell area in the pericarp of one ovary at anthesis increased 5- and 13-fold, respectively, in the developing fruit at 42 DPA. Pericarp cells expanded at similar rates (cell area/day) for 21 DPA. From 21 to 28 DPA, cell expansion in the pericarp was dramatic, with the cell expansion rate doubling up to 42 DPA. Variations in pericarp thickness were linked mainly to variations in mesocarp cell size (Figure 1G–I). Thus, the 128-fold increase in olive fruit weight up to 42 DPA was related to a 7-fold increase in longitudinal diameter, a 5-fold increase in transverse diameter, and a 13-fold increase in mesocarp cell size.

2.2. Ploidy Level during Early Fruit Development in Olive

Here, for the first time, detailed quantitative data is reported at the ploidy level in olive fruit throughout early development. Cell division during early olive fruit development was characterized by flow cytometric analysis of the nuclear-DNA contents from 0 to 42 DPA (Figure 2). At anthesis (0 DPA) and the 7 DPA stage, most of the nuclei were 2C (88% and 65.2% of total nuclei, respectively) and only a low proportion of endoreduplicated nuclei >4C was found (0.8 % and 1.3 % of total nuclei; respectively, Figure 2).

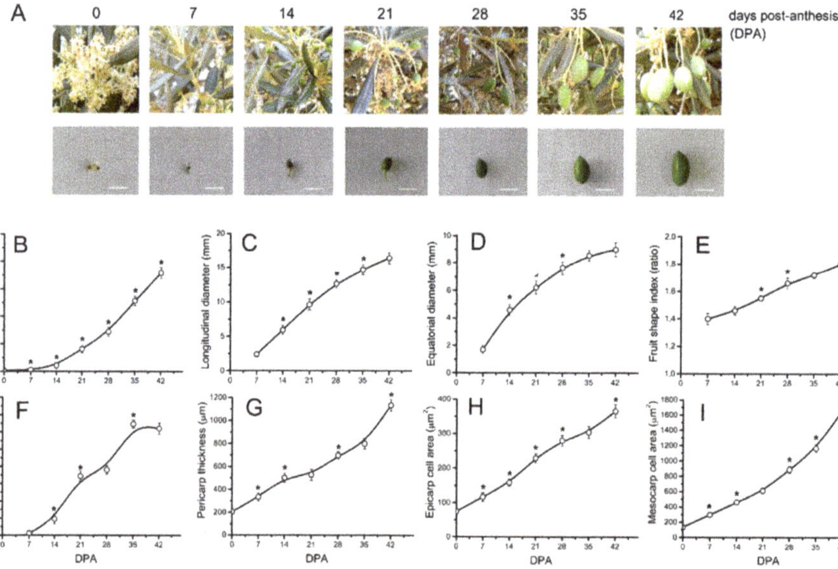

Figure 1. Growth of 'Picual' olive fruits. (**A**) Morphological changes of olive fruit during early fruit development. Increase in 'olive fruit length and diameter as a function of days post anthesis (DPA). (**B**) Changes in fresh weight (FW) (g fruit^{-1}), (**C**) longitudinal diameter (mm), (**D**) transverse diameter (mm), (**E**) fruit-shape index, (**F**) growth rate (mg FW day^{-1}), (**G**) pericarp thickness (mm), (**H**) epicarp cell size (μm^2), and (**I**) mesocarp cell size (μm^2) of developing olive fruit at 0, 7, 14, 21, 28, 35, and 42 DPA. Fruit-shape index is length-to-width ratio of the fruit. The cell area of fruit mesocarp and epicarp cells was measured during early fruit development (staining was with Calcofluor White) using confocal microscopy. Asterisks indicate statistically significant changes based on an unpaired Student's *t*-test ($p < 0.05$) from the preceding point. Data are means ± SE ($n \geq 20$). Scale bar 10 mm. DPA: Days Post-Anthesis.

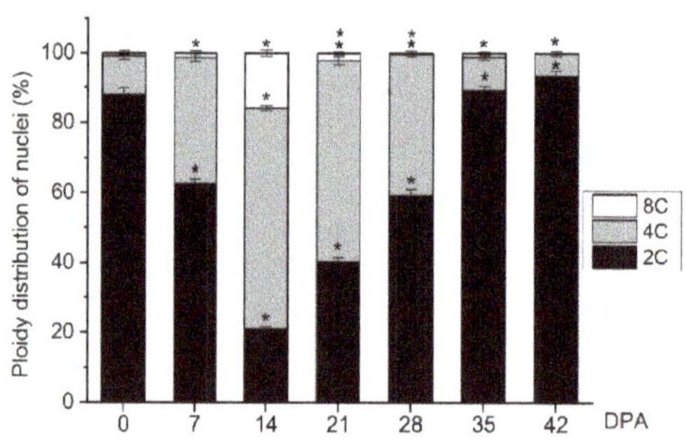

Figure 2. Nuclear ploidy levels from olive fruits during early development. The percentage of nuclei in 2C, 4C, and 8C are shown from 0 to 42 DPA in the developing fruits. Each point is the average of four samples. Asterisks denote significant differences based on unpaired Student's *t*-test ($p < 0.05$) from the preceding point and bars are ± SE.

From 0 to 14 DPA, the proportion of 4C cells increased with time, while the proportion of 2C cells decreased in olive fruits. At 14 DPA, an early stage of fruit development, 4C cells represented the highest proportion of cells in the olive fruits (63.2%), indicating intensive cell division, while the proportion of 8C cells increased in comparison to those in fruits at 7 DPA (1.3% and 15.8% of total nuclei at 7 and 14 DPA, respectively, Figure 2). By contrast, at 28 DPA, cell division was not activated in the fruits based on the increased ratio of 2C to 4C DNA levels relative to that at 21 DPA (Figure 2). The 2C cells represented 40.3% and 59.4% of the cells, and the 4C cells represented 57.6% and 40.4% of the cells in the fruits at 21 and 28 DPA, respectively. The 8C cells represented 2.1% and 0.3% of the cells in the fruits at 21 and 28 DPA, respectively. From 28 to 42 DPA, most of the nuclei were 2C (59.39, 89.3%, and 93.5% of total nuclei of fruits at 28, 35, and 42 DPA, respectively), while only low proportions of 4C and 8C nuclei were detected in these fruits (Figure 2). Thus, the results show that the maximal proportion of 4C cells, which is an indirect estimate for cell division, in olive fruits was reached at 14 DPA, indicating that intense cell division was activated in the fruits at this time.

In olive cultivars, longer durations of the cell division phase have been qualitatively reported through 8 to 10 weeks [42–44]. However, in the present study, flow cytometric analysis revealed that cell division was triggered by pollination in the 'Picual' olive fruit from 0 to 21 DPA (Figure 2). The data here offered no evidence for cell division in fruits from 28 to 42 DPA, and the increased fruit size (weight) in this period apparently resulted from cell expansion, while both cell division and expansion coexisted at gradually increasing rates in the fruits until 21 DPA (Figures 1 and 2). A greater pericarp cell size was evident both in dividing cells (0–21 DPA) and in post-mitotic expanding cells (28–42 DPA) in the 'Picual' olive fruit, which notably contributed to significantly enlarged olive fruit at all the stages examined (Figure 1).

Particularly evident is the transition between 14 and 28 DPA, which corresponds to the shift from intense cell division to cell division arrest in 'Picual' olive fruit. As expected, after the cessation of cell division, pericarp cell size most rapidly increased in 'Picual' olive fruit, and the maximum relative rate of cell expansion in the fruit occurred at 42 DPA (Figure 1); the maximum relative rate of cell division in the fruit was found at 14 DPA (Figure 2). At this stage, a low but significant proportion of endoreduplicated cells of up to 8C (one endocycle) was detected in olive fruits compared with tomato fruits (up to 256C or even 512C) [31,33,64–66]. Endoreduplication increases ploidy in individual cells and reportedly correlates not only with high metabolic activities, cell differentiation, post-mitotic cell growth, and rapid anisotropic cell expansion, but also with the capacity to react under DNA damage [67,68]. In particular, the ploidy increase is strongly correlated with increased cell size [33,67]. In tomato fruit, previous studies have reported that endoreduplication begins in developing ovaries when organogenesis-related cell division ends [31,33]. However, with the exception of some Rosaceae species (e.g., apricot, peach, and plum), endoreduplication does not occur in most of the species where fruit development lasts for a long period (over 14 weeks) of time [66], as is the case of the olive fruit.

2.3. Overall Transcriptional Changes during Early Fruit Development in Olive

Based on the cytological and ploidy analyses of early fruit development (Figures 1 and 2), to ascertain the molecular mechanism of early fruit growth in olive, a comparison was made of the transcriptome, using RNA-seq, of whole 'Picual' olive fruit at selected stages of early fruit development: 0 DPA (P0, anthesis, fruit set), 14 DPA (P14, cell division, mostly dividing cells), and 28 DPA (P28, post-mitotic cell expansion, mostly expanding cells). The use of these three samples (P0, P14, and P28) of developing 'Picual' olive fruit, which represent critical physiological changes during early fruit development, allowed the preferential identification of post-mitotic cell expansion-related genes (Figure 3A).

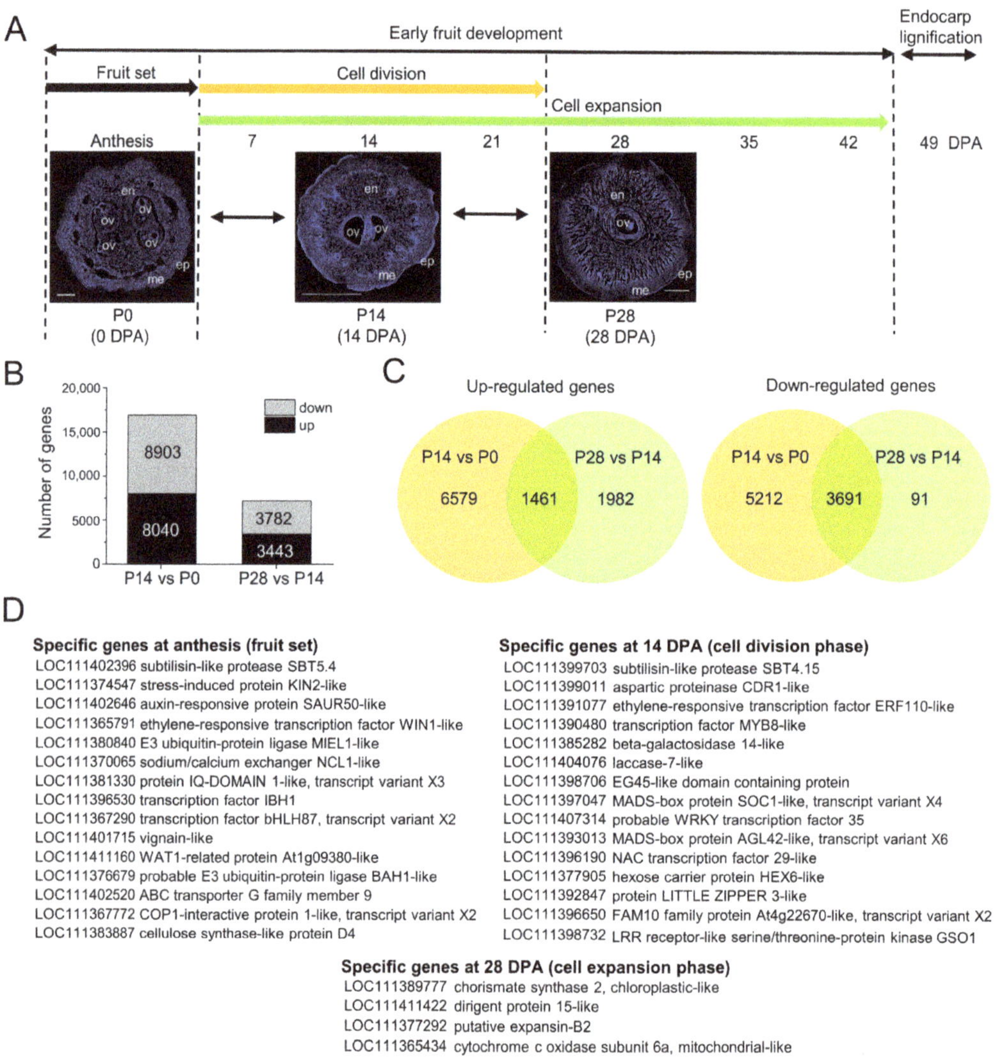

Figure 3. Design of RNA-seq analysis and olive fruit genes during early fruit development. (**A**) Cross-sections of ovaries at anthesis stage (P0) and developing fruits at 14 DPA (P14) and 28 DPA (P28) used in RNA-seq analysis. Micrographs showing changes in fruit surface during early fruit development, stained with Calcofluor White. Calcofluor White staining of cellulose indicated the cell wall; en, endocarp; ep, epicarp; me, mesocarp; ov, ovule. Scale bar 50 µm in P0, and 500 µm in P14 and P28. (**B**) Distribution of genes differentially expressed during early fruit development (P14 vs. P0, and P28 vs. P14 comparisons). (**C**) Overlap of upregulated and downregulated fruit genes during early olive fruit development. This Figure shows that 1461 were upregulated in both comparisons and that 3691 were downregulated in both comparisons. (**D**) Specific genes differentially expressed during early fruit development. Top 15 genes at anthesis (P0), and at 14 DPA (P14). Additional information is presented in Table S8.

The RNA-Seq analysis averaged 40 million reads per sample, of which 78% mapped to the 132,819 annotated transcripts from the *Olea europaea* var. *sylvestris* reference genome [69] (Table S1). From the three olive samples, a PCA was performed using transformed read counts. The three biological replicates for each sample clustered together, thus indicating that the expression levels among replicated samples were closely associated (Figure S1). A total of 24,168 differentially expressed genes (DEGs) related to early fruit development in olive were identified (Figure 3B; Tables S2–S5).

Among the 24,168 DEGs, 16,943 genes were developmentally responsive in the first comparison (between the fruit at 14 DPA and 0 DPA, P14 vs. P0), and 7225 were developmentally responsive in the second comparison (between fruit at 28 DPA and 14 DPA, P28 vs. P14) (Figure 3B; Tables S2–S5). In the first comparison (P14 vs. P0), 8040 genes were upregulated, and 8903 were downregulated in fruit at 14 DPA. In the second comparison (P28 vs. P14), 3443 genes were upregulated, and 3782 were downregulated in fruit at 28 DPA (Figure 3B; Tables S2–S5). A comparison of the genes that were developmentally responsive during early fruit development indicated that 1461 were upregulated in both comparisons, and that 3691 were downregulated in both comparisons (Figure 3C; Tables S6 and S7). Thus, most of the 24,168 DEGs display a distinct temporal expression consistent with the succession of the different phases of early fruit development (Figure 3C; Tables S2–S5).

Additionally, 162 transcripts that were identified could be considered as fruit-set specific (specific genes at anthesis) among the transcripts expressed in the fruit at 0 DPA (fruit set) since they were not detected in the other fruit development stages analyzed (Table S8; Figure 3D), including homologues of stress, cell wall, transport, cell division, transcription factor (TF), and hormone-related genes. An appreciable proportion (almost 30%) of them encode proteins with unknown functions or present no homology with any known genes. Similarly, among the transcripts expressed in the olive fruit at 14 DPA (mostly dividing cells), 67 transcripts could be considered to be cell division specific (specific genes at 14 DPA) (Table S8; Figure 3D). These mainly included genes involved in cell growth processes such as cell wall synthesis and modification, sugar and organic acid transport and metabolism, and hormonal metabolism and signaling. Among the transcripts expressed in the fruit at 28 DPA (post-mitotic cell expansion), only four transcripts (*CS2*, *DIR15*, *EXPB2* and *cox-6A*) could be considered to be cell expansion-specific (specific genes at 28 DPA) (Table S8; Figure 3D): (1) *CS2* coding for chorismate synthase 2 protein controls a key step in the shikimate pathway and catalyzes the transformation of 5-enolpyruvylshikimate 3-phosphate to chorismate, which serves as the initiator metabolite for the synthesis of aromatic amino acids and secondary metabolites (the pathway 'phenylalanine, tyrosine, and tryptophan biosynthesis' and other pathways) [70]; (2) *DIR15* coding for dirigent protein 15-like protein, which is involved in cell wall metabolism [71]; (3) *EXPB2* coding for the expansion of B2 protein, which is involved in cell expansion and other developmental events during which cell wall modification occurs [72]; finally, (4) *cox-6A* coding for cytochrome c oxidase subunit 6a protein is involved in the pathway oxidative phosphorylation, which is part of energy metabolism [73]. Moreover, for validation of the RNA-seq results, a qRT-PCR was performed to determine the levels of *OeCS2*, *OeDIR15*, *OeEXPB2*, and *Oecox-6A* expression across the three samples, which exclusively showed a significant expression level in the fruit at 28 DPA (Figure S2). Thus, they were considered to be prime candidates for further molecular research on the programs of cell division and expansion in olive fruit and their regulation.

2.4. Gene Ontology Functional Enrichment Analysis of Differentially Expressed Genes

For an overall view concerning the functions and processes altered during early olive fruit development, the DEGs were classified using the Gene Ontology (GO) database. Furthermore, GO accessions were assigned to the DEGs based on similar sequences in known proteins available in the UniProt database in addition to InterPro as well as the Pfam domains that these contain. The GO terms 'Transcription', 'ATP binding', and 'Integral component of membrane' were most represented among the categories of biological processes (Figure 4; Tables S9–S12), molecular functions (Figure S3; Tables S9–S12), and cell components (Figure S4; Tables S9–S12), respectively.

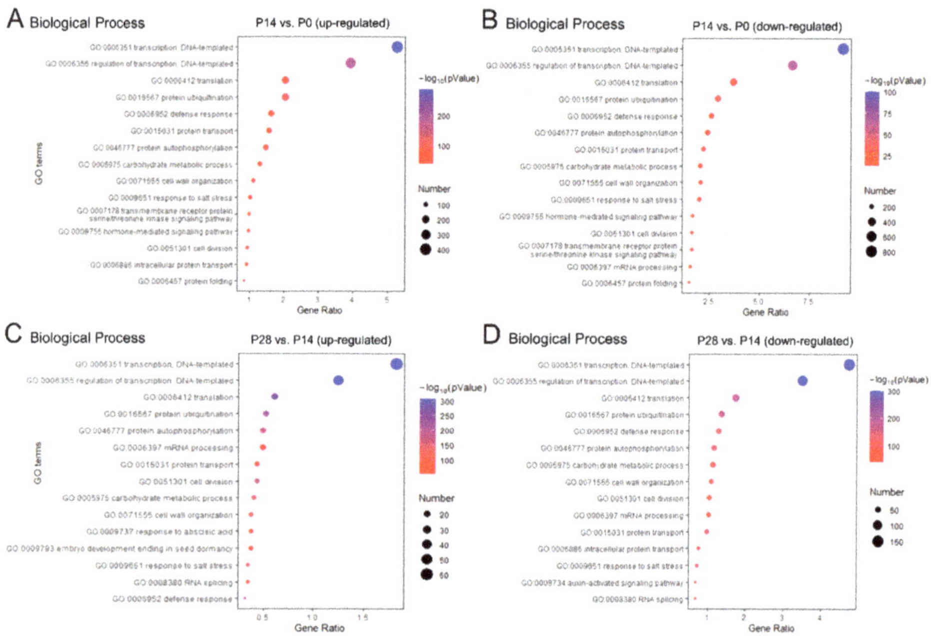

Figure 4. Functional analysis of DEGs during early olive fruit development. The enrichment analysis of GO 'biological process' terms based on DEGs in the olive flowers at anthesis stage (P0), and the developing fruits at 14 (P14) and 28 (P28) DPA. (Top 15). (**A**) Bubble Plot of GO 'biological process' terms in the GO annotations of the genes of the 8040 transcripts with increased transcript accumulation, (**B**) and of the genes of the 8903 transcripts with decreased transcript accumulation in the P14 vs. P0 comparison; (**C**) Bubble Plot of GO 'biological process' terms in the GO annotations of the genes of the 3443 transcripts with increased transcript accumulation, and (**D**) of 3782 transcripts with decreased transcript accumulation in the P28 vs. P14 comparison. The *Y*-axis and *X*-axis denote GO name and gene ratio, respectively. The color of each bubble represents $-\log_{10}$ (*p*-value), and each bubble size represents the count of DEGs. Additional information is presented in Tables S9–S12.

Within the 'Biological process' category, the over-represented group in the dividing olive fruit (14 DPA) with the greatest number among the DEGs was 'Regulation of transcription', 'Translation', 'Protein ubiquitination', 'Defense response', 'Protein transport', 'Protein autophosphorylation', 'Carbohydrate metabolic process', 'Cell wall organization', 'Response to salt stress', 'Transmembrane transport', 'Hormone-mediated signaling pathway', and 'cell division' (Figure 4A). Remarkably, the expanding olive fruit (28 DPA) also bore a significant representation of transcripts associated with 'Regulation of transcription', 'Translation', 'Protein ubiquitination', 'Protein autophosphorylation', 'mRNA processing', 'Protein transport', 'Cell division', 'Carbohydrate metabolic process', and 'Cell wall organization' (Figure 4C), indicating that the same biological processes might necessitate different gene sets in different phases during early fruit development in order to support their activities.

2.5. Characterization of Cell Cycle-Related Genes Associated with Early Fruit Development in Olive

An examination of the current data set corroborates the development-induced accumulation of transcripts presumably participating in the basic cell cycle machinery, including DEGs that encode cyclins (CYC, 48 genes), cyclin-dependent kinases (CDKs, 8 genes), CDK inhibitor proteins (CKIs, 10 genes), homologues of the retinoblastoma (RB) protein

(2 genes), and the E2F TFs (4 genes) during early olive fruit development (Figures 5 and 6; Table S13).

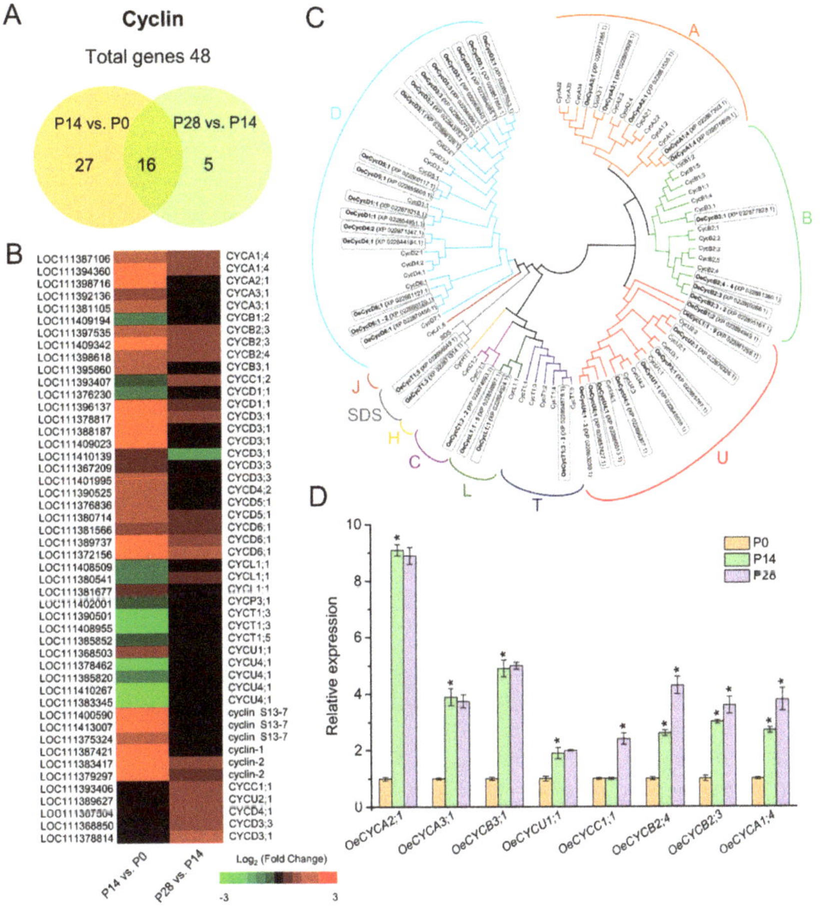

Figure 5. Differential gene expression of *CYC* genes during early olive fruit development. (**A**) Venn diagram showing numbers of overlapping *CYC* genes in both the fruit at 14 DPA (P14) versus fruit at 0 DPA (P0), and in the fruit at 28 DPA (P28) versus fruit at 14 DPA (P14) comparisons. (**B**) Expression values are represented in a heatmap as Log_2 Fold Change in the P14 versus P0 comparison, and the P28 versus P14 comparison, and the color key is indicated at the bottom. (**C**) Phylogenetic analysis of olive *CYC* with other *CYC* genes. The sequences included in this alignment are from olive (*Olea europaea* var. sylvestris Annotation Report; https://www.ncbi.nlm.nih.gov/genome/annotation euk/Olea europaea var. sylvestris/100/, accessed on 10 January 2020), and Arabidopsis (http://www.arabidopsis.org/, accessed on 10 January 2022). The CYC proteins studied from the present work are enclosed in an open box. (**D**) Expression of 8 selected CYC genes in olive fruits at 0 (P0), 14 (P14), and 28 (P28) DPA. Analysis of transcript levels of genes by qRT-PCR. Genes and their primers are shown in Table S20. Relative expression values were normalized to the lowest expression value taken as 1. The data represent the mean values (±SEs) of duplicate experiments from three independent biological samples. Statistical significance compared with the preceding point was determined using Student's *t*-test. * $p < 0.05$. Additional information on the *CYC* genes is presented in Table S13.

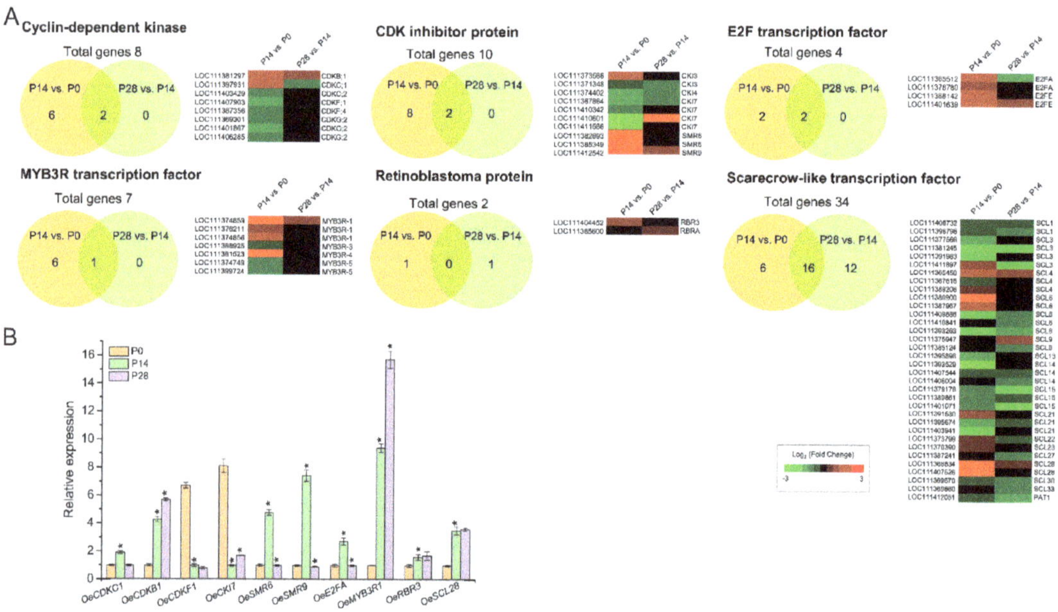

Figure 6. Differential gene expression of candidate cell cycle-related genes during early olive fruit development. (**A**) Venn diagram representing overlap between differential expressed genes *CDK*, *CKI*, *E2F*, *MYB3R*, *RBR*, and *SCL* in both the comparison of fruit at 14 DPA (P14) versus fruit at 0 DPA (P0), and the comparison of fruit at 28 DPA (P28) versus fruit at 14 DPA (P14). Expression values are represented in a heatmap as Log_2 Fold Change in the P14 versus P0, and the P28 versus P14 comparisons, and the color key is indicated at the bottom. (**B**) Expression of cell cycle-related genes in olive fruits at 0 (P0), 14 (P14), and 28 (P28) DPA. Analysis of transcript levels of cell cycle-related genes by qRT-PCR. Genes and their primers are shown in Table S20. Relative expression values were normalized to the lowest expression value taken as 1. The data represent the mean values (±SEs) of duplicate experiments from three independent biological samples. Statistical significance compared with the preceding point was determined using Student's *t*-test. * $p < 0.05$. Additional information on the cell cycle-related genes is presented in Table S13.

CYCs, the regulatory subunits of their respective CDKs, constitute major components in the cell cycle progression machinery [34,74]. In fact, certain cyclins are discontinuously expressed over the cell cycle, as their synthesis as well as their degradation prove to be strictly programmed [75]. This raises the question as to whether the intense cell division phase during early olive fruit development involves the utilization of specific cyclins. A total of 48 cyclins were included in the data set and were retained for further phylogenetic analysis, of which 5 are A-type, 4 are B-type, 2 are C-type, 16 are D-type, 2 are L-type, and 2 are U-type cyclins (Figure 5C; Table S13). Among the 48 genes, 16 (3 A-type, 1 B-type, 6 D-type, 1 L-type, and 1 U-type cyclins) were upregulated exclusively in olive fruit at 14 DPA compared with 0 DPA (P14 vs. P0 comparison), and 5 (one C-type, 3 D-type, and one U-type cyclins) were upregulated only in olive fruit at 28 DPA compared with 14 DPA (P28 vs. P14 comparison), whereas 16 genes (2 A-type, 3 B-type, 1 C-type, 7 D-type, 1 L-type cyclins, and 2 cyclin 2-like) were upregulated in both comparisons (Figure 5A,B). Among the latter, the most abundant proved to be CYCD3;1 and a CYCD6;1 proteins.

Additionally, the present data demonstrated that 11 cyclins genes (1 B-type, 1 D-type, 1 L-type, 1 P-type, 3 T-type, and 4 U-type cyclins) were downregulated in the olive fruit at 14 DPA compared with 0 DPA, and only 1 cyclin gene (D-type) was exclusively downregulated in the olive fruit at 28 DPA compared with 14 DPA, whereas 2 (one C-type and one L-type cyclin) were downregulated exclusively in olive fruit at 28 DPA compared with 14 DPA (P28 vs. P14 comparison) (Figure 5; Table S13). Moreover, differences in expression among different A-type (A1 vs. A2 and A3) cyclins and different B-type (B2 vs. B3) cyclins during early fruit development make it possible to discriminate olive fruits that have an arrested cell division phase. Expressions of A2, A3, B3, and U1 cyclins in olive fruit are restricted to intensive cell division activity during early fruit development. For further analysis of CYC family genes, a qRT-PCR analysis was conducted to assess the expression of eight members of this family in fruits at 0, 14, and 28 DPA (Figure 5D). The qRT-PCR test in the dividing olive fruit corroborated the enrichment in A2, A3, B3, and U1 cyclin genes, as well as the enrichment of A1, B2, and C1 cyclin genes in the expanding olive fruit (Figure 5D). The hypothesis is that the differences in time of expression of cyclin A2 and A3 versus cyclin A1 as well as of cyclin B3 versus cyclin B2 may reflect specific functions during olive fruit development, including A-, B-, and D-type CYCs.

Similarly, only one member of the CDK family (*CDKC1*) was upregulated exclusively in olive fruit at 14 DPA compared with 0 DPA, whereas six of the eight members of the CDK family were downregulated exclusively in olive fruit at 14 DPA compared with 0 DPA (P14 vs. P0 comparison) (Figure 6A; Table S13). Of particular interest is one member of the CDK family, homologous to CDKB1 [76], which proved to be upregulated in both comparisons (Figure 6A; Table S13). In Arabidopsis, the M-specific CDKB1;1 constitutes the plausible candidate kinase that forms part of the mitosis-inducing factor (MIF) when it is bound to the A-type cyclin CYCA2;3, which is capable of inhibiting endoreduplication when it is fully active [77]. In the present study, after fertilization, the cell division phase results mainly from the gene activity of CYCA2, CYCA3, CYCB3, CYCU1, and CDKC1. The post-mitotic cell expansion phase is associated with the gene activity of CYCA1, CYCB2, and CYCC1, whereas transcripts encoding CYCD3;1, CYCD6;1, and CDKB1 proteins are associated with both cell division and cell expansion phases during early fruit growth in olive, in accordance with previous studies [38,47,50,51,57,78]. Therefore, the present data indicate that specific CYC and CDKs are involved in different phases during early olive fruit development.

Regulation of the cell cycle and endocycle by TFs proves to be upstream of the cyclins and CDKs. Furthermore, E2F, OBP1 (DOF TF), and THREE MYB REPEAT (MYB3R) TFs have been found to be primary regulators of the cell cycle [79]. Here, differential expression patterns of other cell cycle-related genes were investigated during early olive fruit development, including *RB*, *E2F*, *MYB3R*, and *SCARECROW-LIKE* (*SCL*) (Figure 6A, Table S13), of which 10 genes were randomly examined and their expression patterns confirmed using qRT-PCR (Figure 6B). E2Fs are known to target genes involved in DNA repair and chromatin dynamics at the transition between the G1 and S phase. In Arabidopsis, E2FA, a protein involved in cell cycle regulation, exerts a dual function, not only maintaining cell proliferation but also stimulating the cell expansion needed in differentiating cells for the growth of organs by promoting endoreduplication [80]. In the present analysis, two members of the E2F TF family homologous to E2FA and E2FE were found to be upregulated in olive fruit at 14 DPA (P14 vs. P0), whereas one other gene homologous to E2FE was upregulated in both comparisons during early olive fruit development (Figure 6; Table S13). In addition, the cyclin D family has a major part in the regulation of the RBR/E2F pathway, activating CDK-mediated RBR phosphorylation and upsetting its interactions with E2F in Arabidopsis [74,81,82]. The results shown here indicate a correlation between the expression of members of the cyclin D family, such as *CYCD3;1* and *CYCD6;1*, and a number of genes, such as *RBR3*, *E2FA*, and *E2FE* (Figure 6A; Table S13), which could affect the balance of mitotic and endoreduplicating cells or the number of endocycles during early olive fruit growth.

By binding to and inhibiting cyclin-dependent kinase complexes, CKIs promote sustained endoreduplication in cells [67]. In addition, one gene homologous to CKI3 and two genes homologous to SIAMESE-RELATED 6 (SMR6) were exclusively transcribed in the dividing olive fruit at 14 DPA, whereas one gene homologous to SMR9 was upregulated in both comparisons. In contrast, 5 of the 10 members of the CKI family were downregulated exclusively in olive fruit at 14 DPA compared with 0 DPA, whereas only one *CKI7* gene was downregulated in both comparisons during early fruit development (Figure 6A, Table S13). Notably, it was found that one gene homologous to AP2-like ethylene-responsive type TF (*SMOS1*), a putative orthologue of the AtSMOS1 (At2g41710) protein in Arabidopsis, is transcriptionally induced during early olive fruit development (Table S6). SMOS1 is involved in the transcription activation of a specific set of SMR family genes, encoding plant-specific CKI and thus inhibiting cell cycle progression at G2 and promoting the onset of endoreplication [67]. In the analysis, *SMR9* exhibited a similar expression pattern as that of *SMOS1* (Figure 6A, Tables S6 and S13). The transcript level of *SMR9* rose in both comparisons during early olive fruit development, leading to the hypothesis that SMOS1/SMR9 is involved in cell cycle control, quickly triggering the repression of cell division during early olive fruit growth. Recently, [83] documented that AtSMOS1 forms a dimer with SCARECROW-LIKE28 (SCL28), a GRAS TF that in Arabidopsis, plays a critical part in regulating cell size as part of a transcriptional network downstream of the central MYB3Rs that regulate the G2 to M phase of cell cycle transition. In the present work, the early fruit growth in olive was found to be associated with upregulated *SCL28* and *MYB3R-1* transcripts (Figure 6A). This suggests that *SCL28* controls cell expansion and differentiation by promoting endoreplication onset during early fruit growth. Thus, the expression of *SCL28*, *SMOS1*, and *MYB3R-1* shows to positively correlated with early fruit growth in olive, indicating that the regulation of cell division and expansion in the olive fruit can be similar to that in other systems.

2.6. *Differing Hormonal Composition and Gene Expression Patterns during Early Fruit Development in Olive*

According to the previous data, GO enrichment identified hormone-mediated signaling pathways that may be key during early olive fruit development (Figure 4). Firstly, in the investigation of a potential relationship between early fruit development and hormonal composition, the hormone profiles of indole-3-acetic acid (IAA), GAs (GA_1 and GA_4), CKs [*trans*-Zeatin (*t*Z)], abscisic acid (ABA), jasmonic acid (JA), and salicylic acid (SA) were examined during early olive fruit development. As shown in Figure 7, all detected hormones decreased during the first 14 days of fruit development in olive fruit, except *t*Z, SA, and IAA. In particular, the results reveal that the *t*Z levels rose during the first 14 days of fruit development, only to disappear later from 14 to 28 DPA in olive fruit. SA levels also rose from the anthesis stage to fruit at 14 DPA, and later fell by 46% in fruit at 28 DPA, while IAA levels rose during early fruit development, reaching a maximum at 28 DPA.

The hormone profiling revealed that the olive flower at anthesis stage (fruit set) is characterized by high levels of JA and GA_1 (Figure 7). In fact, JA was the most abundant in olive flower at anthesis, but JA levels sharply fell during early fruit development (Figure 7). Similarly, upon fertilization, a decrease of GA_4 as well as ABA levels accompanied by a loss of GA_1 were detected in olive fruit at 14 DPA. By contrast, at 28 DPA, the fruit showed the highest peak of hormones with a rise in IAA, GA_1, GA_4, and ABA levels from 14 to 28 DPA, while the *t*Z content was undetected in the fruit at 28 DPA. Notably, GA_1 was absent from the fruit at 14 DPA but was enriched at 28 DPA (Figure 7), suggesting a specific association between this hormone and the cell expansion phase in olive fruit development. Hence, *t*Z, SA, IAA, GA_1, GA_4, and ABA were found to be related to early olive fruit growth. More specifically, this study indicates that, upon fertilization, the olive fruits grow as cell division prevails over expansion associated with high levels of *t*Z and low levels of GA_4 (loss of GA_1). At 28 DPA, the loss of *t*Z and high levels of GA (GA_1 and GA_4) in olive fruit boost fruit growth mainly in terms of cell expansion, concomitantly with a rise in the level of IAA. Thus, IAA, GA_4 and ABA have higher relative importance in the phase when olive fruit growth depends mainly on cell expansion, while CKs (*t*Z) and SA were greater during the phase of more intense cell division.

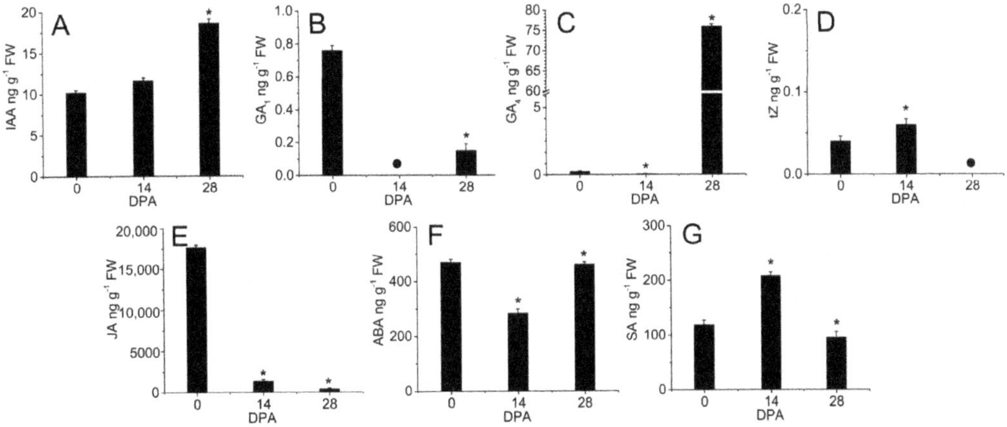

Figure 7. Profiles of (**A**) indole-3-acetic acid (IAA), (**B**) gibberellin 1 (GA_1), (**C**) gibberellin 4 (GA_4), (**D**) *trans*-Zeatin (*t*Z), (**E**) jasmonic acid (JA), (**F**) abscisic acid (ABA), and (**G**) salicylic acid (SA) levels in olive fruits at 0 (P0), 14 (P14), and 28 (P28) DPA during early fruit development. Hormone levels not detected are indicated by a black dot (•). Data are the means ± SD of three biological replicates with three technical repeats each. Statistically significant differences based on unpaired Student's *t*-test at $p < 0.05$ are denoted by an asterisk.

The role of auxins in early fruit growth has been amply reported [46,49,84]. In tomato, auxin signaling appears to be a prerequisite for the expansion of the fruit locular cells [47], for the negative control of the division of the fruit cells [85], and for fruit set [49,86]. A divergence was noted in auxin-related gene expression for cell division versus cell expansion during early olive fruit development (Figure 8; Table S14; Figure S5). Transcripts involved in IAA synthesis, such as a transcript encoding tryptophan aminotransferase protein 4 (TAR4), were upregulated in the dividing olive fruit at 14 DPA (P14 vs. P0), while *YUCCA10* was downregulated during early olive fruit development (Figure 8). In addition, the IAA-amino acid hydrolase genes (*ILR1-like 1*, *ILR1-like 3*, and *ILR1 like 5*), involved in auxin homeostasis are exclusively overexpressed in the dividing olive fruit at 14 DPA (Figure 8). Auxin-amino acid hydrolase may provide local concentrations of auxin within the olive fruit during early development. However, expression levels for the auxin-conjugating enzymes *GH3.6*, *GH3.9*, and *GH3.10* were also found to be boosted in the dividing fruit (P14 vs. P0) during early fruit development. Furthermore, these *GH3.6* and *GH3.9* genes show downregulated expression in the expanding fruit (P28 vs. P14) during early olive fruit development, consistently with the highest IAA levels detected in the fruit at this time (Figure 8; Table S14; Figure S5). In addition, this is consistent with the accepted idea that the release of auxins produced by seeds and/or surrounding fruit tissues triggers fruit growth through cell expansion [29,47]. Nevertheless, the *GH3.1* gene shows downregulated and upregulated expressions in the dividing fruit (P14 vs. P0) and in the expanding fruit (P28 vs. P14), respectively, during early olive fruit development (Figure 8; Table S14; Figure S5). Similarly, during early olive fruit development, genes specifying auxin-signaling components (*TIR1*, *IAA9*, *ARF4*, *ARF8*, *ARF9*, *ARF18*, and *SAUR50)* were upregulated in both comparisons, while some genes (*IAA8*, *IAA26*, *ARF1*, *ARF2*, *ARF3*, *ARF5*, *ARF6 ARF10*, *ARF17*, *ARF19*, and *ARF22)* were exclusively upregulated in the dividing olive fruit (P14 vs. P0), and others (*IAA29* and *SAUR72)* were exclusively upregulated in the expanding olive fruit (P28 vs. P14) during early olive fruit development. Conversely, numerous auxin-regulated genes that were downregulated during early fruit development (*IAA14*, *IAA26*, *SAUR32*, *SAUR36*, *SAUR41*, *SAUR62*, *SAUR66*, *ARGOS*, *DAM1*, *DAM4*, and *PCNT115*, among others) were detected (Figure 8; Table S14; Figure S5). Furthermore, transcript levels for two auxin efflux carriers, *PIN1* and *PIN7*, rose in the dividing fruit and lowered in the expanding fruit, while *PIL2*, *PIL6*, and *PIL7* expressions showed the opposite expression, suggesting that the latter participates in the transition from cell division to expansion during early olive fruit development. However, transcripts encoding auxin influx carrier-like proteins (*LAX2*, *LAX4* and *LAX5*) increased exclusively in the dividing fruit during early development, indicating altered auxin distribution in the fruit at this time. Therefore, the regulation of auxin conjugation, transport, and signaling may differ during early olive fruit development. Rises in endogenous IAA levels in the expanding olive fruit coincided with lower expression levels for the auxin conjugating enzymes, GH3.6 and GH3.9, and for two auxin efflux carriers, PIN1 and PIN7, while the transcript of IAA29, involved in auxin signaling, was upregulated exclusively in the expanding olive fruit, suggesting that *IAA29* may be a key candidate gene of auxin signaling during the cell-expansion phase in early-developing olive fruit.

GA, as well as auxins, reportedly serve as early signals for fruit set as well as for strong fruit growth [87]. A fainter GA stimulation, as well as weakened auxin signaling, has been reported to induce the change from a regular cell cycle to an endocycle [74]. In the case of GA, the present data suggest that GA synthesis is upregulated by *GGPS*, coding geranylgeranyl pyrophosphate synthase (GGPS), and by *GA20ox*, coding GA 20-oxidase; these two genes are active during early olive fruit development. Interestingly, *GA3ox* (GA 3-oxidase), which is a transcript involved in a late GA biosynthetic step, was downregulated in the dividing fruit (P14 vs. P0) and upregulated in the expanding fruit (P28 vs. P14), consistently with the highest GA (GA_1 + GA_4) levels detected in olive fruit at 28 DPA (Figure 8; Table S14; Figure S5). Thus, these results suggest that GA3ox is the major late-step enzyme responsible for the high-level accumulation of GA_4 detected in the expanding olive fruit, whereas the complete absence of GA_1 observed in the dividing olive fruit may be attributed to the expression of the *GA2ox* gene, which is a transcript involved in the deactivation of bioactive GAs. Additionally, the data indicate that GA signaling is negatively regulated by *GAI1* and *RGL1* (DELLA proteins) in the cell-division phase, whereas GA signaling (*GID2* receptor) is upregulated in the expansion phase during early olive fruit development (Figure 8; Table S14; Figure S5).

These results agree with the established notion that GA participates mainly in the cell expansion process in the fruit [34]. Moreover, PACLOBUTRAZOL RESISTANCES 6 (PRE6), belonging to the bHLH TF family, which is involved not only in auxin, GA, and BR signaling, but also in light responses regulating cell expansion in Arabidopsis [88–90], was downregulated during early olive fruit development (Table S8). The downregulated *PRE6* expression is hypothesized to promote cell expansion through GA-, auxin-, and BR-modulating response in olive fruit, as in Arabidopsis. In tomato, SlPRE2 and AtPRE1 proteins are involved in cell elongation through GA modulating response [91]. In this light, it is proposed that an increase in auxin and GA contents as well as GA signaling stimulation trigger the shift from a regular cell cycle to the endocycle in olive fruit, whereas the two hormones, auxin and CK, could activate the cell cycle in the developing fruit, according to absolute levels, cell sensitivity, and other signaling cross-talk. In fact, free CK hydrolyzed from its conjugates by CK riboside 5′-monophosphate phosphoribohydrolases (LOG) exclusively in the dividing fruit could in turn activate and coordinate the expression of cell division-related proteins (via ARR12). In particular, it bares noting that the upregulation of the expression of the three different *LOG3, LOG8,* and *LOG10* genes was found during the first 14 days of olive fruit development (P14 vs. P0), which is consistent with the highest *tZ* levels detected in olive fruit at 14 DPA (Figure 8; Table S14; Figure S5).

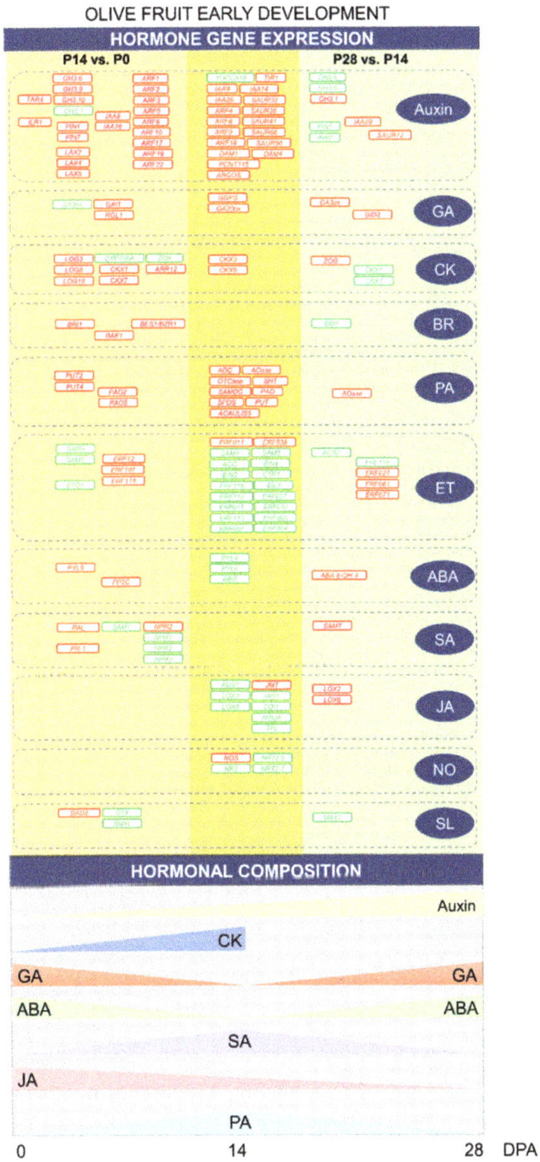

Figure 8. Summary figure including hormonal composition and the most important hormone-related DEGs during early fruit development in olive. DEGs in red represent genes that increased their expression and DEGs in green represent genes that decreased their expression. The middle column of the figure represents DEGs in common in both comparisons. Additional information on the hormone-related genes is presented in Table S14.

By contrast, both *CYP735A* (CK hydroxylase), which catalyzes the biosynthesis of trans-zeatin (*t*Z), and *ZOX*, which encodes zeatin O-xylosyltransferase involved in CK conjugation, were exclusively downregulated in the dividing olive fruit (P14 vs. P0) (Figure 8; Table S14; Figure S5). However, transcript encoding for another zeatin conjugating enzyme *ZOG* (zeatin O-glucosyltransferase) increased exclusively in the expanding olive fruit at

28 DPA, consistently with the loss of *tZ* detected in the olive fruit at 28 DPA (Figure 7). These results suggest local transcriptional control of the CK biosynthesis and conjugation rather than transport from other tissues. Bioactive forms of CKs are, in turn, deactivated by CK oxidase/dehydrogenase (CKX), and thus regulate the amount of bioactive CK. In the present study, *CKX1* and *CKX7* gene expression were upregulated in the P14 versus P0 comparison, and downregulated in the P28 versus P14 comparison, while *CKX3* and *CKX9* gene expression were upregulated in both comparisons, suggesting that the deactivation of CK may actively occur during early olive fruit development (Figure 8; Table S14; Figure S5). Therefore, the upregulated expressions of the three different *LOG3, LOG8,* and *LOG10* genes during the cell-division phase are probably sufficient to maintain *tZ* at levels higher than those that stimulated the fruit growth in olive, suggesting a key role of the LOG-dependent pathway in CK activation during the cell-division phase. Similarly, the upregulation of the expression of CK-responsive *ARR12* TF exclusively in the dividing olive fruit suggests that the *ARR12* gene activates CK-responsive gene expression during the fruit growth phase to ensure a prevalence of cell division over cell expansion.

Furthermore, the BR signaling was clearly stimulated throughout early olive fruit development, since three *BRI1* receptor kinase, two *BAK1 (SERK2)* co-receptors, and three *BES1/BZR1* factor transcriptions were exclusively upregulated in the dividing olive fruit (P14 vs. P0), while only one different receptor *BRI1* gene was exclusively upregulated in the expanding fruit (P28 vs. P14; Figure 8; Table S14; Figure S5), suggesting that the upregulation of receptor *BRI1* may be required for olive fruit development.

In the case of PA, the data imply that early olive fruit growth is mediated by upregulated PA biosynthesis and metabolism, including *ADC* (arginine decarboxylase), *AOase* (acetylornithine deacetylase), *OTCase* (ornithine carbamoyltransferase), *ACAULIS5* (thermospermine synthase), *SAMDC* (S-adenosylmethionine decarboxylase), *SPDS* (spermidine synthase), *SHT* (spermidine hydroxycinnamoyl transferase), *PUT* (polyamine uptake transporter), and *PAO* (polyamine oxidase) genes (Figure 8; Table S14; Figure S5). This agrees with a prior study of the PA levels as well as the induction and spatial expression patterns of *SPDS* and *SAMDC* genes involved in spermidine biosynthesis during early olive fruit development [27]. In fact, both comparisons are apparently characterized by an active PA biosynthesis and conjugation (Figure 8; Table S14; Figure S5), although the upregulation of two different PA transporter *PUT* (*PUT2* At1g31830 and *PUT4* At3g13620) genes, as well as two different *PAO* (*PAO2* and *PAO5*) genes exclusively in the dividing olive fruit, suggest that PA transport and catabolism may significantly influence the role of PA in the cell-division phase. Recently, the importance of PA transporters has been shown in the regulation of flowering and senescence pathways [92]. Substantial information is available on the PA transport systems in bacteria, yeast, and mammals, but the role of PA transporters in plants is only now being recognized [92].

By contrast, a downregulation of genes involved in ethylene (ET) and ABA biosynthesis occurred during early olive fruit development (Figure 8; Table S14; Figure S5). Indeed, five S-adenosylmethionine synthase (*SAM1, SAM2, SAM3, SAM5*) genes, one 1-aminocyclopropane-1 carboxylic acid (ACC) synthase (*ACS*) gene, seven ACC oxidase (*ACO*) genes, and one ethylene-overproduction protein 1 (*ETO1*) gene were downregulated in the dividing olive fruit (P14 vs. P0), while two *SAMS* (*SAM1*, and *SAMS2*) genes, one ACC synthase (*ACS2*) gene, and seven ACC oxidase (*ACO*) genes were downregulated in the expanding olive fruit (P28 vs. P14), suggesting that the early olive fruit development can apparently be characterized by an inactive ET biosynthesis at the transcriptional level. Similarly, other genes related to ET signaling, such as *EIN4, EIN2, CTR1, EIL3* (EIN3-*like, EIL*), and *ERF* (*ERF001, ERF004, ERF005, ERF010, ERF011, ERF027, ERF104, ERF110, ERF113, ERF116, ERF118*), among others, were also downregulated during early olive fruit development (Figure 8; Table S14; Figure S5). Likewise, the upregulation of the genes related to ET signaling during early olive fruit development was also found (Figure 8; Table S14; Figure S5): ERF011 and ERF038 were related to early olive fruit growth, whereas ERF12, ERF107, and ERF118 were likely to be associated with a period of more intense cell division, and ERF027, ERF061, and ERF071 tended to be associated with the cell-expansion period.

Interestingly, the present analysis also found components of ABA signaling repressed during early fruit development, such as two *PYR/PYL (PYL8* and *PYL4,* ABA receptors) genes and one *ABI5* (bZIP TF) gene. By contrast, one *PYR/PYL (PYL9)* and seven *PP2C* genes were exclusively upregulated in the dividing olive fruit (Figure 8; Table S14), suggesting that this PYL9 receptor forms part of an ABA signaling unit that modulates early olive fruit growth. Thus, these results suggest that ABA signaling was downregulated during the cell-expansion phase, and one *ABI5* gene is the major TF controlling ABA-mediated gene expression during olive fruit growth.

Likewise, several genes related to SA were upregulated exclusively during the first 14 days of fruit development, such as *PAL, NPR2,* and *PR-1* genes, consistently with the highest SA levels detected in the fruit at 14 DPA. Meanwhile, other genes such as *SAMT, NPR1, NPR3,* and *NPR5* genes were downregulated in the dividing olive fruit (P14 vs. P0) (Figure 8; Table S14; Figure S5), but one *SAMT* transcript, which encodes increased SA carboxyl methyltransferase (an enzyme responsible for biosynthesis of methyl salicylate) in the expanding olive fruit, was in good agreement with a loss of SA in the expanding olive fruit at 28 DPA (Figure 7; Table S14; Figure S5). Conversely, *JMT* genes encoding a jasmonate-O-methyltransferase, which catalyzes the methylation of JA into methyljasmonate (MeJA), were also upregulated during early fruit development, suggesting that the MeJA could regulate the cell growth rate during early olive fruit development. JA conjugation with isoleucine is reportedly catalyzed by jasmonoyl-isoleucine (JA-Ile) synthetase (JAR1), which is a member of the GH3 family [93]. The most bioactive of the Jas is JA-Ile. The perception of JA-Ile by its coreceptor, the Skp1-Cullin1-F-box-type (SCF) protein ubiquitin ligase complex SCF^{COI1}-JAZ, derepresses the transcriptional repression of target genes in the nucleus. JA-Ile signaling participates in regulating several developmental processes, such as root growth and architecture, tuber and trichome formation, seed germination, and particularly reproductive-organ development [94]. Here, it is reported that *JAR1, COI1, NINJA,* and *TPL* (TOPLESS) genes were downregulated in the olive fruit during early development (Figure 8; Table S14; Figure S5), suggesting that these components may aid in JA-Ile signaling in the nucleus to regulate early fruit development. Notably, these results also suggest that *AOS1,* which encodes an allene oxide synthase (AOS) protein, is the major control point responsible for the drop in JA detected in olive fruit during early development. AOS, the second enzyme in the biosynthesis of the JA, is a regulatory point in the biosynthesis of JA [95]. Interestingly, this analysis showed that *LOX3* (13-LOX member) and *LOX5* (9-LOX member) genes encoding lipoxygenase members were downregulated in the olive fruit during early development, whereas *LOX2* (13-LOX member) and *LOX6* (13-LOX member) were exclusively upregulated in the expanding olive fruit at 28 DPA (Figure 8; Table S14; Figure S5), indicating that these genes modulate the distribution between 13-LOX- and 9-LOX-derived oxylipins during early fruit development.

Finally, the identification of the proteins involved in oxide nitric (NO) formation/ transport and strigolactone (SL) signaling in olive fruit constitutes a promising avenue of research for a better comprehension of NO and SL physiological functions during early fruit development. Here, several transcripts related to NO metabolism were also differentially expressed during early olive fruit development, such as *NOS* genes encoding putative NO synthase, and *NR2* genes encoding putative nitrate reductase (NR), which were upregulated and downregulated during early fruit development, respectively (Figure 8; Table S14; Figure S5). Similarly, *NRT2.5* and *NRT2.7* gene-encoding member nitrate transporters were downregulated in the olive fruit during early development. In addition, genes involved in the SL signaling pathway, such as the receptor *DAD2* (SL esterase), were upregulated exclusively in the dividing olive fruit at 14 DPA (P14 vs. P0), whereas the receptor *D14* (α/β hydrolase DWARF14) and the transcriptional repressor *SMXL* (SUPPRESSOR OF MAX2 LIKE) genes were exclusively downregulated in the dividing olive fruit during early development, suggesting that these components act to regulate karrikins/strigolactone responses during the cell division phase of early olive fruit development. In contrast to this, it is shown here that one *MAX2* gene encoding MORE AXILLARY GROWTH2 (MAX2), an F-box protein that is part of a SKP1-Cullin-F-box (SCF) ubiquitin ligase complex, was exclusively downregulated in the expanding olive fruit at 28 DPA (Figure 8; Table S14; Figure S5).

Therefore, the comprehensive study of varying gene expression, together with analyses of different hormonal composition, unveils complex hormone control underlying early fruit growth in olive, with intricate temporal variations, implying distinct regulatory programs. Key candidate genes involved in CK (*LOG3, LOG8, LOG10, CKX1, CKX7,* and *ARR12*) and auxin (*LAX2, LAX4, LAX5, PIN1, PIN7, GH3.6, GH3.9, GH3.10, IAA8, IAA26, ARF1, ARF2, ARF3, ARF5, ARF6, ARF10, ARF17, ARF19,* and *ARF22)* metabolism and signaling showed a preferential expression in the cell-division phase, while some other candidate genes encoding proteins with roles in auxin (*PIL2, PIL6, PIL7, IAA29, GH3.1,* and *SAUR72)* and GA (*GA3ox* and *GID2*) metabolism and signaling could act in the cell-expansion phase during early olive fruit growth, suggesting that CKs as well as auxins positively affect cell division, whereas the GA and auxin hormones display positive effects on cell expansion during early fruit development in olive. Likewise, cross-talk between CKs, GAs, and auxins stimulate cell growth that is likely in the developing olive fruit where co-expression of the following occurs: *YUCCA10* (auxin synthesis); *TIR1, IAA9, ARF4, ARF8, ARF9, ARF18,* and *SAUR50* (auxin signaling); *PIN1* and *PIN5* (auxin transport); *GGPS* (GA synthesis); *CKX3* and *CKX9* (CK metabolism); *HK3, HK4,* and *ARR5* (CK signaling). In addition, the data imply that both cell-division and cell-expansion phases are mediated by upregulated BR signaling (receptor *BRI1)*, and downregulated ABA (*ABI5*) and ET (*ERF001, ERF004, ERF005, ERF010, ERF011, ERF027, ERF104, ERF110, ERF113, ERF116,* and *ERF118)* signaling, as well as downregulated JA biosynthesis (*AOS*). In contrast, the cell division is mediated primarily by upregulated SA synthesis (*PAL*), auxin (*PIN1* and *PIN7*) and PA (*PUT2* and *PUT4)* transport, GA catabolism (*GA2ox*), and CK signaling (*ARR12*), and the cell expansion is mediated mainly by upregulated GA biosynthesis (*GA3ox*), as well as upregulated auxin (*IAA29*) and GA (*GID2*) signaling during early olive fruit development.

2.7. Signaling Peptides Regulating Early Fruit Development in Olive

The molecular identifications for the signaling peptides during early fruit development are still limited. Signaling peptides in the role of phytohormones control several features of plant growth and development through cell–cell communication networks, e.g., organ abscission, meristem maintenance, gravitropism, cell proliferation and differentiation, cell elongation, and defense [96]. These peptides can be recognized typically by membrane-embedded receptor-like kinases which activate cell signaling to govern plant growth and development. In the present analysis, more than 40 genes encoding putative leucine-rich repeat receptor-like kinases (LRR-RLKs) family proteins, involved in plant peptide signaling, differed in expression level during early olive fruit development (Table S15). Among these, *EMS1* gene encoding for the LRR-RLKs family protein was found to be exclusively detected in the dividing olive fruit. Similarly, one cysteine-rich receptor-like protein kinase 29 (CRK29) and one probable receptor-like protein kinase At5g38990 (MDS1), which are involved in growth adaptation upon exposure to metal ions [97], were found to be detected exclusively in the dividing olive fruit (Table S15). Likewise, more than 400 genes have been found to encode putative proteases (also referred to as peptidases or proteinases), including serine proteases, cysteine proteases, aspartic proteases, and Clp proteases, among others (Table S16). Proteases are enzymes able to participate in almost all stages of plant life [96,98]. In particular, two genes encoding for serine carboxypeptidase-like 45, and serine carboxypeptidase-like 2 were found to be detected only in the dividing olive fruit (Table S16).

Moreover, two genes, *SBT5.4,* and *SBT4.15,* coding for subtilisin-like protease proteins (also known as subtilases; SBT) were found to be detected exclusively in the flower at anthesis (fruit set), and in the dividing fruit at 14 DPA, respectively (Table S15). Although previous studies have indicated that SBT-mediated processing results in the activation of peptide signals regulate stress-induced flower drop, the formation of the embryonic cuticle, and pollen development [98], little has been reported concerning its role in early fleshy fruit development.

The present analysis identified 32 putative SBT genes in olive fruit that differed in expression level during early fruit development (Figure 9A,B), and were retained for further phylogenetic analysis (Figure 9C). Among 32 SBT genes identified in the analysis, 11 were exclusively expressed in the dividing olive fruit at 14 DPA, while 4 were expressed only at 28 DPA (Figure 9; Table S15), indicating that at least some members of SBTs play major roles during early olive fruit development. Overall, 19 SBT genes were upregulated and 9 were downregulated in fruit at 14 DPA compared with 0 DPA, whereas 11 SBT genes were upregulated and 8 were downregulated in the expanding fruit at 28 DPA compared with 14 DPA (Figure 9; Table S15), of which five genes were randomly examined and their expression patterns were confirmed using qRT-PCR (Figure 9D). Notably, it is shown that, after fertilization, transcripts that encode SBT1.1, SBT1.3, SBT1.5, SBT1.6, and SBT1.8 are associated with both cell division and cell expansion phases, whereas the cell division phase involves mainly the gene activity of SBT1.2, SBT1.9, and SBT3.5 proteins. Furthermore, the cell expansion phase is associated with the gene activity of SBT2.5 and SBT3.6 proteins during early olive fruit growth (Figure 9; Table S15). Previously, SBT3.5 and SBT6.1 proteins have been shown to promote cell expansion [99], but cross-talk involving different signaling peptide pathways controlling cell expansion has not been elucidated, nor has their relation to hormonal growth control.

2.8. Transcript Changes in Cell Wall Biosynthesis and Remodeling during Early Olive Fruit Development

In this study, among the DEGs, 895 genes were identified that encode proteins with probable functions in cell wall biosynthesis and remodeling during early olive fruit development (Table S17), of which 300 were upregulated and 258 were downregulated in the P14 versus P0 comparison, and 195 were upregulated and 142 were downregulated in the P28 versus P14 comparison (Table S17). Overall, 119 genes were upregulated, and 75 genes were downregulated in both comparisons during early fruit development (Figure 10; Table S17). The well-represented families included cellulose synthase (CES, 59 genes), arabinogalactan protein (AGP, 59 genes), polygalacturonase (PG, 55 genes), pectin methylesterase (PME) (53 genes), glucan 1,3-β-glucosidase (BG; 43 genes), endo-1,4-β-glucanase or cellulase (EGase/CEL, 41 genes), expansin (EXP, 39 genes), xyloglucan endotransglucosylase/hydrolase (XTH, 37 genes), laccase (LAC, 34 genes), β-galactosidase (βGAL, 28 genes), and extensin (EXT, 28 genes) proteins (Figure 10; Table S17), implying that these cell wall-related enzymes help regulate early olive fruit development.

EGase, XTH, and EXP are cell wall-loosening factors expressed during the stages of maximum growth in fleshy fruit [46,50,53]. An examination of these families identified genes that were upregulated exclusively in the dividing olive fruit (14 DPA), such as genes that encode one EXP (EXPA13), four EGase/CELs, two XTHs (XTH9, XTH10), five CSs (CESA2, CESA3, CESA6, CESE6, CESG2), two AGPs (AGP26, AGP31), three EXTs (EXT2, EXT3, EXT6), and two LAC (LAC15, LAC7) proteins (Figure 10; Table S17). Similarly, genes upregulated only in the expanding olive fruit (28 DPA) were identified, such as genes encoding two EXP (EXPA15, EXPA8), two EGases/CEL (CEL10, CEL25), two XTH (XTH32, XTH33), one CS (CESD4), four AGP (AGP10, AGP19, AGP20, AGP26), and one EXT (EXT1) proteins (Figure 10; Table S17), suggesting that these cell wall-related enzymes regulate cell expansion during early fruit development in the olive.

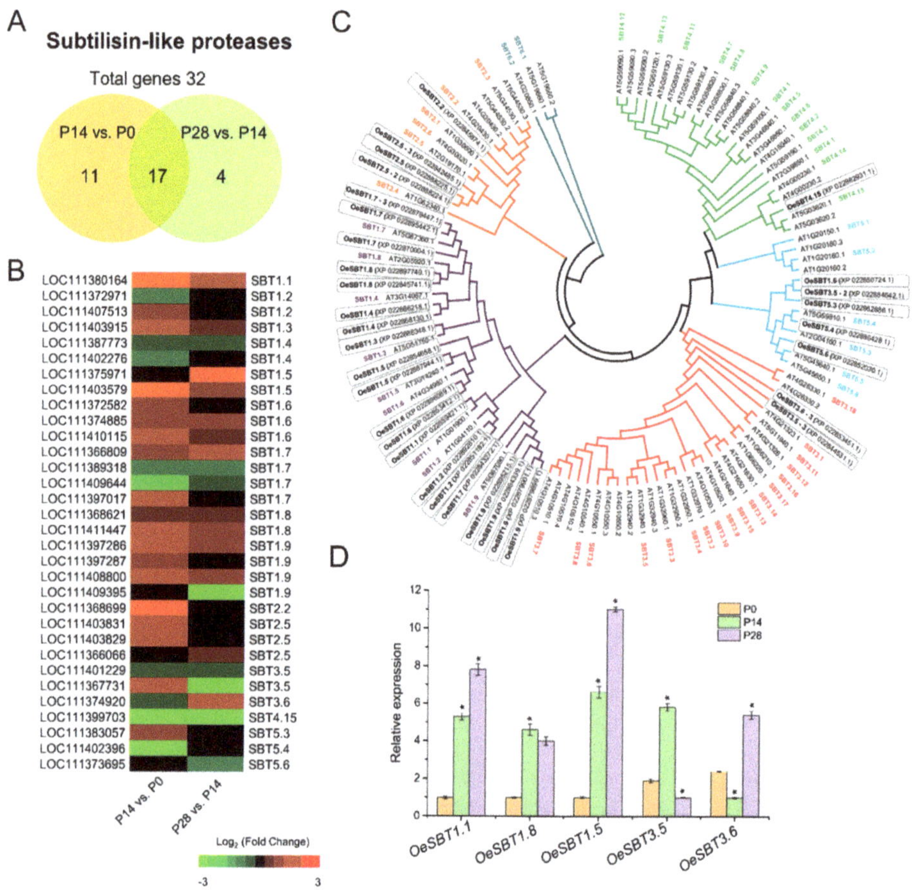

Figure 9. Differential gene expression of subtilisin-like protease (*SBT*) genes during early olive fruit development. (**A**) Venn diagram showing numbers of overlapping *SBT* genes in both the fruit at 14 DPA (P14) versus fruit at 0 DPA (P0), and in the fruit at 28 DPA (P28) versus fruit at 14 DPA (P14) comparisons. (**B**) Expression values are represented in a heatmap as Log_2 Fold Change in the P14 versus P0 comparison, as well as the P28 versus P14 comparison; the color key is indicated at the bottom. (**C**) Phylogenetic analysis of olive SBT (OeSBT) with other SBT genes. The sequences included in this alignment are from olive (*Olea europaea* var. sylvestris Annotation Report; https://www.ncbi.nlm.nih.gov/genome/annotationeuk/Oleaeuropaeaavar.sylvestris/100/, accessed on 10 January 2020) and Arabidopsis (http://www.arabidopsis.org/, accessed on 10 January 2022). The SBT proteins studied from the present work are enclosed in an open box. (**D**) Expression of five selected *SBT* genes in olive fruits at 0 (P0), 14 (P14), and 28 (P28) DPA. Analysis of transcript levels of genes by qRT-PCR. Genes and their primers are shown in Table S20. Relative expression values were normalized to the lowest expression value taken as 1. The data represent the mean values (±SEs) of duplicate experiments from three independent biological samples. Statistical significance compared with the preceding point was determined using a two-sided Student's *t*-test. * $p < 0.05$. Additional information on *SBT* genes is presented in Table S15.

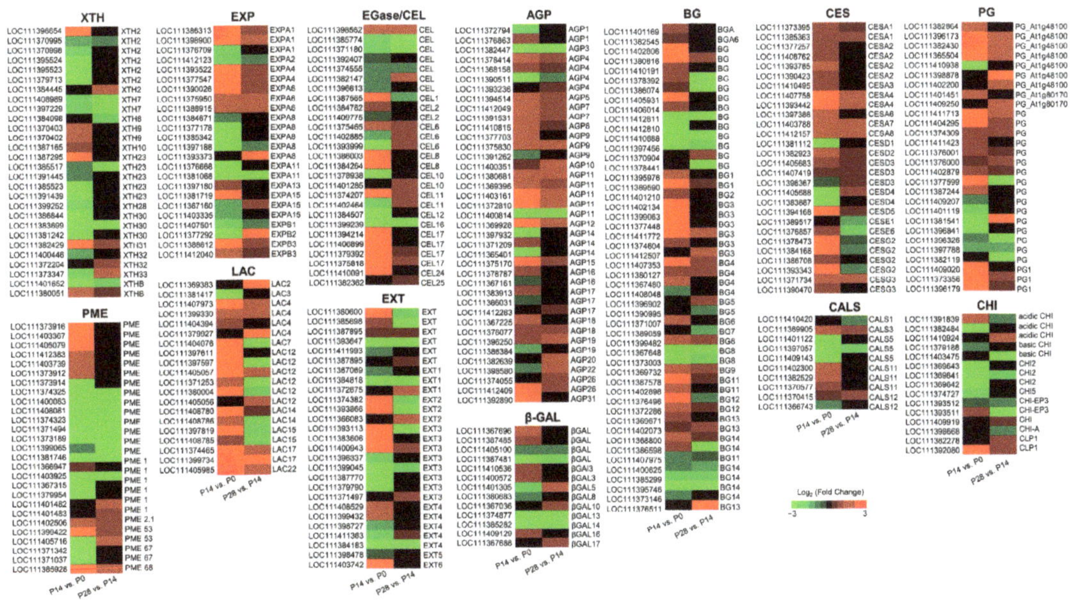

Figure 10. Expression profile of family genes encoding various cell wall proteins during early fruit development in olive. Expression values are represented in a heatmap as Log_2 Fold Change in both the P14 versus P0 (P14 vs. P0) and the P28 versus P14 (P28 vs. P14) comparisons, and the color key is indicated at the bottom. Additional information on the cell wall-related genes is presented in Table S17.

Therefore, the analysis reveals a sequential induction of cell wall biosynthesis and remodeling genes during early olive fruit development, as well as the distinctiveness of candidate genes associated with cell division in comparison with cell expansion during early fruit growth.

2.9. Characterization of Transport-Related Genes Associated with Early Olive Fruit Development

Furthermore, during early fruit development in olive, significant change was identified in the abundance of a subset of 1187 gene-encoding channel and transporter proteins of 127 diverse families (Table S18). Likewise, of the 127 different families of channel and transporter proteins, 26 were expressed only in the dividing olive fruit including the dicarboxylate transporter, GDP-fucose transporter, membrane magnesium transporter, and nucleobase-ascorbate transporter families, 25 families (the protein EXORDIUM, protein YLS3-like, protein NRT1/PTR FAMILY, lipid-transfer protein, sulphite exporter, tetraspanin-2, glycerol-3-phosphate transporter, and glycolipid transfer protein families, among others) are regulated only in the expanding olive fruit, and 76 families (ATP-binding cassette (ABC) transporter, amino acid transporter, aquaporin, glucose-6-phosphate/phosphate translocator, hexose carrier protein, sugar transporter, boron transporter, CMP-sialic acid transporter, copper transport protein, GABA transporter, and magnesium transporter families, among others) are regulated in both developmental stages of olive fruit (Figure 11, Table S18).

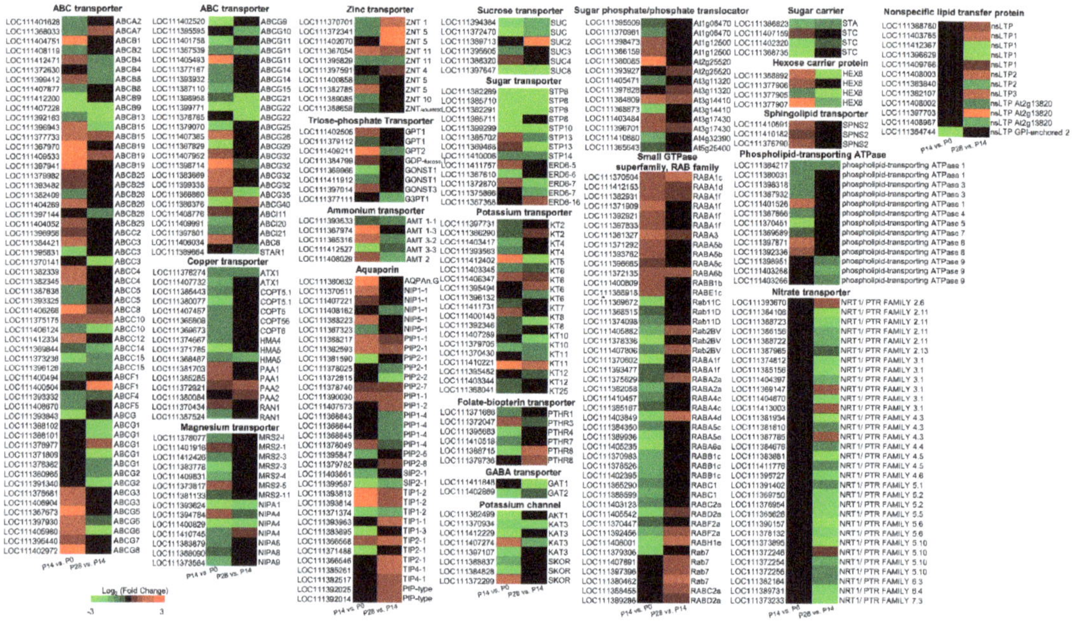

Figure 11. Expression profile of transport-related genes during early fruit development in olive. Expression values are represented in a heatmap as Log_2 Fold Change in both the P14 versus P0 and P28 versus P14 comparisons; the color key is indicated at the bottom. Additional information on the transport-related genes is presented in Table S18.

Among the most abundant upregulated genes in the dividing olive fruit (P14 vs. P0), a homologue of *KT12* (AT1G60160 putative orthologue) encoding a member of the KT/KUP/HAK family of proton-coupled potassium transporters (KT) was identified, these being highly expressed also in the expanding olive fruit (28 DPA). KT12 have potential effects on cell expansion regulated by auxin [100]. Overall, 26 genes that encode KTs fruit (Figure 11; Table S18) were identified. Of these, two genes (*KT5*, *KT8*) were upregulated exclusively in the P14 versus P0 comparison, one gene (*KT2*) was exclusively upregulated in the P28 versus P14 comparison, and two genes (*KT6*, *KT12*) were upregulated in both comparisons. Vigorous cell expansion occurs with greater cell-turgor pressure, which depends on the controlled import of ions to promote water influx. Given that, in plant cells, potassium is the predominant inorganic ion, K transporters and channels constitute the most plausible candidates involved in the production of osmotic gradients through the plasma membrane and thus the molecular mediators responsible for vigorous cell expansion [101]. However, only limited research is available concerning which channels/transporters are involved in K import during early fruit development.

Among the 162 genes related to transport and upregulation in the expanding olive fruit at 28 DPA, the most abundant upregulated genes encode one nonspecific lipid-transfer protein (nsLTP2) (Figure 11; Table S18), which is found in seeds and involved in the transport of the more rigid suberin monomers and sterols [102]. This result may mean that this OeLTP2 protein facilitates the rapid transport of lipids to the plasma membrane in the fruit during early growth, which may then remain at the cell surface to defend against pathogens.

Furthermore, in the present study, three sphingolipid transporter genes encoding SPNS2 proteins were exclusively upregulated in the dividing fruit (Figure 11; Table S18), in agreement with a prior study concerning the sphingolipid levels during early olive fruit development [62], whereas several phospholipid transporting ATPase genes were upregulated in the dividing fruit at 14 DPA, and downregulated at 28 DPA (Figure 11; Table S18), indicating that these lipophilic membrane components are essential for diverse cellular functions during early olive fruit growth. Thus, the metabolism during early fruit development is highly active in this species, including alterations in membrane lipid metabolisms and transport.

2.10. Identifying Transcription Factors Critical for Early Olive Fruit Development

Identifying new TF genes as well as their function in regulating the expression of candidate genes will aid a fuller understanding of the signaling pathways regulating early olive fruit development. Of the DEGs studied, 1150 genes presumably encoding TFs of diverse families were differentially expressed during early fruit development in olive, most of which had a downregulation pattern from 0 to 28 DPA (Figure 12A; Table S19). In particular, 902 DEGs were within the P14 vs. P0 comparison (the set of cell division-related genes), and 564 genes were within the P28 vs. P14 comparison (the set of cell expansion-related genes). The RNA-seq analysis revealed that 385 and 517 genes were upregulated and downregulated in the dividing olive fruit (the P14 vs. P0 comparison), respectively, while 174 and 390 genes were upregulated and downregulated in the expanding fruit (the P28 vs. P14 comparison), respectively (Table S19). These DEGs were shown to be especially related to the TFs MYB, bHLH, ZF, homeobox domain proteins, WRKY, and bZIP families (Figure 12A), implying that TFs from these families may take part in triggering the transcriptional cascade during early fruit development in olive.

The set of cell division-related genes is especially enriched in ZF proteins, whereas set of cell expansion-related genes was found to be rich mainly in the MYB and bHLH families (Figure 12A). Hence, despite the fact that two sets contained members from a number of TF families, clearly significant differences were found in each set regarding the proportion of families (Figure 12A; Table S19). In addition, distinct TF families make up each set: the Whirly, Trihelix, and SNF2 (Helicase-like TF or CHromatin Remodeling Protein) families in the set of cell division-related genes, and the ARID/BRIGHT and VOZ families in the set of cell expansion-related genes (Figure 12A; Table S19). Thus, the results indicate that the cell division phase is potentially controlled by the upregulation of E2F/DP, Whirly, TCP, and homeobox TFs, while the cell expansion phase may be controlled by the upregulation of MADS-box and GATA-binding proteins during early fruit development (Figure 12A; Table S19). However, most members of the zinc finger (ZF), the basic lecine zipper (bZIP), GRAS, WRKY, NAC, PLATZ, and AP2/ERF families were downregulated genes during early fruit development in olive. In different gene groups from each set, the enrichment of sequence elements together with data concerning transcript abundance provide a feasible set of TFs that are able to bind these elements and that may be valuable as a focus of future research.

Within the set of cell division-related genes (P14 vs. P0 comparison), three TFs (TFs) homologous to *AINTEGUMENTA (ANT)*, *bHLH61*, and *MYB3R-1* belonging to the AP2/ERF, bHLH, and MYB families, respectively, have been identified as the most abundant TFs. Indeed, these TF families were among the most widely represented class of proteins in the dividing olive fruit (Figure 12A; Table S19). In addition, three bHLH (bHLH18, bHLH87, and bHLH93), and one ZF CCCH domain-containing (C3H2) protein were the most abundant TFs found in the expanding fruit (the P28 vs. P14 comparison). Of particular interest is ANT, one member of the AP2 family, which was the most abundant TF transcript expressed in the dividing olive fruit (Table S19). ANT is required in the control of cell proliferation in Arabidopsis, and regulates growth and cell numbers during organogenesis, modulating auxin biosynthesis in the ovule via regulation of YUCCA4 [103]. In the present work, on the base of homology and the strong upregulation of *ANT* exclusively in the dividing olive fruit, it is hypothesized that this TF plays a major role in the regulation of cell division and size through its downstream control of auxin biosynthesis in the fruit, consistently with the IAA levels detected in the dividing fruit at 14 DPA (Figure 7). Previously, it has been suggested that *CYCD3;1* is a target of ANT [104]. *CYCD3* genes regulate the length of the temporal period of mitotic cell division during aerial organ development, and are induced by CKs in Arabidopsis [105]. According to the data, the *CYCD3;1*, *ANT*, and CK metabolism, together with signaling gene expression, share a common shift during early olive fruit growth (Figure 8), indicating that these components may act in conserved pathways linking CKs to developmentally regulated cell division in olive fruit enriched mainly in CKs (*tZ*).

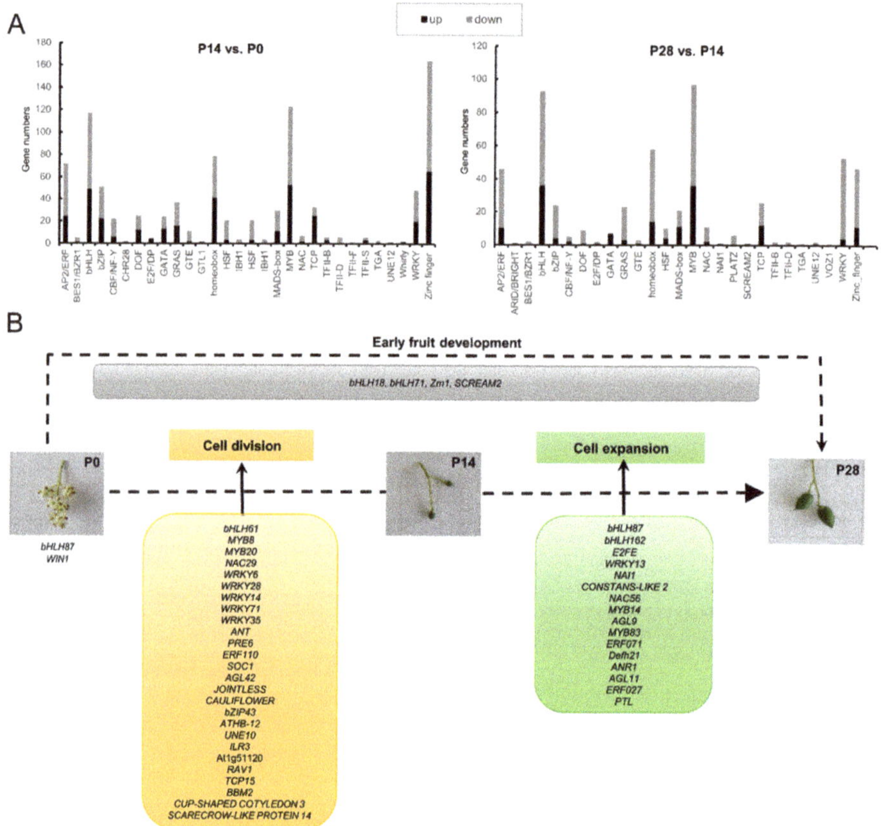

Figure 12. Olive transcription factors (TFs) induced or repressed during early fruit development. (**A**) Summary of the number of significant changes in TF transcripts between the different families during early fruit development in olive. Comparisons of fruit at 14 versus 0 DPA (P14 vs. P0), and 28 versus 14 DPA (P28 vs. P14) for significantly upregulated (black) and downregulated (grey) transcripts revealed differences in the families of TFs during early olive fruit development. (**B**) The TFs involved in regulating cell division and expansion phases during early fruit development in olive. The TFs (log_2FoldChange > 3, $p < 0.01$) were chosen for candidate TFs between the samples.

In Arabidopsis, researchers have identified a small number of organ-size regulators. Notably, among the specific genes in the dividing olive fruit, two *Growth-Regulating Factor* (GRF) genes, GRF1 and GRF3, which encode a transcription activator that plays a role in the regulation of cell proliferation and size in Arabidopsis leaves [106], abundantly accumulated in the dividing olive fruit at 14 DPA (Table S19), indicating the potential roles of these genes in signaling pathways for the regulation of early olive fruit growth. A previous study has shown GRF3 to be a component in a network composed of miR396, GRFs, and their interacting factors (GIFs) contributing to at least partial regulation of meristem function by controlling cell proliferation after infection by cyst nematode [107]. In tomato plants expressing higher levels of *SlGRF1* to *-5*, fruit size and weight are increased as a result of an increased size of the epidermal cells [108]. The present findings provide a valuable tool for unravelling the physiological function of GRF1 and GRF3 in future studies. Conversely, negative effectors of cell proliferation in Arabidopsis, such as BIG BROTHER (BB) [109], was also identified in the present analysis. Here, eight *BB* genes were upregulated in the expanding olive fruit at 28 DPA (Table S19), indicating that at least some members of BBs play a role not only in limiting the cell-proliferation stage, but also in fostering the transition towards cell differentiation and expansion in early olive fruit growth. Nevertheless, the way in which these different effectors act on the regulatory genes of the cell cycle in developing olive fruit has not been elucidated.

Furthermore, the major upregulated differentially expressed TFs (Log_2 Fold Change > 3) were identified during early olive fruit development. As shown in Figure 12B, 26 TFs that might be associated with cell division phase were identified, including 2 DOF-ZF (DOF3.4 and DOF4.7), 2 MYB (MY8 and MYB20), 5 WRKY (WRKY6, WRKY14, WRKY28, WRKY35, and WRKY71), 2 MADS-box (SOC1 and AGL42), 2 NAC (NAC29), and 1 AP2/ERF (ERF110) proteins, among others, while 16 TFs might be involved in cell expansion phase, including 2 bHLH (bHLH87 and bHLH162), E2FE, WRKY13, NAI1, 2 MYB (MYB14 and MYB83), 2 AP2/ERF (ERF071 and ERF027) proteins, among others.

According to their expression profiles, other genes appear to be promising candidates for the main regulations in the early development of olive fruit. In the RNA-seq analysis, four TFs, SCREAM2 (or bHLH33), bHLH18, bHLH71, and Zm1 (R2R3-MYB), were identified within the differentially expressed gene set, and were all upregulated in both the dividing and the expanding olive fruits (P14 vs. P0, and P28 vs. P14 comparisons) and were thus linkable to cell division and cell expansion phases during early fruit development in olive (Figure 12B; Table S19). SCREAM2 (or bHLH33, a putative orthologue At1g12860) encoded a key component of core regulatory units in stomatal development, most likely by controlling successive roles of SPCH, MUTE, and FAMA (bHLH proteins) in Arabidopsis [110]. Additionally, SCREAM is INDUCER of CBF EXPRESSION1 (ICE1), a bHLH TF that positively regulates the cold-induced transcriptome and freezing tolerance [110]. According to the analysis, a homologue of *ICE1* is also upregulated in both comparisons, suggesting that the two proteins prompt sequential steps in the development program (i.e., entry, proliferation, and terminal differentiation) of the stomatal cell during early fruit development in olive. Moreover, here it was found that other members of the bHLH TF family, such as bHLH61 (At5g10570), are exclusively upregulated in the dividing olive fruit, and bHLH87 (At3g21330) and bHLH162 (At4g20970) are exclusively upregulated in the expanding olive fruit. The bHLH proteins represent prime regulatory components in networks of transcription that control several biological processes, such as fruit and flower development [111–113]. Therefore, changes in the temporal distribution of these proteins could lead to unexpected cross-talk in the regulatory network of genes.

Moreover, bHLHs and MYBs or bZIPs bind to DNA as either homodimers or heterodimers, multiplying the feasible regulatory combinations [114]. In this context, MYB and bZIP TFs are members of TF families abundantly represented in the developing olive fruit (Table S20). In the dividing olive fruit, the results demonstrated upregulation of 33 (MYB20, MYBS3, among others) out of 122 MYB genes identified, and 17 (bZIP11, bZIP44, among others) out of 49 bZIP genes identified, while 20 (MYB14, MYB83, among others) MYB genes and 4 bZIP (bZIP6, bZIP11, bZIP18, and RF2b) genes were upregulated in the expanding olive fruit. The possibility that bHLH proteins act as an interaction partner for MYB or bZIP TFs in regulating genes involved in processes downstream during early olive fruit growth should not be disregarded. More research is needed in order to determine whether or not these bHLH TFs participate jointly with MYB or bZIP proteins in the signaling of developing olive fruit. From this perspective, MYB TFs appear to take part in controlling the cell cycle not only in animals, but also in plants and other higher eukaryotes [115]. MYB factors are involved in regulating the transcription of cyclin genes via MYB recognition elements in cyclin promoters [116]. The RNA-Seq analysis showed that the upregulation of MYB3R-4 (3R-MYB, AT5G11510.1), which activates mitotic gene expression and CK response [117], and MYB3R-5 (3R-MYB, AT5G02320.1), which is a member of the DREAM complex [118], are probably associated with the cell division phase, while the upregulation of MYB3R-1 (3R-MYR, AT4G32730.2) is likely associated with the cell expansion phase during early olive fruit development (Table S20). Additionally, members of the R2R3-MYB family have been identified in olive fruit, as in the case for Zm1, which is involved in the regulation of flavonoid biosynthesis [119]. The increased expression of *Zm1* in both the dividing and expanding olive fruits suggests the involvement of Zm1 in early fruit growth. Likewise, many R2R3-MYB genes also act in signal transduction pathways of ABA, GA, JA, and SA [115], supporting the contention that MYB proteins constitute vital components of many hormone-mediated transcriptional cascades, such as ABA, GA, and SA, which regulate early fruit development in olive. Thus, the present study reveals novel regulators associated with early fruit growth and provides a useful resource for further characterization of the physiological functions of these regulators in olive and other plant species.

3. Materials and Methods

3.1. Plant Material and Cytological Study

Olive trees (*Olea europaea* L.) of the 'Picual' cultivar 20 years old in an orchard under drip irrigation (fertigation with suitable liquid fertilizers) near Badajoz (Spain) were studied. From these trees, flowers were collected at the anthesis stage (0 days post-anthesis, DPA) and fruits were sampled at 7, 14, 21, 28, 35, and 42 DPA, spanning a time from the fruit set to the time of endocarp lignification, and 500 flowers or fruits for each developmental stage during the 2018–2019 growing seasons [27,62,63]. Harvested flowers and whole fruits of 'Picual' olive were weighed, and the longitudinal and transverse diameters were measured at different developmental stages (Figure 1). Flowers plus whole fruits at different stages of development were frozen in liquid nitrogen and then stored at −80 °C until analyzed.

In the cytological study, at least three biological replicates were made at each stage. For this, samples were sliced 0.3 to 0.6 mm thick and quickly fixed in ethanol-acetic acid (3:1, *v/v*) for 2 h at an ambient temperature. Then, the samples were rinsed 3 times using 70% ethanol, dried using an ethanol series, and finally embedded in Technovit 7100 (Kulzer). Sections that were 30 μm thick were used for confocal imaging and were stained for 2 min with Calcofluor White M2R (Wyeth, http://www.wyeth.com (accessed on 10 January 2020), 0.1% *w/v* in distilled water), a fluorescence brightener commonly used in the visualization of crystalline cellulose in plant cell walls [120], rinsed with distilled water, mounted with a cover slip in DABCO mounting medium, and sealed with nail polish. Fruit cells and pericarp thickness were examined using a confocal laser-scanning microscope (model Fluoview-BX50, Olympus). For microscopy, the cell size (cell area) and pericarp thickness (cross-section) were determined using a CellProfiler image analysis system [121].

3.2. Flow Cytometry Analysis

The cell division period was precisely determined using flow cytometry with nucleus ploidy profiles taken from ovaries and fruits using the method of Reference [122]. The samples of 0.1–0.2 g fresh weight were diced into 0.5 mL using a blade and placed in an ice-cold buffer (0.2 M Tris-HCl, 4 mM $MgCl_2$, 2 mM EDTA, 86 mM NaCl, 10 mM metabisulphite, pH 7.5, and 1% Triton X-100) then filtered through a 30-μm nylon mesh, and finally stained using 4,6-diamidino-2-phenylindole (DAPI). Next, the distribution of the nuclear-DNA content was analyzed with a FACS CantoII flow cytometer, and the recorded data were analyzed using FACSDiva 6.1.2 software (BD Biosciences, Franklin Lakes, NJ, USA). For each stage, 4 biological replicates, with 10,000 nuclei each, were undertaken. The DNA content (C value) of the olive fruit cells was submitted to flow cytometry with the use of internal calibration standards [122]. The C values were determined following [123]: 2C

DNA (pg) = (mean of the problem sample G1 peak × 2C DNA content of the standard [pg])/mean of the standard G1 peak.

3.3. RNA Extraction, Library Preparation, and Sequencing

The extraction of the total RNA from ovaries and fruits samples was performed as in [27]. For each sample, three biological replicates were collected of the same developmental stage. Tissue frozen at −80 °C was milled into a powder in liquid nitrogen prior to the addition of the buffer (2 mL buffer: 1 g tissue). The RNA quality tests were performed with the Agilent 2100 Bioanalyzer (Agilent Technologies, Santa Clara, CA, USA) employing an RNA 6000 Pico assay kit (Agilent Technologies). This RNA was used in RNA-seq and quantitative RT-PCR (qRT-PCR) analyses. The poly(A)+ mRNA fraction was isolated from the total RNA and obtained cDNA libraries following the recommendations of Illumina (Truseq stranded Illumina kit). In short, poly(A)+ RNA, after being isolated on poly-T oligo attached magnetic beads, was chemically fragmented before reverse transcription and cDNA generation. Next, in the end-repair process of the cDNA fragments, a single 'A' base was added to the 3′ end, and the adapters were ligated afterwards. Finally, the products were purified and enriched with PCR to compile the final indexed double-stranded cDNA library. The quality of the libraries was checked using a Bioanalyzer 2100, High Sensitivity assay. The quantity within the libraries was assessed by real-time PCR in LightCycler 480 (Roche). Before cluster generation in cbot (Illumina), an equimolar pooling of the DNA libraries was performed and the pool was then sequenced using paired-end sequencing (100 × 2) with an Illumina HiSeq 2000 sequencer. The RNA-seq analysis was ordered from Sistemas Genómicos (Valencia, Spain). Three independent biological replicates (three olive trees) per sample were sequenced, each replicate consisting of an equilibrated pool of three RNAs from three different samples per replicate.

Quality control checks of raw sequencing data were performed with FastQC 0.11.7. The generated reads were cleaned using the Trimgalore 0.6.2 method (http://www.bioinformatics.babraham.ac.uk/projects/trimgalore/, accessed on 10 January 2020). Subsequently, all samples were combined, and the complexity of the reads was reduced by removing duplicates and low-quality reads using Picard tools 2.9.2 and Samtools 1.9 algorithms [124] (>Q20). Reads were mapped to the *Olea europaea* var. sylvestris genome as the reference [*Olea europaea* var. sylvestris Annotation Report (nih.gov, accessed on 10 January 2020)] [69]. Each sample was then mapped with the latest bowtie2 2.3.3 version [125]. The reads of good quality (Phred > 20) were selected to increase the resolution of the count expression. Finally, the expression inference was evaluated by means of the count of properly paired end reads by transcript.

3.4. Differential Expression

For the study of differential gene expression and normalization gene expression using library size, the DESeq 1.36 algorithm was used [126]. This method is based on a negative binomial distribution. The genes with a fold change of less than −2 or greater than 2 and a p adjust by FDR < 0.05 were considered differentially expressed [127]. A functional enrichment study was performed using the information obtained from the pFam and uniport databases. This study is based on a hypergeometric distribution using the statistical software R. The principal-component analysis (PCA) was analyzed using the methods described in DESeq using the normalized counts of gene expression obtained from this method. Differential expression was made for each comparison. Then, specific gene expression from each comparison was obtained using a venn diagram (https://bioinfogp.cnb.csic.es/tools/venny/, accessed on 10 January 2020). The GO enrichment was performance using a hypergeometric test based in R package. KEGG data was accessed using AnnotationHub [128,129].

The alignment of the deduced protein sequences and the phylogenetic tree were computed using the ClustalW [130] in MEGA11 software (http://www.megasoftware.net, accessed on 10 January 2022).

3.5. Quantitative RT-PCR

The RNA-seq results by qRT-PCR were validated as indicated in [18]. The primers employed are shown in Table S20. Relative fold differences were calculated based on the comparative Ct method using the *Olea europaea Ubiquitin (OeUb)* as an internal standard [27]. At least two to three independent RNA isolations were used for cDNA synthesis, and each cDNA sample was subjected to real-time PCR analysis in triplicate.

3.6. Quantification of Plant Hormones

A pool of 100-mg fresh weight/sample was used for each measurement split into 3 independent biological replicates per sample. The plant hormones were quantified as described by Reference [131]. Aliquots of lyophilized material were extracted with 80% methanol-1% acetic acid. Deuterium-labeled hormones (purchased from Prof. L Mander-Canberra, OlChemim Ltd-Olomouc, or Cambridge Isotope Lab- Andover) [17,17-^{2}H]GAn, [^{2}H$_{5}$]IAA, and [^{2}H$_{6}$]ABA were added as internal standards for quantification of SA and ABA. For quantification of JA, the compound dhJA was used instead. For collecting the acid fractions containing SA, ABA, and JA, the extracts passed consecutively through HLB (reverse phase), MCX (cationic exchange), and WAX (ionic exchange) columns (Oasis 30 mg, Waters). The final residue was dissolved in 5% acetonitrile-1% acetic acid, and the hormones were separated by reverse phase UPHL chromatography (2.6 µm Accucore RP-MS column, 100 mm length × 2.1 mm i.d., ThermoFisher Scientific) with a 5% to 50% acetonitrile gradient. The hormones were analyzed by electrospray ionization and targeted-SIM using a Q-Exactive spectrometer (Orbitrap detector, ThermoFisher Scientific, Spain). The concentrations of hormones in the extracts were determined using embedded calibration curves and the Xcalibur 4.1 SP1 build 48 and TraceFinder programs.

4. Conclusions

Cytological, ploidy, hormonal, and transcriptional analyses reveal the rate and duration of cell division and cell expansion phases during early olive fruit growth as well as the regulation of these phases. The present study provides evidence that CKs may regulate cell division mainly through the *ARR12* expression gene, and that CKs and SA appear to correlate positively with cell division during early olive fruit development, while GA (via *GA3ox* and *GID2)*, auxin (via *PIL2, PIL6, PIL7, IAA29, GH3.1,* and *SAUR72)*, and ABA (via *ABI5*) have higher relative importance in the period where fruit growth depends mainly on cell expansion. In particular, CK (*t*Z) and GA_4 present different accumulation patterns during early olive fruit development, suggesting that the *tZ* and GA_4 hormones act on early olive fruit development in opposite ways during the transition from cell division to cell expansion. Furthermore, this study offers the first detailed analysis available for an array of cellular responses under the control of gene expression leading to early fruit development in olive. Through gene expression studies, a selection was made of the candidate genes having well-known molecular roles in different biological processes affecting olive fruit growth. This gene expression profile, together with hormonal regulators, will help clarify gene regulatory networks during early olive fruit development. In practical terms, it will aid in designing specific approaches for crop breeding and engineering, facilitating the development of the table olive and olive oil industries.

Supplementary Materials: The following supporting information can be downloaded at: https://www.mdpi.com/article/10.3390/ijms24020961/s1.

Author Contributions: Conceptualization, M.C.G.-J. Funding acquisition, M.C.G.-J. Investigation, M.C.C., B.B., J.C., J.L. and M.G. Methodology, M.C.C., B.B., J.C., J.L. and M.G. Project administration, M.C.G.-J. Resources, M.C.G.-J. Software, M.C.C. and J.C. Supervision, M.C.G.-J. Writing—review and editing, M.C.G.-J. All authors have read and agreed to the published version of the manuscript.

Funding: This work was supported by grants from the Spanish Ministry of Economy and Competitiveness (RTI2018-097244-B-I00), the Junta of Extremadura (Spain), and the European Regional Development Fund (IB18075).

Institutional Review Board Statement: Not applicable.

Informed Consent Statement: Not applicable.

Data Availability Statement: The results of this study are available on request from the corresponding author (Maria C. Gomez-Jimenez).

Acknowledgments: The authors are grateful to the Plant Hormone Quantification Service (IBMCP, Spain) for its contribution to this work. The authors also thank J.L. Grosson for free access to the plant material, D. Nesbitt for correcting the English, and J.C. Triviño (Sistemas Genómicos, Spain) for technical assistance. Maria C. Gomez-Jimenez dedicates this paper to the memory of Miguel A. Paredes (University of Extremadura, Spain).

Conflicts of Interest: The authors declare no conflict of interest.

Abbreviations

ABA	Abscisic acid
ACC	1-Aminocyclopropane-1-carboxylic acid
ABC	ATP-binding cassette transporter
AGP	Arabinogalactan protein
ARFs	Auxin response factors
BG	Glucan 1,3-β-glucosidase
β-GAL	β-galactosidase
BR	Brassinosteroid
BRI1	BR insensitive 1
BAK1	BRI1-associated kinase1
BIN2	BR insensitive 2
BRZ1	Brassinazole resistant 1
CALS	Callose synthase
cDNA	Complementary deoxyribonucleic acid
CDK	Cyclin-dependent kinase
CES	Cellulose synthase
CHI	Chitinase
CK	Cytokinin
CKI	CDK inhibitor protein
CKX	CK oxidase/dehydrogenase
CYC	cyclin
DEG	Differentially expressed gene
DPA	Days post-anthesis
EGase/CEL	Endo-1,4-β-glucanase or cellulase
ERF	Ethylene response factor
ET	Ethylene
EXP	Expansin
EXT	Extensin
GA	Gibberellin
GID1	Gibberellin-insensitive dwarf1
GO	Gene Ontology
IAA	Indole-3-acetic acid
JA	Jasmonic acid
KT	Potassium transporter
LAC	Laccase
LOG	Cytokinin riboside 5′-monophosphate phosphoribohydrolase
PA	Polyamine
PG	Polygalacturonase
PME	Pectin methylesterase

qRT-PCR	Quantitative real-time polymerase chain reaction
RB	Retinoblastoma protein
SA	Salicylic acid
SAUR	Small auxin up RNA
SL	Strigolactone
SBT	Subtilase
TF	Transcription factor
TIR1	Transport inhibitor response 1
tZ	trans-Zeatin
XTH	Xyloglucan endotransglucosylase/hydrolase
ZNT	Zinc transporter

References

1. Baldoni, L.; Belaj, A. Olive. In *Oil Crops. Handbook of Plant Breeding*; Vollmann, J., Rajean, I., Eds.; Springer Science + Business Media: New York, NY, USA, 2009; Volume 4, pp. 397–421. [CrossRef]
2. Rugini, E.; Pannelli, G.; Ceccarelli, M.; Muganu, M. Isolation of triploid and tetraploid olive (*Olea europaea* L.) plants from mixoploid cv. 'Frantoio' and 'Leccino' mutants by in vivo and in vitro selection. *Plant. Breed.* **1996**, *115*, 23–27. [CrossRef]
3. Besnard, G.; García-Verdugo, C.; De Casas, R.; Treier, U.A.; Galland, N.; Vargas, P. Polyploidy in the olive complex (*Olea europaea*): Evidence from flow cytometry and nuclear microsatellite analyses. *Ann. Bot.* **2008**, *101*, 25–30. [CrossRef] [PubMed]
4. Loureiro, J.; Rodriguez, E.; Costa, A.; Santos, C. Nuclear DNA content estimations in wild olive (*Olea europaea* L. ssp. europaea var. sylvestris Brot.) and Portuguese cultivars of *O. europaea* using flow cytometry. *Genet. Resour. Crop. Evol.* **2007**, *54*, 21–25. [CrossRef]
5. Conde, C.; Delrot, S.; Gero, H. Physiological, biochemical and molecular changes occurring during olive development and ripening. *J. Plant Physiol.* **2008**, *165*, 1545–1562. [CrossRef]
6. Breton, C.; Bervillé, A. From the olive flower to the drupe: Flower types, pollination, self and inter-compatibility and fruit set. In *The Mediterranean Genetic Code-Grapevine and Olive*; Sladonja, B., Ed.; InTech: London, UK, 2013. [CrossRef]
7. Alagna, F.; D'Agostino, N.; Torchia, L.; Servili, M.; Rao, R.; Pietrella, M.; Giuliano, G.; Chiusano, M.L.; Baldoni, L.; Perrotta, G. Comparative 454 pyrosequencing of transcripts from two olive genotypes during fruit development. *BMC Genom.* **2009**, *10*, 399. [CrossRef]
8. Alagna, F.; Mariotti, R.; Panara, F.; Caporali, S.; Urbani, S.; Veneziani, G.; Esposto, S.; Taticchi, A.; Rosati, A.; Rao, R.; et al. Olive phenolic compounds: Metabolic and transcriptional profiling during fruit development. *BMC Plant Biol.* **2012**, *12*, 162. [CrossRef]
9. Galla, G.; Barcaccia, G.; Ramina, A.; Collani, S.; Alagna, F.; Baldoni, L.; Cultrera, N.G.; Martinelli, F.; Sebastiani, L.; Tonutti, P. Computational annotation of genes differentially expressed along olive fruit development. *BMC Plant Biol.* **2009**, *9*, 128. [CrossRef]
10. Martinelli, F.; Tonutti, P. Flavonoid metabolism and gene expression in developing olive (*Olea europaea* L.) fruit. *Plant Biosyst.* **2012**, *146*, 164–170. [CrossRef]
11. Parra, R.; Paredes, M.A.; Sanchez-Calle, I.M.; Gomez-Jimenez, M.C. Comparative transcriptional profiling analysis of olive ripe fruit pericarp and abscission zone tissues shows expression differences and distinct patterns of transcriptional regulation. *BMC Genom.* **2013**, *9*, 866. [CrossRef]
12. Parvini, F.; Zeinanloo, A.A.; Ebrahimie, E.; Tahmasebi-Enferadi, S.; Hosseini-Mazinani, M. Differential expression of fatty acid desaturases in Mari and Shengeh olive cultivars during fruit development and ripening. *Eur. J. Lipid Sci. Technol.* **2015**, *117*, 523–531. [CrossRef]
13. Inês, C.; Parra-Lobato, M.C.; Paredes, M.A.; Labrador, J.; Gallardo, M.; Saucedo-García, M.; Gavilanes-Ruiz, M.; Gomez-Jimenez, M.C. Sphingolipid distribution, content and gene expression during olive-fruit development and ripening. *Front. Plant Sci.* **2018**, *9*, 28. [CrossRef] [PubMed]
14. Mougiou, N.; Trikka, F.; Trantas, E.; Ververidis, F.; Makris, A.; Argiriou, A. Expression of hydroxytyrosol and oleuropein biosynthetic genes are correlated with metabolite accumulation during fruit development in olive, *Olea europaea*, cv. Koroneiki. *Plant Physiol. Biochem.* **2018**, *128*, 41–49. [CrossRef] [PubMed]
15. Bruno, L.; Picardi, E.; Pacenza, M.; Chiappetta, A.; Muto, A.; Gagliardi, O.; Muzzalupo, I.; Pesole, G.; Bitonti, M.B. Changes in gene expression and metabolic profile of drupes of *Olea europaea* L. cv Carolea in relation to maturation stage and cultivation area. *BMC Plant Biol.* **2019**, *19*, 428. [CrossRef] [PubMed]
16. Carbone, F.; Bruno, L.; Perrotta, G.; Bitonti, M.B.; Muzzalupo, I.; Chiappetta, A. Identification of miRNAs involved in fruit ripening by deep sequencing of *Olea europaea* L. transcriptome. *PLoS ONE* **2019**, *14*, e0223354. [CrossRef] [PubMed]
17. Xiaoxia, L.; Jianguo, Z.; Ying, L.; Guodong, R. Metabolome and transcriptome analyses reveal tissue-specific variations in gene expression and metabolites of olive. *J. Plant Biol.* **2020**, *63*, 73–82. [CrossRef]
18. Briegas, B.; Corbacho, J.; Parra-Lobato, M.C.; Paredes, M.A.; Labrador, J.; Gallardo, M.; Gomez-Jimenez, M.C. Transcriptome and hormone analyses revealed insights into hormonal and vesicle trafficking regulation among *Olea europaea* fruit tissues in late development. *Int. J. Mol. Sci.* **2020**, *21*, 4819. [CrossRef]

19. Liu, X.; Guo, L.; Zhang, J.; Xue, L.; Luo, Y.; Rao, G. Integrated analysis of fatty acid metabolism and transcriptome involved in olive fruit development to improve oil composition. *Forests* **2021**, *12*, 1773. [CrossRef]
20. Rao, G.; Zhang, J.; Liu, X.; Li, X.; Wang, C. Combined metabolome and transcriptome profiling reveal optimal harvest strategy model based on different production purposes in olive. *Foods* **2021**, *10*, 360. [CrossRef]
21. Haralampidis, K.; Milioni, D.; Sanchez, J.; Baltrusch, M.; Heinz, E.; Hatzopoulos, P. Temporal and transient expression of stearoyl-ACP carrier protein desaturase gene during olive fruit development. *J. Exp. Bot.* **1998**, *49*, 1661–1669. [CrossRef]
22. Martsinkovskaya, A.I.; Poghosyan, Z.P.; Haralampidis, K.; Murphy, D.J.; Hatzopoulos, P. Temporal and spatial gene expression of cytochrome B5 during flower and fruit development in olives. *Plant Mol. Biol.* **1999**, *40*, 79–90. [CrossRef]
23. Hatzopoulos, P.; Banilas, G.; Giannoulia, K.; Gazis, F.; Nikoloudakis, N.; Milioni, D.; Haralampidis, K. Breeding, molecular markers and molecular biology of the olive tree. *Eur. J. Lipid Sci. Technol.* **2002**, *104*, 574–586. [CrossRef]
24. Banilas, G.; Moressis, A.; Nikoloudakis, N.; Hatzopoulos, P. Spatial and temporal expressions of two distinct oleate desaturases from olive (*Olea europaea* L.). *Plant Sci.* **2005**, *168*, 547–555. [CrossRef]
25. Banilas, G.; Karampelias, M.; Makariti, I.; Kourti, A.; Hatzopoulos, P. The olive *DGAT2* gene is developmentally regulated and shares overlapping but distinct expression patterns with *DGAT1*. *J. Exp. Bot.* **2011**, *62*, 521–532. [CrossRef]
26. Poghosyan, Z.P.; Giannoulia, K.; Katinakis, P.; Murphy, D.J.; Hatzopoulos, P. Temporal and transient expression of olive enoyl-ACP reductase gene during flower and fruit development. *Plant Physiol. Biochem.* **2005**, *43*, 37–44. [CrossRef] [PubMed]
27. Gomez-Jimenez, M.C.; Paredes, M.A.; Gallardo, M.; Fernandez-Garcia, N.; Olmos, E.; Sanchez-Calle, I.M. Tissue-specific expression of olive S-adenosyl methionine decarboxylase and spermidine synthase genes and polyamine metabolism during flower opening and early fruit development. *Planta* **2010**, *232*, 629–647. [CrossRef] [PubMed]
28. Alagna, F.; Cirilli, M.; Galla, G.; Carbone, F.; Daddiego, L.; Facella, P.; Lopez, L.; Colao, C.; Mariotti, R.; Cultrera, N.; et al. Transcript analysis and regulative events during flower development in olive (*Olea europaea* L.). *PLoS ONE* **2016**, *17*, e0263101. [CrossRef] [PubMed]
29. Gillaspy, G.; Ben-David, H.; Gruissem, W. Fruits: A developmental perspective. *Plant Cell* **1993**, *5*, 1439–1451. [CrossRef]
30. Bertin, N. Analysis of the tomato fruit growth response to temperature and plant fruit load in relation to cell division, cell expansion and DNA endoreduplication. *Ann. Bot.* **2005**, *295*, 439–447. [CrossRef]
31. Cheniclet, C.; Rong, W.Y.; Causse, M.; Frangne, N.; Bolling, L.; Carde, J.P. Cell expansion and endoreduplication show a large genetic variability in pericarp and contribute strongly to tomato fruit growth. *Plant Physiol.* **2005**, *139*, 1984–1994. [CrossRef]
32. Chevalier, C.; Nafati, M.; Mathieu-Rivet, E.; Bourdon, M.; Frangne, N.; Cheniclet, C.; Renaudin, J.P.; Gvaudant, F.; Hernould, M. Elucidating the functional role of endoreduplication in tomato fruit development. *Ann. Bot.* **2011**, *107*, 1159–1169. [CrossRef]
33. Chevalier, C.; Bourdon, M.; Pirrello, J.; Cheniclet, C.; Gévaudant, F.; Frangne, N. Endoreduplication and fruit growth in tomato: Evidence in favour of the karyoplasmic ratio theory. *J. Exp. Bot.* **2014**, *65*, 2731–2746. [CrossRef] [PubMed]
34. Mauxion, J.P.; Chevalier, C.; Gonzalez, N. Complex cellular and molecular events determining fruit size. *Trends Plant Sci.* **2021**, *26*, 1023–1038. [CrossRef] [PubMed]
35. Quinet, M.; Angosto, T.; Yuste-Lisbona, F.J.; Blanchard-Gros, R.; Bigot, S.; Martinez, J.P.; Lutts, S. Tomato fruit development and metabolism. *Front. Plant Sci.* **2019**, *10*, 1554. [CrossRef] [PubMed]
36. Pećinar, I.; Quarrie, S.P.; Bertin, N.; Rančić, D., Savić, S.; Jovanović, Z.; Stikić, R. Tomato fruit development in response to different irrigation practices: Developmental study of pericarp cell layers. *Biol. Life Sci. Forum.* **2021**, *4*, 105. [CrossRef]
37. Pirrello, J.; Deluche, C.; Frangne, N.; Gevaudant, F.; Maza, E.; Djari, A.; Djari, A.; Bourge, M.; Renaudin, J.P.; Brown, S.; et al. Transcriptome profiling of sorted endoreduplicated nuclei from tomato fruits: How the global shift in expression ascribed to DNA ploidy influences RNA-Seq data normalization and interpretation. *Plant J.* **2018**, *93*, 387–398. [CrossRef] [PubMed]
38. Farinati, S.; Forestan, C.; Canton, M.; Galla, G.; Bonghi, C.; Varotto, S. Regulation of fruit growth in a peach slow ripening phenotype. *Genes* **2021**, *12*, 482. [CrossRef] [PubMed]
39. Edgar, B.A.; Orr-Weaver, T.L. Endoreplication cell cycles. *Cell* **2001**, *105*, 297–306. [CrossRef] [PubMed]
40. Frawley, L.E.; Orr-Weaver, T.L. Polyploidy. *Curr. Biol.* **2015**, *25*, R353–R358. [CrossRef]
41. Lang, L.; Schnittger, A. Endoreplication—A means to an end in cell growth and stress response. *Curr. Opin. Plant Biol.* **2020**, *54*, 85–92. [CrossRef]
42. Rallo, P.; Rapoport, H. Early growth and development of the olive fruit mesocarp. *J. Hort. Sci. Biotechnol.* **2001**, *76*, 408–412. [CrossRef]
43. Rosati, R.; Caporali, S.; Hammami, S.; Moreno-Alías, I.; Paoletti, A.; Rapoport, H.F. Tissue size and cell number in the olive (*Olea europaea*) ovary determine tissue growth and partitioning in the fruit. *Funct. Plant Biol.* **2012**, *39*, 580–587. [CrossRef] [PubMed]
44. Trentacoste, E.; Puertas, C.; Sadras, V.O. Modelling the intraspecific variation in the dynamics of fruit growth, oil and water concentration in olive (*Olea europaea* L.). *Eur. J. Agron.* **2012**, *38*, 83–93. [CrossRef]
45. Hernández-Santana, V.; Perez-Arcoiza, A.; Gomez-Jimenez, M.C.; Diaz-Espejo, A. Disentangling the link between leaf photosynthesis and turgor in fruit growth. *Plant J.* **2021**, *107*, 1788–1801. [CrossRef] [PubMed]
46. Catalá, C.; Rose, J.K.; Bennett, A.B. Auxin-regulated genes encoding cell wall-modifying proteins are expressed during early tomato fruit growth. *Plant Physiol.* **2000**, *122*, 527–534. [CrossRef]
47. Lemaire-Chamley, M.; Petit, J.; Garcia, V.; Just, D.; Baldet, P.; Germain, V.; Fagard, M.; Mouassite, M.; Cheniclet, C.; Rothan, C. Changes in transcriptional profiles are associated with early fruit tissue specialization in tomato. *Plant Physiol.* **2005**, *139*, 750–769. [CrossRef]

48. Mounet, F.; Moing, A.; Garcia, V.; Petit, J.; Maucourt, M.; Deborde, C.; Bernillon, S.; Le Gall, G.; Colquhoun, I.; Defernez, M.; et al. Gene and metabolite regulatory network analysis of early developing fruit tissues highlights new candidate genes for the control of tomato fruit composition and development. *Plant Physiol.* **2009**, *149*, 1505–1528. [CrossRef]
49. Wang, H.; Schauer, N.; Usadel, B.; Frasse, P.; Zouine, M.; Hernould, M.; Latché, A.; Pech, J.C.; Fernie, A.R.; Bouzayen, M. Regulatory features underlying pollination-dependent and -independent tomato fruit set revealed by transcript and primary metabolite profiling. *Plant Cell* **2009**, *21*, 1428–1452. [CrossRef]
50. Pattison, K.; Carr, K.M.; Grumet, R. Transcriptome analyses of early cucumber fruit growth identifies distinct gene modules associated with phases of development. *BMC Genom.* **2012**, *13*, 518. [CrossRef]
51. Kang, C.Y.; Darwish, O.; Geretz, A.; Shahan, R.; Alkharouf, N.; Liu, Z.C. Genome-scale transcriptomic insights into early-stage fruit development in woodland strawberry *Fragaria vesca*. *Plant Cell* **2013**, *25*, 1960–1978. [CrossRef]
52. McAtee, P.; Karim, S.; Schaffer, R.; David, K. A dynamic interplay between phytohormones is required for fruit development, maturation, and ripening. *Front. Plant. Sci.* **2013**, *4*, 79. [CrossRef]
53. Pattison, R.J.; Csukasi, F.; Zheng, Y.; Fei, Z.; van der Knaap, E.; Catalá, C. Comprehensive tissue-specific transcriptome analysis reveals distinct regulatory programs during early tomato fruit development. *Plant Physiol.* **2015**, *168*, 1684–1701. [CrossRef] [PubMed]
54. Fenn, M.A.; Giovannoni, J.J. Phytohormones in fruit development and maturation. *Plant J.* **2021**, *105*, 446–458. [CrossRef] [PubMed]
55. Atta-Aly, M.A.; Riad, G.S.; Lacheene, Z.E.S.; El-Beltagy, A.S. Early ethrel applications extend tomato fruit cell division and increase fruit size and yield with ripening delay. *J. Plant. Growth Regul.* **1999**, *18*, 15–24. [CrossRef] [PubMed]
56. Fos, M.; Proaño, K.; Nuez, F.; García-Martínez, J.L. Role of gibberellins in parthenocarpic fruit development induced by the genetic system pat-3/pat-4 in tomato. *Physiol. Plant.* **2001**, *111*, 545–550. [CrossRef] [PubMed]
57. Fu, F.Q.; Mao, W.H.; Shi, K.; Zhou, Y.H.; Asami, T.; Yu, J.Q. A role of brassinosteroids in early fruit development in cucumber. *J. Exp. Bot.* **2008**, *59*, 2299–2308. [CrossRef] [PubMed]
58. Mariotti, L.; Picciarelli, P.; Lombardi, L.; Ceccarelli, N. Fruit-set and early fruit growth in tomato are associated with increases in indoleacetic acid, cytokinin, and bioactive gibberellin contents. *J. Plant Growth Regul.* **2011**, *30*, 405–415. [CrossRef]
59. Wang, Y.; Clevenger, J.P.; Illa-Berenguer, E.; Meulia, T.; van der Knaap, E.; Sun, L. A comparison of sun, ovate, fs8.1 and auxin application on tomato fruit shape and gene expression. *Plant Cell Physiol.* **2019**, *60*, 1067–1081. [CrossRef]
60. Shulman, Y.; Lavee, S. Endogenous cytokinins in maturing Manzanillo olive fruits. *Plant Physiol.* **1976**, *57*, 490–492. [CrossRef]
61. Shulman, Y.; Lavee, S. Gibberellin-like substances during ripening of olive fruit. *Sci. Hortic.* **1980**, *12*, 169–175. [CrossRef]
62. Corbacho, J.; Inês, C.; Paredes, M.A.; Labrador, J.; Cordeiro, A.M.; Gallardo, M.; Gomez-Jimenez, M.C. Modulation of sphingolipid long-chain base composition and gene expression during early olive-fruit development, and putative role of brassinosteroid. *J Plant Physiol.* **2018**, *231*, 383–392. [CrossRef]
63. Inês, C.; Corbacho, J.; Paredes, M.A.; Labrador, J.; Cordeiro, A.M.; Gomez-Jimenez, M.C. Regulation of sterol content and biosynthetic gene expression during flower opening and early fruit development in olive. *Physiol. Plant.* **2019**, *167*, 526–539. [CrossRef] [PubMed]
64. Bergervoet, J.H.W.; Verhoeven, H.A.; Gilissen, L.J.W.; Bino, R.J. High amounts of nuclear DNA in tomato (*Lycopersicon esculentum* Mill.) pericarp. *Plant Sci.* **1996**, *116*, 141–145. [CrossRef]
65. Joubès, J.; Phan, T.-H.; Just, D.; Rothan, C.; Bergounioux, C.; Raymond, P.; Chevalier, C. Molecular and biochemical characterization of the involvement of cyclin-dependent linase CDKA during the early development of tomato fruit. *Plant Physiol.* **1999**, *121*, 857–869. [CrossRef]
66. Bourdon, M.; Frangne, N.; Mathieu-Rivet, E.; Nafati, M.; Cheniclet, C.; Renaudin, J.P.; Chevalier, C. Endoreduplication and growth of fleshy fruits. *Prog. Bot.* **2010**, *71*, 101–132.
67. Robinson, D.O.; Coate, J.E.; Singh, A.; Hong, L.; Bush, M.; Doyle, J.J.; Roeder, A.H.K. Ploidy and Size at Multiple Scales in the Arabidopsis Sepal. *Plant Cell* **2018**, *30*, 2308–2329. [CrossRef]
68. Tsukaya, H. Re-examination of the role of endoreduplication on cell-size control in leaves. *J. Plant Res.* **2019**, *132*, 571–580. [CrossRef]
69. Unver, T.; Wu, Z.; Sterck, L.; Turktas, M.; Lohaus, R.; Li, Z.; Yang, M.; He, L.; Deng, T.; Escalante, F.J.; et al. Genome of wild olive and the evolution of oil biosynthesis. *Proc. Natl. Acad. Sci. USA* **2017**, *114*, E9413–E9422. [CrossRef]
70. Macheroux, P.; Schmid, J.; Amrhein, N.; Schaller, A. A unique reaction in a common pathway: Mechanism and function of chorismate synthase in the shikimate pathway. *Planta* **1999**, *207*, 325–334. [CrossRef]
71. Paniagua, C.; Pačínková, A.; Jackson, P.; Dabravolski, S.; Riber, W.; Didi, V.; Houser, J.; Gigli-Bisceglia, N.; Wimmerová, M.; Budinská, E.; et al. Dirigent proteins in plants: Modulating cell wall metabolism during abiotic and biotic stress exposure. *J. Exp. Bot.* **2017**, *68*, 3287–3301. [CrossRef]
72. Kende, H.; Bradford, K.J.; Brummell, D.A.; Cho, H.T.; Cosgrove, D.J.; Fleming, A.J.; Gehring, C.; Lee, Y.; McQueen-Mason, S.; Rose, J.K.C.; et al. Nomenclature for members of the expansin superfamily of genes and proteins. *Plant Mol. Biol.* **2004**, *55*, 311–314. [CrossRef]
73. Fontanesi, F.; Soto, I.C.; Barrientos, A. Cytochrome c oxidase biogenesis: New levels of regulation. *IUBMB Life* **2008**, *60*, 557–568. [CrossRef] [PubMed]

74. De Veylder, L.; Larkin, J.C.; Schnittger, A. Molecular control and function of endoreplication in development and physiology. *Trends Plant. Sci.* **2011**, *16*, 624–634. [CrossRef] [PubMed]
75. Bulankova, P.; Akimcheva, S.; Fellner, N.; Riha, K. Identification of *Arabidopsis* meiotic cyclins reveals functional diversification among plant cyclin genes. *PLoS Genet.* **2013**, *9*, e1003508. [CrossRef] [PubMed]
76. Weimer, A.K.; Biedermann, S.; Harashima, H.; Roodbarkelari, F.; Takahashi, N.; Foreman, J.; Guan, Y.; Pochon, G.; Heese, M.; Van Damme, D.; et al. The plant-specific CDKB1-CYCB1 complex mediates homologous recombination repair in Arabidopsis. *EMBO J.* **2016**, *35*, 2068–2086. [CrossRef] [PubMed]
77. Boudolf, V.; Lammens, T.; Boruc, J.; Van Leene, J.; Van Den Daele, H.; Maes, S.; Van Isterdael, G.; Russinova, E.; Kondorosi, E.; Witters, E.; et al. CDKB1;1 forms a functional complex with CYCA2;3 to suppress endocycle onset. *Plant Physiol.* **2009**, *150*, 1482–1493. [CrossRef] [PubMed]
78. Luo, W.; Li, Y.; Sun, Y.; Lu, L.; Zhao, Z.; Zhou, J.; Li, X. Comparative RNA-seq analysis reveals candidate genes associated with fruit set in pumpkin. *Sci. Hortic.* **2021**, *288*, 110255. [CrossRef]
79. Komaki, S.; Sugimoto, K. Control of the plant cell cycle by developmental and environmental cues. *Plant Cell Physiol.* **2012**, *53*, 953–964. [CrossRef]
80. Magyar, Z.; Horváth, B.; Khan, S.; Mohammed, B.; Henriques, R.; De Veylder, L.; Bakó, L.; Scheres, B.; Bögre, L. Arabidopsis E2FA stimulates proliferation and endocycle separately through RBR-bound and RBR-free complexes. *EMBO J.* **2012**, *31*, 1480–1493. [CrossRef]
81. Ramirez-Parra, E.; Gutierrez, C. E2F regulates FASCIATA1, a chromatin assembly gene whose loss switches on the endocycle and activates gene expression by changing the epigenetic status. *Plant Physiol.* **2007**, *144*, 105–120. [CrossRef]
82. Sablowski, R.; Gutierrez, C. Cycling in a crowd: Coordination of plant cell division, growth, and cell fate. *Plant Cell* **2022**, *34*, 193–208. [CrossRef]
83. Nomoto, Y.; Takatsuka, H.; Yamada, K.; Nomoto, Y.; Takatsuka, H.; Yamada, K.; Suzuki, T.; Suzuki, T.; Huang, Y.; Latrasse, D.; et al. A hierarchical transcriptional network activates specific CDK inhibitors that regulate G2 to control cell size and number in Arabidopsis. *Nat. Commun.* **2022**, *13*, 1660. [CrossRef] [PubMed]
84. Pandolfini, T.; Molesini, B.; Spena, A. Molecular dissection of the role of auxin in fruit initiation. *Trends Plant Sci.* **2007**, *12*, 327–329. [CrossRef] [PubMed]
85. de Jong, M.; Wolters-Arts, M.; Schimmel, B.C.; Stultiens, C.L.; de Groot, P.F.; Powers, S.J.; Tikunov, Y.M.; Bovy, A.G.; Mariani, C.; Vriezen, W.H.; et al. *Solanum lycopersicum* AUXIN RESPONSE FACTOR 9 regulates cell division activity during early tomato fruit development. *J. Exp. Bot.* **2015**, *66*, 3405–3416. [CrossRef] [PubMed]
86. de Jong, M.; Wolters-Arts, M.; Feron, R.; Mariani, C.; Vriezen, W.H. The *Solanum lycopersicum* auxin response factor 7 (SlARF7) regulates auxin signaling during tomato fruit set and development. *Plant J.* **2009**, *57*, 160–170. [CrossRef] [PubMed]
87. Pattison, R.J.; Csukasi, F.; Catalá, C. Mechanisms regulating auxin action during fruit development. *Physiol. Plant.* **2014**, *151*, 62–72. [CrossRef]
88. Zhang, L.Y.; Bai, M.Y.; Wu, J.Y.; Zhu, J.Y.; Wang, H.; Zhang, Z.; Wang, W.; Sun, Y.; Zhao, J.; Sun, X.; et al. Antagonistic HLH/bHLH transcription factors mediate brassinosteroid regulation of cell elongation and plant development in rice and Arabidopsis. *Plant Cell* **2009**, *21*, 3767–3780. [CrossRef]
89. Mara, C.D.; Huang, T.; Irish, V.F. The Arabidopsis floral homeotic proteins APETALA3 and PISTILLATA negatively regulate the BANQUO genes implicated in light signaling. *Plant Cell* **2010**, *22*, 690–702. [CrossRef]
90. Zheng, K.; Wang, Y.; Zhang, N.; Jia, Q.; Wang, X.; Hou, C.; Chen, J.-C.; Wang, S. Involvement of PACLOBUTRAZOL RESISTANCE6/KIDARI, an atypical bHLH transcription factor, in auxin responses in Arabidopsis. *Front. Plant Sci.* **2017**, *8*, 1813. [CrossRef]
91. Zhu, Z.; Liang, H.; Chen, G.; Li, F.; Wang, Y.; Liao, C.; Hu, Z. The bHLH transcription factor SlPRE2 regulates tomato fruit development and modulates plant response to gibberellin. *Plant Cell Rep.* **2019**, *38*, 1053–1064. [CrossRef]
92. Pál, M.; Szalai, G.; Gondor, O.K.; Janda, T. Unfinished story of polyamines: Role of conjugation, transport and light-related regulation in the polyamine metabolism in plants. *Plant Sci.* **2021**, *308*, 110923. [CrossRef]
93. Li, M.; Yu, G.; Cao, C.; Liu, P. Metabolism, signaling, and transport of jasmonates. *Plant Commun.* **2021**, *2*, 100231. [CrossRef] [PubMed]
94. Huang, H.; Liu, B.; Liu, L.; Song, S. Jasmonate action in plant growth and development. *J. Exp. Bot.* **2017**, *68*, 1349–1359. [CrossRef] [PubMed]
95. Farmer, E.E.; Goossens, A. Jasmonates: What ALLENE OXIDE SYNTHASE does for plants. *J. Exp. Bot.* **2019**, *70*, 3373–3378. [CrossRef] [PubMed]
96. Ghorbani, S.; Fernandez, A.; Hilson, P.; Beeckman, T. Signaling peptides in plants. *Cell Dev. Biol.* **2014**, *3*, 141. [CrossRef]
97. Richter, J.; Watson, J.M., Stasnik, P.; Borowska, M.; Neuhold, J.; Berger, M.; Stolt-Bergner, P.; Schoft, V.; Hauser, M.T. Multiplex mutagenesis of four clustered CrRLK1L with CRISPR/Cas9 exposes their growth regulatory roles in response to metal ions. *Sci Rep.* **2018**, *8*, 12182. [CrossRef] [PubMed]
98. Stintzi, A.; Schaller, A. Biogenesis of post-translationally modified peptide signals for plant reproductive development. *Curr. Opin. Plant Biol.* **2022**, *69*, 102274. [CrossRef]
99. Ghorbani, S.; Hoogewijs, K.; Pečenková, T.; Fernandez, A.; Inzé, A.; Eeckhout, D. The SBT6.1 subtilase processes the GOLVEN1 peptide controlling cell elongation. *J. Exp. Bot.* **2016**, *67*, 4877–4887. [CrossRef]

100. Tenorio-Berrío, R.; Pérez-Alonso, M.-M.; Vicente-Carbajosa, J.; Martín-Torres, L.; Dreyer, I.; Pollmann, S. Identification of two auxin-regulated potassium transporters involved in seed maturation. *Int. J. Mol. Sci.* **2018**, *19*, 2132. [CrossRef]
101. Dreyer, I.; Uozumi, N. Potassium channels in plant cells. *FEBS J.* **2011**, *278*, 4293–4303. [CrossRef]
102. Samuel, D.; Liu, Y.J.; Cheng, C.S.; Lyu, P.C. Solution Structure of Plant Nonspecific Lipid Transfer Protein-2 from Rice (*Oryza sativa*). *J. Biol. Chem.* **2002**, *277*, 35267–35273. [CrossRef]
103. Li, Y.J.; Yu, Y.; Liu, X.; Zhang, X.S.; Su, Y.H. The Arabidopsis MATERNAL EFFECT EMBRYO ARREST45 protein modulates maternal auxin biosynthesis and controls seed size by inducing *AINTEGUMENTA*. *Plant Cell* **2021**, *33*, 1907–1926. [CrossRef]
104. Wu, S.C.; Zhang, Y. Cyclin-dependent kinase 1 (CDK1)-mediated phosphorylation of Enhancer of Zeste 2 (Ezh2) regulates its stability. *J. Biol. Chem.* **2011**, *286*, 28511–28519. [CrossRef]
105. Randall, R.S.; Miyashima, S.; Blomster, T.; Zhang, J.; Elo, A.; Karlberg, A.; Immanen, J.; Nieminen, K.; Lee, J.-Y.; Kakimoto, T.; et al. *AINTEGUMENTA* and the D-type cyclin CYCD3;1 regulate root secondary growth and respond to cytokinins. *Biol. Open* **2015**, *4*, 1229–1236. [CrossRef]
106. Kim, J.H.; Tsukaya, H. Regulation of plant growth and development by the GROWTH-REGULATING FACTOR and GRF-INTERACTING FACTOR duo. *J. Exp. Bot.* **2015**, *66*, 6093–6107. [CrossRef]
107. Hewezi, T.; Maier, T.R.; Nettleton, D.; Baum, T.J. The Arabidopsis microRNA396-GRF1/GRF3 regulatory module acts as a developmental regulator in the reprogramming of root cells during cyst nematode infection. *Plant Physiol.* **2012**, *159*, 321–335. [CrossRef]
108. Cao, D.; Wang, J.; Ju, Z.; Liu, Q.; Li, S.; Tian, H.; Fu, D.; Zhu, H.; Luo, Y.; Zhu, B. Regulations on growth and development in tomato cotyledon, flower and fruit via destruction of miR396 with short tandem target mimic. *Plant Sci.* **2016**, *247*, 1–12. [CrossRef]
109. Vanhaeren, H.; Nam, Y.J.; De Milde, L.; Chae, E.; Storme, V.; Weigel, D.; Gonzalez, N.; Inzé, D. Forever Young: The role of ubiquitin receptor DA1 and E3 Ligase BIG BROTHER in controlling leaf growth and development. *Plant Physiol.* **2017**, *173*, 269–1282. [CrossRef]
110. Kanaoka, M.M.; Pillitteri, L.J.; Fujii, H.; Yoshida, Y.; Bogenschutz, N.L.; Takabayashi, J.; Zhu, J.K.; Torii, K.U. SCREAM/ICE1 and SCREAM2 specify three cell-state transitional steps leading to Arabidopsis stomatal differentiation. *Plant Cell* **2008**, *20*, 1775–1785. [CrossRef]
111. Liljegren, S.J.; Roeder, A.H.; Kempin, S.A.; Gremski, K.; Østergaard, L.; Guimil, S.; Reyes, D.K.; Yanofsky, M.F. Control of fruit patterning in *Arabidopsis* by INDEHISCENT. *Cell* **2004**, *116*, 843–853. [CrossRef]
112. Tani, E.; Tsaballa, A.; Stedel, C.; Kalloniati, C.; Papaefthimiou, D.; Polidoros, A.; Darzentas, N.; Ganopoulos, I.; Flemetakis, E.; Katinakis, P.; et al. The study of a SPATULA-like bHLH transcription factor expressed during peach (*Prunus persica*) fruit development. *Plant Physiol. Biochem.* **2011**, *49*, 654–663. [CrossRef]
113. Wang, L.; Tang, W.; Hu, Y.; Zhang, Y.; Sun, J.; Guo, X.; Lu, H.; Yang, Y.; Fang, C.; Niu, X.; et al. A MYB/bHLH complex regulates tissue-specific anthocyanin biosynthesis in the inner pericarp of red-centered kiwifruit *Actinidia chinensis* cv. Hongyang. *Plant J.* **2019**, *99*, 359–378. [CrossRef] [PubMed]
114. Ezer, D.; Shepherd, S.J.K.; Brestovitsky, A.; Dickinson, P.; Cortijo, S.; Charoensawan, V.; Box, M.S.; Biswas, S.; Jaeger, K.E.; Wigge, P.A. The G-Box transcriptional regulatory code in Arabidopsis. *Plant Physiol.* **2017**, *175*, 628–640. [CrossRef] [PubMed]
115. Ambawat, S.; Sharma, P.; Yadav, N.R.; Yadav, R.C. MYB transcription factor genes as regulators for plant responses: An overview. *Physiol. Mol. Biol. Plants.* **2013**, *19*, 307–321. [CrossRef] [PubMed]
116. Ito, M.; Araki, S.; Matsunaga, S.; Itoh, T.; Nishihama, R.; Machida, Y.; Doonan, J.H.; Watanabe, A. G2/M-phase-specific transcription during the plant cell cycle is mediated by c-Myb-like transcription factors. *Plant Cell* **2001**, *13*, 1891–1905. [CrossRef]
117. Furuya, T.; Saito, M.; Uchimura, H.; Satake, A.; Nosaki, S.; Miyakawa, T.; Shimadzu, S.; Yamori, W.; Tanokura, M.; Fukuda, H.; et al. Gene co-expression network analysis identifies BEH3 as a stabilizer of secondary vascular development in Arabidopsis. *Plant Cell* **2021**, *33*, 2618–2636. [CrossRef]
118. Chen, P.; Takatsuka, H.; Takahashi, N.; Kurata, R.; Fukao, Y.; Kobayashi, K.; Ito, M.; Umeda, M. Arabidopsis R1R2R3-Myb proteins are essential for inhibiting cell division in response to DNA damage. *Nat. Commun.* **2017**, *8*, 635. [CrossRef]
119. Franken, P.; Schrell, S.; Peterson, P.A.; Saedler, H.; Wienand, U. Molecular analysis of protein domain function encoded by the myb homologous maize genes C1, Zm1 and Zm38. *Plant J.* **1989**, *6*, 21–30. [CrossRef]
120. Gahan, P.B. Plant Histochemistry and Cytochemistry. Academic Press: London. UK, 1984.
121. Lamprecht, M.R.; Sabatini, D.M.; Carpenter, A.E. CellProfiler: Free, versatile software for automated biological image analysis. *Biotechniques* **2007**, *42*, 71–75. [CrossRef]
122. Loureiro, J. Flow cytometric approaches to study plant genomes. *Ecosistemas* **2009**, *18*, 103–108.
123. Dolezel, J.; Greilhuber, J.; Suda, J. Estimation of nuclear DNA content in plants using flow cytometry. *Nature Protocols* **2007**, *2*, 2233–2244. [CrossRef]
124. Li, H.; Handsaker, B.; Wysoker, A.; Fennell, T.; Ruan, J.; Homer, N.; Marth, G.; Abecasis, G.; Durbin, R.; 1000 Genome Project Data Processing Subgroup. The Sequence Alignment/Map format and SAM tools. *Bioinformatics* **2009**, *25*, 2078–2079. [CrossRef]
125. Langmead, B.; Salzberg, S.L. Fast gapped-read alignment with Bowtie 2. *Nat. Methods* **2012**, *9*, 357–359. [CrossRef] [PubMed]
126. Love, M.I.; Huber, W.; Anders, S. Moderated estimation of fold change and dispersion for RNA-seq data with DESeq2. *Genome Biol.* **2014**, *15*, 550. [CrossRef]

127. Benjamini, Y.; Hochberg, Y. Controlling the false discovery rate: A practical and powerful approach to multiple testing. *J. R. Stat. Soc.* **1995**, *57*, 289–300. [CrossRef]
128. Yu, G.; Wang, L.-G.; Han, Y.; He, Q.-Y. ClusterProfler: An R package for com- paring biological themes among gene clusters. *Omics J. Integr. Biol.* **2012**, *16*, 284–287. [CrossRef]
129. Morgan, M.; Shepherd, L. AnnotationHub: Client to Access Annotation- Hub Resources. R Package Version 3.2.2. 2019. Available online: https://bioconductor.org/packages/release/bioc/html/AnnotationHub.html (accessed on 10 January 2020).
130. Thompson, J.; Higgins, D.; Gibson, T. CLUSTAL W: Improving the sensitivy of progressive multiple sequence through sequence weighting, position-specific gap penalties and weight matrix choice. *Nucl. Acids Res.* **1994**, *22*, 4673–4680. [CrossRef]
131. Seo, M.; Jikumaru, Y.; Kamiya, Y. Profiling of hormones and related metabolites in seed dormancy and germination studies. *Methods Mol. Biol.* **2011**, *773*, 99–111.

International Journal of
Molecular Sciences

Article

Genome-Wide Identification of the *SAUR* Gene Family in Wax Gourd (*Benincasa hispida*) and Functional Characterization of *BhSAUR60* during Fruit Development

Chen Luo [1,2], Jinqiang Yan [1,2], Changxia He [1,2], Wenrui Liu [1,2], Dasen Xie [1,2] and Biao Jiang [1,2,*]

1 Vegetable Research Institute, Guangdong Academy of Agricultural Sciences, Guangzhou 510640, China
2 Guangdong Key Laboratory for New Technology Research of Vegetables, Guangzhou 510640, China
* Correspondence: jiangbiao@gdaas.cn; Tel.: +86-020-38469441

Abstract: The wax gourd (*Benincasa hispida*) is an important vegetable crop whose fruits contain nutrients and metabolites. *Small auxin upregulated RNA* (*SAUR*) genes constitute the largest early auxin-responsive gene family and regulate various biological processes in plants, although this gene family has not been studied in the wax gourd. Here, we performed genome-wide identification of the *SAUR* gene family in wax gourds and analyzed their syntenic and phylogenetic relationships, gene structures, conserved motifs, *cis*-acting elements, and expression patterns. A total of 68 *SAUR* (*BhSAUR*) genes were identified, which were distributed on nine chromosomes with 41 genes in two clusters. More than half of the *BhSAUR* genes were derived from tandem duplication events. The BhSAUR proteins were classified into seven subfamilies. *BhSAUR* gene promoters contained *cis*-acting elements involved in plant hormone and environmental signal responses. Further expression profiles showed that *BhSAUR* genes displayed different expression patterns. *BhSAUR60* was highly expressed in fruits, and overexpression led to longer fruits in *Arabidopsis*. In addition, the plants with overexpression displayed longer floral organs and wavy stems. In conclusion, our results provide a systematic analysis of the wax gourd *SAUR* gene family and facilitate the functional study of *BhSAUR60* during wax gourd fruit development.

Keywords: wax gourd; genome-wide identification; *SAUR*; gene family; fruit development

1. Introduction

The wax gourd (*Benincasa hispida* (Thunb.) Cogn., 2n = 2*x* = 24) is an important vegetable crop of the Cucurbitaceae family [1,2]. The fruits of wax gourds contain nutrients and metabolites, and can be used as both a food and a medicine [3]. There are several characteristics of these fruits, such as the size, shape, and peel color, which are important agronomic traits in breeding [4–9]. Previously, quantitative trait locus (QTL) mapping identified size-associated traits in wax gourd fruits, including fruit weight, fruit length, fruit diameter, as well as flesh thickness [5]. The fine-mapping results showed that *Benincasa hispida Fruit Shape* (*BFS*) was the candidate gene for wax gourd fruit shape determination [4]. Recently, *Arabidopsis Pseudo-Response Regulator2* (*APRR2*) was identified as the candidate gene controlling peel color in wax gourds [6].

Auxin is the first identified plant hormone, and controls numerous aspects of plant growth and development by regulating cell division, expansion, differentiation, and patterning [10,11]. Auxin can induce the expression of various genes [12,13]. Currently, three major classes of early auxin-responsive gene families have been identified in plants, including the *Auxin/Indole-3-Acetic Acid* (*Aux/IAA*) family, the *Small Auxin Up-regulated RNA* (*SAUR*) family, and the *Gretchen Hagen3* (*GH3*) family [14], which play important roles in auxin signaling and homeostasis.

Among the early auxin-responsive gene families, the *SAUR* gene family is the largest [15,16]. In most plant genomes, the number of *SAUR* genes ranges from 60 to 140, and the genes are usually arranged in clusters [14,17–19]. The SAUR proteins are characterized by a conserved core of about 60 residues, named the SAUR domain [15,16]. *SAUR* genes were originally identified as a class of small RNAs that were rapidly induced by auxin in the soybean [18]. In the model plant *Arabidopsis*, the *SAUR* gene family contains 79 members and participates in a wide range of cellular, physiological, and developmental processes [15,16,20]. For example, *SAUR19*, *SAUR41*, and *SAUR63* promote hypocotyl elongation by increasing cell expansion [21–23]. PHYTOCHROME RAPIDLY REGULATED1 (PAR1) and PAR2 repress the expression of *SAUR15* and *SAUR68* to inhibit shade-avoidance responses [24]. The *SAUR19* subfamily genes function downstream of PHYTOCHROME INTERACTING FACTOR4 (PIF4) to regulate hypocotyl growth in response to high temperatures [25]. SAUR17 and SAUR50 proteins differentially regulate Protein Phosphatase 2C D-clade 1 (PP2C-D1) during apical hook development and cotyledon opening [26]. Moreover, the *SAUR19* subfamily genes positively regulate leaf growth [22], while *SAUR36* negatively regulates leaf cell expansion and positively regulates leaf senescence [27]. It was also shown that some SAUR proteins, such as SAUR19, SAUR41, and SAUR63, are capable of modulating IAA transport [21–23]. In addition to auxin, brassinosteroid (BR) [28], gibberellin (GA) [29], jasmonate (JA) [30], abscisic acid (ABA) [31], and cytokinin (CK) [32] also regulate the expression of some *SAUR* genes.

Furthermore, it has been claimed that the *SAUR* genes are also involved in fruit development and ripening. In *Arabidopsis*, overexpression of *SAUR8*, *SAUR10*, and *SAUR16* led to longer siliques when compared with the wild-type (WT) version [32,33]. Tomato *SlSAUR69* could repress auxin transport and enhance sensitivity to ethylene at the onset of fruit ripening [34]. In grapes, *SAUR* members were identified and *SAUR041* was proposed to be a candidate regulator of berry size [35]. In cucurbit crops, the *SAUR* gene family had been characterized in watermelon, cucumber, and melon [36–38], and some *SAUR* genes were significantly accumulated in the fruits, suggesting their potential roles in controlling fruit development.

Previously, we identified a *SAUR* gene as the candidate domestication gene conferring large fruit size in wax gourds [7], while no comprehensive analysis of the *SAUR* gene family has been reported in wax gourds. In this study, we performed genome-wide identification of the *SAUR* gene family in wax gourds and analyzed their syntenic and phylogenetic relationships. We determined their gene structures, conserved motifs, *cis*-acting elements, and expression patterns. Then, we selected *BhSAUR60* for further characterization. Our results provide systematic information on the wax gourd *SAUR* gene family, which will facilitate the study of functional and regulatory mechanisms of *SAUR* family members during wax gourd fruit development.

2. Results

2.1. Identification of the SAUR Gene Family in Wax Gourds

To identify the *SAUR* gene family in wax gourds, HMMER searches were performed in the genome of wax gourd inbred line B227 using the *SAUR* gene family characteristic domain (Pfam: PF02519). A total of 68 *SAUR* genes were obtained (Table S1). A multiple sequence alignment analysis identified the conserved SAUR domain of BhSAUR proteins (Figure S1), which was consistent with the previous report [15]. According to their location on the chromosomes, the 68 wax gourd *SAUR* genes were named, from *BhSAUR1* to *BhSAUR68* (Figure 1). Among the 68 genes, 66 were distributed on the nine chromosomes, with *BhSAUR67* and *BhSAUR68* not mapped to the existing chromosomes (Figure 1). Moreover, *BhSAUR10*–*BhSAUR28* and *BhSAUR40*–*BhSAUR61* constituted two gene clusters on chromosome 5 and chromosome 10, respectively (Figure 1).

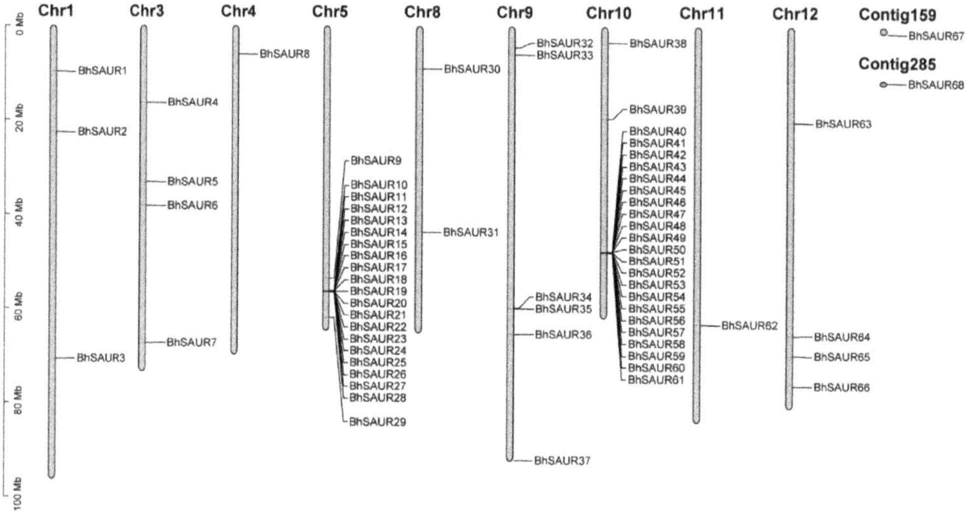

Figure 1. Chromosomal location of the *BhSAUR* genes.

The open reading frame (ORF) length of the 68 *BhSAUR* genes ranged from 225 to 573 bp, encoding polypeptides of 74–190 amino acids (aa) (Table S1). The molecular weight (Mw) and theoretical isoelectric point (pI) of their proteins were 8.27–22.29 kDa and 4.60–10.67, respectively (Table S1). Subcellular localization prediction results showed that most BhSAUR proteins were located in the chloroplast or nucleus, and only a few of them were located in the mitochondrion, cytoplasm, cell membrane, or peroxisome (Table S1).

2.2. Gene Duplication and Synteny Analysis of the BhSAUR Gene Family

As the *SAUR* gene family is a large family, we analyzed the gene duplication modes of *BhSAUR* genes. We found that *BhSAUR* genes were derived from five types of duplication events, including whole-genome duplication (WGD), tandem duplication (TD), dispersed duplication (DSD), proximal duplication (PD), and transposed duplication (TRD) (Table S2). Most genes (36 members) were derived from TD (Table S2), indicating that the TD event was the main driver of *SAUR* gene family expansion in wax gourds. In addition, all tandem-duplicated genes were located in the two *SAUR* gene clusters.

To detect and evolutionarily analyze the synteny of *BhSAUR* genes, we performed an intra-species synteny analysis of the wax gourd genome. Finally, 13 paralogous *BhSAUR* gene pairs were identified, among which seven *BhSAUR* gene pairs were found between the two gene clusters (Figure 2A). In addition, we calculated the non-synonymous substitutions/synonymous substitutions (Ka/Ks) values of the *BhSAUR* gene pairs to evaluate sequence variations, and all Ka/Ks values were below 1 (Table S3), indicating that these genes had undergone purifying selection throughout their evolution.

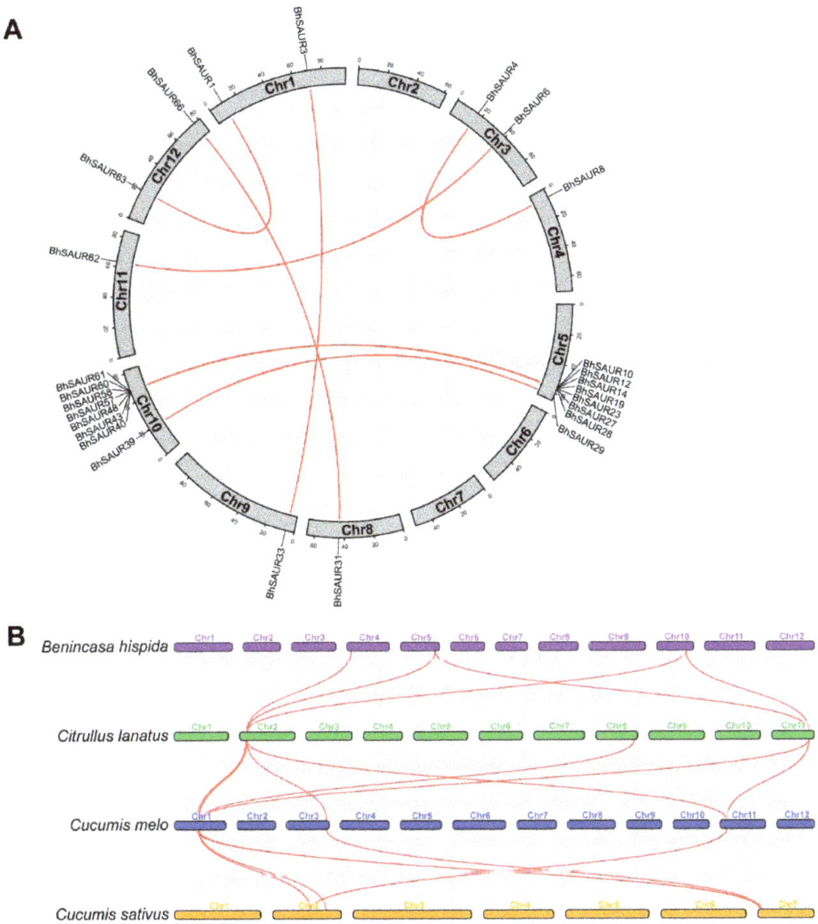

Figure 2. Synteny analysis of the *SAUR* genes. (**A**) Synteny analysis of the *SAUR* genes in wax gourds. Grey lines display the collinear gene pairs in the wax gourd genome. Red lines indicate the collinear *BhSAUR* gene pairs; (**B**) gene synteny analysis across the wax gourd (*Benincasa hispida*), watermelon (*Citrullus lanatus*), melon (*Cucumis melo*), and cucumber (*Cucumis sativus*). Grey lines display the collinear gene pairs across four cucurbit crop genomes. Red lines indicate the collinear *SAUR* gene pairs in the gene clusters.

The *SAUR* gene family had been identified in other cucurbit crops, including watermelon, melon, and cucumber, which also contained two *SAUR* gene clusters [36–38]. To explore the evolutionary trajectory of the *SAUR* family, we constructed a syntenic map and generated collinear *SAUR* gene pairs across four cucurbit crops (Figure S2; Table S4), which showed that most *SAUR* genes were mapped to these syntenic blocks (Figure S2). Furthermore, we analyzed the collinear gene pairs in the *SAUR* gene clusters. The obtained results showed that genes in the *SAUR* gene clusters displayed one-to-one, one-to-many, or many-to-one syntenic relationships between any two cucurbit crops (Figure 2B).

2.3. Phylogenetic Analysis of the BhSAUR Gene Family

To understand the phylogenetic relationships among the identified SAUR proteins in cucurbit crops, we constructed a neighbor-joining phylogenetic tree using the complete amino acid sequences of SAURs from the wax gourd, watermelon, melon, cucumber, and

Arabidopsis. The obtained phylogenetic tree showed that these SAURs were divided into seven subfamilies (Figure 3).

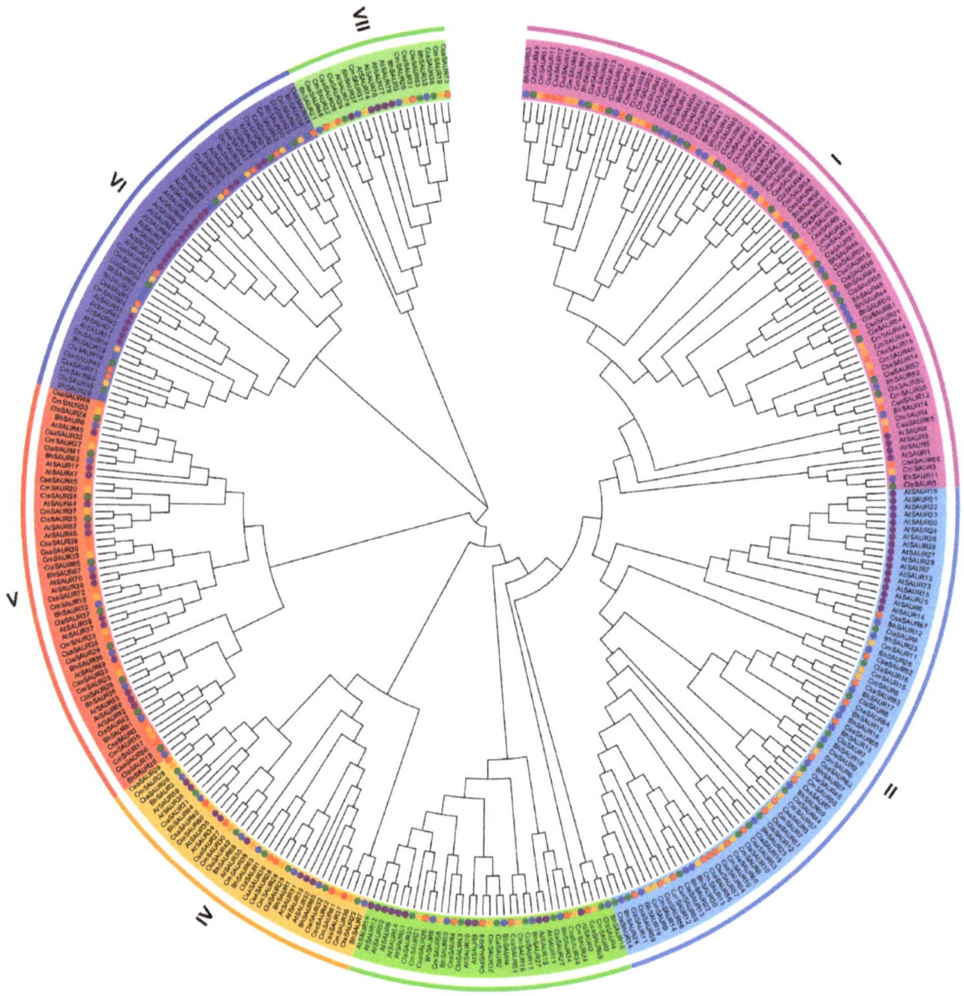

Figure 3. Phylogenetic tree of the SAUR family. The unrooted tree was constructed from all SAUR proteins of *Arabidopsis* (At), wax gourd (Bh), watermelon (Cla), melon (Cm), and cucumber (Csa). Different color circles represent different species. I–VII represent the seven subfamilies.

The SAUR proteins were randomly grouped into their subfamilies except for subfamilies I and II, as the SAUR proteins of cucurbit crops were branched together while the *Arabidopsis* SAUR proteins were branched together in these two subfamilies (Figure 3). In addition, most clustered BhSAUR proteins on chromosome 10 and chromosome 5 were classified into subfamily I and subfamily II, respectively (Figure 3).

2.4. Gene Structure and Conserved Motif Analysis of the BhSAUR Gene Family

To investigate the exon and intron structures of *BhSAUR* genes, their coding sequences and genomic sequences were extracted for comparison. The obtained results showed that most *BhSAUR* genes did not contain introns, and only two genes (*BhSAUR14* and

BhSAUR67) contained one intron (Figure 4B). We then used the MEME online software to analyze the conserved motifs of BhSAUR proteins. Four conserved motifs were identified, and the closely related BhSAUR proteins had similar motifs (Figure 4A,C). Motifs 1–3 constituted the SAUR domain, and 61 BhSAUR proteins contained all three of these motifs, while the other seven contained only one or two motifs (Figure 4C). In addition, motif 4 was present only in some BhSAUR proteins (Figure 4C), most of which were in the two gene clusters on chromosome 5 and chromosome 10, indicating that motif 4 may contribute to the special function of these proteins.

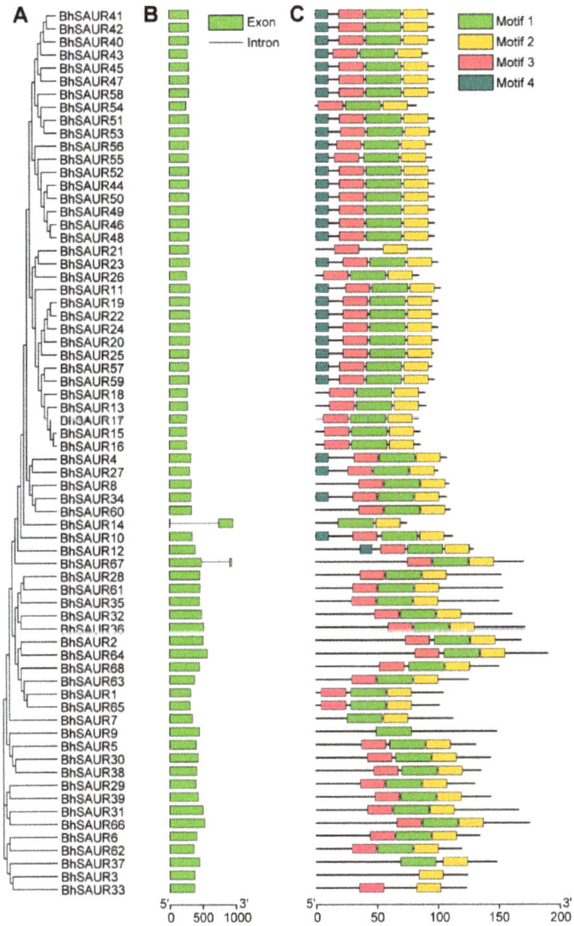

Figure 4. Gene structure and conserved motifs of the *BhSAUR* genes. (**A**) Phylogenetic tree of the BhSAUR proteins; (**B**) gene structure of the *BhSAUR* genes; (**C**) conserved motifs of the BhSAUR proteins.

2.5. *Cis-Acting Element Analysis in the Promoters of the BhSAUR Gene Family*

To identify the *cis*-acting elements in the promoters of *BhSAUR* genes, their 2000 bp promoter regions upstream of the start codon were obtained and analyzed with the PlantCARE online software. Almost all *BhSAUR* gene promoters contained one or multiple *cis*-acting elements involved in plant hormone responses (Figure 5A), including abscisic acid-responsive element (ABRE), auxin-responsive elements (AuxRE, TGA-element, AuxRR-core), gibberellin-responsive elements (GARE-motif, P-box, TATC-box), methyl jasmonate (MeJA)-responsive elements (CGTCA-motif, TGACG-motif), and salicylic acid-responsive element (TCA-element). In addition, *cis*-acting elements associated with environmental signal response were also found (Figure 5A), such as a defense and stress-responsive element (TC-rich repeats), drought-inducibility element (MBS), and low-temperature-responsive element (LTR). Therefore, plant hormones and environmental conditions might be important for regulating the expression of *BhSAUR* genes.

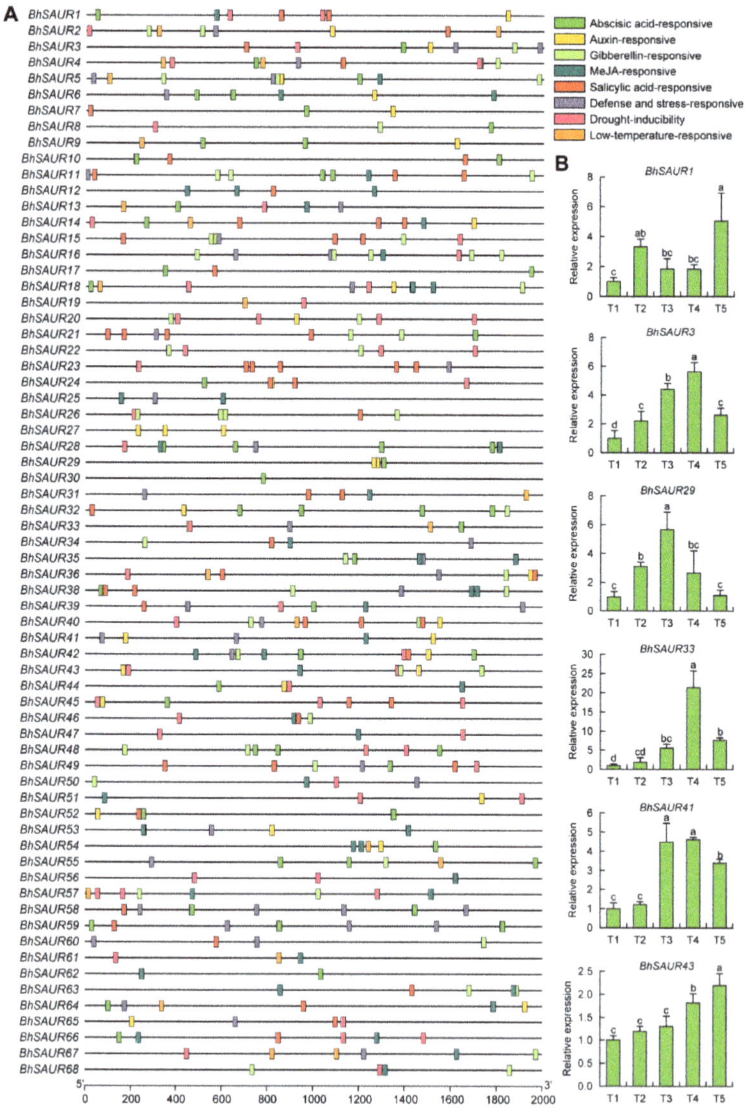

Figure 5. *Cis*-acting element analysis of the *BhSAUR* gene promoters. (**A**) *Cis*-acting elements in the promoters of the *BhSAUR* genes; (**B**) expression analysis of the *BhSAUR* genes under IAA treatment. T1, T2, T3, T4, and T5 represent 0, 10, 30, 60, and 120 min after IAA treatment, respectively. Values are means ± SDs ($n = 3$). Different lowercase letters indicate significant differences determined by one-way ANOVA with Duncan's test ($p < 0.05$).

As *SAUR* genes are major early auxin-responsive genes and putative auxin-responsive *cis*-acting elements were found in some *BhSAUR* gene promoters, we performed an IAA-treatment experiment to test whether these genes could respond to auxin. Then, six *BhSAUR* genes with auxin-responsive *cis*-acting elements in their promoters were randomly selected for validation. Quantitative real-time PCR (qRT-PCR) results showed that all selected genes responded to IAA treatment (Figure 5B). *BhSAUR1*, *BhSAUR3*, and *BhSAUR29* were upregulated at 10 min after IAA treatment, while the expression of the other three genes increased significantly at 30 or 60 min (Figure 5B). Moreover, the expression of *BhSAUR3*, *BhSAUR29*, *BhSAUR33*, and *BhSAUR41* decreased at 60 or 120 min (Figure 5B). Our results implied that these *BhSAUR* genes were likely to play critical roles in the IAA signaling transduction pathway.

2.6. *Expression Characterization of the BhSAUR Gene Family*

To further explore the function of *BhSAUR* genes, we investigated their expression patterns through previously published transcriptome data [7,39]. Expression profiles of the *BhSAUR* genes in different tissues showed that these genes displayed different expression patterns (Figure 6A). For example, *BhSAUR3* was highly expressed in every tissue, with higher expression in the flower. *BhSAUR33* showed higher expression in the root and flower than in other tissues. *BhSAUR60* and *BhSAUR66* were specially expressed in the fruit and root, respectively. Moreover, *BhSAUR60* exhibited the highest expression in the fruit among the 68 *BhSAUR* genes. We then performed a qRT-PCR analysis to validate the expression patterns of these four genes, and the results were generally consistent with the transcriptome data (Figure 6C).

In addition, we analyzed the expression profiles of *BhSAUR* genes in fruits at different developmental stages and found that several of them were expressed at one to three stages (Figure 6B). Our qRT-PCR results also verified the expression patterns of five *BhSAUR* genes (Figure 6D). Consistently, the expression of *BhSAUR1*, *BhSAUR2*, and *BhSAUR66* increased with the development of fruit. *BhSAUR33* showed higher expression in 0- and 20-days after pollination (DAP) fruits than in 10-DAP fruits. *BhSAUR60* was highly expressed in 0-DAP fruits, and its expression decreased as the fruit developed. Together, these analyses suggested that *BhSAUR60* might be crucial in wax gourd fruit development, and we chose *BhSAUR60* for further characterization.

2.7. *Subcellular Localization Analysis of the BhSAUR60 Protein*

To determine the subcellular localization of BhSAUR60 protein, the green fluorescent protein (GFP) tag was fused to the C terminus of *BhSAUR60* under the control of the super promoter. The control construct and *Super:BhSAUR60–GFP* recombinant construct were transferred into *Nicotiana benthamiana* leaves using *Agrobacterium tumefaciens*-mediated transient expression. The result showed that the BhSAUR60–GFP fusion protein was localized to both the nucleus and cytoplasm (Figure 7).

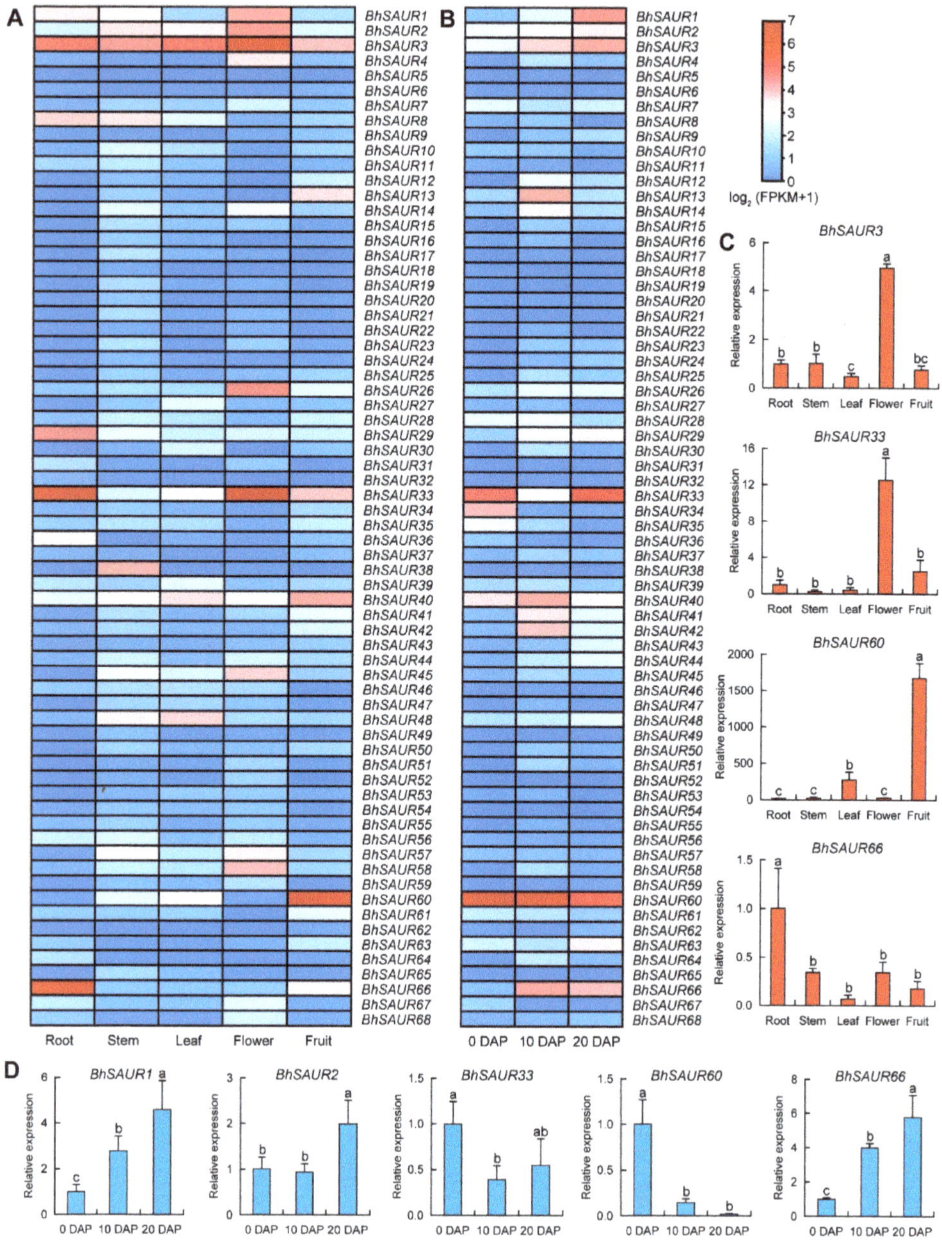

Figure 6. Expression characterization of the *BhSAUR* genes. (**A**) Expression profile of the *BhSAUR* genes in different tissues; (**B**) expression profile of the *BhSAUR* genes in fruits at different developmental stages. DAP, days after pollination; (**C**) validation of the expression of some *BhSAUR* genes in different tissues. Values are means ± SDs ($n = 3$); (**D**) validation of the expression of some *BhSAUR* genes in fruits at different developmental stages. Values are means ± SDs ($n = 3$). Different lowercase letters in (**C**,**D**) indicate significant differences determined by one-way ANOVA with Duncan's test ($p < 0.05$).

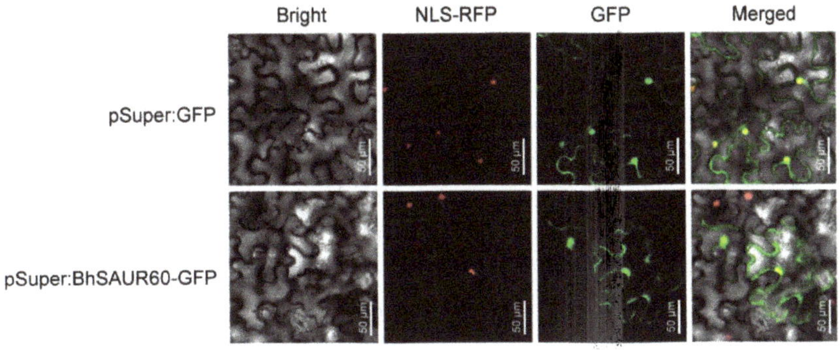

Figure 7. Subcellular localization of the BhSAUR60 Protein.

2.8. Functional Analysis of the BhSAUR60 Gene

Phylogenetic analysis showed that BhSAUR60 was closely related to the *Arabidopsis* SAUR10 (AtSAUR10) protein and grouped into the subfamily III (Figure 3), indicating that BhSAUR60 may have similar functions as AtSAUR10. In *Arabidopsis*, overexpression of *AtSAUR10* showed pleiotropic growth-related phenotypes, including longer hypocotyls, cauline leaves, sepals, filaments, pistils, and siliques, and wavier stems [32,33]. To further investigate the biological function of *BhSAUR60*, we generated a *35S:BhSAUR60* construct and transformed it into *Arabidopsis*. All transgenic lines showed similar phenotypes, and we selected three representative lines for further characterization (Figure 8A). The expression of *BhSAUR60* increased significantly in the transgenic lines (Figure S3). Overexpression of *BhSAUR60* in *Arabidopsis* displayed wavy main stems and side branches (Figure 8A,B). The overexpression lines also displayed longer floral organs, including longer sepals, filaments, and pistils (Figure S4). Due to the long pistil size, the overexpression lines often showed reduced fertility, and long siliques were only reached in pollinated pistils (Figure 8B). Measurement of the well-developed siliques showed that the silique lengths of overexpression lines were significantly increased when compared with those of WT (Figure 8C,D). These results indicated that *BhSAUR60* could promote fruit elongation.

Furthermore, we performed an expression analysis of fruit development related genes in WT and *35S:BhSAUR60* siliques. Our qRT-PCR results showed that the expression levels of *AtARF6*, *AtARF8*, *AtSTK*, and *AtCKX7* decreased significantly in the *35S:BhSAUR60* plants (Figure 8E–H). These genes regulate auxin signaling or cytokinin degradation to influence fruit elongation in *Arabidopsis* [40,41], which implied that *BhSAUR60* might regulate fruit elongation of *Arabidopsis* through the auxin and cytokinin pathways.

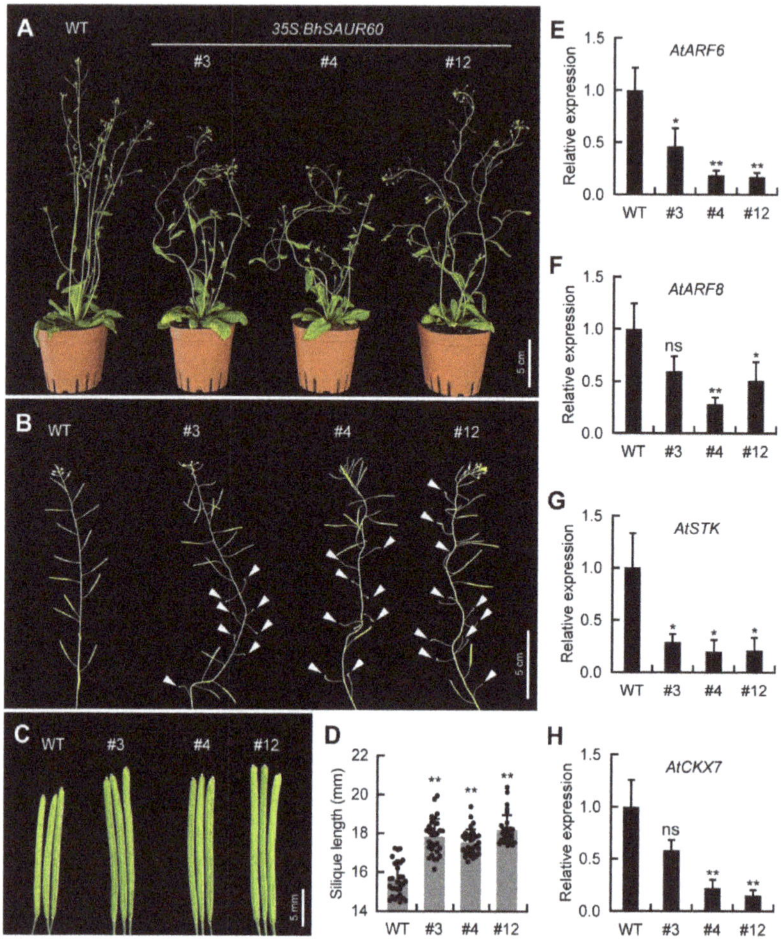

Figure 8. Overexpression of the *BhSAUR60* gene in *Arabidopsis*. (**A**) Whole-plant phenotypes of wild-type (WT) and *35S:BhSAUR60* plants around 10 days after bolting (DAB); (**B**) main-stem phenotypes of WT and *35S:BhSAUR60* plants at 18 DAB. Triangles indicate unfertilized pistils; (**C**) mature siliques of WT and *35S:BhSAUR60* plants; (**D**) silique lengths of WT and *35S:BhSAUR60* plants. Values are means ± SDs ($n = 30$), **, $p < 0.01$ (Student's *t*-test); (**E–H**) expression analysis of fruit development-related genes in WT and *35S:BhSAUR60* siliques. Values are means ± SDs ($n = 3$), ns, not significant, *, $p < 0.05$, **, $p < 0.01$ (Student's *t*-test).

3. Discussion

The *SAUR* family is the largest family of early auxin-responsive genes and participates in a variety of processes of plant growth and development [15,16]. Previously, *SAUR* genes have been identified in the model plant *Arabidopsis* and other diverse plant species [15,20]. In cucurbit crops, there are 65, 73, and 66 *SAUR* genes in watermelon [37], cucumber [36], and melon [38], respectively. In this study, we identified 68 *SAUR* genes in the wax gourd genome (Figure 1; Table S1), a number of *SAURs* similar to those of cucurbit crops. Together with evolutionary relationships (Figures 2 and 3), bioinformatic analysis (Figures 4 and 5), and expression profiles (Figure 6), the comprehensive characteristics of *BhSAUR* genes were also discovered.

The 68 *BhSAUR* genes identified were distributed on the nine chromosomes and two contigs, with *BhSAUR10–BhSAUR28* and *BhSAUR40–BhSAUR61* constituting two gene clusters on chromosomes 5 and chromosomes 10, respectively (Figure 1). Previous studies indicated that tandem duplicates have been found to play significant roles in plant adaptations to rapidly changing environments [42,43]. Our results showed that most *BhSAUR* genes were derived from TD events (Table S2), and all tandem-duplicated genes were located in the two *SAUR* gene clusters (Figure 1), indicating that TD events led to the expansion of the *SAUR* gene family and increased gene diversity in the wax gourd.

In cucurbit crops, the basic helix–loop–helix (bHLH) transcription factor family had over 100 members, and over 10 orthologous *bHLH* genes formed a superblock and had perfect syntenic relationships among seven cucurbit crop genomes [44]. There were two *SAUR* gene clusters in the wax gourd, watermelon, melon, and cucumber [36–38]. In this study, we detected the syntenic *SAUR* genes across four cucurbit crops (Table S4), while no perfectly matched syntenic block was found in the *SAUR* clusters across four cucurbit crops (Figure 2B). Therefore, we speculate that the conserved syntenic block across cucurbit crops may be lost due to gene duplication during the evolution and divergence of *SAUR* genes.

It was shown that *SAUR* genes integrate hormonal and environmental signals to regulate plant growth and development [15,16]. In the present study, we identified *cis*-acting elements associated with plant hormone responses in the *BhSAUR* gene promoters, including ABA, auxin, GA, MeJA, and salicylic acid responses (Figure 5A). Additionally, *cis*-acting elements in response to environmental signals were also found, such as defense- and stress-responsive elements, drought-inducibility elements, and low-temperature-responsive elements (Figure 5B). These results indicate that plant hormones and environmental signals may be important for regulating the expression of *BhSAUR* genes. As *SAUR* genes comprise the largest family of early auxin-responsive genes, we performed an IAA-treatment experiment to test whether these genes could respond to auxin. The obtained results showed that selected *BhSAUR* genes, including *BhSAUR1*, *BhSAUR3*, *BhSAUR29*, *BhSAUR33*, *BhSAUR41*, and *BhSAUR43* responded to auxin (Figure 5C).

Phylogenetic analysis of the SAUR proteins from four cucurbit crops and *Arabidopsis* showed that these SAURs were classified into seven subfamilies (Figure 3). In addition, most of the clustered BhSAUR proteins on chromosomes 10 and 5 were classified into subfamilies I and II, respectively (Figure 3). Although *BhSAUR60* was a member of the *BhSAUR* gene cluster on chromosome 10, we found that the BhSAUR60 protein lacked motif 4 and did not group with the clustered BhSAUR proteins into subfamily I (Figures 3 and 4), which implies that BhSAUR60 may have distinct functions from these proteins. In *Arabidopsis*, eight closely related SAUR proteins constitute the SAUR10-clade, and their overexpression lines all showed similar growth-related phenotypes, including longer hypocotyls, cauline leaves, sepals, filaments, pistils, and siliques, as well as wavy stems [32,33]. Our phylogenetic analysis showed that BhSAUR60 was more closely related to the members of *Arabidopsis* SAUR10-clade in the subfamily III (Figure 3), and overexpression of *BhSAUR60* in *Arabidopsis* also displayed wavy stems and longer organs (Figure 8A–C; Figure S4), suggesting that it had the conserved functions of the SAUR10-clade proteins in *Arabidopsis*, especially the regulating of cell elongation. Previously, we also demonstrated the function of BhSAUR60 in cell expansion in wax gourds [7]. Together, these results imply that BhSAUR60 may have the ability to promote fruit elongation in the wax gourd.

Fruits are the main, edible part of cucurbits, which have diverse fruit sizes and shapes [2,45]. Fruit size affects the appearance quality and the yield of wax gourds, while the molecular mechanisms regulating fruit development remain largely unknown [7]. In this study, we analyzed the expression profiles of *BhSAUR* genes using previously published transcriptome data, and found that several of them were expressed in fruits at one to three stages (Figure 6A,B). Previously, a study had identified several regions/genes potentially selected during wax gourd domestication, and *Bhi10G001538* (*BhSAUR60* in this study) was proposed to be an important candidate domestication gene conferring large fruit size in the wax gourd [7]. In the present study, we demonstrated that *BhSAUR60* was highly expressed in the fruits specifically (Figure 6C,D), and that overexpression led to longer fruits in *Arabidopsis* (Figure 8C,D), suggesting the crucial role of *BhSAUR60* during fruit development.

4. Materials and Methods

4.1. Plant Materials and Treatment

Wax gourd inbred lines B227 (reference genome-based variety) and P109 were used as the experimental materials in this study. The original seeds were provided by the Vegetable Research Institute of the Guangdong Academy of Agricultural Sciences. The B227 variety was grown in the research-experimental field of the Vegetable Research Institute, Guangdong Academy of Agricultural Sciences under the natural sunlight conditions with standard agricultural practices. Roots, stems, leaves, male flowers (indicated as flowers), and fruits of the B227 variety were collected with three biological replicates at 0 DAP, 10 DAP, and 20 DAP. For auxin treatment, the P109 variety was grown in a greenhouse of the Vegetable Research Institute, Guangdong Academy of Agricultural Sciences under a cycle of 16-h light and 8-h dark at 25 °C. The fully expanded leaves of 3-week-old seedlings were sprayed with 100 μM IAA, and leaf samples were collected at 0, 10, 30, 60, and 120 min after treatment with three biological replicates. All samples were immediately frozen in liquid nitrogen and stored at −80 °C for subsequent use.

4.2. Identification of the Wax Gourd SAUR Gene Family

The SAUR proteins of wax gourds were identified from the B227 genome database using the HMMER software (version 3.3.2, Howard Hughes Medical Institute, Chevy Chase, MD, USA) based on the SAUR Hidden Markov Model (HMM) profile from the Pfam database (PF02519, the characteristic domain of auxin-inducible *SAUR* gene family) [46,47]. The identified BhSAUR proteins with an E-value of less than 10^{-5} were reserved for further analysis. Multiple sequence alignment of BhSAUR proteins was performed on WebLogo [48], and the sequence logo of the SAUR domain was then generated. The chromosomal location of the *BhSAUR* genes was visualized using the TBtools software (version 1.0987663, South China Agricultural University, Guangzhou, China) based on the gene annotation file of the wax gourd [49]. The molecular weight, theoretical isoelectric point, and other physical and chemical properties of BhSAUR proteins were analyzed with the ExPASY Server [50]. The subcellular localization of BhSAUR proteins was predicted by the Plant-mPLoc [51]. The basic information of *BhSAUR* genes is shown in Table S1.

4.3. Gene Duplication and Synteny Analysis

DupGen_finder was used to identify the different modes of gene duplication in wax gourds [42]. Gene duplication modes of *BhSAUR* genes are listed in Table S2. Gene synteny analysis was performed using the *MCScanX* toolkit with default parameters [52]. Based on the syntenic blocks, intra-species paralogous genes and inter-species orthologous genes were identified. Paralogous gene pairs and orthologous gene pairs are listed in Table S3 and Table S4, respectively. Then, the orthologous gene pairs among four cucurbit crops were merged. Visualization of collinear gene pairs was performed using the Advanced Circos or Multiple Synteny Plot function of TBtools software (version 1.0987663) [49]. The non-synonymous substitutions (Ka), synonymous substitutions (Ks), and Ka/Ks values of *BhSAUR* paralogous gene pairs were calculated by the KaKs Calculator and are listed in Table S3 [53].

4.4. Phylogenetic Analysis

Sequences of identified SAUR proteins in the wax gourd, watermelon, melon, and cucumber were obtained from the Cucurbit Genomics Database (http://cucurbitgenomics.org/, accessed on 14 December 2021). Sequences of *Arabidopsis* SAUR proteins were obtained from the Arabidopsis Information Resource (https://www.arabidopsis.org, accessed on 14 December 2021). Alignment of all SAUR protein sequences was carried out using the

ClustalW tool in the MEGA X software (version 10.2.6, The Pennsylvania State University, University Park, PA, USA) [54]. Based on the results of multiple sequence alignment, the phylogenetic tree was constructed by the MEGA X software (version 10.2.6) using the neighbor-joining method with 1000 bootstrap value and 50% partial deletion for gap treatment [54]. Then, the resulting phylogeny was visualized using the Evolview tool [55]. The accession numbers of genes are listed in Table S5.

4.5. Gene Structure Analysis and Conserved Motif Prediction

Comparison of the gene structure was carried out using the Gene Structure Display Server [56]. To predict the conserved motifs of BhSAUR proteins, the complete amino acid sequences of BhSAURs were analyzed using the MEME Suite (https://meme-suite.org/meme/meme_5.4.1/tools/meme, accessed on 22 June 2022) [57]. The number of motifs was set to 4, and the other parameters were default. Visualization of gene structure and conserved motifs was performed using the TBtools software (version 1.0987663) [49].

4.6. Cis-Acting Element Analysis

The promoter sequences (2000 bp genomic sequence upstream of the start codon) of the *BhSAUR* genes were extracted from the wax gourd genome database. *Cis*-acting elements of the promoter regions were analyzed using the PlantCARE online software (https://bioinformatics.psb.ugent.be/webtools/plantcare/html/, accessed on 20 May 2022) [58]. Visualization of *cis*-acting elements was performed using the TBtools software (version 1.0987663) [49].

4.7. Expression Profile Analysis

The transcriptome data of five wax gourd tissues (root, stem, leaf, flower, and fruit) and fruits at three developmental stages (0 DAP, 10 DAP, and 20 DAP) were obtained from previously published studies [7,39]. The average fragments per kilobase of transcript per million mapped reads (FPKM) values of each gene in different samples were used for further expression analysis. Heatmaps of the expression patterns of *BhSAUR* genes were created using the TBtools software (version 1.0987663) [49].

4.8. RNA Extraction and qRT-PCR

Total RNA of different tissues was extracted using the *TransZol* Up Plus RNA Kit (TransGen Biotech, Beijing, China). Then, a FastKing gDNA Dispelling RT SuperMix Kit (Tiangen Biotech, Beijing, China) was used to synthesize the cDNAs. The qRT-PCR was performed on the CFX96™ Real-Time PCR Detection System (Bio-Rad, Hercules, CA, USA) using the *PerfectStart*® Green qPCR SuperMix (TransGen Biotech, Beijing, China). All reactions were performed with three biological replicates and three technical replicates. The wax gourd ubiquitin gene (*Bhi10G000739*) and *Arabidopsis* actin gene (*AT5G09810*) were used as internal references. The relative expression levels were calculated by the $2^{-\Delta\Delta CT}$ method [59]. All primers specific to the templates were designed using the NCBI Primer-BLAST tool. Primers used for qRT-PCR are listed in Table S6.

4.9. Subcellular Localization Analysis

The full-length coding sequence (CDS), without the stop codon of *BhSAUR60*, was cloned and inserted into the pSuper-1300 GFP vector to fuse with the GFP tag. The recombinant construct and empty vector were transformed into *A. tumefaciens* strain GV3101. The *N. benthamiana* plants were grown in a greenhouse under the 16-h light/8-h dark cycle at 25 °C for five weeks. Then, bacterial suspensions were infiltrated into fully expanded young leaves of *N. benthamiana* plants. The RFP-labeled nuclear marker (NLS-RFP) was used to mark the nucleus position. After 2 d of infiltration, fluorescence signals of the leaf epidermal cells were visualized using a confocal laser scanning microscope (Zeiss LSM710, Jena, Germany). Primers used for vector construction are listed in Table S6.

4.10. Transformation of Arabidopsis and Trait Measurement

The CDS of *BhSAUR60* was inserted into the pBI121 vector under the control of the 35S promoter. The recombinant construct was transformed into *A. tumefaciens* strain GV3101. *Arabidopsis* Columbia ecotype (Col-0) was used as the WT. The *Arabidopsis* plants were grown in a growth chamber under the 16-h light/8-h dark cycle at 22 °C. *Agrobacterium*-mediated transformation of *Arabidopsis* was performed using the floral dip method [60]. Transgenic plants were selected by kanamycin resistance and further confirmed by PCR. T_2-generation plants were used for functional identification. The silique length (from stigma to gynophore) of three individual plants was measured for each line, with ten well-developed siliques from the middle of the main inflorescence per individual plant. Finally, the average length of 30 siliques was measured to represent each transgenic line. Primers used for vector construction and genotyping are listed in Table S6.

5. Conclusions

In the present study, we identified 68 *SAUR* genes in the wax gourd, which were distributed on nine chromosomes with 41 genes in two clusters. More than half of the *BhSAUR* genes were derived from tandem-duplication events. The BhSAUR proteins were classified into seven subfamilies, and members grouped into the same subfamily shared similar conserved motifs and arrangements. The *BhSAUR* gene promoters contained *cis*-acting elements involved in plant hormone responses, including the ABA, auxin, GA, MeJA, and salicylic acid responses. The IAA-treatment experiment and qRT-PCR analysis showed that *BhSAUR1, BhSAUR3, BhSAUR29, BhSAUR33, BhSAUR41*, and *BhSAUR43* could respond to auxin. In addition, *cis*-acting elements in response to environmental signals, such as defense- and stress-responsive elements, drought-inducibility elements, and low-temperature-responsive elements, were also found in the *BhSAUR* gene promoters. Further expression profiles showed that *BhSAUR* genes displayed different expression patterns. *BhSAUR3, BhSAUR60*, and *BhSAUR66* were predominantly expressed in the flower, fruit, and root, respectively. Furthermore, we demonstrated that the BhSAUR60 protein was localized to the nucleus and cytoplasm, and that its overexpression led to increased fruit length in *Arabidopsis*. These results enhanced our understanding of the *SAUR* gene family in wax gourds and revealed the potential role of *BhSAUR60* in regulating fruit elongation. In future studies, the detailed functions of *BhSAUR60* will be elucidated through genome editing and overexpression technologies, thereby revealing the regulatory mechanisms of *BhSAUR60* in determining fruit length and facilitating the molecular breeding of wax gourds.

Supplementary Materials: The following supporting information can be downloaded at: https://www.mdpi.com/article/10.3390/ijms232214021/s1.

Author Contributions: Conceptualization, B.J. and C.L.; methodology, C.L.; formal analysis, C.L., J.Y. and C.H.; resources, B.J., W.L. and D.X.; writing—original draft preparation, C.L.; writing—review and editing, B.J. and C.L.; supervision, B.J. All authors have read and agreed to the published version of the manuscript.

Funding: This research was funded by the Guangdong Basic and Applied Basic Research Foundation (2021A1515111214), the National Natural Science Foundation of China (32202504), the Training Plan for Young and Middle-Aged Discipline Leaders of Guangdong Academy of Agricultural Sciences (R2020PY-JG003), the Special Fund for Scientific and Technological Talents Introduction of Guangdong Academy of Agricultural Sciences (R2021YJ-YB2004), and the Discipline Team Construction Project of Guangdong Academy of Agricultural Sciences (202114TD).

Institutional Review Board Statement: Not applicable.

Informed Consent Statement: Not applicable.

Data Availability Statement: The sequence data of the wax gourd, watermelon, melon, and cucumber can be found in the Cucurbit Genomics Database (http://cucurbitgenomics.org/, accessed on 14 December 2021) under accession numbers in Tables S1 and S5. The sequence data of *Arabidopsis* can be found in the Arabidopsis Information Resource (https://www.arabidopsis.org, accessed on 14 December 2021) under accession numbers in Table S5. Transcriptome data of the wax gourd were downloaded from the National Center for Biotechnology Information (NCBI) SRA database under accession number SRP224600 and SRA074508.

Conflicts of Interest: The authors declare no conflict of interest.

References

1. Guo, J.; Xu, W.; Hu, Y.; Huang, J.; Zhao, Y.; Zhang, L.; Huang, C.-H.; Ma, H. Phylotranscriptomics in Cucurbitaceae reveal multiple whole-genome duplications and key morphological and molecular innovations. *Mol. Plant* **2020**, *13*, 1117–1133. [CrossRef] [PubMed]
2. Ma, L.; Wang, Q.; Zheng, Y.; Guo, J.; Yuan, S.; Fu, A.; Bai, C.; Zhao, X.; Zheng, S.; Wen, C.; et al. Cucurbitaceae genome evolution, gene function and molecular breeding. *Hortic. Res.* **2022**, *9*, uhab057. [CrossRef] [PubMed]
3. Gu, M.; Fan, S.; Liu, G.; Guo, L.; Ding, X.; Lu, Y.; Zhang, Y.; Ji, G.; Huang, C. Extract of wax gourd peel prevents high-fat diet-induced hyperlipidemia in C57BL/6 mice via the inhibition of the PPARγ pathway. *Evid. Based Compl. Alt.* **2013**, *2013*, 342561. [CrossRef] [PubMed]
4. Cheng, Z.; Liu, Z.; Xu, Y.; Ma, L.; Chen, J.; Gou, J.; Su, L.; Wu, W.; Chen, Y.; Yu, W.; et al. Fine mapping and identification of the candidate gene *BFS* for fruit shape in wax gourd (*Benincasa hispida*). *Theor. Appl. Genet.* **2021**, *134*, 3983–3995. [CrossRef]
5. Liu, W.; Jiang, B.; Peng, Q.; He, X.; Lin, Y.; Wang, M.; Liang, Z.; Xie, D.; Hu, K. Genetic analysis and QTL mapping of fruit-related traits in wax gourd (*Benincasa hispida*). *Euphytica* **2018**, *214*, 136. [CrossRef]
6. Ma, L.; Liu, Z.; Cheng, Z.; Gou, J.; Chen, J.; Yu, W.; Wang, P. Identification and application of BhAPRR2 controlling peel colour in wax gourd (*Benincasa hispida*). *Front. Plant Sci.* **2021**, *12*, 716772. [CrossRef] [PubMed]
7. Xie, D.; Xu, Y.; Wang, J.; Liu, W.; Zhou, Q.; Luo, S.; Huang, W.; He, X.; Li, Q.; Peng, Q.; et al. The wax gourd genomes offer insights into the genetic diversity and ancestral cucurbit karyotype. *Nat. Commun.* **2019**, *10*, 5158. [CrossRef] [PubMed]
8. Jiang, B.; Liu, W.; Xie, D.; Peng, Q.; He, X.; Lin, Y.; Liang, Z. High-density genetic map construction and gene mapping of pericarp color in wax gourd using specific-locus amplified fragment (SLAF) sequencing. *BMC Genom.* **2015**, *16*, 1035. [CrossRef]
9. Yan, J.; Sun, P.; Liu, W.; Xie, D.; Wang, M.; Peng, Q.; Sun, Q.; Jiang, B. Metabolomic and transcriptomic analyses reveal association of mature fruit pericarp color variation with chlorophyll and flavonoid biosynthesis in wax gourd (*Benincasa hispida*). *Agronomy* **2022**, *12*, 2045. [CrossRef]
10. Du, M.; Spalding, E.P.; Gray, W.M. Rapid auxin-mediated cell expansion. *Annu. Rev. Plant Biol.* **2020**, *71*, 379–402. [CrossRef]
11. Zhao, Y. Essential roles of local auxin biosynthesis in plant development and in adaptation to environmental changes. *Annu. Rev. Plant Biol.* **2018**, *69*, 417–435. [CrossRef] [PubMed]
12. Chapman, E.J.; Estelle, M. Mechanism of auxin-regulated gene expression in plants. *Annu. Rev. Genet.* **2009**, *43*, 265–285. [CrossRef] [PubMed]
13. Weijers, D.; Wagner, D. Transcriptional responses to the auxin hormone. *Annu. Rev. Plant Biol.* **2016**, *67*, 539–574. [CrossRef] [PubMed]
14. Hagen, G.; Guilfoyle, T. Auxin-responsive gene expression: Genes, promoters and regulatory factors. *Plant Mol. Biol.* **2002**, *49*, 373–385. [CrossRef]
15. Ren, H.; Gray, W.M. SAUR proteins as effectors of hormonal and environmental signals in plant growth. *Mol. Plant* **2015**, *8*, 1153–1164. [CrossRef]
16. Stortenbeker, N.; Bemer, M. The *SAUR* gene family: The plant's toolbox for adaptation of growth and development. *J. Exp. Bot.* **2019**, *70*, 17–27. [CrossRef]
17. Wu, J.; Liu, S.; He, Y.; Guan, X.; Zhu, X.; Cheng, L.; Wang, J.; Lu, G. Genome-wide analysis of *SAUR* gene family in Solanaceae species. *Gene* **2012**, *509*, 38–50. [CrossRef]
18. McClure, B.A.; Guilfoyle, T. Characterization of a class of small auxin-inducible soybean polyadenylated RNAs. *Plant Mol. Biol.* **1987**, *9*, 611–623. [CrossRef]
19. Chen, Y.; Hao, X.; Cao, J. *Small auxin upregulated RNA* (*SAUR*) gene family in maize: Identification, evolution, and its phylogenetic comparison with *Arabidopsis*, rice, and sorghum. *J. Integr. Plant Biol.* **2014**, *56*, 133–150. [CrossRef]
20. Gil, P.; Liu, Y.; Orbović, V.; Verkamp, E.; Poff, K.L.; Green, P.J. Characterization of the auxin-inducible *SAUR-AC1* gene for use as a molecular genetic tool in *Arabidopsis*. *Plant Physiol.* **1994**, *104*, 777–784. [CrossRef]
21. Chae, K.; Isaacs, C.G.; Reeves, P.H.; Maloney, G.S.; Muday, G.K.; Nagpal, P.; Reed, J.W. *Arabidopsis SMALL AUXIN UP RNA63* promotes hypocotyl and stamen filament elongation. *Plant J.* **2012**, *71*, 684–697. [CrossRef] [PubMed]
22. Spartz, A.K.; Lee, S.H.; Wenger, J.P.; Gonzalez, N.; Itoh, H.; Inzé, D.; Peer, W.A.; Murphy, A.S.; Overvoorde, P.J.; Gray, W.M. The *SAUR19* subfamily of *SMALL AUXIN UP RNA* genes promote cell expansion. *Plant J.* **2012**, *70*, 978–990. [CrossRef] [PubMed]

23. Kong, Y.; Zhu, Y.; Gao, C.; She, W.; Lin, W.; Chen, Y.; Han, N.; Bian, H.; Zhu, M.; Wang, J. Tissue-specific expression of *SMALL AUXIN UP RNA41* differentially regulates cell expansion and root meristem patterning in *Arabidopsis*. *Plant Cell Physiol.* **2013**, *54*, 609–621. [CrossRef]
24. Roig-Villanova, I.; Bou-Torrent, J.; Galstyan, A.; Carretero-Paulet, L.; Portolés, S.; Rodríguez-Concepción, M.; Martínez-García, J.F. Interaction of shade avoidance and auxin responses: A role for two novel atypical bHLH proteins. *Embo J.* **2007**, *26*, 4756–4767. [CrossRef]
25. Franklin, K.A.; Lee, S.H.; Patel, D.; Kumar, S.V.; Spartz, A.K.; Gu, C.; Ye, S.; Yu, P.; Breen, G.; Cohen, J.D.; et al. PHYTOCHROME-INTERACTING FACTOR 4 (PIF4) regulates auxin biosynthesis at high temperature. *Proc. Natl. Acad. Sci. USA* **2011**, *108*, 20231–20235. [CrossRef] [PubMed]
26. Wang, J.; Sun, N.; Zhang, F.; Yu, R.; Chen, H.; Deng, X.W.; Wei, N. SAUR17 and SAUR50 differentially regulate PP2C-D1 during apical hook development and cotyledon opening in *Arabidopsis*. *Plant Cell* **2020**, *32*, 3792–3811. [CrossRef] [PubMed]
27. Hou, K.; Wu, W.; Gan, S.-S. *SAUR36*, a small auxin up RNA gene, is involved in the promotion of leaf senescence in *Arabidopsis*. *Plant Physiol.* **2013**, *161*, 1002–1009. [CrossRef]
28. Oh, E.; Zhu, J.Y.; Bai, M.Y.; Arenhart, R.A.; Sun, Y.; Wang, Z.Y. Cell elongation is regulated through a central circuit of interacting transcription factors in the *Arabidopsis* hypocotyl. *Elife* **2014**, *3*, e03031. [CrossRef]
29. Stamm, P.; Kumar, P.P. Auxin and gibberellin responsive *Arabidopsis SMALL AUXIN UP RNA36* regulates hypocotyl elongation in the light. *Plant Cell Rep.* **2013**, *32*, 759–769. [CrossRef]
30. Nemhauser, J.L.; Hong, F.; Chory, J. Different plant hormones regulate similar processes through largely nonoverlapping transcriptional responses. *Cell* **2006**, *126*, 467–475. [CrossRef]
31. Kodaira, K.-S.; Qin, F.; Tran, L.-S.P.; Maruyama, K.; Kidokoro, S.; Fujita, Y.; Shinozaki, K.; Yamaguchi-Shinozaki, K. *Arabidopsis* Cys2/His2 zinc-finger proteins AZF1 and AZF2 negatively regulate abscisic acid-repressive and auxin-inducible genes under abiotic stress conditions. *Plant Physiol.* **2011**, *157*, 742–756. [CrossRef]
32. Van Mourik, H.; van Dijk, A.D.J.; Stortenbeker, N.; Angenent, G.C.; Bemer, M. Divergent regulation of *Arabidopsis SAUR* genes: A focus on the *SAUR10*-clade. *BMC Plant Biol.* **2017**, *17*, 245. [CrossRef] [PubMed]
33. Bemer, M.; van Mourik, H.; Muino, J.M.; Ferrandiz, C.; Kaufmann, K.; Angenent, G.C. FRUITFULL controls *SAUR10* expression and regulates *Arabidopsis* growth and architecture. *J. Exp. Bot.* **2017**, *68*, 3391–3403. [CrossRef] [PubMed]
34. Shin, J.-H.; Mila, I.; Liu, M.; Rodrigues, M.A.; Vernoux, T.; Pirrello, J.; Bouzayen, M. The RIN-regulated Small Auxin-Up RNA SAUR69 is involved in the unripe-to-ripe phase transition of tomato fruit via enhancement of the sensitivity to ethylene. *New Phytol.* **2019**, *222*, 820–836. [CrossRef]
35. Li, M.; Chen, R.; Gu, H.; Cheng, D.; Guo, X.; Shi, C.; Li, L.; Xu, G.; Gu, S.; Wu, Z.; et al. Grape small auxin upregulated RNA (*SAUR*) 041 is a candidate regulator of berry size in grape. *Int. J. Mol. Sci.* **2021**, *22*, 11818. [CrossRef] [PubMed]
36. Wang, H.; Shang, Q. Genome-wide identification and expression analysis of the *SAUR* gene family in *Cucumis sativus*. *Acta Hortic. Sin.* **2019**, *46*, 1093–1111.
37. Zhang, N.; Huang, X.; Bao, Y.; Wang, B.; Zeng, H.; Cheng, W.; Tang, M.; Li, Y.; Ren, J.; Sun, Y. Genome-wide identification of *SAUR* genes in watermelon (*Citrullus lanatus*). *Physiol. Mol. Plant Pathol.* **2017**, *23*, 619–628. [CrossRef]
38. Tian, Z.; Han, J.; Che, G.; Hasi, A. Genome-wide characterization and expression analysis of *SAUR* gene family in Melon (*Cucumis melo* L.). *Planta* **2022**, *255*, 123. [CrossRef]
39. Jiang, B.; Xie, D.; Liu, W.; Peng, Q.; He, X. *De novo* assembly and characterization of the transcriptome, and development of SSR markers in wax gourd (*Benicasa hispida*). *PLoS ONE* **2013**, *8*, e71054. [CrossRef]
40. Jose Ripoll, J.; Bailey, L.J.; Mai, Q.A.; Wu, S.L.; Hon, C.T.; Chapman, E.J.; Ditta, G.S.; Estelle, M.; Yanofsky, M.F. microRNA regulation of fruit growth. *Nat. Plants* **2015**, *1*, 15036. [CrossRef]
41. Di Marzo, M.; Herrera-Ubaldo, H.; Caporali, E.; Novak, O.; Strnad, M.; Balanza, V.; Ezquer, I.; Mendes, M.A.; de Folter, S.; Colombo, L. SEEDSTICK controls *Arabidopsis* fruit size by regulating cytokinin levels and *FRUITFULL*. *Cell Rep.* **2020**, *30*, 2846–2857.e3. [CrossRef] [PubMed]
42. Qiao, X.; Li, Q.; Yin, H.; Qi, K.; Li, L.; Wang, R.; Zhang, S.; Paterson, A.H. Gene duplication and evolution in recurring polyploidization-diploidization cycles in plants. *Genome Biol.* **2019**, *20*, 38. [CrossRef] [PubMed]
43. Hanada, K.; Zou, C.; Lehti-Shiu, M.D.; Shinozaki, K.; Shiu, S.-H. Importance of lineage-specific expansion of plant tandem duplicates in the adaptive response to environmental stimuli. *Plant Physiol.* **2008**, *148*, 993–1003. [CrossRef]
44. Xu, Y.; Zhang, H.; Zhong, Y.; Jiang, N.; Zhong, X.; Zhang, Q.; Chai, S.; Li, H.; Zhang, Z. Comparative genomics analysis of *bHLH* genes in cucurbits identifies a novel gene regulating cucurbitacin biosynthesis. *Hortic. Res.* **2022**, *9*, uhac038. [CrossRef] [PubMed]
45. Pan, Y.; Wang, Y.; McGregor, C.; Liu, S.; Luan, F.; Gao, M.; Weng, Y. Genetic architecture of fruit size and shape variation in cucurbits: A comparative perspective. *Theor. Appl. Genet.* **2020**, *133*, 1–21. [CrossRef] [PubMed]
46. Mistry, J.; Chuguransky, S.; Williams, L.; Qureshi, M.; Salazar, G.A.; Sonnhammer, E.L.L.; Tosatto, S.C.E.; Paladin, L.; Raj, S.; Richardson, L.J.; et al. Pfam: The protein families database in 2021. *Nucleic Acids Res.* **2021**, *49*, D412–D419. [CrossRef]
47. Mistry, J.; Finn, R.D.; Eddy, S.R.; Bateman, A.; Punta, M. Challenges in homology search: HMMER3 and convergent evolution of coiled-coil regions. *Nucleic Acids Res.* **2013**, *41*, e121. [CrossRef]
48. Crooks, G.E.; Hon, G.; Chandonia, J.-M.; Brenner, S.E. WebLogo: A sequence logo generator. *Genome Res.* **2004**, *14*, 1188–1190. [CrossRef]

49. Chen, C.; Chen, H.; Zhang, Y.; Thomas, H.R.; Frank, M.H.; He, Y.; Xia, R. TBtools: An integrative toolkit developed for interactive analyses of big biological data. *Mol. Plant* **2020**, *13*, 1194–1202. [CrossRef]
50. Duvaud, S.; Gabella, C.; Lisacek, F.; Stockinger, H.; Ioannidis, V.; Durinx, C. Expasy, the Swiss Bioinformatics Resource Portal, as designed by its users. *Nucleic Acids Res.* **2021**, *49*, W216–W227. [CrossRef]
51. Chou, K.-C.; Shen, H.-B. Plant-mPLoc: A top-down strategy to augment the power for predicting plant protein subcellular localization. *PLoS ONE* **2010**, *5*, e11335. [CrossRef] [PubMed]
52. Wang, Y.; Tang, H.; Debarry, J.D.; Tan, X.; Li, J.; Wang, X.; Lee, T.-h.; Jin, H.; Marler, B.; Guo, H.; et al. *MCScanX*: A toolkit for detection and evolutionary analysis of gene synteny and collinearity. *Nucleic Acids Res.* **2012**, *40*, e49. [CrossRef] [PubMed]
53. Wang, D.; Zhang, Y.; Zhang, Z.; Zhu, J.; Yu, J. KaKs_Calculator 2.0: A toolkit incorporating gamma-series methods and sliding window strategies. *Genom. Proteom. Bioinf.* **2010**, *8*, 77–80. [CrossRef]
54. Kumar, S.; Stecher, G.; Li, M.; Knyaz, C.; Tamura, K. MEGA X: Molecular evolutionary genetics analysis across computing platforms. *Mol. Biol. Evol.* **2018**, *35*, 1547–1549. [CrossRef] [PubMed]
55. Subramanian, B.; Gao, S.; Lercher, M.J.; Hu, S.; Chen, W.-H. Evolview v3: A webserver for visualization, annotation, and management of phylogenetic trees. *Nucleic Acids Res.* **2019**, *47*, W270–W275. [CrossRef] [PubMed]
56. Hu, B.; Jin, J.; Guo, A.-Y.; Zhang, H.; Luo, J.; Gao, G. GSDS 2.0: An upgraded gene feature visualization server. *Bioinformatics* **2015**, *31*, 1296–1297. [CrossRef]
57. Bailey, T.L.; Johnson, J.; Grant, C.E.; Noble, W.S. The MEME Suite. *Nucleic Acids Res.* **2015**, *43*, W39–W49. [CrossRef]
58. Lescot, M.; Déhais, P.; Thijs, G.; Marchal, K.; Moreau, Y.; Van de Peer, Y.; Rouzé, P.; Rombauts, S. PlantCARE, a database of plant cis-acting regulatory elements and a portal to tools for *in silico* analysis of promoter sequences. *Nucleic Acids Res.* **2002**, *30*, 325–327. [CrossRef]
59. Livak, K.J.; Schmittgen, T.D. Analysis of relative gene expression data using real-time quantitative PCR and the $2^{-\Delta\Delta CT}$ Method. *Methods* **2001**, *25*, 402–408. [CrossRef]
60. Clough, S.J.; Bent, A.F. Floral dip: A simplified method for *Agrobacterium*-mediated transformation of *Arabidopsis thaliana*. *Plant J.* **1998**, *16*, 735–743. [CrossRef]

International Journal of
Molecular Sciences

MDPI

Article

Solanum tuberosum Microtuber Development under Darkness Unveiled through RNAseq Transcriptomic Analysis

Eliana Valencia-Lozano [1,†], Lisset Herrera-Isidrón [2,†], Jorge Abraham Flores-López [2], Osiel Salvador Recoder-Meléndez [2], Aarón Barraza [3] and José Luis Cabrera-Ponce [1,*]

1 Departamento de Ingeniería Genética, Centro de Investigación y de Estudios Avanzados del IPN, Unidad Irapuato, Irapuato 36824, Guanajuato, Mexico
2 Unidad Profesional Interdisciplinaria de Ingeniería Campus Guanajuato (UPIIG), Instituto Politécnico Nacional, Av. Mineral de Valenciana 200, Puerto Interior, Silao de la Victoria 36275, Guanajuato, Mexico
3 CONACYT-Centro de Investigaciones Biológicas del Noreste, SC. IPN 195, Playa Palo de Santa Rita Sur, La Paz 23096, Baja California Sur, Mexico
* Correspondence: jlcabre@yahoo.com.mx; Tel.: +52-462-6239600 (ext. 9421)
† These authors contributed equally to this work.

Abstract: Potato microtuber (MT) development through in vitro techniques are ideal propagules for producing high quality potato plants. MT formation is influenced by several factors, i.e., photoperiod, sucrose, hormones, and osmotic stress. We have previously developed a protocol of MT induction in medium with sucrose (8% *w*/*v*), gelrite (6g/L), and 2iP as cytokinin under darkness. To understand the molecular mechanisms involved, we performed a transcriptome-wide analysis. Here we show that 1715 up- and 1624 down-regulated genes were involved in this biological process. Through the protein–protein interaction (PPI) network analyses performed in the STRING database (v11.5), we found 299 genes tightly associated in 14 clusters. Two major clusters of up-regulated proteins fundamental for life growth and development were found: 29 ribosomal proteins (RPs) interacting with 6 PEBP family members and 117 cell cycle (CC) proteins. The PPI network of up-regulated transcription factors (TFs) revealed that at least six TFs–MYB43, TSF, bZIP27, bZIP43, HAT4 and WOX9–may be involved during MTs development. The PPI network of down-regulated genes revealed a cluster of 83 proteins involved in light and photosynthesis, 110 in response to hormone, 74 in hormone mediate signaling pathway and 22 related to aging.

Keywords: transcriptome-wide analysis; microtubers; potato; *Solanum tuberosum*; darkness; cell cycle; ribosomal proteins; PEBP family genes; cytokinin

1. Introduction

Potato (*Solanum tuberosum* L.) is the most important staple food worldwide just after maize, rice, and wheat with an approximate production of around 360 million tons in 2020 on a surface area of 16.5 million hectares [1]. Potato is a geophyte plant with a dual reproduction system, forming vegetative storage organs called tubers and a sexual propagation that is tightly regulated to ensure fitness in adverse climatic conditions [2].

The potato tubers are swollen underground storage organs differentiated from the subapical meristem of the stolon (modified stem) mediated by radial divisions, expanding to further differentiate into starch accumulating parenchyma. Tubers function in vegetative propagation (eyes; axillary meristems) and as storage enabling survival during winter months and/or prolonged periods of abiotic stress (perennialism) and are a means to bypass sexual reproduction.

The molecular mechanisms of the potato tuberization process have been well documented and depend on light (day length and light quality), temperature, and nutrient supply [3–5].

Plant biotechnology techniques have been applied to produce potato tubers in vitro called microtubers (MTs). The MTs process occurs when axillary buds are cultured in medium containing high content of sucrose, hormones, and different light quality/darkness. MTs have advantages in terms of storage, transport, and cultivation, due to their reduced size and weight, requiring less space compared to seedlings, with a higher multiplication rate producing seed potato faster and cheaper than other methods [6–9].

We developed a protocol for MTs induction in potato *S. tuberosum* var Alpha culturing stolon explants in MS medium supplemented with 8% SUC, high content of gelrite 6.0 g/L, and cytokinin (CK) 2iP 10 mg/L under darkness. The rationale of this protocol was that two-component and cytokinin (CK) signaling interacts with homeobox TFs; ribosomal proteins (RPs), cell cycle (CC), carbon metabolism, auxin responsive factors, and stem cell maintenance genes were involved in this process [10].

Several transcriptomic analyses during potato tuberization have revealed conserved molecular machinery during this process: PEBP members (florigen and tuberigen genes), photoperiod, transcriptional regulation, protein synthesis, cell cycle regulation, metabolic pathways, secondary metabolism, starch and sucrose metabolism, hormone signaling, among others [11–16].

Moreover, a few papers have published reliable gene ID in order to compare results. Sharma et al. [15] published the targets of StBEL5 in tuber development by a transcriptome analysis. A set of more than 2000 up-regulated genes can be read from the Supplementary Data Table S2. A PPI network in STRING data base v11.5 with the highest confidence (0.900) revealed that the up-regulated genes are involved in fundamental processes for life: RPs, CC, spliceosome, ribonucleoprotein complex, and basal TFs interacting with several clusters of genes involved in light, stress and metabolic pathways.

Light modulates the molecular mechanisms during potato tuberization (photomorphogenesis); little attention has been given to the molecular events in the dark. To elucidate the molecular mechanisms of MTs development under these conditions, we performed a transcriptomic-wide analysis.

Differential expressed genes (DEG) obtained from the analysis were screened in a bioinformatic tool STRING database v11.5 (https://string-db.org/cgi/input?sessionId=bbmDlM5pze5B&input_page_active_form=multiple_identifiers) with the highest confidence (0.860).

The PPI network of up-regulated TFs revealed that at least six TFs–MYB43, TSF, bZIP27, bZIP43, HAT4, and WOX9–may be involved during MTs development. These TFs were involved in vascularization, proliferation process, cell fate commitment, epigenetic regulation, meristem organization, cell cycle, and ribosomes.

The PPI network of down-regulated TFs demonstrated several TCP TFs families known as repressor of shoot morphogenesis interacting with MYB TFs involved in flower development.

The PPI network of up-regulated genes revealed two clusters–one made of RPs, that play an important role in the control of cell growth, division, and development [17].

PEBP family members, StSP6A, StSP3D, StSP5G, CEN1/BFT, ATC, and FD TF interact with RPL11. The second cluster belongs to CC, whose function is to correctly duplicate the DNA present in the genome and to segregate the resulting copies correctly into the two daughter cells [18]. The CC cluster interacts tightly with several metabolic pathways: one and carbon metabolism, CK response, citrate cycle, fatty acid metabolism, thylakoid, sulfur metabolism, disulfide isomerase, oxidative stress, and amino sugar and nucleotide sugar metabolism. Down-regulated genes were associated with photosynthesis, light harvesting complex, response to light, hormone signal transduction, response to hormone, flowering, and plant organ development.

In this manuscript we have made a deep analysis by text mining, using the STRING data base from which we proposed a new model of MTs development under darkness conditions.

2. Results

2.1. MTs Transcriptomic-Wide Analysis

In order to identify the genes involved in the development of MTs under dark conditions, a transcriptomic-wide analysis was performed. We sequenced cDNA libraries constructed from two treatments: MR8-G6-2iP inducing MTs and MR1-G3-2iP, and non-inducing MTs using the Illumina HiSeq 2000 platform. This produced a total of 397,834,274 reads, from all six cDNA libraries from MR8-G6-2iP and MR1-G3-2iP. A summary of mapping statistics obtained for each sample is given in Table 1.

Table 1. Mapping statistics for quality filtered reads generated for lines with MTs and non-MTs induction.

	MR8G62IP (MTs)	MR1G32IP (non-MTs)
Read -1	58,784,802.00	71,588,702.00
Read -2	59,487,917.00	73,230,643.00
Read -3	59,185,434.00	75,556,776.00
Yield (Mbases)	177,458,153.00	220,376,121.00

Q40: Error probability 0.0001 (1 in 10.000 base calling).

Overall, 4839 DEG were found in MR8G62iP (MTs inducing treatment) versus control line (non-MTs); 2896 were up-regulated and 1843 down-regulated.

Gene Ontology (GO) enrichment analysis of up-regulated genes identified 2713 genes with annotated GO terms.

GO biological process includes 440 significantly enriched GO terms. Both up- and down-regulated biological process were related to protein phosphorylation, regulation of transcription, transmembrane transport, carbohydrate metabolic process, and proteolysis.

Specific up-regulated biological processes include: microtubule-based movement, translation, fatty acid biosynthetic process, lipid transport, DNA replication, gycolitic process, positive regulation of transcription, cell wall modification, DNA repair, defense response, carboxylic acid metabolic process, protein dephosphorylation, gluthatione metabolic process, and the amino acid metabolic process. Specific down-regulated biological processes include: photosynthesis, light harvesting, cell wall biogenesis, cellular glucan, and the xyloglucan metabolic process (Figure 1).

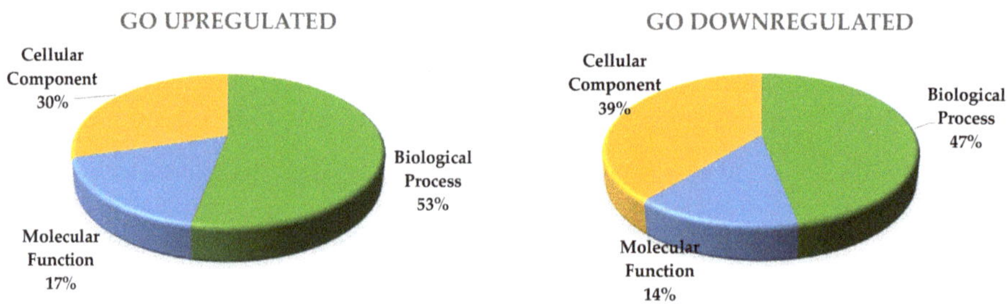

Figure 1. Biological process regulated during MTs development of potato induced in dark.

GO cellular components include 82 significantly enriched GO-terms; 64 were up-regulated and those specific of MTs were: ribosome, nucleosome, microtubule, extracellular region, and cytoplasm. Specific down-regulated cellular components include: cell wall, extrinsic component of membrane, photosystem I and II (Figure 2).

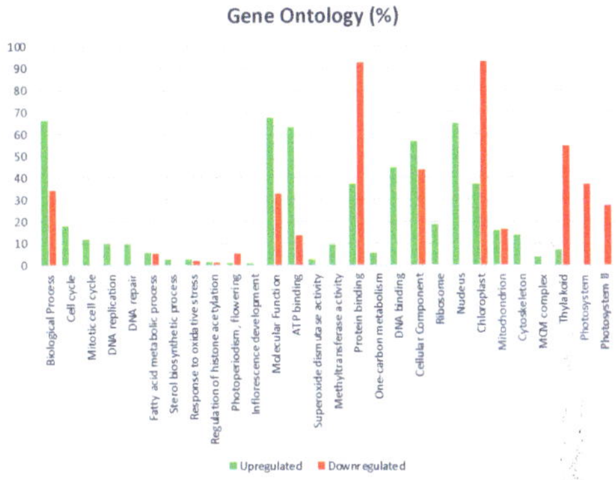

Figure 2. Cellular components regulated during MTs development of potato induced in the dark.

GO molecular function includes 132 significantly enriched GO terms; the seven main functional subgroups are: GO:0003824 catalytic activity (963), GO:0005488 binding (998), GO:00116491 oxidoreductase activity (243), GO:0043167 ion binding (604), GO:0016787 hydrolase activity, GO:0098772 molecular function regulator, and GO:0008092 cytoskeletal protein binding (67).

2.2. *TFs*

Overall, 112 up-regulated and 145 down-regulated TFs were found in the transcriptomic-wide analysis of MTs induced under dark conditions.

2.2.1. Up-Regulated TFs

Up-regulated TFs are related to C2H2-type zinc finger (20), F-box (15), bHLH (13), AP2 ERF (13), MYB (10), GATA (6), ZF-HD protein (3), Basic leucine zipper (5), bZiP (5), B3 DNA binding d2ggomain (5), NF-Y (4), WRKY (3), TCP (2).

The PPI network of up-regulated TFs made on the STRING data base v11.5 with medium confidence revealed that 23 out of 112 TFs interact (Figure 3), while 14 interact with at least 1 TF. The list of up-regulated TFs interacting in the PPI network and levels of gene regulation are shown in the heat-map (Figure 4).

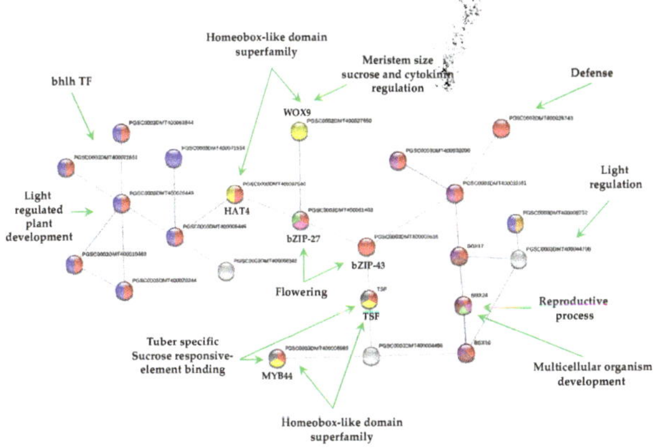

Figure 3. PPI network of up-regulated TFs derived using STRING database from transcriptomic-wide analysis of potato *S. tuberosum* with medium confidence (0.400). The figure represents a full network; the edges indicate both functional and physical protein associations.

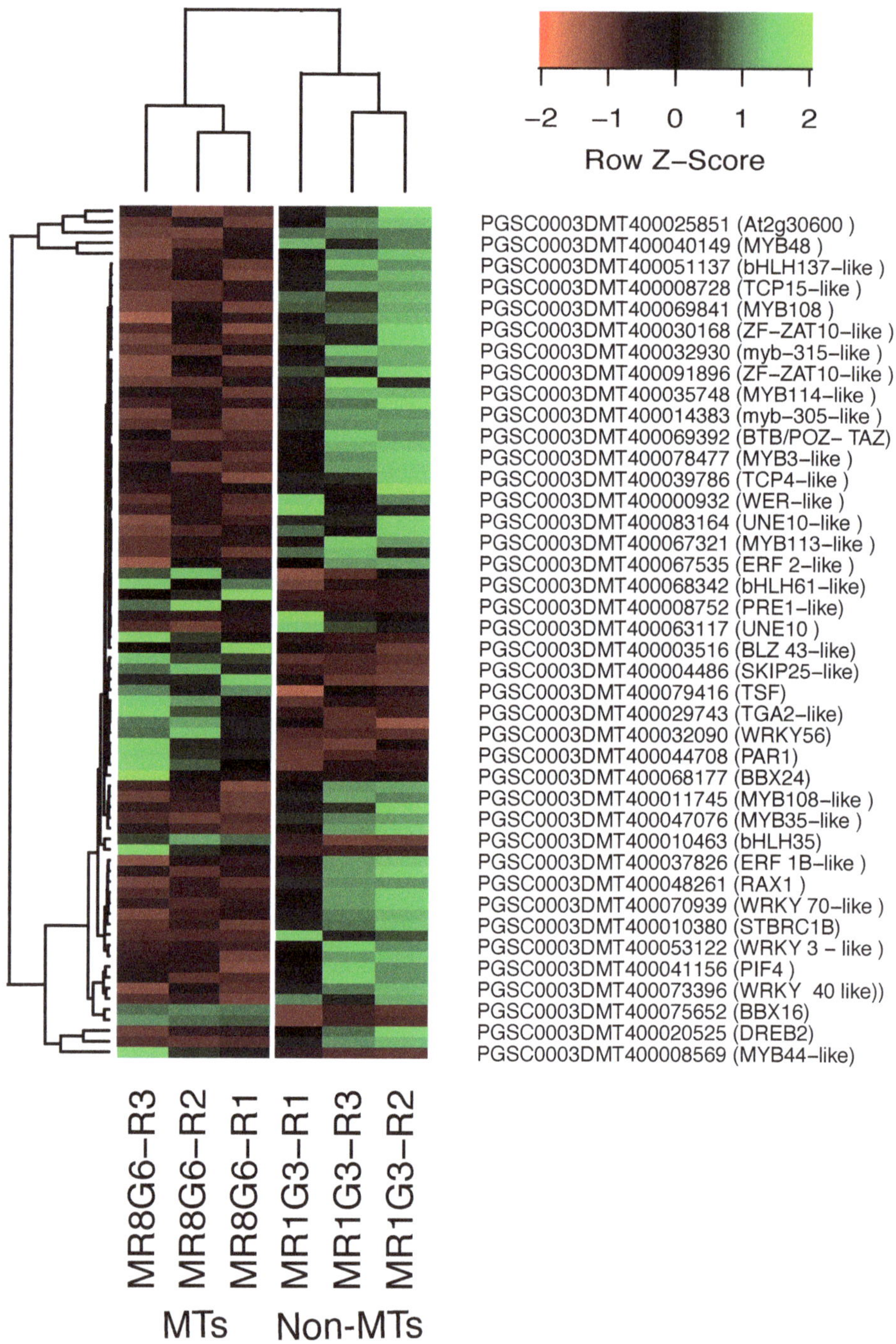

Figure 4. Hierarchical clustering analyses (HCA) and heat map of up- and down-regulated TFs during MTs development under darks conditions, levels of up-regulation are present in Log2.

2.2.2. Down-Regulated TFs

Overall, 145 TFs were found to be down-regulated in MTs development. These TFs are related to bHLH (38), Homeobox (30), AP2/ERF (21), SANT/Myb domain (19), Myb domain (19), Zinc finger C2H2 superfamily (18), F-box (14), B-box-type zinc finger (11), WRKY (9), CCT domain (8), B3 DNA binding domain (7), Transcription factor TCP (6), BTB/POZ domain (6), and SKP1/BTB/POZ (6).

The PPI network of down-regulated TFs revealed that 58 out of 112 TFs interact (Figure 5), while 9 interact with at least 1 TF.

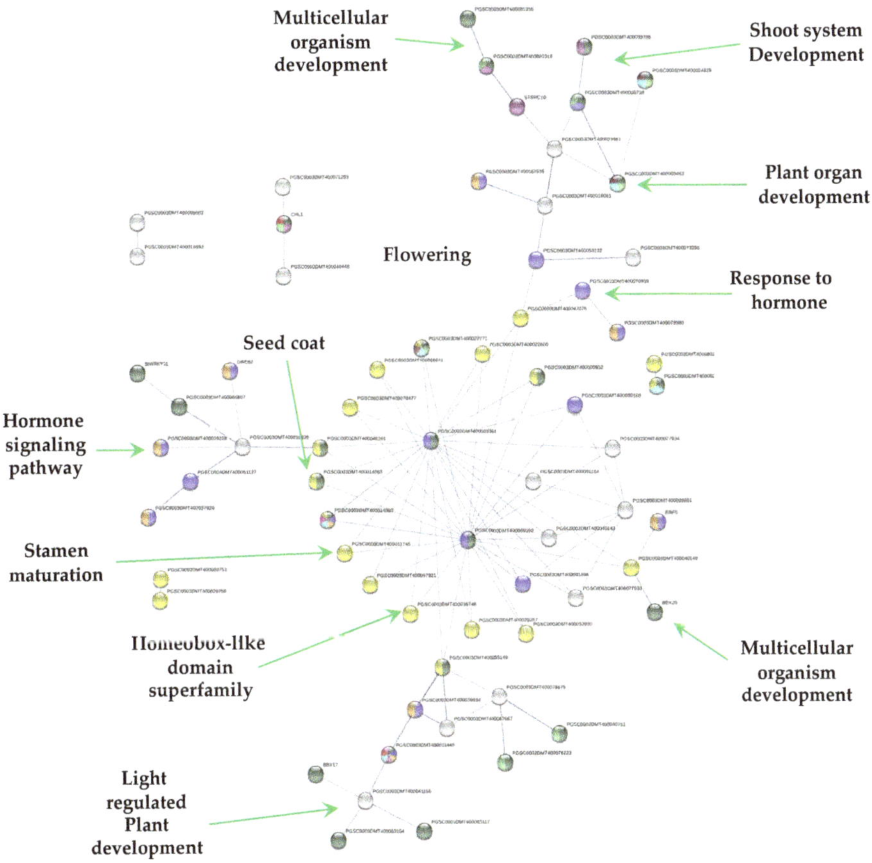

Figure 5. The PPI network of down-regulated TFs derived using the STRING database from the transcriptomic-wide analysis of potato *S. tuberosum* with medium confidence (0.400). The figure represents a full network, the edges indicate both functional and physical protein associations.

The list of down-regulated TFs interacting in the network and levels of gene regulation are shown in the heat-map (Figure 4).

2.2.3. PPI Network of Up- and Down-Regulated TFs

The PPI network of up- and down-regulated TFs revealed both sets of TFs found interact in separated clusters–one forming MTs and the another corresponding to down-regulation TFs involved in shoot-stolon differentiation, gametophytes, plant organ development, and hormone signaling (Supplementary Figure S1).

2.3. Up-Regulated Genes

From 1699 up-regulated genes, 299 were tightly associated in two main clusters (Figure 6). Two fundamental processes for life were detected: RPs (29) and CC (117), interacting directly with proteins that sense the environment: osmotic stress (21), oxidative stress (9), CK response (23), one carbon metabolism (6), carbon metabolism (38), TCA cycle (16), acyl carrier proteins (6), fatty acid metabolism (14), thylakoid (13), redoxins (9), sulfur metabolism (8), disulfide isomerase activity (5), and immunophilins (12). Three clusters comprising 37 genes do not interact: sterol and terpenoid biosynthesis (16) and amino sugar and nucleotide sugar metabolism (7) (Figure 6).

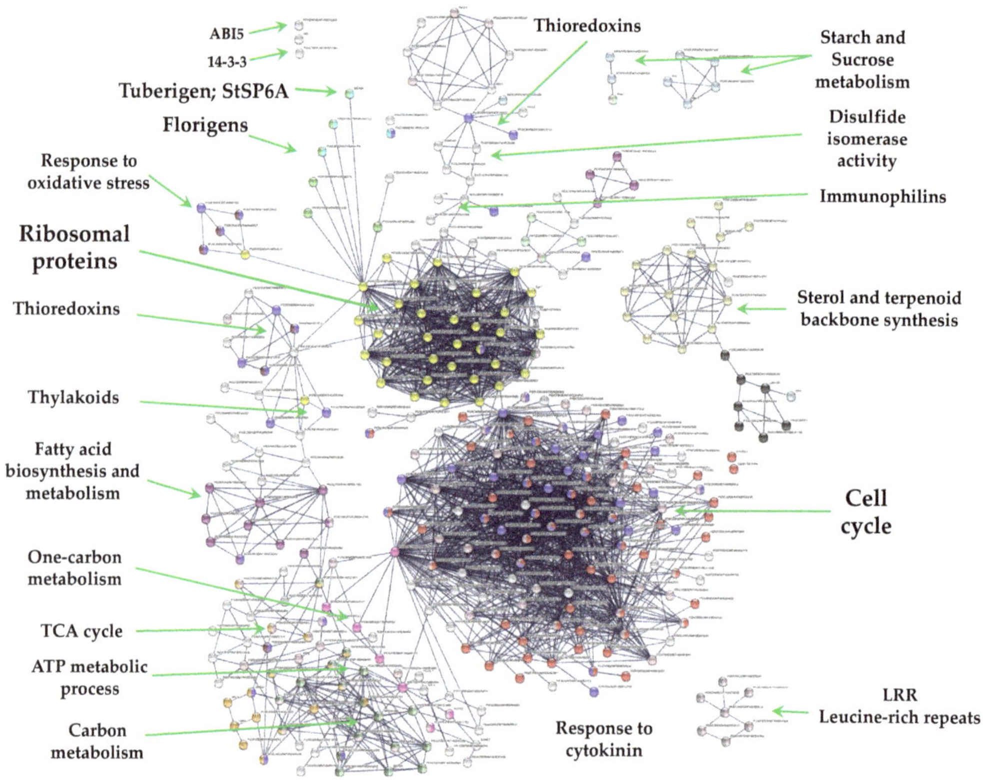

Figure 6. PPI network of up-regulated genes derived from the STRING database of potato *S. tuberosum* from the transcriptomic-wide analysis with high confidence (0.860). Clusters are highlighted with the name of the function. The figure represents a full network, the edges indicate both functional and physical protein associations.

2.4. Down-Regulated Genes

Overall, 271 out of 1616 down-regulated genes were present in the PPI network with high confidence (0.700). Since MTs were induced in darkness, the network consists in one cluster related to photosynthesis (69) tightly interacting with several functions (Figure 7). These functions include; light harvesting complex (26), chloroplast (134), biosynthesis of secondary metabolites (65), carbohydrate derivative binding (27), cellular lipid metabolic process (17), starch and sucrose metabolism and glycosyl hydrolases (14), carbohydrate metabolic process (38), carbon metabolism (28), cellular lipid metabolic process (17), homeobox and calmodulin binding motif (2), response to hormone (11), flowering (1), plant hormone signal transduction (2), and response to blue light (10) (Figure 7).

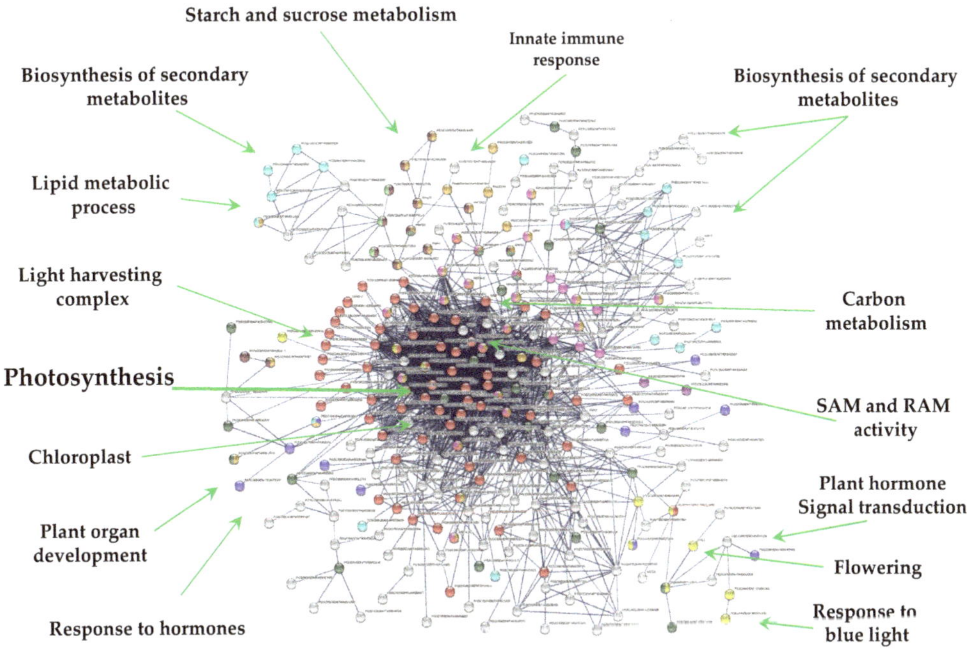

Figure 7. PPI network of down-regulated genes derived from the STRING database of potato *S. tuberosum* from the transcriptomic-wide analysis with high confidence (0.700). Cluster is highlighted with the name of the function. The figure represents a full network, the edges indicate both functional and physical protein associations.

2.5. DEG Related to PEBP Family Members

Overall, six PEBP family members were present in MTs development, three FT orthologs of PEBP family are present in potato, StSP6A, StSP5G (PGSC0003DMT400041726), and StSP3D (PGSC0003DMT400041725), two CEN1-BFT (PGSC0003DMT400030575), CEN1-BFT (PGSC0003DMT400037143), and a self-pruning ATC (PGSC0003DMT400018307). The TF FD (PGSC0003DMT400061403) a bZIP transcription factor-27 self-pruning was found interacting with ATC (Figure 8).

Gene ontology (GO) from functional annotation revealed a relationship between florigen genes present in the PPI network and RPs. bZIP27 FD (PGSC0003DMT400061403) was associated in 35 out of 59 GO-terms in which RPs were present (GO:0009889, GO:0009891, GO:0009893, GO:0010468, GO:0010556, GO:0010557, GO:0010604, GO:0010628, GO:0019222, GO:0031323, GO:0031325, GO:0031326, GO:0031328, GO:0048522, GO:0050789, GO:0050794, GO:0051171, GO:0051173, GO:0060255, GO:0065007, GO:0065009, GO:0080090, GO:2000112, GO:0005488, GO:0097159, GO:0098772, GO:0140110, GO:1901363, GO:0005622, GO:0005634, GO:0043226, GO:0043227, GO.0043229, GO:0043231, GO:0110165).

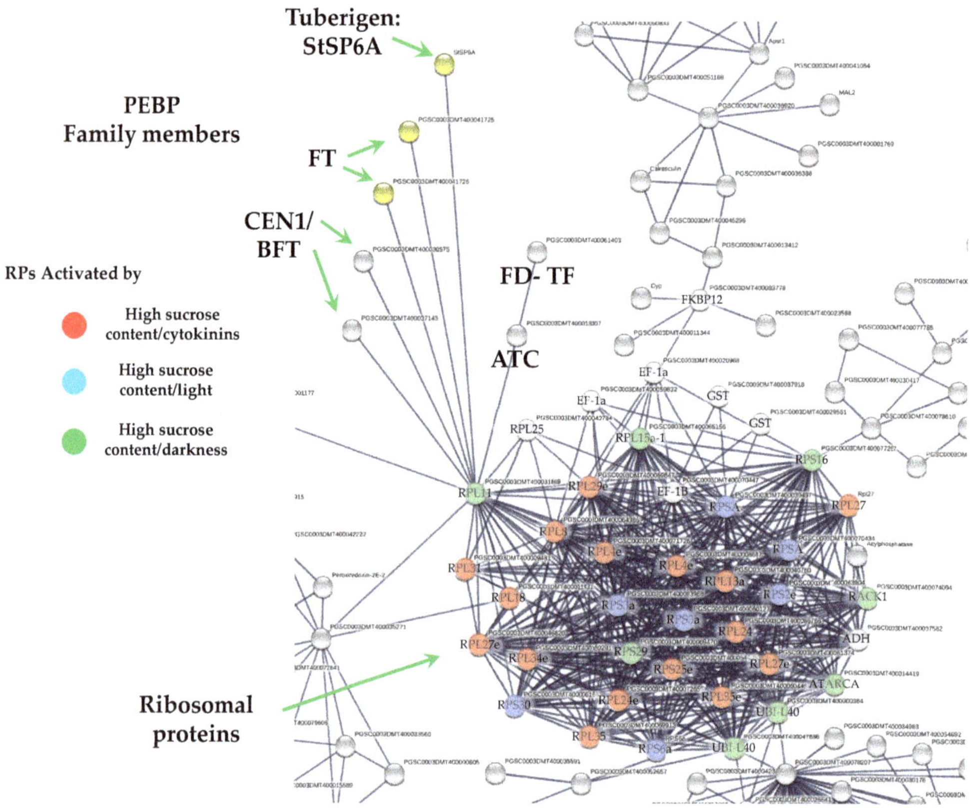

Figure 8. PPI network of up-regulated PEBP family members and RPs cluster of potato S. tuberosum from the transcriptomic-wide analysis with high confidence (0.860). Selected proteins are highlighted with the name of the gene. RPs highlighted in red, blue, and green correspond to activated proteins by sucrose and CKs. The list of RPs and levels of gene regulation are shown in the heat map (Figure 9).

In the case of StSP6A, it was associated in 12 out of 51 GO-terms in which RPs were present (GO:0009987, GO:0050789, GO:0065007, GO:0005488, GO:0005622, GO:0005634, GO:0005737, GO:0043226, GO:0043227, GO:0043229, GO:0043231, GO:0110165).

StSP3D (PGSC0003DMT400041725 and PGSC0003DMT40004172a) were associated in 12 out of 51 GO-terms in which RPs were present (GO:0009987, GO:0050789, GO:0065007, GO:0005488, GO:0005622, GO:0005634, GO:0005737, GO:0043226, GO:0043227, GO:0043229, GO:0043231, GO:0110165).

CEN1/BFT/TFL1C genes (PGSC0003DMT400030575 and PGSC0003DMT400037143) were associated in 3 out of 17 GO-terms in which RPs were present (GO:0005622, GO:0005737, GO:0110165). The list of PEBP family members and levels of gene regulation are shown in the heat map (Figure 9).

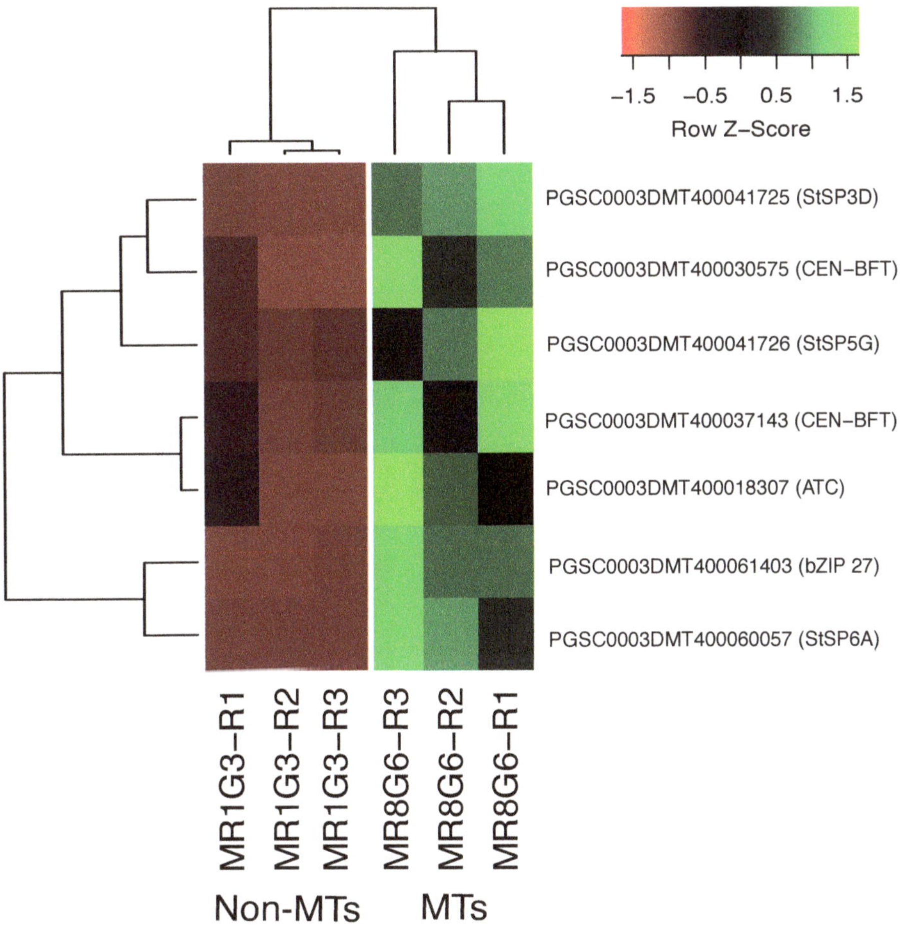

Figure 9. Hierarchical clustering analyses (HCA) and heat map of PEBP family members during MTs development under darks conditions; levels of up-regulation are present in Log2.

2.6. DEG Related to RPs

The PPI network analysis for the up-regulated transcripts corresponding to RPs in the established MTs developmental conditions yielded a cluster of 31 RPs interacting proteins, from which 11 proteins corresponded for the small subunit and 18 for the large subunit, three for eukaryotic elongation factors, two for WD40, RACK1, and ATARCA, and two for glutathione S-transferase genes (Figure 10). The architecture of the RPs cluster yielded in this analysis is quite similar to the evolutionary traits previously determined in PPI network analyses present in the Eukarya domain [19]. Moreover, in the PPI network analysis we were able to disentangle the protein interactions in four main categories. The RPs transcripts that were activated under high sucrose content conditions were highlighted in yellow. Those RPs transcripts that were activated under high sucrose content/darkness conditions, according to Gamm et al. [20], were highlighted in green. The RPs transcripts that were activated under high sucrose content/light conditions were highlighted in blue [20] as well as the RPs transcripts that were activated under high sucrose content/CKs according to Brenner and Schmülling [21]. We also found that among the RPs transcripts was present RPL29e (PGSC0003DMT400069847) that directly interacts with FKBP12 (PGSC0003DMT400083778) and RPL11 (PGSC0003DMT400031869), which in turn interacts with florigens and the thioredoxin-dependent protein (PGSC0003DMT400035271), the ubiquitin fusion protein, and the proliferating cell nuclear antigen (PGSC0003DMT400078207).

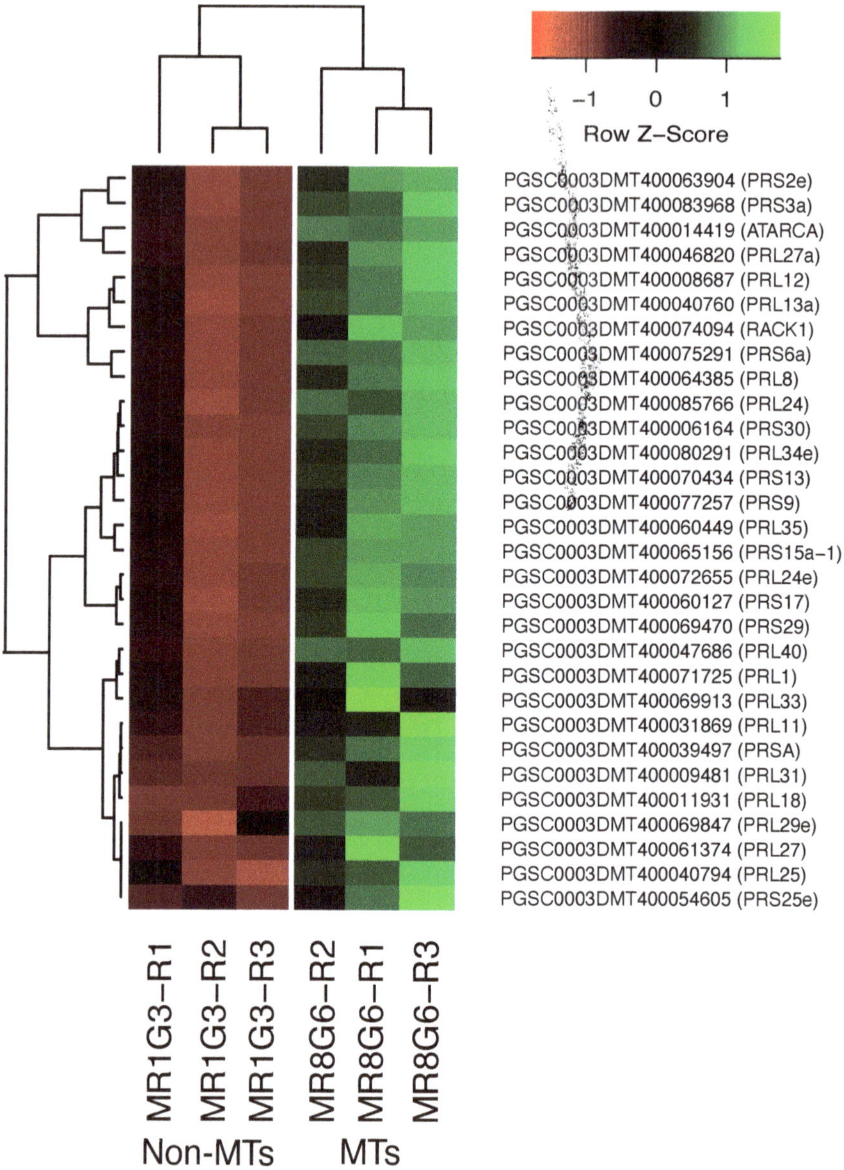

Figure 10. Hierarchical clustering analyses (HCA) and heat map of RPs during MTs development under darkness conditions from levels of up-regulation are present in Log2.

2.7. DEG Related to CC

We found 117 up-regulated transcript genes directly related to the CC of potato MTs under darkness condition, from which 36 transcripts were directly related with mitotic regulation, 31 transcripts corresponded to E2F factors, 25 transcripts were related with cytoskeleton polymerization, 5 transcripts were related with motor proteins, 13 with epigenetic regulation through DNA/Histone methylation, and 5 with histone/nucleosome binding (Figures 11 and S2, Table S1).

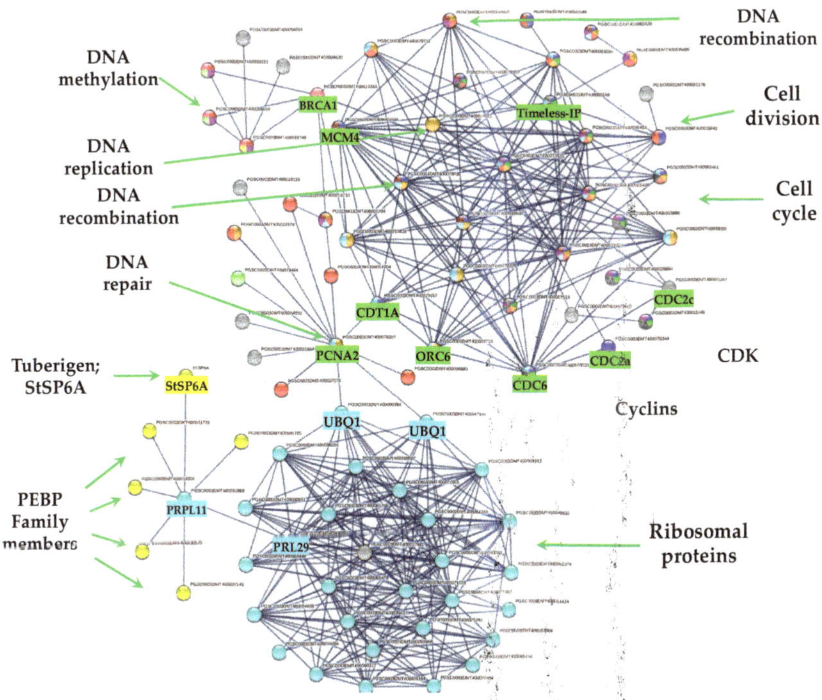

Figure 11. PPI physical network of up-regulated transcripts genes of potato S. tuberosum from the transcriptomic-wide analysis with high confidence (0.860). Selected proteins are highlighted with the name of the gene.

2.7.1. Cyclins/CDKs

Two CDKs: CDC2-CDKA (PGSC0003DMT400075349) and CDC20-CDKA (PGSC0003DMT400071215) were found among the up-regulated transcripts in this work. The plant CDC2 is homologous to CDKA and in mammals to CDK1 and CDK2. CDKA function has been characterized as a regulatory protein at the G1/S and G2/M checkpoints. In addition, among the up-regulated transcripts related to Cyclins, we found 10 cyclins, which correspond to cyclin-A (four transcripts), cyclin-B (three transcripts), and cyclin-D (three transcripts). Furthermore, in the PPI network analysis the CDKA (PGSC0003DMT400071215) protein interacts with all the cyclins A and B in the network (Figures 11 and S1, Table S1). The list of CC genes and levels of gene regulation are shown in the heat map (Figure 12).

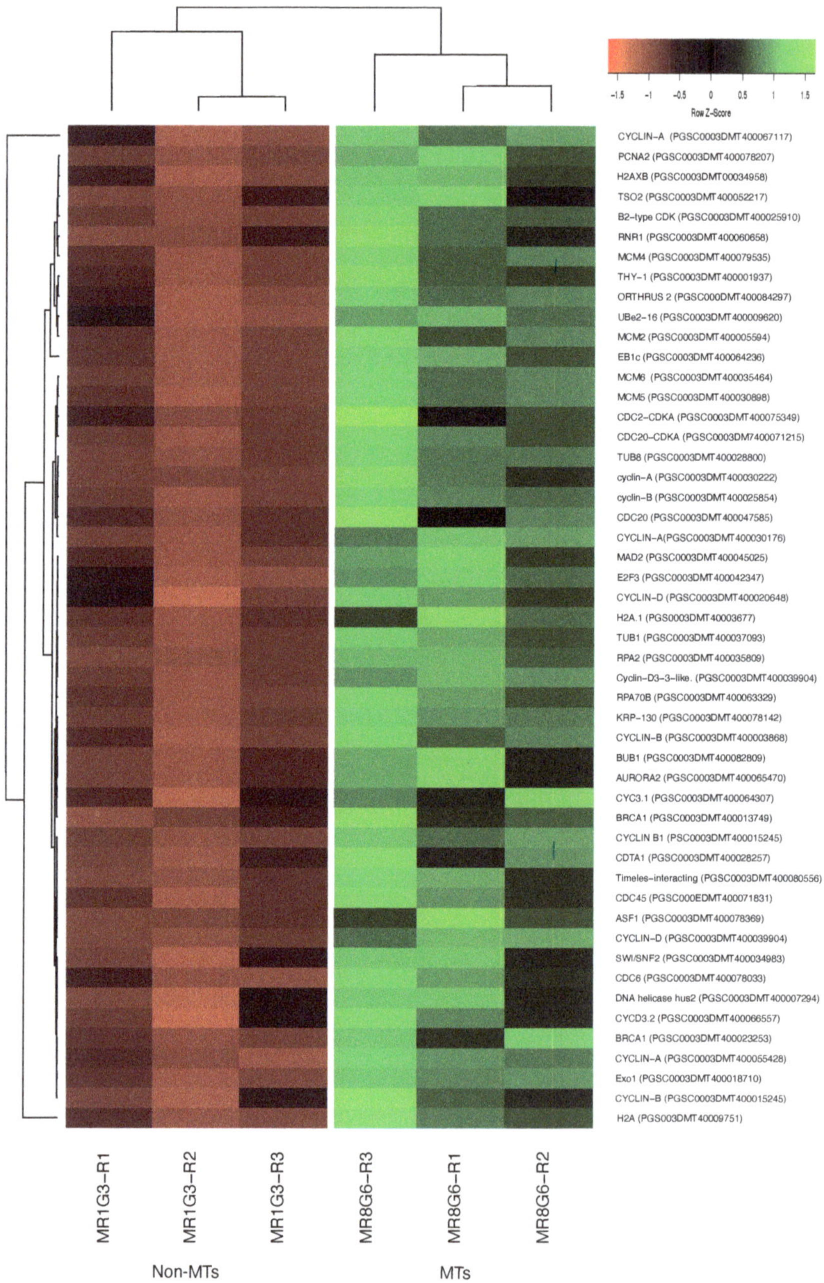

Figure 12. Heat map of CC regulation proteins during MTs development under dark conditions; levels of up-regulation are present in Log2.

2.7.2. E2F Factors

In this work, we found up-regulated transcripts belonging to the E2F TFs family, which are directly involved in the regulation of the DNA replication in the onset of S phase. Among those transcripts, the PCNA2 (PGSC0003DMT400078207) interacts with E2F3

(PGSC0003DMT400042347), which in turn interacts with *THY-1* (PGSC0003DMT400001937). The THY-1 interacts with LOG10, PGSC0003DMT400057413) which their transcriptional activation is CK-dependent. Further, PCNA2 interacts with SWI/SNF2 (PGSC0003DMT400034983), RNR1 (PGSC0003DMT400060658), and TSO2 (PGSC0003DMT400052217). The E2F3 transcription factor also interacts with BRCA1 (PGSC0003DMT400023253) (Figure S1, Table S1).

2.7.3. Histone Binding Proteins

Among the proteins present in the CC cluster, we also found three histone proteins: H2A (PGS003DMT40009751), H2A.1 (PGS0003DMT40003677), and H2AXB (PGSC0003DMT00034958). Those three proteins interact with the plant protein BRCA1 that in turn interacts with ORTHRUS 2 (PGSC000DMT400084297) and PCNA2 (PGSC0003DMT400078207). Moreover, ASF1 (PGSC0003DMT400078369) interacts with PCNA2 and with THY-1.

2.7.4. PPI Network Interaction of CC with RPs and Florigens

The PPI network analyses shed light on to protein interactions and allow a glimpse as to their biological and functional relationship. In this regard, we found that PEBP family members including the tuberigen StSP6A interact with RPs via RPL11 (PGSC0003DMT400031869), then with RPL29e (GSC0003DMT400069847) and UBQ1 (PGSC0003DMT400000364) within the cluster of RPs. UBQ1 interacts with the first protein of CC, PCNA2 (PGSC0003DMT400078207) (Figure 11). PCNA2 interacts with CDT1A (PGSC0003DMT00028257). CDT1A interacts with CDC6 (PGSC0003DMT400078033), then with CDC2a (PGSC0003DMT400075349), followed with CDC45 (PGSC000EDMT400071831) and with B1-type cyclin-dependent kinase (PGSC0003DMT400071215) interacting with CYCLIN B1 (PSC0003DMT400015245) and with B2-type cyclin-dependent kinase (PGSC0003DMT400025910) (Figure 11). In addition, CDT1A interacts with Exonuclease 1 (PGSC0003DMT400018710), with HUS2 (PGSC0003DMT400007294), then with BRCT domain BRCA1 (PGSC0003DMT400023253), with BRCA1 (PGSC0003DMT400013749), with Ubiquitin-conjugating enzyme e2-16 (PGSC0003DMT400009620), MCM4 (PGSC0003DMT400079535), then with MCM2 (PGSC0003DMT400005594), and with Timeless-interacting (PGSC0003DMT400080556) and two replication proteins RPA2 (PGSC0003DMT400035809) and RPA70B (PGSC0003DMT400063329).

2.8. DEG Transcripts with Biological Functions

The list of the transcripts genes found in the DEG analysis under MTs developmental darkness condition were directly related to cytokinin response, osmotic stress, one-carbon, carbon metabolism, glycolysis/gluconeogenesis, citrate cycle, fatty acid metabolism, thylakoids, acyl-carrier proteins, redoxins, response to oxidative stress, immunophilins, ribosomal proteins, sulfur metabolism, mitotic cell cycle, E2F activation, cytoskeleton, motor proteins, methylation, histone binding, sterol biosynthetic process and terpenoid backbone biosynthesis, amino sugar and nucleotide sugar metabolism, disulfide isomerase activity, LRR; leucine rich repeats, and sucrose metabolism, which are available in Table S1.

2.9. Validation of the Transcriptome-Wide Analysis

Validation of the transcriptomic-wide analysis was made by selecting 19 DEG and analyzing their regulation by quantitative reverse transcription PCR (qRT-PCR), using the primers described in Supplementary Table S1. The results are shown in Figure 13, indicating that the values are consistent with those obtained in the transcriptomic-wide analysis.

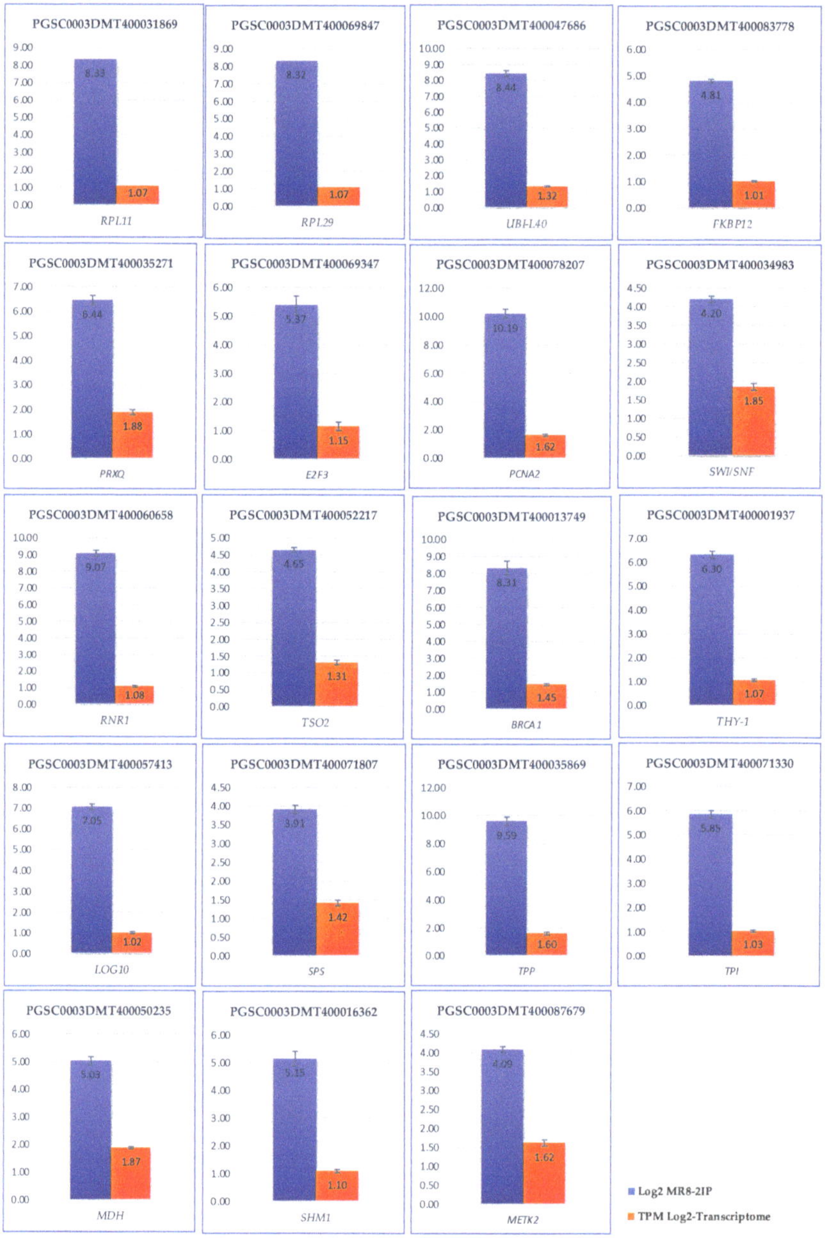

Figure 13. Validation of the transcriptomic-wide analysis by quantitative reverse transcription PCR (qRT-PCR) of 19 DEG up-regulated genes involved in RPs, PEBP, CC, carbon metabolism of potato *S. tuberosum*. Blue columns correspond to absolute gene expression derived from the genome-wide analysis. The orange bars represent the expression of the transcriptome expressing the number of transcripts per million.

The genes selected to validate de transcriptomic-wide analysis of MTs development were: StSP6, ATC, UBQ1, FKBP12, RPL29, RPL11, PRXQ, MDH1, SHM1, METK2, TPI, TPP,

SPP, BRCA1, RNR1, TSO2, THY-1, LOG10, SWI/SNF, PCNA2, and E2F3. EFa1, SEC3 were used as endogens.

3. Discussion

MTs development were induced from a stolon in medium containing sucrose 8%, gelrite 6 g/L, a CK 2iP 10 mg/L under darkness. In this environment, light modulating/repressing signals from leaves to underground were absent.

The molecular mechanisms of tuber development depend on genes related to light perception, temperature, and nutrient supply. These mechanisms have been associated mainly with phytochromes (PHYB and PHYF), homeobox TFs (BEL5 and POTH1), florigens belonging to the PEBP gene family (FT, StSP3D, StSP6A, StSP5G, TFL1), St14-3-3, StFDL1, TAC complex, StABI5 (StABL1), and several molecular regulators controlled under length day conditions (for review see [3–5]).

3.1. PPI Network of Up-Regulated TFs

TFs are proteins that control the transcription of genes by binding to a defined region of the genome. TFs are essential for the regulation of biological process, such as body planning, development, differentiation, and responses to various environmental signals [22].

MTs development showed 23 up-regulated TFs interacting in a PPI network (Figure 3). The interpretation of the molecular mechanisms from a stolon to a tuber development in high content of sucrose and gelrite, 2iP as CK under darkness is as follows:

The MYB44 TF (PGSC000DMT400008569) a tuber-specific and sucrose-responsive element binding factor. MYB44 confers resistance to abiotic stresses dependent on ABA. Enrichment analysis with 50 interactors revealed that MYB44 is involved in auxin signaling pathways and tissue vascularization. MYB44 interacts with an AFR protein (PGSC0003DMT400004486), F-box/kelch-repeat skip25-like. AFR protein interacts with another tuber-specific and sucrose-responsive element binding factor TSF (PGSC0003DMT000794169). Enrichment analysis with 50 interactors revealed that TSF is involved in cell fate specification and commitment, multicellular organism development, cell cycle regulation, spliceosomal and ribonucleoprotein complex. TSF interacts with bZIP43 (PGSC0003DMT400003516); it is expressed in inflorescence meristem during petal differentiation and expansion stage. An enrichment analysis with 50 interactors revealed that bZIP43 is involved in histone modification, h3-k9 methylation, ribonucleoprotein complex, 14-3-3 interaction, and amino acid biosynthesis. bZIP43 interacts with a bZIP27-like, self-pruning TF (PGSC0003DMT400061403), required for positive regulation of flowering. Enrichment analysis with 50 interactors revealed that bZIP27 is involved in structural component of ribosomes, cell cycle regulation, cyrcadian rhythm, protein transport, methylation. bZIP43 interacts with a wuschel-related homeobox 9-like (WOX9), COMPOUND INFLORESCENCE (PGSC0003DMT400027850). WOX9 is required for meristem growth and development, and acts through a positive regulation of wuschel. Mutant phenotypes include embryo lethality, reduced root development, and smaller meristems. Phenotypes can be rescued by the addition of sucrose in the growth media. Overexpression can partially rescue the triple mutants of CK receptor phenotype, as a downstream effector of CK signaling. Enrichment analysis with 50 interactors revealed that WOX9 is involved in ribosome, cell cycle regulation, cell fate commitment, meristem growth and maintenance, translation initiation factor 4f, and flower development.

bZIP43 also interacts with a homeobox-leucine zipper protein HAT4 (PGSC0003DMT400067540). HAT4 acts as mediator of the red/far-red light effects on leaf cell expansion in the shading response. It is also involved in the negative regulation of cell elongation and proliferation processes, such as root formation and secondary growth of the vascular system. HAT4 TF interacts with 6 bhlh TFs. Enrichment analysis with 50 interactors revealed that HAT4 is involved in flower development, meristem maintenance, photoreception, sugar signaling, and abiotic stimulus.

The above information about PPI network of up-regulated TFs during MTs development may explain partially some molecular events required to develop MTs: vascularization (MYB44), proliferation process (HAT4), cell cycle, ribosomes (bZIP27), cell fate commitment and meristems (WOX9), light regulation (HAT4), and epigenetic regulation (bZIP43) present during MTs differentiation. The rest of TFs in the PPI network may participate during MTs development (Figure 3). Further analysis of each component is required to elucidate it.

3.2. PPI Network of Down-Regulated TFs

MTs development revealed that 57 down-regulated TFs interact in a PPI network (Figure 5). The morphological transformation from a stolon explant (modified stem) to a tuber must involve the down-regulation of TFs involved in stem, leaf, root, and flower development. From 57 down-regulated TFs, 14 TFs were related to shoot system development, 7 to flower and gametophyte development, 6 to phyllome development, 32 in response to hormone, and 23 in hormone signaling pathways.

The PPI network revealed a cluster of 18 homedomain TF superfamily/16 MYB interacting with two BTI TF (PGSC0003DMT400039384) and (PGSC0003DMT400069392) involved in gametophyte development. The MYB TF are involved in multicellular organism development, stamen maturation and dehiscence, seed coat, lateral organ boundaries, and root and xylem differentiation (Figure 5). The gametophyte cluster interacts with a light regulated plant development consisting in BBX17, PIL6 (phytochrome interacting factor), two FAMA TFs, involved in differentiation of stomatal guard cells, derepresses stem cell gene expression, and 10 TF required during fertilization of ovules by pollen.

Another cluster interacting with the gametophyte is composed by TCP TFs. These TFs are involved in leaf development, branching, circadian rhythm, embryonic growth, floral organ morphogenesis and hormone signaling. Recently, Nicolas et al. [23], demonstrated that the TCP TF BRC1b is a tuber repressor in aerial axillary buds, limits sucrose accumulation, promotes dormancy, abscisic acid response, and a reduced number of plasmodesmata. In addition, it interacts with StSP6A, and inhibits its tuber-inducing activity in aerial nodes. In our work, StBRC1b (PGSC0003DMT400010380) was found to be down-regulated and interacting within the network (Figure 5). StBRC1b interacts with TCP23-like and F-box/lrr-repeat MAX2 homolog a-like. A combined up- and down-regulated TFs networks revealed that StBRC1b interacts close to one up-regulated TF, the tuber-specific and sucrose-responsive element TSF (PGSC0003DMT000794169). STBRC1b interacts with MAX2 (PGSC0003DMT400020318) and KUF1 (PGSC0003DMT400004486), a karrikin protein that interacts with TSF TF, up-regulated in this work (Supplementary Figure S2).

Tang et al. [24], reported that the gene Soltu.DM.06GO25210, named by authors IT1, is identical to StBRC1b. When silenced by CRISPR/Cas, stolons were converted to branches instead of swelling with the capability of developing small tubers within time. StSP6A interacts with IT1 (StBRC1b), and authors suggest that it forms a complex during tuber initiation. They also found that there was an impaired interaction between the StSP6A gene from Etuberosum, a non-tuber producer Solanum plant species, due to the deleted fourth exon in Etuberosum. MAX2 is required for responses to strigolactones and karrikins as well as the repression of axillary shoots outgrowth. MEE1 is required during endosperm development in embryos.

3.3. Up-Regulated DEG PPI Network

3.3.1. PEBP Family Members and FD TF Interacting with RPs

The PEBP gene family is well-conserved across species from animals to plants, which play central roles in several biological processes. In *Homo sapiens*, *PEBP1 (RKIP)* has been implicated in several cancers, probably acting as a metastasis suppressor [25]. In addition, it has been proposed to be an active regulator for growth, transformation, differentiation [26], and a cell cycle modulator [27]. In flowering plants, PEBP gene family members have undergone duplication. PEBP are involved in determination of plant architecture. PEBP family consists of three groups: the *FLOWERING LOCUS* (*FT*)-like, the *TERMINAL FLOWER1* (*TFL1*)-like, and *CENTRORADIALS* (*CEN*)-like. The *MOTHER OF FT AND TFL1*-like (*MFT*), and the *BROTHER OF FT AND TFL1* (*BFT*)-like are present as a separate subclade within the CEN group [28]. In potato, multiple *FT* paralogues are related to both sexual and vegetative reproduction. SP3D, an FT orthologue, controls flowering, while SP5G and SP6A, two other PEBP genes, act as a repressor and inducer of tuber formation, respectively [29]. The balance between *FT*/*TFL1* defines the plant growth habit as indeterminate or determinate by modulating the pattern of formation of vegetative and reproductive structures in the apical and axillary meristems [30]. In our study, six PEBP family members were transcriptionally active in MTs development in darkness conditions, such as three orthologs *SP6A*, *StSP5G* (PGSC0003DMT400041726), *SP3D* (PGSC0003DMT400041725), and two orthologs of *CEN1-BFT* (PGSC0003DMT400030575). In the PPI network analysis, we found that all PEBP protein family members interact with the universal plastid RPL11 (PGSC0003DMT400031869). PEBPs lack a DNA-binding domain but can exert the function as transcriptional co-regulators through complexing with TFs, regulating the expression of downstream target genes [31]. The TF FD (PGSC0003DMT400061403), a bZIP transcription factor-27 self-pruning was also found interacting with ATC (Figure 8).

Purwestri et al. [32] and Wang et al. [33], through a yeast two-hybrid screening approach, found that SP6A can interact with RPs and other proteins involved in protein synthesis, RNA and DNA binding proteins, histones, initiation factors, signaling and carbon metabolism, and regulate flowering. In addition, the PPI network analysis unveiled 77 interactors, widely distributed throughout the cells in nearly all subcellular components, including plastids, ribosomes, thylakoids, cytosol, and chloroplasts. SP6A functions as positive regulator of tuber development under high contents of sucrose [30,34] and represses bud flower formation under short day conditions [35]. It is involved in several pathways during tuberization, including temperature, microRNA, and photoperiod [36–38]. Several transcriptomic analysis of potato *S. tuberosum* have revealed the presence of RPs during the tuberization process [12,15,16,39,40]. In Sharma et al. [15], we found through PPI network analysis with the highest confidence (0.900) that RPs proteins interacted with SP6A, RP40SA, RP60S, RPS8, RPS4A, RPL10, BEL5, and RPL14.

The ABI5-like 1 (StABL1) bind to SP6A in a 14-3-3 manner forming an alternative tuberigen activation complex (aTAC), thereby resulting in early tuberization [41]. In our study, the transcript for abscisic acid-insensitive 5-like protein (PGSC0003DMT400072260) was up-regulated; it did not interact directly with SP6A in the PPI network analysis, even using the lower confidence (0.150) value. This suggest that a 14-3-3 protein is essential to establish a protein interacting complex with SP6A. In our study, the 14-3-3, (34G; PGSC0003DMT400079172) transcript was up-regulated, and its protein interacts with SP6A, using a high confidence (0.780) value. It is very likely that SP6A and 14-3-3 might be forming a complex under MTs darkness conditions. 34G (14-3-3) is involved in cell cycle regulation, mitotic cycle, ribosome biogenesis, response to stress, starch and sucrose metabolism, reproductive process, and basal transcription factor [42,43].

CENTRORADIALS (*CEN*) represses the initiation of floral meristems in several species. The *DHvCEN5* barley plants developed spikelet initiation in both photoperiods; GO enrichment analysis demonstrate that *HvCEN* had a negative impact on to chromatin modification, cell cycle regulation, CK signaling, and cell growth through RPs [44]. A member of the *TERMINAL FLOWER-1/CENTRORADIALS* gene family (*CEN*) has shown to function as a negative regulator of tuberization. At the protein level, CEN competes with SP6A for the interaction with other members of the TAC complex [45]. Moreover, in potato plants, *CEN* overexpression induced a lower rate of sprout growth, lower content of CKs, and higher content of abscisic acid (ABA) [46].

Another transcript for a *FT* was found in this work–the *SP5G* (PGSC0003DMT400041726) which was up-regulated. The function of this *FT* is to blockade the tuberization process. In the absence of light/darkness, SP5G might not repress *SP6A*. In addition, it has been shown that PHYB and PHYF co-stabilized CONSTANS-like 1 protein (COL1), binds SP5G, and represses *SP6A* and tuberization [47]. We also found that the transcript for *SP3D* was up-regulated and that this protein is involved in flowering, and it has been shown that florigens can promote tuberization; therefore, the SP3D might be able to promote this process [29,48].

3.3.2. RPs Cluster

Several transcriptomic analyses in *S. tuberosum* have revealed the transcriptional activation of the RPs coding genes in the tuberization process [12,15,16,20,39,40]. Specifically, the work of Taylor et al. [49] found a 15-20-fold increase in transcript levels of two RPs, S19 and L7, during the early stages of tuberization in stolon tips. Moreover, the RP proteins have been directly related with growth, development, embryo viability, flowering, and leaf development [50–56].

Sucrose Activation of RPs

The up-regulation of RPs in Arabidopsis is activated by different biotic and abiotic factors [57–63] such as sucrose [20,64–67] and CKs [21].

In our study, a high content of sucrose (8%), a CK 2iP 10 mg/L, and osmotic stress conditions were applied under darkness conditions. Hummel et al. [67] found that in response to 6% sucrose feeding, 166 out of 204 RPs were up-regulated. In our study, 31 RPs were detected, and according to the data obtained by Gamm et al. [20] who applied 5% of sucrose, we obtained 14 RPs activated in dark conditions and 12 with similar in light and dark RPs activated by sucrose.

CK Activation of RPs

In our study, 12 transcripts for RPs were directly related to CKs activation [21], 12 transcripts for RPs were transcriptionally activated by both sucrose and CK treatment. In *Arabidopsis,* 87 transcripts for RP of the 474 identified responded specifically to CKs; CK-responsive RPs related genes encode proteins for both plastidic and cytoplasmatic localization [21].

CKs binds directly to ribosomes in higher plants with high affinity [68,69]. CKs alter the polyribosome formation, ribosomal protein synthesis, number of ribosomes, and the phosphorylation status of ribosomal proteins [70–77]. In *Arabidopsis,* Karunadasa et al. [74] demonstrated that protein synthesis is induced by CKs and CK signaling and requires isoforms of the RP L4. The loss of function of RPL4 increases the tolerance to osmotic stress and decreases sensitivity to CK-induced growth. In our study, 12 RPs were activated by CKs [21], and 12 RPs were activated by both sucrose and CKs treatment. In *Arabidopsis,* 87 out of 474 respond to CKs [21]. The CK-responsive genes encode proteins of both plastidic and cytoplasmatic localization.

Stress Activation of RPs

The expression levels of RPs are altered under stress conditions [57,58,61–63,75], suggesting that it participates in signal detection and stress response. In rice, RPs large subunits are water and salt stress responsive proteins [61–63]. In maize, ribosome footprints from chloroplast and cytosol have been analyzed [76] and provided insights into the dynamic translation mechanisms during drought [77] and viral infection [78]. In cotton, RPL14-2 interact with several RPs to activate drought and salt tolerance [79]. In tea (*Camellia sinensis* L.), RPL32 is up-regulated in drought stress [80].

RPs and Their Relationship with CC

In *Saccharomyces cerevisiae,* the repression of nine RPs 60S genes (L4, L7, L18, L3, L17, L28, L35, L37, and L40) resulted in G2/M phase arrest. The repression of nine other 60S RPs (L1, L3, L9, L16, L19, L21, L25, L30, and L43) and 22 40S genes (S0, S1, S2, S3, S4, S5, S6, S9, S10, S13, S14, S15, S17, S19, S20, S21, S22, S26, S27, S29, and S30) caused arrest in the G1 phase [81].

Overall, 14 of 38 proteins analyzed in yeast were present in the potato RPs network: RPL1, RPL18, RPL25, RPL35, RPL40, RPS2, RPS3, RPS6, RPS9, RPS13, RPS15, RPS17, RPS29, and RPS30. Accordingly, they could contribute to the activation of CC during potato MTs development.

Maitra et al. [82] demonstrated a relationship between RPs, CC, and one-carbon metabolism. Yeasts regulate the one-carbon flow balance between donors by controlling the activity of cytoplasmic serine hydroxymethyl transferase (SHM2P) [83]. In cells lacking certain ribosomal proteins, SHM2 translation efficiency and Gly:Ser ratios were found to be low. These observations are consistent with low flow through the 1C pathways.

We proposed that RPs present in our analysis, in addition to being essential in the progression of CC in MTs development, could also participate in the one-carbon pathways that ultimately comprise a series of interrelated metabolic pathways that include the methionine and folate cycles–critical for cell function–providing 1C units (methyl groups) for the synthesis of DNA, polyamines, amino acids, creatine, and phospholipids [84].

3.3.3. CC Cluster

Tuber development relies on the temporal and spatial control of cell proliferation and cell growth to correctly develop an organ that will help the potato plant to survive under adverse conditions. Potato tubers develop from stolons, from which swelling growth occurs in the subapical region of the meristem. Initial expansion and radial cell division of cells in the pith and cortex, in combination with restricted longitudinal growth is followed by cell division and enlargement in random orientations in the perimedullary region once the swelling has reached a diameter of 2 to 4 mm, continuing until the tuber reaches its full size. In this study, genes involved in CC regulation were present, cyclins/CDKs, E2F factors and histone binding proteins.

Cyclins/CDKs

Overall, 2 cyclin-dependent kinase (CDKs); CDC2 (PGSC0003DMT400075349) and CDKB1;2 (PGSC0003DMT400071215), 10 cyclins: 4 cyclin-A, 3 cyclins-B and 3 cyclin-D were found. CDKB1;2 (PGSC0003DMT400071215) interacts with cyclins A and B present in our network. CKs in the presence of sucrose activate the expression of cyclins-D that have been shown to be essential in CC regulation. Yang et al. [85] demonstrated that CKs (Zeatin, BAP, and 2iP) regulate cell cycle by promoting the nuclear shuttling of the MYB-domain protein 3R4. MYB3R4 binds 5′-AACGG-3′ motifs of promoters of G2/M-specific genes and E2F targets. In our analysis, a gene homologous of MYB3R4 was found (PGSC0003DMT00047825). In accordance, it is very feasible that the potato MYB3R4 activate the cell cycle regulation genes during MTs development.

Rhee et al. [86] analyzed cyclins-D gene expression in potato exposed to different light regimes, hormones (CKs and auxins), and carbohydrates (SUC and glucose) and found that *StCYCD3;1* and StCYCD3;3 were more expressed in above-ground organs whereas StCYCD3;2 was expressed more abundant in undergrounds organs. The above expression levels were when cytokinins like zeatin and BAP combined with sucrose induced higher expression compared to 2,4-D. In our work, StCYCD3;2 was found interacting within the PPI network with a cyclin-dependent kinase B2 (CDKB) (PGSC0003DMT400025910). The levels of expression found were similar; *StCYCD3;2* was 3.5, while *StCYCD3;1* was 3.03, and *StCYCD3;3* was a 2.85 fold change.

In *H. sapiens*, *PEBP1* (*RKIP*) is a metastasis suppressor in aggressive cancers [25] and cell cycle modulator [27]. Silenced-*RKIP* accelerate DNA synthesis and G1/S transition entry by inducing the expression of *CDC6*, *MCM2,4,6,7*, *cyclinD2*, *D1*, and *E2*. It also enhances G2/M transition and down-regulates G2/M checkpoints like *AURORA-B*, cyclin G1, and slow G2/M transition time. This led to a higher proliferation rate compared to control cells [27].

In our work, the components of G2/M checkpoints were present, showing similarities with the function of PEBP1(RKIP) in *H. sapiens*.

The CC regulation of mitosis is summarized as follows: The CYCB-CDKA complex (PGSC0003DMT400071215) interacts with MAD2 (PGSC0003DMT400045025) (mitotic arrest deficient), MAD2 interacts with CDC20 protein (PGSC0003DMT400047585) (APC metaphase/anaphase complex). MAD2 and CDC20 are checkpoint proteins locally catalyzed by kinetochores, the central player in the distribution of genetic material to the daughter cells during mitosis [87].

MAD2 interacts with EB1c (PGSC0003DMT400064236) (microtubule stability). MAD2 interacts with TUB1 (PGSC0003DMT400037093) (tubulin formation). In Arabidopsis, *TUB1* expression is linked to light induction [88].

In hypocotyl explants, the *TUB1* highest expression levels are found in dark conditions. The *TUB1* transcriptional behavior in previous reports is in accordance with our data, since our work was carried out in dark conditions [88].

MAD2 interacts with TUB8 (PGSC0003DMT400028800). During the initiation of tuber formation, there are changes in the expression of tubulins at the tip of the stolons [87].

MAD2 interacts with AURORA2 (PGSC0003DMT400065470), a serine/threonine protein kinase. It is abundant in tissues with high division, such as roots and flowers, but low in leaves and stems. In mitosis it is intimately linked with the cytoskeleton (pre-prophase band, phragmoplast, metaphase plate) required for cytokinesis [89].

MAD2 interacts with KRP-130 (PGSC0003DMT400078142). KRP-130 are involved in microtubule dynamics, morphogenesis, chromosome segregation, organelle transport, and vesicles [90].

KRP-130 interacts with BUB1 (PGSC0003DMT400082809); BUB1 binds to the kinetochore and has an important role in establishing the mitotic spindle checkpoint and chromosome aggregation [91].

BUB1 interacts with the CDKA-CYCA complex (PGSC0003DMT400071215). BUB1 interacts in the same way with the complex of CDKA-CYCA. CDKA (PGSC0003DMT400071215) interacts with four B cyclins. CDKA has been identified at the G1/S and G2/M checkpoints [92] and is modulated by cyclins A and B [93].

E2F Factors

The E2F proteins form a family of TFs that regulate the transition from the G_1 to the S phase in the cell cycle. We found 31 E2F factors in this analysis. The activation of genes such as *PCNA2, RNR1, TSO2, THY-1, SWI/SNF2, ORC6, ORC4, ORC5, SPT16, MCM6, SMC3, MCM2, CHR11, POLA2,* and *PSF2* demonstrates that DNA replication was carried out correctly, validating that the conditions evaluated are optimal for the development of MTs. Abnormal induction of the S phase re-entry in mature leaf cells is observed in overexpression of E2F in Arabidopsis [94]. Unregulated activity of this family of TFs has been characterized broadly in human cancers [95].

Abiotic stress is perceived through SWI/SNF2 and its interaction with *PCNA2*. CKs are perceived through the activation of *LOG10* to the synthesis of dihydrofolate reductase. Key proteins were found during mitosis, such as those of the cytoskeleton, microtubules, tubulins, motor proteins, and chromosome stability, as well as the proteins of the synthesis phase. Like PCNA2, RNR1, TSO2, THY-1, SWI/SNF2.

Histone Binding Proteins

Three histone proteins are present in the CC cluster; histone H2A (PGS003DMT40009751), histone H2A.1 (PGS0003DMT40003677), and histone H2AXB (PGSC0003DMT00034958).

Chromatin structure is essential for normal cell cycle progression, chromosome segregation, and centromere function [96].

Loss of function mutations in *H2A* cause an increase in ploidy phenotype, and an increase rate of chromosome loss and defects in completing the G2/M phase of the cell cycle. Altered centromeric chromatin structure and mutations in genes encoding kinetochore components revealed the role for *H2A* in the proper centromere-kinetochore function [97].

Histones also regulate a DNA damage response by epigenetic modifications. The expression of a loss of function mutant of *H2AX* Ser139Ala, that partially disrupts the phosphorylation site, suppressed the normal G2/M cell cycle arrest following DNA damage by ionizing radiation. The G2/M cell cycle arrest by DNA damage is promoted by checkpoints such as *BRCA1* [98]. Taken together, *H2AXB* (PGSC0003DMT00034958) regulates the DNA damage response via *BRCA1* (PGSC0003DMT400023253) to arrest mitotic cell cycle and ensure an appropriate cell division during MTs development.

These three proteins interact with BRCA1, a heterodimeric ubiquitin E3 ligase, required for the accumulation of ubiquitin conjugates at sites of DNA damage and for silencing at DNA satellite repeat regions. BRCA1 interact with E3 ubiquitin-protein ligase ORTHRUS 2 (PGSC000DMT400084297) and PCNA2.

Histone chaperone ASF1 (PGSC0003DMT400078369) interacts with PCNA2 with a combined score of 0.970 and with THY-1 with a score (0.860). ASF1 play an important role in chromatin replication, maintenance of genome integrity, and cell proliferation during plant development. Histones prevent DNA from becoming tanged, protect it from DNA damage, and play important roles in gene regulation and DNA replication.

MTs development under high content of sucrose/gelrite, 2iP and darkness is activated by PEBP members (StSP6A, StSP5G, StSP3D, CEN1/BFT, ATC) and the TF FD. PEBP members interact with RPs to modulate CC regulation (DNA synthesis and mitosis) (Figure 14).

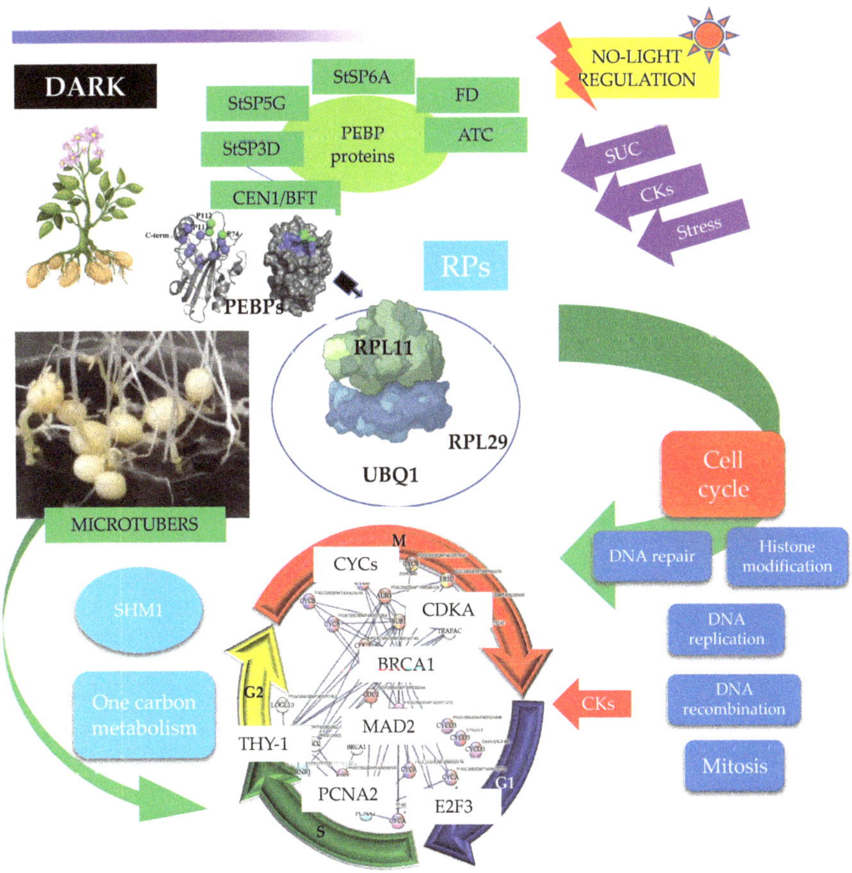

Figure 14. PPI network of up-regulated CC proteins involved in mitosis (M), S, G2, G1 phase of potato S. tuberosum from transcriptomic-wide analysis with high confidence (0.860).

3.4. Validation of Transcriptomic-Wide Analysis by RT-qPCR

To validate the results of the transcriptomic-wide analysis and the conclusions reached, we performed an analysis by RT-qPCR of 21 genes interacting within the PPI network.

Figure 13 shows the results of validation by RT-qPCR. For all cases, the correct validation is observed as they are up-regulated both in the transcriptome and in the relative quantification.

4. Materials and Methods

4.1. Plant Material

4.1.1. Potato Shoot Micropropagation

Potato cv. Alpha plantlets were propagated by shoot proliferation under in vitro conditions in MS medium [99], supplemented with sucrose 30 g/L (CAT 57-50-1 Sigma-Aldrich, St. Louis, MO, USA), activated charcoal 3 g/L (CAT: 242276 Sigma-Aldrich, St. Louis, MO, USA), pH 5.8 and solidified with gelrite 3 g/L (GELZAN CAT. G1910 Sigma-Aldrich, St. Louis, MO, USA). Shoots were incubated at 25/17 °C under fluorescent light at 25 μmol/m^2/s of irradiance. Stolon explants derived from propagated shoots were used for MTs induction.

4.1.2. Potato MTs Induction

Potato MTs induction protocol was made according to Herrera-Isidron et al. [10]. Stolon explants were cultured in flasks containing MR8-G6-2iP medium: MS medium, supplemented with 2iP 10 mg/L, 8% sucrose, 6 g/L gelrite, 3 g/L activated charcoal, pH 5.8, osmotic potential of 2.02 mPa. As control, stolon explants were cultured in MR1-G3-2iP: MS medium supplemented with 10 mg/2iP, 1% sucrose, 3 g/L gelrite, 3 g/L activated charcoal, pH 5.8, osmotic potential of −0.88 Mpa.). Flasks were sealed with plastic and incubated in dark at 25/17 °C for 15 days of incubation.

4.1.3. Transcriptome Sequencing and Assembly

To isolate RNA from MTs, tissue samples were collected after 15 days of incubation; at this time, MTs initiate to develop [10].

Total RNA was isolated by using Trizol (Invitrogen, Carlsbad, CA, USA), RNA concentration was measured by its absorbance at 260 nm, ratio 260 nm/280 nm was assessed, and its integrity confirmed by electrophoresis in agarose 1% (*w*/*v*) gels.

Samples of cDNA were amplified by PCR using SYBRTM Green (ThermoFisher CAT: 4312704, Waltham, MA, USA) in Real-Time PCR Systems (CFX96 BioRad, Herules, CA, USA).

Final product cDNA samples for sequencing were sent to GENEWIZ, Plainfield, NJ, USA. Illumina HiSeq 2500 (Illumina, San Diego, CA, USA) was used for sequencing. Sequenced reads were tested for quality using the FastQC software package (http://www.bioinformatics.babraham.ac.uk/projects/fastqc/) and preprocessed to remove sequence adapters and low-quality bases using the software Trimmomatics [100].

4.1.4. Analysis of DEG

Adapter removal was performed using the Trimmomatic v0.3.6 program [100]. RNA-seq reads were aligned to the *Solanum tuberosum* reference genome available in Phytozome v12.1. (https://phytozome.jgi.doe.gov/pz/portal.html) with the STAR aligner v.2.5.2b. [101]. In this step, the BAM (Binary Alignment/Map) files were generated. Subsequently, a count and set of transcripts were made using the featureCounts program of the Subread v.1.5.2 package [102]. A quantification and differential analysis of the transcripts was performed using the DESeq2 v1.12.4 program. Finally, an ontology analysis was performed using Blast2GO.

4.1.5. PPI Analysis of Microtuberization

A gene network with high confidence (0.860) was performed with the STRING database v11.5 [103], based on *S. tuberosum*, homologous genes present in the *S. tuberosum* genome in Sol genomics network [104]. Gene identifier (Id) was made according to the UNIPROT [105], NCBI [106] database. Homologous in *S. tuberosum* greater than 60% in protein sequence with *A. thaliana* were considered. Oligonucleotides were designed to qPCR ($2-\Delta\Delta CT$ method analysis) [107] gene expression or transcriptional analysis (Table S1).

5. Conclusions

1. A PPI network of up-regulated TFs revealed that at least six TFs–MYB43, TSF, bZIP27, bZIP43, HAT4, WOX9–may be involved during MTs development.
2. Two fundamental biological process essential for life and highly conserved through organisms were found interacting tightly: RPs comprising 29 and CC 117 proteins.
3. PEBP members interact with RPs and CC process to activate MTs development under high content of sucrose and gelrite, 2iP under darkness.
4. Further experiments by yeast two-hybrid screening approach of genome edited up- and down-regulated PEBP members with RPL11 are required to demonstrate our model of MTs development under darkness.

Supplementary Materials: The following supporting information can be downloaded at: https://www.mdpi.com/article/10.3390/ijms232213835/s1. References [55,108–142] are cited in Supplementary Materials Table S1.

Author Contributions: E.V.-L. and L.H.-L.: Conceptualization, methodology, validation, formal analysis, investigation, data curation, writing original draft preparation, review and editing, funding acquisition; J.A.F.-L. and O.S.R.-M.: Methodology, software, validation, investigation; A.B.: Software, data curation, formal analysis, supervision, writing, review and editing; J.L.C.-P.: Conceptualization, methodology, validation, formal analysis, investigation, data curation, writing original draft preparation, review and editing, funding acquisition, writing, review and editing. All authors have read and agreed to the published version of the manuscript.

Funding: This research was funded, by SIP 20222061 IPN, proyectos de desarrollo tecnológico o inovación para alumnos del IPN # 120.

Data Availability Statement: Not applicable.

Acknowledgments: We thank Rosalia Vuelvas-Nolasco for her excellent technical assistance of plant tissue culture of potato and medium preparation.

Conflicts of Interest: The authors declare no conflict of interest.

References

1. Food and Agriculture Organization Corporate Statistical Database. FAOSTAT. 2022. Available online: https://www.fao.org/faostat/es/#data/QCL/visualize (accessed on 26 July 2022).
2. Khosa, J.; Bellinazzo, F.; Kamenetsky, R.; Macknight, R.; Immink, R.G.H. PHOSPHATIDYLETHANOLAMONE-BINDING PROTEINS: The conductors of dual reproduction in plants with vegetative storage organs. *J. Exp. Bot.* **2021**, *72*, 2845–2856. [CrossRef] [PubMed]
3. Dutt, S.; Manjul, A.S.; Raigond, P.; Singh, B.; Siddappa, S.; Bhardwaj, V.; Kawar, P.G.; Patil, V.U.; Kardile, H.B. Key Players Associated with Tuberization in Potato: Potential Candidates for Genetic Engineering. *Crit. Rev. Biotechnol.* **2017**, *37*, 942–957. [CrossRef] [PubMed]
4. Kondhare, K.R.; Malankar, N.N.; Devani, R.S.; Banerjee, A.K. Genome-Wide Transcriptome Analysis Reveals Small RNA Profiles Involved in Early Stages of Stolon-to-Tuber Transitions in Potato under Photoperiodic Conditions. *BMC Plant Biol.* **2018**, *18*, 284. [CrossRef] [PubMed]
5. Zierer, W.; Rüscher, D.; Sonnewald, U.; Sonnewald, S. Tuber and Tuberous Root Development. *Annu. Rev. Plant Biol.* **2021**, *72*, 551–580. [CrossRef] [PubMed]
6. Donnelly, D.J.; Coleman, W.K.; Coleman, S.E. Potato Microtuber Production and Performance: A Review. *Am. J. Pot. Res.* **2003**, *80*, 103–115. [CrossRef]

7. Hannapel, D.J. Signalling the Induction of Tuber Formation. In *Potato Biology and Biotechnology*; Elsevier: Amsterdam, The Netherlands, 2007; pp. 237–256. [CrossRef]
8. Vinterhalter, D.; Dragicevic, I.; Vinterhalter, B. Potato in vitro culture techniques and biotechnology. *Fruit Veg. Cereal Sci. Biotechnol. (Spec. Issue 1)* **2008**, *2*, 16–45.
9. Coleman, W.K.; Donnelly, D.J.; Coleman, S.E. Potato Microtubers as Research Tools: A Review. *Am. J. Pot. Res.* **2001**, *78*, 47–55. [CrossRef]
10. Herrera-Isidron, L.; Valencia-Lozano, E.; Rosiles-Loeza, P.Y.; Robles-Hernández, M.G.; Napsuciale-Heredia, A.; Cabrera-Ponce, J.L. Gene Expression Analysis of Microtubers of Potato Solanum Tuberosum L. Induced in Cytokinin Containing Medium and Osmotic Stress. *Plants* **2021**, *10*, 876. [CrossRef]
11. Crookshanks, M.; Emmersen, J.; Welinder, K.G.; Lehmann-Nielsen, K. The potato tuber transcriptome analysis of 6067 expressed sequence tags. *FEBS Lett.* **2001**, *506*, 123–126. [CrossRef]
12. Shan, J.; Song, W.; Zhou, J.; Wang, X.; Xie, C.; Gao, X.; Xie, T.; Liu, J. Transcriptome analysis reveals novel genes potentially involved in photoperiodic tuberization in potato. *Genomics* **2013**, *102*, 388–396. [CrossRef]
13. Kumar, J.; Sapna, T.; Sundaresha, D.S.; Chandel, P.; Ali, N.; Singh, B.; Bhardwaj, V.; Singh, B.P. Microarray analysis of gene expression patterns in the leaf during potato tuberization in the potato somatic hybrid *Solanum tuberosum* and *Solanum etuberosum*. *Genome* **2014**, *58*, 305–313. [CrossRef]
14. Gong, L.; Zhang, H.; Gan, X.; Zhang, L.; Chen, Y.; Nie, F.; Shi, L.; Li, M.; Guo, Z.; Zhang, G.; et al. Transcriptome Profiling of the Potato (*Solanum tuberosum* L.) Plant under Drought Stress and Water-Stimulus Conditions. *PLoS ONE* **2015**, *10*, e0128041. [CrossRef]
15. Sharma, P.; Lin, T.; Hannapel, D.J. Targets of the StBEL5 Transcription Factor Include the FT Ortholog StSP6A. *Plant Physiol.* **2016**, *170*, 310–324. [CrossRef]
16. Vulavala, V.K.R.; Fogelman, E.; Faigenboim, A.; Shoseyov, O.; Ginzberg, I. The Transcriptome of Potato Tuber Phellogen Reveals Cellular Functions of Cork Cambium and Genes Involved in Periderm Formation and Maturation. *Sci. Rep.* **2019**, *9*, 10216. [CrossRef]
17. Wilson, D.N.; Doudna Cate, J.H. The Structure and Function of the Eukaryotic Ribosome. *Cold Spring Harb. Perspect. Biol.* **2012**, *4*, a011536. [CrossRef]
18. Inzé, D.; De-Veylder, L. Cell Cycle Regulation in Plant Development. *Annu. Rev. Genet.* **2006**, *40*, 77–105. [CrossRef]
19. Timsit, Y.; Sergeant-Perthuis, G.; Bennequin, D. Evolution of ribosomal protein network architectures. *Sci. Rep.* **2021**, *11*, 625. [CrossRef]
20. Gamm, M.; Peviani, A.; Honsel, A.; Snel, B.; Smeekens, S.; Hanson, J. Increased Sucrose Levels Mediate Selective mRNA Translation in Arabidopsis. *BMC Plant Biol.* **2014**, *14*, 306. [CrossRef]
21. Brenner, W.G.; Schmülling, T. Transcript Profiling of Cytokinin Action in Arabidopsis Roots and Shoots Dcovers Largely Similar but Also Organ-Specific Responses. *BMC Plant Biol.* **2012**, *12*, 112. [CrossRef]
22. Latchman, D.S. Transcription factors: An overview. *Int. J. Biochem. Cell Biol.* **1997**, *29*, 1305–1312. [CrossRef]
23. Nicolas, M.; Torres-Pérez, R.; Wahl, V.; Cruz-Oró, E.; Rodríguez-Buey, M.L.; Zamarreño, A.M.; Martín-Jouve, B.; García-Mina, J.M.; Oliveros, J.C.; Prat, S.; et al. Spatial control of potato tuberization by the TCP transcription factor BRANCHED1b. *Nat. Plants* **2022**, *8*, 281–294. [CrossRef] [PubMed]
24. Tang, D.; Jia, Y.; Zhang, J.; Li, H.; Cheng, L.; Wang, P.; Bao, Z.; Liu, Z.; Feng, S.; Zhu, X.; et al. Genome evolution and diversity of wild and cultivated potatoes. *Nature* **2022**, *606*, 535–541. [CrossRef] [PubMed]
25. Ahmed, M.; Lai, T.H.; Kim, D.R. A functional Network Model of the Metastasis Supressor PEBP1/RKIP and its Regulators in Breast Cancer Cells. *Cancers* **2021**, *13*, 6098. [CrossRef] [PubMed]
26. Yeung, K.; Seitz, T.; Li, S.; Janosch, P.; McFerran, B.; Kaiser, C.; Fee, F.; Katsanakis, K.D.; Rose, D.W.; Mischak, H.; et al. Suppression of Raf-1 kinase activity and MAP kinase signalling by RKIP. *Nature* **1999**, *401*, 173–177. [CrossRef]
27. Al-Mulla, F.; Bitar, M.S.; Taqi, Z.; Rath, O.; Kolch, W. RAF kinase activity signalling by (RKIP) modulates cell cycle kinetics and motility. *Mol. Biosyst.* **2011**, *7*, 928–941. [CrossRef]
28. Voogd, C.; Brian, L.A.; Wang, T.; Allan, A.C.; Varkony-Gasic, E. Three FT and multiple CEN and BFT genes regulate maturity, flowering, and vegetative phenology in kiwifruit. *J. Exp. Bot.* **2017**, *68*, 1539–1553. [CrossRef]
29. Navarro, C.; Abelenda, J.A.; Cruz-Oró, E.; Cuéllar, C.A.; Tamaki, S.; Silva, J.; Shimamoto, K.; Prat, S. Control of Flowering and Storage Organ Formation in Potato by FLOWERING LOCUS T. *Nature* **2011**, *478*, 119–122. [CrossRef]
30. Moraes, T.S.; Dornelas, M.C.; Martinelli, A.P. FT/TFL1: Calibrating Plant Architecture. *Front. Plant Sci.* **2019**, *10*, 97. [CrossRef]
31. Collani, S.; Neumann, M.; Yant, L.; Schmid, M. FT Modulates Genome-wide DNA-Binding of the bZIP Transcription Factor FD. *Plant Physiol.* **2019**, *180*, 367–380. [CrossRef]
32. Purwestri, Y.A.; Susanto, F.A.; Tsuji, H. Hd3a florigen recruits different proteins to reveal its function in plant growth and development. *Intech* **2017**, *4*, 49–67. [CrossRef]
33. Wang, E.; Liu, T.; Sun, X.; Jing, S.; Zhou, T.; Liu, T.; Song, B. Profiling of the Candidate Interacting Proteins of SELF-PRUNING 6A (SP6A) in *Solanum tuberosum*. *Int J. Mol. Sci.* **2022**, *15*, 9126. [CrossRef] [PubMed]
34. Abelenda, J.A.; Bergonzi, S.; Oortwijn, M.; Sonnewald, S.; Du, M.; Visser, R.G.; Bachem, C.W. Source-sink regulation is mediated by interaction of an FT homolog with a SWEET protein in potato. *Curr. Biol.* **2019**, *29*, 1178–1186. [CrossRef] [PubMed]

35. Plantenga, F.D.M.; Bergonzi, S.; Abelenda, J.A.; Bachem, C.W.B.; Visser, R.G.F.; Heuvelink, E.; Marcelis, L.F.M. The tuberization signal StSP6A represses flower bud development in potato. *J. Exp. Bot.* **2019**, *70*, 937–948. [CrossRef]
36. Navarro, C.; Cruz-Oró, E.; Pratt, S. Conserved function of flowering (FT) homologues as signals for storage organ differentiation. *Curr. Opin. Plant Biol.* **2015**, *23*, 45–53. [CrossRef] [PubMed]
37. Hastilestari, B.R.; Lorenz, J.; Reid, S.; Hofman, J.; Pscheidt, D.; Sonnewald, U.; Sonnewald, S. Deciphering source and sink responses of potato plants (*Solanum tuberosum* L.) to elevated temperatures. *Plant Cell Envirom.* **2018**, *41*, 2600–2616. [CrossRef] [PubMed]
38. Lehretz, G.G.; Sonnewald, S.; Hornyik, C.; Corral, J.M.; Sonnewald, U. Post-transcriptional Regulation of FLOWERING LOCUS T Modulates Heat-Dependent Source-Sink Development in Potato. *Curr. Biol.* **2019**, *29*, 1614–1624.e3. [CrossRef] [PubMed]
39. Salvato, F.; Havelund, J.F.; Chen, M.; Rao, R.S.P.; Rogowska-Wrzesinska, A.; Jensen, O.N.; Gang, D.R.; Thelen, J.J.; Møller, I.M. The Potato Tuber Mitochondrial Proteome. *Plant Physiol.* **2014**, *16*, 637–653. [CrossRef] [PubMed]
40. Van Dijk, J.P.; Cankar, K.; Scheffer, S.J.; Beenen, H.G.; Shepherd, L.V.T.; Stewart, D.; Davies, H.V.; Wilkockson, S.J.; Leifert, C.; Gruden, K.; et al. Transcriptome Analysis of Potato Tubers Effects of Different Agricultural Practices. *J. Agric. Food Chem.* **2009**, *57*, 1612–1623. [CrossRef] [PubMed]
41. Jing, S.; Sun, X.; Yu, L.; Wang, E.; Cheng, Z.; Liu, H.; Jiang, P.; Qin, J.; Begum, S.; Song, B. Transcription factor StABI5-like 1 binding to the FLOWERING LOCUS T homologs promotes early maturity in potato. *Plant Physiol.* **2022**, *18*, 1677–1693. [CrossRef]
42. Mhawech, P. 14-3-3 proteins-an update. *Cell Res.* **2005**, *15*, 228–236. [CrossRef]
43. Huang, Y.; Wang, W.; Yu, H.; Peng, J.; Hu, Z.; Chen, L. The role of 14-3-3 proteins in plant growth and response to abiotic stress. *Plant Cell Rep.* **2022**, *41*, 833–852. [CrossRef] [PubMed]
44. Bi, X.; van Esse, W.; Mulki, M.A.; Kirschner, G.; Zhong, J.; Simon, R.; von Korff, M. CENTRORADIALIS Interacts with *FLOWERING LOCUS T*-Like Genes to Control Floret Development and Grain Number. *Plant Physiol.* **2019**, *180*, 1013–1030. [CrossRef] [PubMed]
45. Zhang, X.; Campbell, R.; Ducreux, L.J.M.; Morris, J.; Hedley, P.E.; Mellado-Ortega, E.; Roberts, A.G.; Stephens, J.; Bryan, G.J.; Torrance, L.; et al. TERMINAL FLOWER-1/CENTRORADIALIS inhibits tuberization via protein interaction with the tuberigen activation complex. *Plant J.* **2020**, *103*, 2263–2278. [CrossRef] [PubMed]
46. Morris, W.L.; Alamar, M.C.; Lopez-Cobollo, R.M.; Castillo Cañete, J.; Bennett, M.; Van der Kaay, J.; Stevens, J.; Kumar Sharma, S.; McLean, K.; Thompson, A.J.; et al. A member of the TERMINAL FLOWER 1/CENTRORADIALIS gene family controls sprout growth in potato tubers. *J. Exp. Bot.* **2019**, *70*, 835–843. [CrossRef] [PubMed]
47. Abelenda, J.A.; Cruz-Oró, E.; Franco-Zorrilla, J.M.; Prat, S. Potato StCONSTANS-Like1 Suppresses Storage Organ Formation by Directly Activating the FT-like StSP5G Repressor. *Curr. Biol.* **2016**, *26*, 872–881. [CrossRef]
48. Teo, C.J.; Takahashi, K.; Shimizu, K.; Shimamoto, K.; Taoka, K.I. Potato Tuber Induction is Regulated by Interactions Between Components of a Tuberigen Complex. *Plant Cell Physiol.* **2017**, *58*, 365–374. [CrossRef]
49. Taylor, M.A.; Arif, S.A.M.; Pearce, S.R.; Davies, H.V.; Kumar, A.; George, L.A. Differential Expression and Sequence Analysis of Ribosomal Protein Genes Induced in Stolon Tips of Potato (*Solanum Tuberosum* L.) during the Early Stages of Tuberization. *Plant Physiol.* **1992**, *100*, 1171–1176. [CrossRef]
50. Van Lijsebettens, M.; Vanderhaeghen, R.; De Block, M.; Bauw, G.; Villarroel, R.; Van Montagu, M. An S18 Ribosomal Protein Gene Copy at the Arabidopsis PFL Locus Affects Plant Development by Its Specific Expression in Meristems. *EMBO J.* **1994**, *13*, 3378–3388. [CrossRef]
51. Tsugeki, R.; Kochieva, E.Z.; Fedoroff, N.V. A Transposon Insertion in the Arabidopsis SSR16 Gene Causes an Embryo-Defective Lethal Mutation. *Plant J.* **1996**, *10*, 479–489. [CrossRef]
52. Zinn, A.R.; Ross, J.L. Turner Syndrome and Haploinsufficiency. *Curr. Opin. Genet. Dev.* **1998**, *8*, 322–327. [CrossRef]
53. Lambertsson, A. The Minute Genes in Drosophila and Their Molecular Functions. *Adv. Genet.* **1998**, *38*, 69–134. [CrossRef] [PubMed]
54. Volarević, S.; Stewart, M.J.; Ledermann, B.; Zilberman, F.; Terracciano, L.; Montini, E.; Grompe, M.; Kozma, S.C.; Thomas, G. Proliferation, But Not Growth, Blocked by Conditional Deletion of 40 S Ribosomal Protein S6. *Science* **2000**, *288*, 2045–2047. [CrossRef] [PubMed]
55. Ito, T.; Kim, G.-T.; Shinozaki, K. Disruption of an Arabidopsis Cytoplasmic Ribosomal Protein S13-Homologous Gene by Transposon-Mediated Mutagenesis Causes Aberrant Growth and Development. *Plant J.* **2000**, *22*, 257–264. [CrossRef] [PubMed]
56. Martinez-Seidel, F.; Beine-Golovchuk, O.; Hsieh, Y.-C.; Kopka, J. Systematic Review of Plant Ribosome Heterogeneity and Specialization. *Front. Plant Sci.* **2020**, *11*, 948. [CrossRef]
57. Ferreyra, M.L.F.; Pezza, A.; Biarc, J.; Burlingame, A.L.; Casati, P. Plant L10 Ribosomal Proteins Have Different Roles during Development and Translation under Ultraviolet-B Stress. *Plant Physiol.* **2010**, *153*, 1878–1894. [CrossRef]
58. Wang, J.; Lan, P.; Gao, H.; Zheng, L.; Li, W.; Schmidt, W. Expression Changes of Ribosomal Proteins in Phosphate- and Iron-Deficient Arabidopsis Roots Predict Stress-Specific Alterations in Ribosome Composition. *BMC Genom.* **2013**, *14*, 783. [CrossRef]
59. Cheong, B.E.; Beine-Golovchuk, O.; Gorka, M.; Ho, W.W.H.; Martinez-Seidel, F.; Firmino, A.A.P.; Skirycz, A.; Roessner, U.; Kopka, J. Arabidopsis REI-LIKE Proteins Activate Ribosome Biogenesis during Cold Acclimation. *Sci. Rep.* **2021**, *11*, 2410. [CrossRef]
60. Zhang, C.; Li, H.; Yuan, C.; Liu, S.; Li, M.; Zhu, J.; Lin, X.; Lu, Y.; Guo, X. CKB 1 Regulates Expression of Ribosomal Protein L10 Family Gene and Plays a Role in UV -B Response. *Plant Biol. J.* **2020**, *22*, 143–152. [CrossRef]

61. Moin, M.; Bakshi, A.; Saha, A.; Dutta, M.; Madhav, S.M.; Kirti, P.B. Rice Ribosomal Protein Large Subunit Genes and Their Spatio-Temporal and Stress Regulation. *Front. Plant Sci.* **2016**, *7*, 1284. [CrossRef]
62. Moin, M.; Bakshi, A.; Madhav, M.S.; Kirti, P.B. Expression Profiling of Ribosomal Protein Gene Family in Dehydration Stress Responses and Characterization of Transgenic Rice Plants Overexpressing RPL23A for Water-Use Efficiency and Tolerance to Drought and Salt Stresses. *Front. Chem.* **2017**, *5*, 97. [CrossRef]
63. Moin, M.; Saha, A.; Bakshi, A.; Divya, D.; Madhav, M.S.; Kirti, P.B. Study on Transcriptional Responses and Identification of Ribosomal Protein Genes for Potential Resistance against Brown Planthopper and Gall Midge Pests in Rice. *Curr. Genom.* **2021**, *22*, 98–110. [CrossRef] [PubMed]
64. Gonzali, S.; Loreti, E.; Solfanelli, C.; Novi, G.; Alpi, A.; Perata, P. Identification of Sugar-Modulated Genes and Evidence for in Vivo Sugar Sensing in Arabidopsis. *J. Plant Res.* **2006**, *119*, 115–123. [CrossRef] [PubMed]
65. Baena-González, E.; Rolland, F.; Thevelein, J.M.; Sheen, J. A Central Integrator of Transcription Networks in Plant Stress and Energy Signalling. *Nature* **2007**, *44*, 938–942. [CrossRef] [PubMed]
66. Usadel, B.; Bläsing, O.E.; Gibon, Y.; Retzlaff, K.; Höhne, M.; Günther, M.; Stitt, M. Global Transcript Levels Respond to Small Changes of the Carbon Status during Progressive Exhaustion of Carbohydrates in Arabidopsis Rosettes. *Plant Physiol.* **2008**, *146*, 1834–1861. [CrossRef]
67. Hummel, M.; Cordewener, J.H.G.; de Groot, J.C.M.; Smeekens, S.; America, A.H.P.; Hanson, J. Dynamic Protein Composition of *Arabidopsis Thaliana* Cytosolic Ribosomes in Response to Sucrose Feeding as Revealed by Label Free MSE Proteomics. *Proteomics* **2012**, *12*, 1024–1038. [CrossRef]
68. Berridge, M.V.; Ralph, R.K.; Letham, D.S. The Binding of Kinetin to Plant Ribosomes. *Biochem. J.* **1970**, *119*, 75–84. [CrossRef]
69. Fox, J.E.; Erion, J.L.A. Cytokinin Binding Protein from Higher Plant Ribosomes. *Biochem. Biophys. Res. Commun.* **1975**, *64*, 694–700. [CrossRef]
70. Tepfer, D.A.; Fosket, D.E. Hormone-Mediated Translational Control of Protein Synthesis in Cultured Cells of Glycine Max. *Dev. Biol.* **1978**, *62*, 486–497. [CrossRef]
71. Feierabend, J.; de Boer, J. Comparative Analysis of the Action of Cytokinin and Light on the Formation of Ribulosebisphosphate Carboxylase and Plastid Biogenesis. *Planta* **1978**, *142*, 75–82. [CrossRef]
72. Yakovleva, L.; Kulaeva, O. The effect of phytohormones on phosphorylation of ribosomal proteins in detached pumpkin cotyledons. *Biochem. Physiol. Pflanz.* **1987**, *182*, 359–365. [CrossRef]
73. Cherepneva, G.N.; Schmidt, K.-H.; Kulaeva, O.N.; Oelmüller, R.; Kusnetsov, V.V. Expression of the Ribosomal Proteins S14, S16, L13a and L30 Is Regulated by Cytokinin and Abscisic Acid. *Plant Sci.* **2003**, *165*, 925–932. [CrossRef]
74. Karunadasa, S.S.; Kurepa, J.; Shull, T.E.; Smalle, J.A. Cytokinin-induced Protein Synthesis Suppresses Growth and Osmotic Stress Tolerance. *New Phytol.* **2020**, *227*, 50–64. [CrossRef] [PubMed]
75. Hulm, J.L.; McIntosh, K.B.; Bonham-Smith, P.C. Variation in Transcript Abundance among the Four Members of the *Arabidopsis thaliana* RIBOSOMAL PROTEIN S15a Gene Family. *Plant Science* **2005**, *169*, 267–278. [CrossRef]
76. Chotewutmontri, P.; Stiffler, N.; Watkins, K.P.; Barkan, A. Ribosome Profiling in Maize. In *Maize*; Methods in Molecular, Biology; Lagrimini, L.M., Ed.; Springer: New York, NY, USA, 2018; Volume 1676, pp. 165–183. [CrossRef]
77. Lei, L.; Shi, J.; Chen, J.; Zhang, M.; Sun, S.; Xie, S.; Li, X.; Zeng, B.; Peng, L.; Hauck, A.; et al. Ribosome Profiling Reveals Dynamic Translational Landscape in Maize Seedlings under Drought Stress. *Plant J.* **2015**, *84*, 1206–1218. [CrossRef] [PubMed]
78. Xu, T.; Lei, L.; Shi, J.; Wang, X.; Chen, J.; Xue, M.; Sun, S.; Zhan, B.; Xia, Z.; Jiang, N.; et al. Characterization of Maize Translational Responses to Sugarcane Mosaic Virus Infection. *Virus Res.* **2019**, *259*, 97–107. [CrossRef] [PubMed]
79. Shiraku, M.L.; Magwanga, R.O.; Cai, X.; Kirungu, J.N.; Xu, Y.; Mehari, T.G.; Hou, Y.; Wang, Y.; Agong, S.G.; Peng, R.; et al. Functional Characterization of GhACX3 Gene Reveals Its Significant Role in Enhancing Drought and Salt Stress Tolerance in Cotton. *Front. Plant Sci.* **2021**, *12*, 658755. [CrossRef]
80. Mehdi, R.; Mojtaba, K.; Mojtaba, M.; Sanam, S.C. Identification of Drought-Responsive Proteins of Sensitive and Tolerant Tea (*Camellia sinensis* L.) Clones under Normal and Drought Stress Conditions. *Curr. Prot.* **2020**, *17*, 227–240. [CrossRef]
81. Thapa, M.; Bommakanti, A.; Shamsuzzaman, M.; Gregory, B.; Samsel, L.; Zengel, J.M.; Lindahl, L. Repressed Synthesis of Ribosomal Proteins Generates Protein-Specific Cell Cycle and Morphological Phenotypes. *MBoC.* **2013**, *24*, 3620–3633. [CrossRef]
82. Maitra, N.; He, C.; Blank, H.M.; Tsuchiya, M.; Schilling, B.; Kaeberlein, M.; Aramayo, R.; Kennedy, B.K.; Polymenis, M. Translational Control of One-Carbon Metabolism Underpins Ribosomal Protein Phenotypes in Cell Division and Longevity. *eLife* **2020**, *9*, e53127. [CrossRef]
83. Piper, M.D.; Hong, S.-P.; Ball, G.E.; Dawes, I.W. Regulation of the Balance of One-Carbon Metabolism in *Saccharomyces Cerevisiae*. *J. Biol. Chem.* **2000**, *275*, 30987–30995. [CrossRef]
84. Clare, C.E.; Brassington, A.H.; Kwong, W.Y.; Sinclair, K.D. One-Carbon Metabolism: Linking Nutritional Biochemistry to Epigenetic Programming of Long-Term Development. *Annu. Rev. Anim. Biosci.* **2019**, *7*, 263–287. [CrossRef] [PubMed]
85. Yang, W.; Cortijo, S.; Korsbo, N.; Roszak, P.; Schiessl, K.; Gurzadyan, A.; Wightman, R.; Jönsson, H.; Meyerowitz, E. Molecular mechanism of cytokinin-activated cell division in *Arabidopsis*. *Science* **2021**, *26*, 1350–1355. [CrossRef] [PubMed]
86. Rhee, Y.; Hwang, K.; Cho, S.; Lee, M.; Kil, E.J.; Choi, S.; Hahn, B.S.; Kim, D.; Auh, C.K.; Lee, S. Expression analysis of D-type cyclin in potato (*Solanum tuberosum* L.) under different culture conditions. *Acta Physiol. Plant.* **2016**, *38*, 36. [CrossRef]
87. Lara-Gonzalez, P.; Kim, T.; Oegema, K.; Corbett, K.; Desai, A. A tripartite mechanism catalyzes Mad2-Cdc20 assembly at unattached kinetochores. *Science* **2021**, *374*, 64–67. [CrossRef]

88. Leu, W.M.; Cao, X.L.; Wilson, T.J.; Snustad, D.P.; Chua, N.H. Phytochrome A and phytochrome B mediate the hypocotyl-specific down-regulation of TUB1 by light in arabidopsis. *Plant Cell* **1995**, *7*, 2187–2196. [CrossRef]
89. Demidov, D.; Van Damme, D.; Geelen, D.; Blattner, F.R.; Houben, A. Identification and dynamics of two classes of aurora-like kinases in Arabidopsis and other plants. *Plant Cell* **2005**, *17*, 836–848. [CrossRef]
90. Cai, G.; Cresti, M. Are kinesins required for organelle trafficking in plant cells? *Front. in Plant Sci.* **2012**, *3*, 170. [CrossRef]
91. Weaver, R.L.; Limzerwala, J.F.; Naylor, R.M.; Jeganathan, K.B.; Baker, D.J.; van Deursen, J.M. BubR1 alterations that reinforce mitotic surveillance act against aneuploidy and cancer. *Elife* **2016**, *5*, e16620. [CrossRef]
92. Hemerly, A.; Engler, J.D.A.; Bergounioux, C.; Van Montagu, M.; Engler, G.; Inzé, D.; Ferreira, P. Dominant negative mutants of the Cdc2 kinase uncouple cell division from iterative plant development. *EMBO J.* **1995**, *14*, 3925–3936. [CrossRef]
93. Boruc, J.; Van den Daele, H.; Hollunder, J.; Rombauts, S.; Mylle, E.; Hilson, P.; Russinova, E. Functional modules in the Arabidopsis core cell cycle binary protein–protein interaction network. *Plant Cell* **2010**, *22*, 1264–1280. [CrossRef]
94. Rossignol, P.; Stevens, R.; Perennes, C.; Jasinski, S.; Cella, R.; Tremousaygue, D.; Bergounioux, C. AtE2F-a and AtDP-a, members of the E2F family of transcription factors, induce Arabidopsis leaf cells to re-enter S phase. *Mol. Genet. Genom.* **2002**, *266*, 995–1003. [CrossRef] [PubMed]
95. Chen, H.Z.; Tsai, S.Y.; Leone, G. Emerging roles of E2Fs in cancer: An exit from cell cycle control. *Nat. Rev. Cancer* **2009**, *11*, 785–797. [CrossRef]
96. Schulman, I.; Bloom, K.S. Centromeres: An integrated protein/DNA complex required for chromosome movement. *Annu. Rev. of Cell Biol.* **1991**, *7*, 311–336. [CrossRef] [PubMed]
97. Pinto, I.; Winston, F. Histone H2A is required for normal centromere function in *Saccharomyces cerevisiae*. *EMBO J.* **2002**, *19*, 1598–1612. [CrossRef] [PubMed]
98. Rios-Doria, J.; Velkova, A.; Dapic, V.; Galán-Caridad, J.M.; Dapic, V.; Carvalho, M.A.; Melendez, J.; Monteiro, A.N.A. Ectopic expression of histone H2AX mutants reveals a role for its post-translational modifications. *Cancer Biol. Ther.* **2009**, *8*, 422–434. [CrossRef] [PubMed]
99. Murashige, T.; Skoog, F. A revised medium for rapid growth and bioassays with tobacco tissue cultures. *Physiol. Plant.* **1962**, *15*, 473–497. [CrossRef]
100. Bolger, A.M.; Lohse, M.; Usadel, B. Trimmomatic: A flexible trimmer for Illumina sequence data. *Bioinformatics* **2014**, *30*, 2114–2120. [CrossRef]
101. Dobin, A.; Davis, C.A.; Schlesinger, F.; Drenkow, J.; Zaleski, C.; Jha, S.; Gingeras, T.R. STAR: Ultrafast universal RNA-seq aligner. *Bioinformatics* **2013**, *29*, 15–21. [CrossRef]
102. Liao, Y.; Smyth, G.K.; Shi, W. featureCounts: An efficient general-purpose program for assigning sequence reads to genomic features. *Bioinformatics* **2014**, *30*, 923–930. [CrossRef]
103. Szklarczyk, D.; Franceschini, A.; Wyder, S.; Forslund, K.; Heller, D.; Huerta-Cepas, J.; Von Mering, C. STRING v10: Protein–protein interaction networks, integrated over the tree of life. *Nucleic Acids Res.* **2015**, *43*, D447–D452. [CrossRef]
104. Bombarely, A.; Menda, N.; Tecle, I.Y.; Buels, R.M.; Strickler, S.; Fischer-York, T.; Mueller, L.A. The Sol Genomics Network (solgenomics. net): Growing tomatoes using Perl. *Nucleic Acids Res.* **2010**, *39*, D1149–D1155. [CrossRef] [PubMed]
105. The UniProt Consortium. UniProt. The universal protein knowledgebase in 2021. *Nucleic Acids Res.* **2021**, *49*, D480–D489. [CrossRef] [PubMed]
106. Sherry, S.T.; Ward, M.H.; Kholodov, M.; Baker, J.; Phan, L.; Smigielski, E.M.; Sirotkin, K. dbSNP: The NCBI database of genetic variation. *Nucleic Acids Res.* **2001**, *29*, 308–311. [CrossRef] [PubMed]
107. Livak, K.J.; Schmittgen, T.D. Analysis of relative gene expression data using real-time quantitative PCR and the $2^{-\Delta\Delta CT}$ Method. *Methods* **2001**, *25*, 402–408. [CrossRef]
108. Chen, G.-H.; Liu, M.-J.; Xiong, Y.; Sheen, J.; Wu, S.-H. TOR and RPS6 Transmit Light Signals to Enhance Protein Translation in Deetiolating *Arabidopsis* Seedlings. *Proc. Natl. Acad. Sci. USA* **2018**, *115*, 12823–12828. [CrossRef]
109. Kim, Y.-K.; Kim, S.; Shin, Y.; Hur, Y.-S.; Kim, W.-Y.; Lee, M.-S.; Cheon, C.-I.; Verma, D.P.S. Ribosomal Protein S6, a Target of Rapamycin, Is Involved in the Regulation of RRNA Genes by Possible Epigenetic Changes in Arabidopsis. *J. Biol. Chem.* **2014**, *289*, 3901–3912. [CrossRef]
110. Tzafrir, I. The Arabidopsis SeedGenes Project. *Nucleic Acids Res.* **2003**, *31*, 90–93. [CrossRef]
111. Wang, R.; Zhao, J.; Jia, M.; Xu, N.; Liang, S.; Shao, J.; Qi, Y.; Liu, X.; An, L.; Yu, F. Balance between Cytosolic and Chloroplast Translation Affects Leaf Variegation. *Plant Physiol.* **2018**, *176*, 804–818. [CrossRef]
112. Creff, A.; Sormani, R.; Desnos, T. The Two Arabidopsis RPS6 Genes, Encoding for Cytoplasmic Ribosomal Proteins S6, Are Functionally Equivalent. *Plant Mol. Biol.* **2010**, *73*, 533–546. [CrossRef]
113. Sun, L.; Yu, Y.; Hu, W.; Min, Q.; Kang, H.; Li, Y.; Hong, Y.; Wang, X.; Hong, Y. Ribosomal Protein S6 Kinase1 Coordinates with TOR-Raptor2 to Regulate Thylakoid Membrane Biosynthesis in Rice. *Biochim. Biophys. Acta (BBA) Mol. Cell Biol. Lipids* **2016**, *1861*, 639–649. [CrossRef]
114. Kim, K.-Y. Molecular Cloning of Low-Temperature-Inducible Ribosomal Proteins from Soybean. *J. Exp. Bot.* **2004**, *55*, 1153–1155. [CrossRef] [PubMed]
115. Romani, I.; Tadini, L.; Rossi, F.; Masiero, S.; Pribil, M.; Jahns, P.; Kater, M.; Leister, D.; Pesaresi, P. Versatile Roles of Arabidopsis Plastid Ribosomal Proteins in Plant Growth and Development: Different Functions of Plastid Ribosome Subunits. *Plant J.* **2012**, *72*, 922–934. [CrossRef] [PubMed]

116. Szick-Miranda, K.; Zanial, A.S.; Zanial, A.S.; Abidayo, S.; Slater, K.L.C. Analysis of RPS15aE, an Isoform of a Plant-Specific Evolutionarily Distinct Ribosomal Protein in Arabidopsis Thaliana, Reveals Its Potential Role as a Growth Regulator. *Plant Mol. Biol. Rep.* **2010**, *28*, 239–252. [CrossRef]
117. Lu, C.; Xie, Z.; Yu, F.; Tian, L.; Hao, X.; Wang, X.; Chen, L.; Li, D. Mitochondrial Ribosomal Protein S9M Is Involved in Male Gametogenesis and Seed Development in *Arabidopsis*. *Plant Biol. J.* **2020**, *22*, 655–667. [CrossRef]
118. Ma, Z.; Dooner, H.K. A Mutation in the Nuclear-Encoded Plastid Ribosomal Protein S9 Leads to Early Embryo Lethality in Maize. *Plant J.* **2004**, *37*, 92–103. [CrossRef]
119. Qiu, Z.; Chen, D.; He, L.; Zhang, S.; Yang, Z.; Zhang, Y.; Wang, Z.; Ren, D.; Qian, Q.; Guo, L.; et al. The Rice White Green Leaf 2 Gene Causes Defects in Chloroplast Development and Affects the Plastid Ribosomal Protein S9. *Rice* **2018**, *11*, 39. [CrossRef]
120. Tzafrir, I.; Pena-Muralla, R.; Dickerman, A.; Berg, M.; Rogers, R.; Hutchens, S.; Sweeney, T.C.; McElver, J.; Aux, G.; Patton, D.; et al. Identification of Genes Required for Embryo Development in Arabidopsis. *Plant Physiol.* **2004**, *135*, 1206–1220. [CrossRef]
121. Schultes, N.P.; Sawers, R.J.H.; Brutnell, T.P.; Krueger, R.W. Maize High Chlorophyll Fluorescent 60 Mutation Is Caused by an Ac Disruption of the Gene Encoding the Chloroplast Ribosomal Small Subunit Protein 17. *Plant J.* **2000**, *21*, 317–327. [CrossRef]
122. Chan, Y.L.; Paz, V.; Olvera, J.; Wool, I.G. The Primary Structure of L37—A Rat Ribosomal Protein with a Zinc Finger-like Motif. *Biochem. Biophys. Res. Commun.* **1993**, *192*, 590–596. [CrossRef]
123. Finkelshtein, A.; Khamesa, H.; Tuan, L.A.; Rabanim, M.; Chamovitz, D.A. Overexpression of the Ribosomal S30 Subunit Leads to Indole-3-carbinol Tolerance in *Arabidopsis Thaliana*. *Plant J.* **2021**, *105*, 668–677. [CrossRef]
124. Nagaraj, S.; Senthil-Kumar, M.; Ramu, V.S.; Wang, K.; Mysore, K.S. Plant Ribosomal Proteins, RPL12 and RPL19, Play a Role in Nonhost Disease Resistance against Bacterial Pathogens. *Front. Plant Sci.* **2016**, *6*, 1192. [CrossRef] [PubMed]
125. Zhao, D.; Zhang, C.; Li, Q.; Yang, Q.; Gu, M.; Liu, Q. A Residue Substitution in the Plastid Ribosomal Protein L12/AL1 Produces Defective Plastid Ribosome and Causes Early Seedling Lethality in Rice. *Plant Mol. Biol.* **2016**, *91*, 161–177. [CrossRef] [PubMed]
126. Yang, L.; Xie, C.; Li, W.; Zhang, R.; Jue, D.; Yang, Q. Expression of a Wild Eggplant Ribosomal Protein L13a in Potato Enhances Resistance to Verticillium Dahliae. *Plant Cell Tissue Organ Cult* **2013**, *115*, 329–340. [CrossRef]
127. Song, J.; Wei, X.; Shao, G.; Sheng, Z.; Chen, D.; Liu, C.; Jiao, G.; Xie, L.; Tang, S.; Hu, P. The Rice Nuclear Gene WLP1 Encoding a Chloroplast Ribosome L13 Protein Is Needed for Chloroplast Development in Rice Grown under Low Temperature Conditions. *Plant Mol. Biol.* **2014**, *84*, 301–314. [CrossRef] [PubMed]
128. Lee, J.; Jang, S.; Ryu, S.; Lee, S.; Park, J.; Lee, S.; An, G.; Park, S.K. Mutation of Plastid Ribosomal Protein L13 Results in an Albino Seedling-Lethal Phenotype in Rice. *Plant Breed. Biotechnol.* **2019**, *7*, 395–404. [CrossRef]
129. Zhang, H.; Luo, M.; Day, R.C.; Talbot, M.J.; Ivanova, A.; Ashton, A.R.; Chaudhury, A.M.; Macknight, R.C.; Hrmova, M.; Koltunow, A.M. Developmentally Regulated *HEART STOPPER*, a Mitochondrially Targeted L18 Ribosomal Protein Gene, Is Required for Cell Division, Differentiation, and Seed Development in *Arabidopsis*. *J. Exp. Bot.* **2015**, *66*, 5867–5880. [CrossRef]
130. Yan, H.; Chen, D.; Wang, Y.; Sun, Y.; Zhao, J.; Sun, M.; Peng, X. Ribosomal Protein L18aB Is Required for Both Male Gametophyte Function and Embryo Development in Arabidopsis. *Sci. Rep.* **2016**, *6*, 31195. [CrossRef]
131. Nishimura, T.; Wada, T.; Yamamoto, K.T.; Okada, K. The *Arabidopsis* STV1 Protein, Responsible for Translation Reinitiation, Is Required for Auxin-Mediated Gynoecium Patterning. *Plant Cell* **2005**, *17*, 2940–2953. [CrossRef]
132. Li, S.; Liu, K.; Zhang, S.; Wang, X.; Rogers, K.; Ren, G.; Zhang, C.; Yu, B. STV1, a Ribosomal Protein, Binds Primary MicroRNA Transcripts to Promote Their Interaction with the Processing Complex in *Arabidopsis*. *Proc. Natl. Acad. Sci. USA* **2017**, *114*, 1424–1429. [CrossRef]
133. Park, S.-H.; Chung, M.-S.; Lee, S.; Lee, K.-H.; Kim, C.S. Loss of Ribosomal Protein L24A (RPL24A) Suppresses Proline Accumulation of Arabidopsis Thaliana Ring Zinc Finger 1 (Atrzf1) Mutant in Response to Osmotic Stress. *Biochem. Biophys. Res. Commun.* **2017**, *494*, 499–503. [CrossRef]
134. Szakonyi, D.; Byrne, M.E. Ribosomal Protein L27a Is Required for Growth and Patterning in Arabidopsis Thaliana: Ribosomal Protein-Mediated Patterning. *Plant J.* **2011**, *65*, 269–281. [CrossRef] [PubMed]
135. Zsögön, A.; Szakonyi, D.; Shi, X.; Byrne, M.E. Ribosomal Protein RPL27a Promotes Female Gametophyte Development in a Dose-Dependent Manner. *Plant Physiol.* **2014**, *165*, 1133–1143. [CrossRef] [PubMed]
136. Oristian, D.S.; Sloofman, L.G.; Zhou, X.; Wang, L.; Farach-Carson, M.C.; Kirn-Safran, C.B. Ribosomal Protein L29/HIP Deficiency Delays Osteogenesis and Increases Fragility of Adult Bone in Mice. *J. Orthop. Res.* **2009**, *27*, 28–35. [CrossRef] [PubMed]
137. Aseev, L.V.; Boni, I.V. Extraribosomal Functions of Bacterial Ribosomal Proteins. *Mol. Biol.* **2011**, *45*, 739–750. [CrossRef]
138. Finley, D.; Bartel, B.; Varshavsky, A. The Tails of Ubiquitin Precursors Are Ribosomal Proteins Whose Fusion to Ubiquitin Facilitates Ribosome Biogenesis. *Nature* **1989**, *338*, 394–401. [CrossRef]
139. Chen, J.-G. RACK1 Mediates Multiple Hormone Responsiveness and Developmental Processes in Arabidopsis. *J. Exp. Bot.* **2006**, *57*, 2697–2708. [CrossRef]
140. Guo, J.; Chen, J.-G. RACK1 Genes Regulate Plant Development with Unequal Genetic Redundancy in Arabidopsis. *BMC Plant Biol.* **2008**, *8*, 108. [CrossRef]
141. Cheng, Z.; Li, J.-F.; Niu, Y.; Zhang, X.-C.; Woody, O.Z.; Xiong, Y.; Djonović, S.; Millet, Y.; Bush, J.; McConkey, B.J.; et al. Pathogen-Secreted Proteases Activate a Novel Plant Immune Pathway. *Nature* **2015**, *521*, 213–216. [CrossRef]
142. Kakehi, J.-I.; Kawano, E.; Yoshimoto, K.; Cai, Q.; Imai, A.; Takahashi, T. Mutations in Ribosomal Proteins, RPL4 and RACK1, Suppress the Phenotype of a Thermospermine-Deficient Mutant of Arabidopsis Thaliana. *PLoS ONE* **2015**, *10*, e0117309. [CrossRef]

International Journal of
Molecular Sciences

Article

Integrated Metabolome and Transcriptome Analysis Reveals a Regulatory Network of Fruit Peel Pigmentation in Eggplant (*Solanum melongena* L.)

Xiaohui Zhou, Songyu Liu, Yan Yang, Jun Liu and Yong Zhuang *

Institute of Vegetable Crops, Jiangsu Academy of Agricultural Sciences, Jiangsu Key Laboratory for Horticultural Crop Genetic Improvement, Nanjing 210014, China
* Correspondence: 19960033@jaas.ac.cn; Tel.: +86-25-8439-1752

Abstract: The color of fruit peel is an economically important character of eggplant, and black-purple eggplant has received much attention for being rich in anthocyanin. However, the reason why different fruit peel colors form in eggplant is not well understood. In the present study, an integrative analysis of the metabolome and transcriptome profiles was performed in five eggplant varieties with different fruit colors. A total of 260 flavonoids were identified, and most of them showed significantly higher abundance in black-purple varieties than in other varieties. The transcriptome analysis indicated the activation of early phenylpropanoid biosynthesis genes (*SmPAL*, *SmC4H*, and *Sm4CL*) was more responsible for anthocyanin accumulation, while *SmF3′5′H* was the key factor for the formation of a purple color. Furthermore, two transcription factors, *SmGL2* and *SmGATA26*, were identified as new hub genes associated with anthocyanin accumulation. The silencing of *SmGL2* and *SmGATA26* reduced anthocyanin accumulation in eggplant fruit peels, suggesting the possible involvement of *SmGL2* and *SmGATA26* in regulating anthocyanin biosynthesis. In addition, the pathway of plant hormone signal transduction was significantly enriched, indicating that phytohormones may cooperatively interact to modulate flavonoid biosynthesis. This study provides comprehensive information of flavonoid metabolites and new insights into the regulatory network of fruit coloration, which might be useful for the molecular breeding of eggplant.

Keywords: eggplant; fruit peel color; flavonoids; anthocyanin; metabolome; transcriptome

1. Introduction

Flavonoids, as important secondary metabolites, are widely distributed in plants. They are involved in many biological functions and play crucial roles in plant pigmentation and protection against biotic and abiotic stress [1]. Flavonoids mainly contain seven groups, including flavonols, flavones, flavanones, flavanols, isoflavones, chalcones, and anthocyanins. Meanwhile, various modifications (glycosylation, acylation, and others) and molecular polymerization lead to the formation of a great number of flavonoid compounds [2]. Anthocyanins are the most important water-soluble pigments in horticultural plants and confer a red-to-blue coloration to plant tissues [3]. Moreover, the copigmentation initiated by flavones and flavonols as well as glycosides increases the stability of the colored structural forms of anthocyanins and consequently enhances the final coloration [4,5].

Flavonoids/anthocyanins are synthesized via the phenylpropanoid and flavonoid pathways, which involve at least two types of genes: structural genes and regulatory genes [2]. Structural genes encode the enzymes that directly participate in flavonoid biosynthesis, including phenylalanine ammonia lyase (*PAL*), cinnamate 4-hydroxylase (*C4H*), 4-coumarate-CoA ligase (*4CL*), chalcone synthase (*CHS*), chalcone isomerase (*CHI*), flavanone 3-hydroxyl enzyme (*F3H*), dihydroflavonol reductase (*DFR*), anthocyanin synthase (*ANS*), and flavonoid 3-O-glucosyltransferase (*UFGT*) [6]. Transcription factors have been reported to regulate the expression of structural genes, and the MYB-bHLH-WD40 (MBW) transcriptional complex is the main regulator involved in flavonoid biosynthesis [7]. Moreover, other transcription factors were also proven to participate in regulating anthocyanin biosynthesis, such as WRKY, bZIP, and ERF [8–10].

Eggplant (*Solanum melongena* L.) belongs to the Solanaceae family and is an important vegetable crop worldwide [6,11]. Eggplant fruits are widely consumed due to their generous composition of nutraceuticals [12]. The fruit peel color of eggplant has a wide range and mainly includes white, green, reddish purple, and dark purple, which is one of the primary exterior quality and economic traits. The commercial value of eggplant peel color has strong regional characteristics, and consumers in different regions have different preferences for it, especially in China. The color of eggplant fruit peel is mainly determined by the composition and contents of anthocyanins and chlorophyll [6]. The breeding of the full range of possible fruit colors is valued in diverse world markets [13].

Although many studies have been conducted on the fruit color of eggplant, most of them have mainly focused on purple fruit peel and light-induced anthocyanin biosynthesis [14–16]. The metabolic pathways and candidate genes that contribute to fruit peel coloration are still not fully understood. In addition, there are few reports on the non-anthocyanin flavonoid composition of eggplants with different fruit peel colors. Recently, the combined analyses of the transcriptome and metabolome have been widely used and have become a powerful tool to identify key genes and metabolites related to phenotypic traits, such as color variation and quality formation, in many horticultural plants [17,18]. Thus, the characteristic metabolic components and candidate regulatory genes underlying fruit peel coloration were investigated by integrated metabolome and transcriptome analyses with five different color varieties of eggplant in the present study. The regulatory networks and some new key genes involved in flavonoid synthesis were obtained through a weighted gene coexpression network analysis (WGCNA). The findings not only provide new insights into the molecular network of flavonoid synthesis but also provide valuable information for the future molecular breeding of eggplant.

2. Results

2.1. Variation in Fruit Peel Color and Total Anthocyanin and Chlorophyll Contents

The fruit colors of the five varieties used in this study were representative of the Chinese varieties, with A1 and A3 possessing black-purple fruit peels but different calyx colors, A2 possessing a green peel, A4 possessing a white peel, and A5 possessing a reddish-purple fruit peel (Figure 1A). The lightness values (L^*) for A1 (black-purple), A2 (green), A3 (black-purple), A4 (white), and A5 (reddish-purple) were 24.21, 53.75, 24.61, 89.47, and 33.05, respectively. The black-purple cultivars A1 and A3 had the lowest lightness values. In other words, the fruit peel colors of A1 and A3 were deeper than those of the other varieties. Meanwhile, the redness values (a^*) ranged from −20.54 to 27.23, and A5, with a reddish-purple fruit peel, exhibited the highest redness value. The yellowness parameters (b^*) ranged from −8.55 to 37.88, and A2, with a green peel, possessed the highest yellowness value (Figure 1B).

The total anthocyanin and chlorophyll contents were also measured for the five varieties. The total anthocyanin content was highest in A1, followed by A3 and A5, while the total contents of anthocyanin in A2 and A4 were significantly lower (Figure 1C). The total chlorophyll content of A2 was slightly higher than those of A1 and A3. Very low amounts of total chlorophyll were detected in A4 and A5 (Figure 1D). Both total the anthocyanin and chlorophyll contents were lowest in A4. In general, the color values and pigment contents were in line with the fruit peel color variation.

Figure 1. Phenotypes of the five eggplant varieties with different fruit colors. (**A**) Morphologies of the fruits of the five eggplant varieties; (**B**) Color parameters of fruit peels of the five eggplant varieties; (**C**) Total anthocyanin contents of the five eggplant varieties; (**D**) Total chlorophyll contents of the five eggplant varieties. The different letters on top of each bar indicate significant differences among the five eggplant varieties at $p < 0.05$.

2.2. Flavonoid Metabolic Differences

A widely targeted metabolite analysis was performed for the comprehensive profiling of flavonoids in colored eggplant fruit based on LC-MS/MS. A total of 260 flavonoid metabolites were identified, including 32 anthocyanins, 10 chalcones, 5 flavanols, 25 flavnones, 11 flavanonols, 58 flavones, 11 flavonoid carbonoside, 86 flavonols, 10 isoflavones, proanthocyanidin A6, and 11 tannins. A hierarchical cluster analysis (HCA) was carried out to demonstrate the phenotypic variations in terms of the relative contents of the five eggplant cultivars (Figure 2A).

A total of 32 anthocyanins were identified, including 10 delphinidin glycosides, 14 cyanidin glycosides, 3 petunidin glycosides, 2 malvidin glycosides, 2 peonidin glycosides, and 1 pelargonidin glycoside. The relative contents of the identified anthocyanins varied greatly in the five eggplant cultivars (Figure 2B). Delphinidin glycosides, petunidin glycosides, and malvidin glycosides were not detected in A2 and A4 with green and white fruit peels. Among the identified anthocyanins in the reddish to dark purple group (A1, A3, and A5), del-3-O-rutinoside (D3R), del-3-O-(2-O-p-coumaroyl)rutinoside-7-O-glu, and del-3-O-(2-O-p-coumaroyl)rutinoside-5-O-glu (nasunin) were present at the highest levels in A1 and A5. Interestingly, D3R and nasunin, which were previously reported to be the most representative anthocyanin compounds in purple eggplant, were not detected in A3, whereas the levels of del-3-O-(6-O-p-coumaroyl)glu and cya-3-O-(6-O-p-coumaroyl)glu were the highest in A3. In general, most of the identified anthocyanins were more abundant in A1, followed by A3 and A5.

In this study, remarkable differences in flavonoid metabolites were observed among the five varieties. Various flavonoid glycosides possessing rhamnose, galactose, rutinose, and glucose as major sugar ligands were detected. Kaempferol, quercetin, myricetin, luteolin, and their derivatives were the major flavonols and flavones present in the eggplants. The majority of the flavonoids accumulated at the lowest level in A4, followed by A2. As for the reddish to black-purple group (A1, A3, and A5), A1 and A5 seemed to have more similar flavonoid accumulation patterns compared to A3. Typically, most of the chalcones, tannins, and flavonols were more abundant in A3 than in A1 and A5.

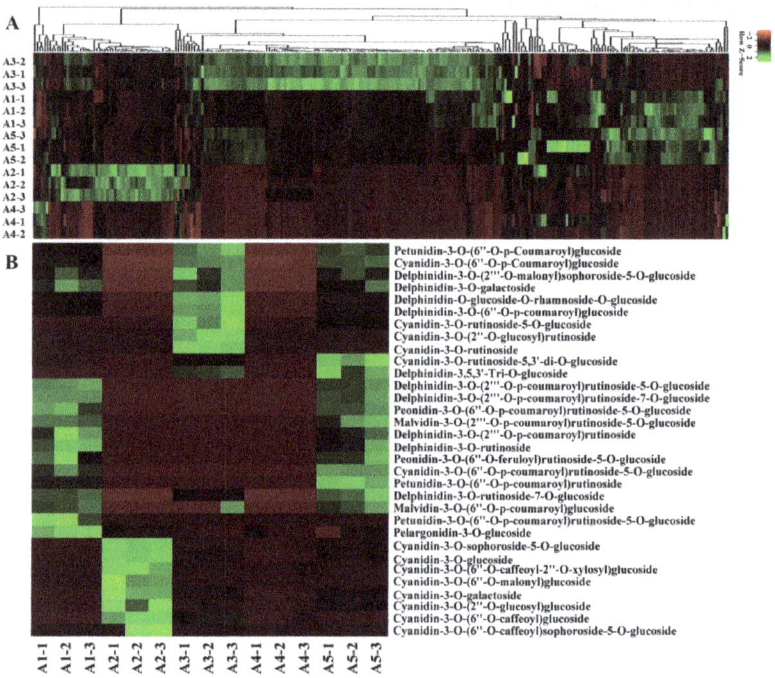

Figure 2. Metabolome analyses of different-colored eggplant fruits. (**A**) Hierarchical clustering of flavonoid metabolites in the five eggplant varieties; (**B**) The anthocyanins detected in the five eggplant varieties.

A principal component analysis was also used to illustrate the internal structures of multiple variables in the five varieties (Figure 3). In the PCA plot, the quality control (QC) samples were grouped together, indicating the stability and repeatability of our analysis. Consistent with the HCA results, A1 and A5 showed the closet distance in the plot, indicating similarities in their metabolic accumulation pattern. Although A1 and A5 gathered near each other, the five varieties could be distinguished easily.

2.3. *Comparison of Differentially Accumulated Metabolites among the Five Eggplant Varieties*

Score plots of the orthogonal partial least-squares discriminant analysis (OPLS-DA) were created to analyze the differences between pairs of the five varieties. High predictability (Q2 > 0.9) and strong goodness of fit (R2X and R2Y) of the OPLS-DA models were observed in pairwise comparisons among the five varieties (Figure 4 and Supplementary Figure S1), suggesting the reliability of the metabolic data. Although in the PCA plots A1and A5 gathered near each other, in the OPLS-DA models, A1, A2, A3, A4, and A5 were clearly separated from each other, suggesting major distinctions in the flavonoid metabolic profiles between the different peel color groups.

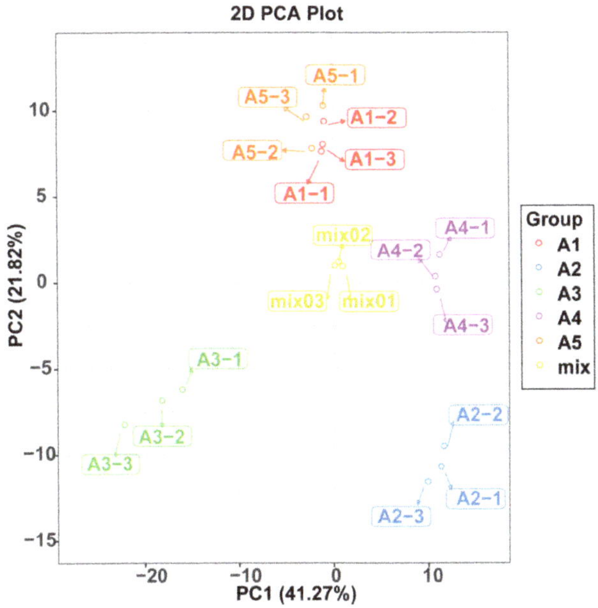

Figure 3. PCA score plot of flavonoid profiles from the five eggplant varieties.

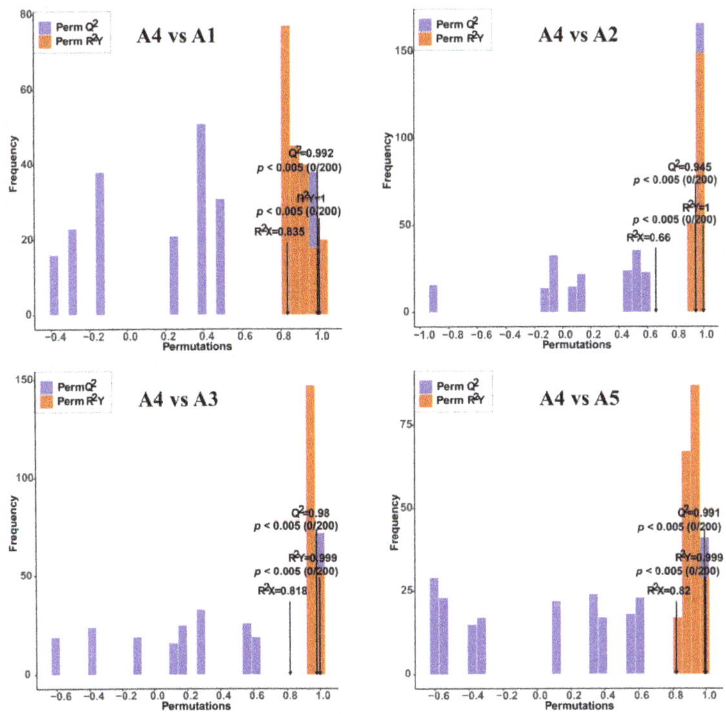

Figure 4. Orthogonal partial least-squares discriminant analysis (OPLS-DA) model of comparison groups, including A4 vs. A1, A4 vs. A2, A4 vs. A3, and A4 vs. A5.

The differentially accumulated metabolites (DAMs) were further identified in pairwise comparisons using the criteria of a fold change ≥2 or ≤0.5 and a VIP (variable importance in project) ≥1. Compared with A4, possessing a white fruit peel color, a total of 156, 101, 145, and 150 DAMs were identified in A1 (black purple), A2 (green), A3 (black purple), and A5 (reddish purple), respectively (Figure 5A). As expected, most of the DAMs were upregulated in the other four varieties as the peel color changed from white to black purple. A total of 48 DAMs were observed among the comparison groups A4 vs. A1, A4 vs. A2, A4 vs. A3, and A4 vs. A5 (Figure 5B).

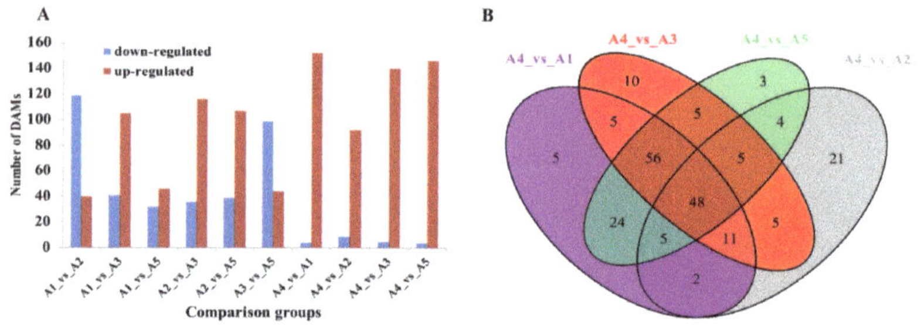

Figure 5. Differential flavonoid analysis among the five eggplant varieties. (**A**) Number of differentially accumulated flavonoids among the five eggplant varieties; (**B**) Venn diagram showing the overlapping and cultivar-specific differential metabolites among the comparison groups, including A4 vs. A1, A4 vs. A2, A4 vs. A3, and A4 vs. A5.

For the reddish to black-purple group (A1 vs. A3, A1 vs. A5, and A3 vs. A5), the lowest number of DAMs was found for A1 vs. A5 (46 upregulated and 32 downregulated DAMs in A5). Compared with A3, a total of 146 DAMs (105 downregulated and 41 upregulated) as well as 143 DAMs (99 downregulated and 44 upregulated) were observed in A1 and A5, respectively (Figure 5A). Most of the DAMs, including chalcones, flavanonols, flavonols, and tannins, were upregulated in A3.

The differential metabolites among the five varieties were mapped to the Kyoto Encyclopedia of Genes and Genomes (KEGG) database, and a KEGG enrichment analysis was performed for the DAMs of each comparison group. The most enriched pathways detected for all compared groups were flavonoid biosynthesis, anthocyanin biosynthesis, flavone and flavonol biosynthesis, isoflavonoid biosynthesis, and the biosynthesis of secondary metabolites.

2.4. Comparative Transcriptome of Fruit Peels among the Five Eggplant Varieties

A total of fifteen libraries were established using the five varieties of samples with three biological replicates for each variety. After removing the adaptor and low-quality sequences, a total of 20.66~21.4 Gb of clean reads were obtained for each variety. The Q20 and Q30 values for each library were equal to or greater than 97.43 and 92.92%, respectively (Table 1). These clean reads were mapped to the reference genome with match ratios in the range of 96.64~97.91 %. Differentially expressed genes (DEGs) were identified with an expression fold change of $\log_2$ ratio ≥ 1 and a false discovery rate ≤ 0.01 as the thresholds, and a total of 10110 DEGs were obtained among all samples (Supplementary Table S1). There were 4265, 3987, 4005, 4253, 3837, 3629, 4456, 4095, 2881, and 2152 DEGs in the comparison groups, respectively, including A1 vs. A2, A1 vs. A3, A1 vs. A5, A2 vs. A3, A2 vs. A5, A3 vs. A5, A4 vs. A1, A4 vs. A2, A4 vs. A3, and A4 vs. A5 (Table 2). Among these comparison groups, the A4 vs. A5 group had the smallest number of upregulated and downregulated DEGs.

Table 1. Overview of the transcriptome sequencing data and quality check.

Samples	Clean Reads	Reads Mapped	Q20 (%)	Q30 (%)	GC Content (%)
A1–1	47409416	45993726 (97.01%)	97.69	93.49	42.23
A1–2	46685604	45365789 (97.17%)	97.77	93.64	42.46
A1–3	48592694	47236677 (97.21%)	97.8	93.68	42.69
A2–1	48679876	47284281 (97.13%)	97.75	93.59	42.33
A2–2	43769020	42395918 (96.86%)	97.67	93.51	41.91
A2–3	49045732	47596182 (97.04)	97.68	93.41	42.18
A3–1	46608676	45126433 (96.82)	97.76	93.63	41.66
A3–2	45840606	44446440 (96.96%)	97.8	93.73	41.9
A3–3	49665704	47998296 (96.64%)	97.68	93.5	42.34
A4–1	47772226	46365463 (97.06%)	97.79	93.67	42.50
A4–2	42518600	41148256 (96.78%)	97.43	92.92	42.12
A4–3	49236290	47743262 (96.97%)	97.77	93.64	42.22
A5–1	46807752	45518684 (97.25%)	97.88	93.9	42.42
A5–2	47101278	46077478 (97.83%)	98.15	94.37	42.34
A5–3	43808476	42891632 (97.91%)	98.23	94.57	42.44

Table 2. The number of DEGs identified in the pairwise comparisons.

Comparison Groups	Total DEGs	Downregulated DEGs	Upregulated DEGs
A1 vs. A2	4265	1936	2329
A1 vs. A3	3987	2236	1751
A1 vs. A5	4005	2431	1574
A2 vs. A3	4253	2507	1746
A2 vs. A5	3837	2540	1297
A3 vs. A5	3629	2087	1542
A4 vs. A1	4456	1679	2777
A4 vs. A2	4095	1314	2781
A4 vs. A3	2881	1176	1705
A4 vs. A5	2152	1146	1006

A GO enrichment analysis of these DEGs showed that the genes related to "cell", "cell part", and "organelle" were predominant in the cellular component category. In the molecular function category, the most enriched terms were binding and catalytic activity. In the biological process category, the genes were mostly involved in metabolic and cellular processes. For the KEGG annotation results, the DEGs among all samples were mapped to 141 KEGG pathways, and the KEGG pathways of the biosynthesis of secondary metabolites, plant hormone signal transduction, phenylpropanoid biosynthesis, flavonoid biosynthesis, anthocyanin biosynthesis, photosynthesis, photosynthesis-antenna proteins, and porphyrin and chlorophyll metabolism were significantly enriched.

2.5. Identification of DEGs Involved in Flavonoid Biosynthesis

To explore the mechanism of accumulating flavonoids in the different fruit peel colors of eggplant, the gene expression patterns involved in the phenylpropanoid and flavonoid pathways were analyzed. As expected, most of the structural genes involved in flavonoid biosynthesis, including *PAL*, *C4H*, *4CL*, *CHS*, *CHI*, *F3H*, *DFR*, *ANS*, and *UFGT* had significantly lower levels of transcription in A2 and A4, which may be one of the reasons for lower flavonoids in A2 and A4 (Figure 6). *F3. wH* was found to have no or very weak expression levels in A2 and A4, suggesting that *F3thaH* may be a critical factor that directly determines the purple color of fruit peels in eggplant.

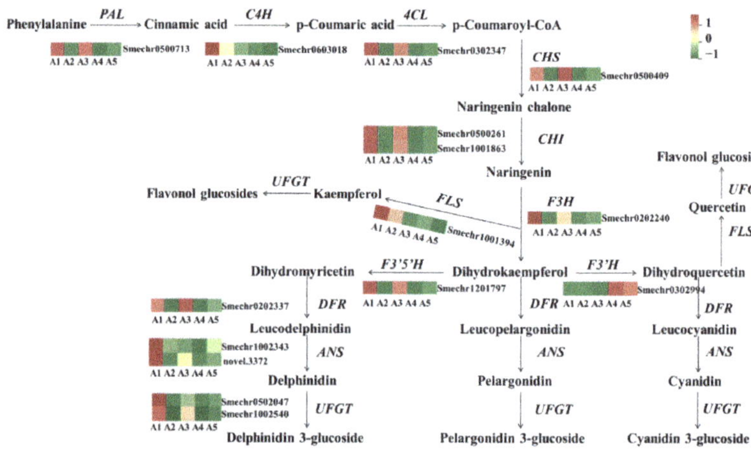

Figure 6. The expression profiles of genes involved in the anthocyanin/flavonoid pathway. PAL, phenylalanine ammonia lyase; C4H, cinnamate 4-hydroxylase; 4CL, 4-coumarate-CoA ligase; CHS, chalcone synthase; CHI, chalcone isomerase; F3H, flavanone 3-hydroxyl enzyme; FLS, flavonol synthase; DFR, dihydroflavonol reductase; ANS, anthocyanin synthase; UFGT, flavonoid 3-O-glucosyltransferase.

The structural genes involved in the flavonoid biosynthesis pathway were also differently expressed among the comparison groups of A1 vs. A3, A1 vs. A5, and A3 vs. A5. In the comparisons of A1 vs. A5 and A3 vs. A5, both early and late biosynthesis genes were downregulated in A5, including *PAL*, *4CL*, *C4H*, *CHS*, *CHI*, *F3H*, *F3 3HH*, *DFR*, *ANS*, and *UFGT*. Notably, in the A1 vs. A3 comparison, the early biosynthesis genes, including *CHS*, *CHI*, *F3H*, and *F3nd H*, were not differently expressed, and most of the DEGs were late biosynthesis genes, such as *DFR*, *ANS*, and anthocyanin-modified genes (Smechr0502047, Smechr0400421, and Smechr1102789).

In general, most of the structural genes involved in flavonoid biosynthesis were highly expressed in A1 and A3 with black-purple fruit peels. With decreasing purple color, the expression of most annotated flavonoid genes decreased in A5, while the lowest expression levels were found in A2 and A4.

2.6. Identification of DEGs Involved in Chlorophyll Metabolism

The green color of eggplant fruit peels is mainly affected by chlorophylls. In this study, a total of 27 DEGs related to chlorophyll metabolism were identified (Figure 7). Compared to A2 with green fruit peels, most of the DEGs showed lower expression in A4 and A5, including *HEMA* (Smechr0402030 and Smechr0100518), *HEMC* (Smechr0702754), *HEME* (Smechr0400257 and Smechr0601135), *HEMF* (Smechr0400064), *HEMG* (Smechr0102139), *CHLH* (Smechr0400430), *CHLM* (Smechr0303264), *CRD* (Smechr1002947), *POR* (Smechr0702049), *DVR* (Smechr0100030), *CAO* (Smechr0601739 and Smechr1200635), *NOL* (Smechr0701042), and *HCAR* (Smechr0902459).

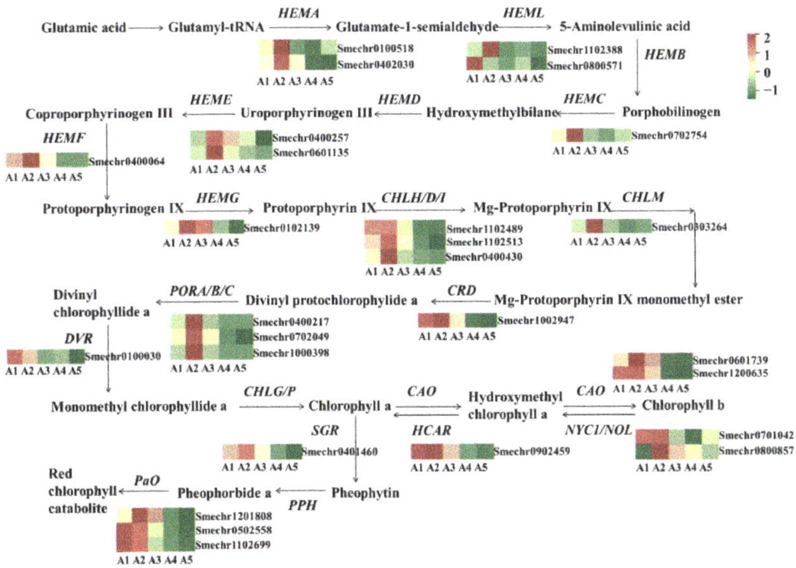

Figure 7. Differential expression of genes related to the chlorophyll metabolism pathway. HEMA, Glutamyl-RNA reductase; HEMB, Delta-aminolevulinic acid dehydratase; HEMC, Porphobilinogen deaminase; HEME, Uroporphyrinogen decarboxylase; HEMF, Coproporphyrinogen-III oxidase; HEMG, Protoporphyrinogen III oxidase; CHLH/D/I, Magnesium-chelatase subunit ChlH/ChlD/ChlI; CHLM, Magnesium protoporphyrin IX methyltransferase; CRD, Magnesium-protoporphyrin IX monomethyl ester cyclase; DVR, Diviyl chlorophyllide a 8-vinyl-reductase; POR, Protochlorophyllide oxidoreductase; CHLG, Chlorophyll synthase; NYC1/NOL: Chlorophyll (ide) b reductase; CAO, chlorophyllide a oxygenase; HCAR, 7-hydroxymethyl chlorophyll a reductase; SGR, Stay-green; PPH, Pheophytinase; PaO, Pheophorbide a oxygenase; RCCR, Red Chl catabolite reductase.

As for the comparison of A2 vs. A1, only four DEGs related to chlorophyll biosynthesis were detected, while in A2 vs. A3 there were eight DEGs exhibiting higher expression levels in A2, including *HEMA*, *HEML*, *HEMC*, *CHLH*, *CHLM*, and *POR*. Notably, except for A2 vs. A1, the pathway of photosynthesis-antenna proteins was also enriched, based on a KEGG analysis of the comparison of A2 vs. A3, A2 vs. A4, and A2 vs. A5. A total of 15 DEGs related to the light-harvesting chlorophyll protein complex (LHC), including 7 DEGs related to LHCI and 8 DEGs related to LHCII, exhibited declined expression in A4 and A5 (Supplementary Table S2).

2.7. Identification of DEGs Involved in Hormone-Mediated Signaling Pathways

The KEGG enrichment analysis showed that the plant hormone signal transduction pathway was one of the most significantly enriched pathways. Therefore, to better investigate hormonal regulation in fruit peel color variation, the DEGs enriched in the signaling pathways of auxin, gibberellin, ethylene, brassinosteroid (BR), abscisic acid (ABA), salicylic acid (SA), and jasmonic acid (JA) were analyzed (Supplementary Figure S2).

In the auxin signaling pathway, the expression levels of five DEGs were correlated with color variation, including *AUX/IAA* (Smechr0303465), *ARF* (Smechr0101326), *GH3* (Smechr0400187), and *SAUR* (Smechr0100144 and Smechr1100127). In the gibberellin signaling pathway, the expression of *GID1* (Smechr1000150) and *DELLA* (Smechr0500162 and Smechr0500354) exhibited positive correlations with fruit peel color variation. Similarly, the expression of *ETR* (Smechr0902323) and *EBF1/2* (Smechr0700262 and Smechr1200241), which are involved in the ethylene signaling pathway, as well as *BRI1* (Smechr0702406) and *BKI1* (Smechr0502339), which are involved in the BR signaling pathway, were associated with color variation among the five varieties. *ABF* (Smechr0100651) and *SnRK2* (Smechr0401004 and Smechr0802058) in the ABA signaling pathway showed higher expression levels in A1 and A3 with black-purple fruit skins. Meanwhile, the expression of *TGA* (Smechr1002700) in the SA signaling pathway was correlated with color variation among the five varieties. Specifically, two bHLH transcription factors involved in the JA signaling pathway had different expression levels among the five varieties, of which *SmTT8* (Smechr0901701) was also involved in flavonoid biosynthesis.

2.8. Identification of Coexpressed Gene Networks and Key Candidates

To investigate the gene regulatory network of peel coloration of eggplant fruit, a weighted gene coexpression network analysis (WGCNA) using 6241 non-redundant DEGs was performed, and 20 distinct modules that were labeled in different colors were obtained (Figure 8). The total anthocyanin contents and total chlorophyll contents as well as delphinidin-type anthocyanins were used as phenotypic data for the analysis. The analysis revealed that the purple and violet modules had positive correlations with the anthocyanin contents and chlorophyll contents, respectively (Figure 9A). The KEGG enrichment analysis of genes grouped into different modules showed that genes in the purple module were enriched in flavonoid biosynthesis, while genes in the violet module were enriched in photosynthesis-antenna proteins and porphyrin and chlorophyll metabolism.

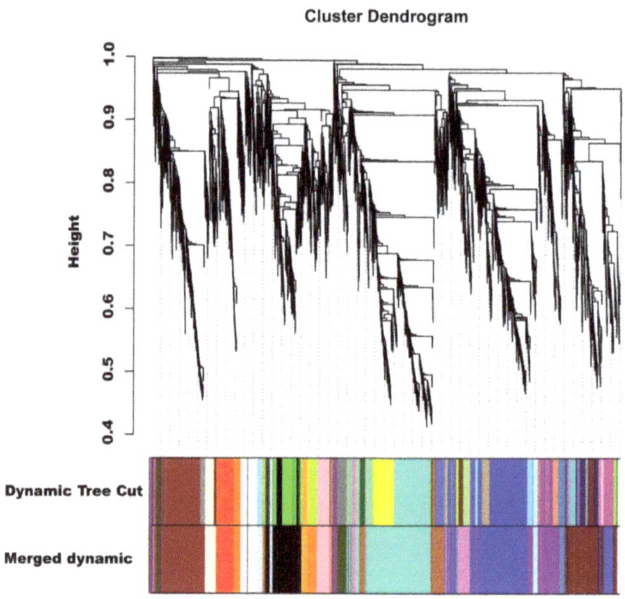

Figure 8. Hierarchical clustering presenting twenty modules with coexpressed genes.

In the correlation network of the purple module, four transcription factors, including *SmGL2* (Smechr0303487), *SmGATA26* (Smechr0601814), *SmWRKY44* (Smechr1002420), and *SmTT8* (Smechr0901701) as well as genes involved in the flavonoid biosynthesis pathway, such as *SmF3mFtH* (Smechr1201797), *SmUFGT* (Smechr1002540), *SmPAL* (Smechr0500713), and *Sm4CL* (Smechr0302347), which had the highest degree of connectivity, were identified as hub genes (Table 3, Figure 9B). SmWRKY44 and SmTT8 were previously reported to be involved in anthocyanin biosynthesis. Thus, to further investigate the function of SmGL2 and SmGATA26, the TRV2-*SmGL2*, TRV2-*SmGATA26*, and TRV2-*SmWRKY44* recombinants were constructed to silence *SmGL2*, *SmGATA26*, and *SmWRKY44* in the purple peels of eggplants. Compared to TRV2, the *SmGL2*-silenced, *SmGATA26*-silenced, and *SmWRKY44*-silenced eggplants had decreased anthocyanin accumulation in the fruit peels, suggesting that SmGL2 and SmGATA26 may be involved in anthocyanin biosynthesis (Figure 9C).

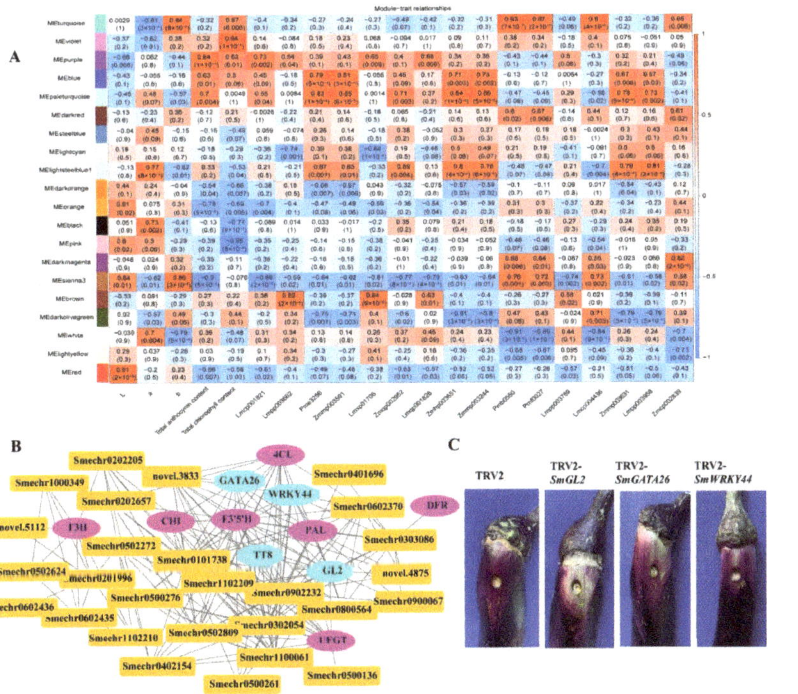

Figure 9. Gene network and key candidate genes involved in anthocyanin biosynthesis, identified by WGCNA. (**A**) Module–trait associations based on Pearson correlations, (**B**) Gene network of the hub genes in the purple module, which were positively correlated with the total anthocyanin content (r^2 = 0.84, $p = 1 \times 10^{-4}$); (**C**) The effect of *SmGL2*, *SmGATA26*, and *SmWRKY44* silencing in VIGS eggplant fruits.

Table 3. The hub genes detected in the purple and violet modules.

Module	Gene	Annotation	Connectivity
Purple	Smechr0303487	Homobox-leucine zipper protein GLABRA2	164.2872382
Purple	Smechr1201797	Flavonoid 3′5′-hydroxylase	163.1378627
Purple	Smechr0302054	Uncharacterized protein	163.0477793
Purple	Smechr0601814	GATA transcription factor 26	162.7337461
Purple	Smechr1002540	Anthocyanin 3-O-glucosyltransferase	159.0606892
Purple	Smechr0500713	Phenylalanine ammonia-lyase 2	157.7681631
Purple	Smechr1002420	WRKY transcription factor 44	157.7222576
Purple	Smechr1102209	Kirola	157.0941481
Purple	Smechr0901701	Anthocyanin-related transcription factor TT8	156.9502973
Purple	Smechr0302347	4-coumarate-CoA ligase 2	156.3548796
Violet	Smechr1000379	Chlorophyll a-b binding protein	103.30002
Violet	Smechr0902639	30S ribosomal protein S1	102.53807
Violet	Smechr0901976	Uncharacterized protein	101.83604
Violet	Smechr0502558	Protein TIC55, chloroplastic	101.50027

Table 3. *Cont.*

Module	Gene	Annotation	Connectivity
Violet	Smechr1100719	ABC transporter F family member 5	98.330714
Violet	Smechr0900664	Chlorophyll a-b binding protein	97.539107
Violet	Smechr1001174	Uncharacterized protein	94.912891
Violet	Smechr1200635	Chlorophyllide a oxygenase	90.176381

In the correlation network of the violet module, two genes belonged to the LHCB family, including *SmLhcb3* (Smechr1000379) and *SmLhcb4* (Smechr0900664) as well as *SmTIC55* (Smechr0502558) and *SmCAO* (Smechr1200635), which are involved in porphyrin and chlorophyll metabolism, had the highest degree of connectivity and were identified as hub genes (Supplementary Figure S3).

2.9. Validation of Differentially Expressed Genes by Quantitative Real-Time PCR

To validate the accuracy of the RNA-Seq data, a total of 16 differentially expressed genes, including 8 hub genes, were selected for a quantitative real-time PCR (qRT-PCR) analysis. As illustrated in Figure 10, the qRT-PCR results showed that the expression patterns of these genes were consistent with the RNA-seq results, indicating that the transcriptome sequencing results were reliable.

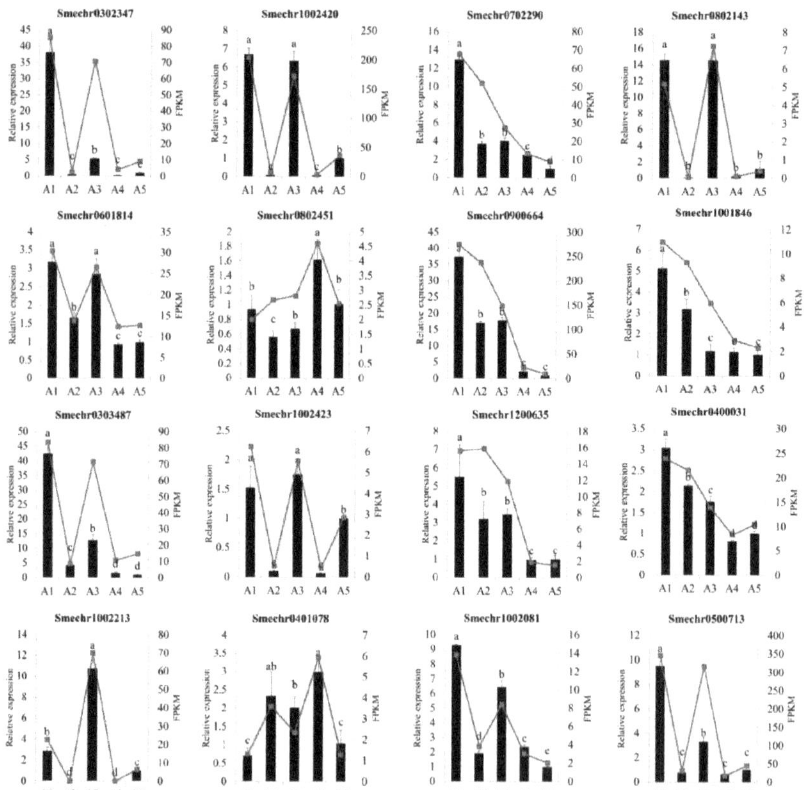

Figure 10. Quantitative real-time PCR analysis of the sixteen DEGs among the five eggplant cultivars. The different letters on top of each bar indicate significant differences among the five eggplant varieties at $p < 0.05$.

3. Discussion

3.1. Pigment Accumulation in Different Colors of Eggplant Fruit Peels

Anthocyanins were recognized as the major pigments in red to black-purple eggplant fruit peels. Previous studies suggested that delphinidin-3-O-(2-O-p-coumaroyl)rutinoside-5-O-glucoside (nasunin) and delphinidin-3-O-rutinoside (D3R) were the main types of anthocyanins in the Japanese-type and non-Japanese-type eggplants, respectively [19,20]. In our study, delphinidin and cyanidin derivatives were the major detected anthocyanins, and delphinidin was the most abundant. Cyanidin derivatives, but not delphinidin derivatives, were identified in A2 and A4, which supported that delphindin-type anthocyanins were more responsible for the red to black-purple pigmentation of eggplant fruit peels. Interestingly, both nasunin and D3R were highly accumulated in A1 and A5 but not in A3, suggesting that the types of anthocyanins are genotype-dependent. This variation may be due to the different origins and genetic backgrounds, as A3 belongs to the European type of eggplant, exhibiting more adaption to low temperatures, while A1 and A5 belong to the Chinese type, which have better fruit taste.

Previous studies have suggested that flavonoid metabolites, including tannins, flavones, and flavonols, are important copigment factors, contributing to the color of most flowers, fruits, and seeds [4,5]. The hydrophobic interactions between anthocyanin and other organic components form molecular associations and stabilize their final color appearance [21]. In this study, higher levels of tannins and flavonol glycosides were detected in the reddish to black-purple group, especially in A3. Despite no significant differences being found in the appearance of fruit peels and the total anthocyanin contents between A1 and A3, big differences in the compositions and contents of flavonoid metabolites still existed, suggesting that the pigment accumulation in eggplant fruit peels is a complex process and that the final purple color appearance can be attributed to a combination of multiple factors.

It has been reported that the presence of both anthocyanins and chlorophylls resulted in a black-purple fruit color, and dark purple fruits had higher levels of chlorophyll than dark-green ones [19]. However, the total content of chlorophyll of A2 was slightly higher than those of A1 and A3, but the total chlorophyll contents of A1, A2, and A3 did not vary significantly. The presence of chlorophylls may have a minor darkening effect on the black-purple appearance [6].

3.2. The Activation of Early Phenylpropanoid/Flavonoid Biosynthesis Genes for Anthocyanin Accumulation in Fruit Peels

Anthocyanins are produced by a branch of the phenylpropanoid and flavonoid pathway, and differences in the expression pattern of genes involved in phenylpropanoid/flavonoid biosynthesis result in diverse anthocyanin profiles [22]. Our study provides a comprehensive profile of the flavonoid biosynthesis pathway of eggplant fruit peels with different colors. The results revealed by the WGCNA showed that several structural genes, including *SmPAL*, *Sm4CL*, *SmF3mFCH*, and *SmUFGT* were identified as hub genes associated with anthocyanin accumulation. In the initial steps of the flavonoid pathway, enzymes encoded by *PAL*, *C4H*, and *4CL* sequentially catalyze the metabolism of phenylalanine to coumaroyl-CoA, which provides the precursor compounds for the synthesis of flavonoids [2]. Considering that all of the structural genes involved in anthocyanin biosynthesis had higher expression in A1 and A3, and the higher expression of *SmPAL*, *SmC4H*, and *Sm4CL* provided sufficient precursor compounds for the synthesis of flavonoids, we concluded that the activation of early phenylpropanoid/flavonoid biosynthesis genes was more responsible for anthocyanin accumulation in eggplant.

F3′5′H is an important enzyme that catalyzes the hydroxylation of flavonoids at C3′ and C5′ of ring B, resulting in the production of delphinidin-based anthocyanins [23]. In this study, undetectable or very low expression of *SmF3mr H* were observed in A2 and A4 with green and white fruit peel colors, but high expression was found in A1 and A3, followed by A5. Thus, we concluded that *SmF3mFuH* was the key factor for the purple color formation of fruit peels at the transcriptional level. This was consistent with a recent report by Yang et al. [24]. Another key gene was anthocyanin 3-O glucose transferase (*UFGT*), which is the key enzyme for anthocyanin stability and water solubility [2]. Previous studies showed that *UFGT* was the key gene involved in the accumulation of anthocyanins in jujube fruit peels [25,26]. *UFGT* was also found to be present but not expressed in white skin apples [27]. Their findings were in line with our results that the different expression levels of *SmUFGT* coincided with the different anthocyanin accumulations and colorations among the five eggplant cultivars.

3.3. Candidate Transcription Factors Modulated Anthocyanin Accumulation in Eggplant Fruit Peels

The MYB-bHLH-WD40 transcriptional complex is considered to be an important and conserved model in the regulation of anthocyanin biosynthesis [7]. In eggplant, previous studies have reported that several MYB and bHLH TFs, including SmMYB113, SmMYB86, SmMYB75, SmMYB35, SmbHLH1, SmbHLH13, and SmTT8, are involved in anthocyanin biosynthesis, of which SmMYB113 and SmTT8 were proven to be the key factors responsible for anthocyanin accumulation [24,28–33]. In our study, very low transcription levels of *SmMYB113* were detected in A2 and A4, and compared to the reddish to black-purple group (A1, A3, and A5), a 6 bp deletion in the first exon of *SmMYB113* was found in white and green fruit peels of eggplants (Supplementary Figure S4). A previous report on grapes revealed that a retrotransposon-induced mutation in *VvmybA1* is associated with the loss of pigmentation in white cultivars [34]. Whether the different gene expression levels or the difference in sequence variation in *SmMYB113* affects anthocyanin accumulation in white and green eggplant fruit peels needs further investigation. Interestingly, although SmMYB113 and SmTT8 were proven to participate in the regulation of anthocyanin biosynthesis, only the transcript levels of *SmTT8* correlated with those of anthocyanin biosynthesis structural genes among different-colored fruits. Additionally, four transcription factors, including *SmGL2*, *SmGATA26*, *SmWRKY44*, and *SmTT8*, were also considered as hub genes associated with anthocyanin accumulation based on the WGCNA analysis. Thus, we speculated that SmTT8 is the backbone of the anthocyanin biosynthetic regulation network. This finding was in line with a previous study on mulberry fruit colors, which showed that bHLH3 is indispensable for the regulation of the flavonoid pathway [3]. However, the reason for the differences in the expression levels of *SmTT8* among different-colored fruits remains unknown.

In addition to the MYB-bHLH-WD40 regulatory complex, which is involved in anthocyanin biosynthesis, many transcription factors, such as WRKY, NAC, and HD-ZIP, have been reported to regulate anthocyanin biosynthesis [8,9,35]. Most of them exert their function either by regulating the expression of the MBW complex components at the transcriptional level or by destabilizing the MBW complex through protein–protein interactions [36]. The different expression levels of *SmTT8* among different-colored fruit peels may be caused by other TFs, such as SmGL2, SmGATA26, or SmWRKY44. SmGL2 is a homolog of Arabidopsis GL2. In Arabidopsis, AtGL2 is a negative regulator of anthocyanin synthesis that directly binds the TAAATGTT/A L1 box in the promoters of *AtPAP2* and *AtMYB113*, repressing their expression at the transcriptional level [37]. SmWRKY44 was previously reported as a hub gene involved in light-induced anthocyanin biosynthesis in eggplant [38]. The transcript levels of *SmGL2*, *SmGATA26*, and *SmWRKY44* were positively correlated with the expression of *SmTT8* in the connection network. These genes may have upstream or downstream regulatory relationships with SmTT8. In addition, the TRV2-*SmGL2* and TRV2-*SmGATA26* silencing vector effectively inhibited anthocyanin accumulation in fruit peels, indicating that SmGL2 and SmGATA26 are involved in anthocyanin biosynthesis. The exact functions of SmGL2 and SmGATA26 in the anthocyanin biosynthetic regulation network remain unclear and will be the focus of future research. Nevertheless, the hub genes and regulation network presented in our study broaden our understanding of anthocyanin biosynthesis in eggplant.

3.4. Plant Hormone Signaling Pathway for Color Variation in Eggplant Fruit Skins

Phytohormones play important roles in regulating anthocyanin accumulation [39]. However, few studies have been reported on eggplant. In the present study, plant hormone signal transduction was one of the most significantly enriched pathways. The signaling pathway from hormone perception to anthocyanin regulation is relatively clear for jasmonate, auxin, strigolactone, and gibberellic acid [36]. The basic logic of the signaling mechanisms for these four hormones is similar. They all involve an Skp1/Cullin/F-box (SCF) E3 ubiquitin ligase complex that targets a protein substrate for poly-ubiquitination and degradation by the 26S proteasome [36]. In Arabidopsis, JAZ proteins directly bound the MYB and bHLH TFs of the anthocyanin-activating MBW complex to regulate jasmonate-mediated anthocyanin accumulation [40]. In apples, MdJAZ1 interacted with MdTRB1 and interfered with the interaction between MdTRB1 and MdMYB9, thereby negatively modulating MdTRB1-promoted anthocyanin biosynthesis [41]. In this study, several bHLH TFs, including SmTT8 in the JA signaling pathway, were also found to be differently expressed among different-colored fruits, indicating a possible involvement of JA-mediated regulation of anthocyanin biosynthesis. In addition, some of the genes involved in the phytohormone signaling pathway, such as ethylene, abscisic acid, salicylic acid, and BR, also exhibited different expression levels correlated with color variation in this study. Whether these hormones have a positive or negative effect on anthocyanin accumulation seems to vary from species to species [36]. In apples, MdERF1B bound to the promoters of *MdMYB9* and *MdMYB11* and regulated anthocyanin and proanthocyanidin accumulation [42]. ABI5 regulated ABA-induced anthocyanin biosynthesis by modulating the MdMYB1-MdbHLH3 complex [43]. However, the signaling mechanisms that regulate anthocyanin biosynthesis are still poorly understood. The crosstalk of these signaling pathways might be extensive to respond to different developmental and environmental cues. The elucidation of the molecular nature of these interactions will be valuable for augmenting the understanding of the regulation of anthocyanin biosynthesis.

4. Materials and Methods

4.1. Plant Materials

Five eggplant varieties, A1 (cv. Suqie 6), A2 (cv. Suqie 801), A3 (cv. Bulita), A4 (Suqie 11), and A5 (cv. Hangqie 1), with different fruit peel colors were used in this study. The fruit peel color of A1 and A3 is black-purple. A1 is the Chinese type with long fruit and a purple calyx, while A3 is the European type with ovate fruit and a green calyx. A2, A4, and A5 are all Chinese types, and their fruit peel colors are green, white, and reddish purple, respectively. The seeds of A1, A2, and A4 were collected from Jiangsu Academy of Agricultural Sciences, while the seeds of A3 and A5 were bought from a Chinese market. All of the samples were grown in a completely randomized block design in a greenhouse at the Liuhe experimental station (32°29′17″N and 118°37′17″E) of Jiangsu Academy of Agricultural Sciences from March to August 2021.

The fruit peels of the five varieties were collected at the commercial stage (30 days after anthesis). Six fruit peels from six individual plants were pooled as a sample, and three samples were set as biological replicates for each variety. Before the fruit peels were collected, the fruit peel color intensity was measured for each variety. The fruit peel samples of each cultivar were divided into three sample sets and stored at −80 °C for subsequent analysis. One sample set was used for total anthcoyanin and chlorophyll content measurement, and the other two sample sets were used for metabolite analysis and transcriptome sequencing (qRT PCR), respectively.

4.2. Total Anthocyanin and Chlorophyll Analysis

The total anthocyanin contents of the five varieties were measured using the spectrophotometric differential pH method [6]. First, 100 mg samples were ground into powder and extracted with 2 mL of a pH 1.0 buffer (containing 50 mM KCl and 150 mM HCl) and a pH 4.5 buffer (containing 400 mM sodium acetate and 240 mM HCl). The extraction solutions were vortexed and stored at 4 °C for 12 h. The solutions were then centrifuged at 10,625× g for 20 min at 4 °C. The absorption at 510 nm was measured, and the total anthocyanin content was calculated according to Zhang et al. [44]. The extraction of chlorophyll of the five varieties was performed as previously described [6]. Approximately 200 mg samples were ground into powder and extracted with 5 mL of 80% acetone. The extraction was then centrifuged at 3000× g for 20 min. The supernatants were collected and measured at the absorption wavelengths of 663 nm and 645 nm.

4.3. Fruit Peel Color Intensity Measurement

The fruit peel colors of the five varieties were analyzed as previously described [6]. The values of L^* (lightness), a^* (redness and greenness), and b^* (yellowness and blueness) were measured using a handheld spectrophotometer (CR-400, Minolta, Japan). The L^* value represents the fruit peel color intensity, and a lower L^* value generally means a deeper color.

4.4. Flavonoid Identification and Quantification

The sample preparation and flavonoid identification and quantification were performed at Wuhan MetWare Biotechnology Co., Ltd. (www.metware.cn, accessed on 12 June 2021) following the standard procedures. The sample extracts were analyzed using an LC-ECI-MS/MS system (UPLC, SHIMADZU Nexera X2, www.shimadzu.com.cn, accessed on 12 June 2021; MS, Applied Biosystems 4500 Q TRAP, www.appliedbiosystems.com.cn, accessed on 12 June 2021). Normalized metabolite data from the five varieties were used to compare the metabolites. A hierarchical clustering analysis (HCA), a principal component analysis (PCA), and an orthogonal partial least-squares discriminant analysis (OPLS-DA) were processed by the R-package to study the metabolite differences among the five samples. The OPLS-DA model plots were created using the MetaboAnalystR (v1.0.1) package. The significantly regulated metabolites between groups were determined by the criteria of VIP $\geq$ 1 and a fold change $\geq$2 or $\leq$0.5. The Kyoto Encyclopedia of Genes and Genomes (KEGG) compound database (http://www.kegg.jp/kegg/compound, accessed on 12 June 2021) was used for the annotation of the identified metabolites, and annotated metabolites were then mapped to the KEGG pathway database (http://www.kegg.jp/kegg/pathway.html, accessed on 12 June 2021).

4.5. RNA-Seq Analysis

The total RNA was extracted from fruit peels of the five varieties using the RNeasy Plant Mini Kit (Qiagen, Hilden, Germany). The RNA quantity, integrity, and purity were analyzed using a Nanodrop and an Agilent Bioanalyzer 2100 (Agilent Technologies, Palo Alto, CA, USA). The sequencing of the libraries was performed on an Illumina HiSeq platform. The clean reads for each library were 6.38–7.45 Gb after filtration. All clean reads were then mapped to the eggplant reference genome (http://eggplant-hq.cn, accessed on 21 June 2021), with match ratios in the range of 96.64–97.91%.

4.6. Bioinformatic Analysis

FPKM (fragments per kb per million fragments) was used for gene expression level quantification [45]. Differentially expressed genes (DEGs) were identified by DESeq2 [46]. The threshold of the p value was calculated using the false discovery rate (FDR) control method. The criterion for the judgment of significance of the DEGs was set at an FDR < 0.01 and the absolute value of $\log_2$ ratio ≥ 1. The gene ontology (GO) and KEGG pathway enrichment analyses were performed using the topGO method and KOBAS 2.0 (http://www.biostars.org/p/200126, accessed on 12 June 2021), respectively [47].

4.7. Coexpression Network Analysis

A weighted gene coexpression network analysis (WGCNA) was performed for all differentially expressed genes using the WGCNA R package [48]. The adjacency matrix was generated by calculating the Pearson correlations among all the genes. A correlation analysis between the identified modules and sample traits was performed with all genes in each module. The hub genes related to the total anthocyanin content and chlorophyll content in the significant positive module were used to construct coexpression networks with Cytoscape 3.7.2 (http://www.cytoscape.org, accessed on 12 June 2021).

4.8. Quantitative Real-Time PCR Analysis

The quantitative reverse-transcription PCR (qRT-PCR) was applied to analyze the expression levels of 16 DEGs among the five varieties. First-strand cDNA was syntheized by the ReverTran Ace qPCR RT kit (Toyobo). The primers used for qRT-PCR were designed by Primer 3 (http://primer3.ut.ee, accessed on 12 June 2021). All primers are listed in Supplementary Table S3. qRT-PCR was performed as previously described [6]. The gene expression levels were normalized with *SmGAPDH* as an internal control and were calculated using the $2^{-\Delta\Delta Ct}$ method [49].

4.9. Vector Construction and Transformation of Eggplant

The tobacco rattle virus (TRV)-based VIGS system was employed for SmGL2, SmGATA26, and SmWRKY44 silencing. The 300 bp best target regions of SmGL2, SmGATA26, and SmWRKY44 predicted by SGN VIGS Tool (https://vigs.solgenomics.net, accessed on 12 June 2021) were cloned into a pTRV2 vector using specific primers. The TRV1, empty TRV2, TRV2-SmGL2, TRV2-SmGATA26, and TRV2-SmWRKY44 plasmids were then transformed into Agrobacterium tumefaciens strain GV3101. An equal volume of TRV1 Agrobacterium culture was mixed with empty TRV2 or TRV2-SmGL2, TRV2-SmGATA26, and TRV2-SmWRKY44 cultures before infiltration. The mixtures were then infiltrated into eggplant peels via needleless syringes. The treated eggplant fruits were kept in the dark for 24 h at 22 °C and then cultivated in regular light conditions for 5 days. Three biological replicates were conducted for each combination.

4.10. Statistical Analysis

The statistical analysis was performed using Microsoft Excel 2013 and SPSS statistics 19 (IBM SPSS, Chicago, IL, USA). Significant differences were determined using a one-way ANOVA with Duncan's test at $p < 0.05$.

5. Conclusions

In this study, delphinidin-type anthocyanins were the main anthocyanin components in reddish to black-purple eggplant fruit peels. The activation of early phenylpropanoid biosynthesis genes was more responsible for anthocyanin accumulation in eggplant, while *SmF3mFnH* was the key factor for the formation of a purple color. Moreover, four TFs, including *SmGL2*, *SmGATA26*, *SmWRKY44*, and *SmTT8*, were identified as hub genes associated with anthocyanin accumulation. In addition, the pathway of plant hormone signal transduction was significantly enriched. It is possible that phytohormones cooperatively interact to modulate flavonoid biosynthesis. The results of this study provide valuable information and new insights into the regulatory network of fruit coloration and will therefore contribute to the molecular breeding of eggplant.

Supplementary Materials: The following supporting information can be downloaded at: https://www.mdpi.com/article/10.3390/ijms232113475/s1, Figure S1: Orthogonal partial least squares discriminant analysis discriminant analysis (OPLS-DA) model of comparison groups including A1 vs. A2, A1 vs. A3, A1 vs. A5, A2 vs. A3, A2 vs. A5, and A3 vs. A5.; Figure S2: Differential expressed genes involving in plant hormone transduction; Figure S3: Gene network of the hub genes in the violet module, which positively correlated with total chlorophyll content ($r^2 = 0.94$, $p = 1 \times 10^{-7}$); Figure S4: Sequence variation of *SmMYB113* among white, green, and purple fruit of eggplant. Table S1: DEGs identified in the pair-wise comparisons; Table S2: DEGs related to light-harvesting chlorophyll protein complex; Table S3: Primers used for quantitative real-time PCR (qRT-PCR).

Author Contributions: Conceptualization, Y.Z. and X.Z.; methodology, Y.Y. and J.L.; investigation, X.Z. and S.L.; writing—original draft, X.Z.; writing—review and editing, X.Z. and Y.Y. All authors have read and agreed to the published version of the manuscript.

Funding: This work was partially supported by the Key Research and Development Program of Jiangsu Province of China (grant number: BE2021377).

Institutional Review Board Statement: Not applicable.

Informed Consent Statement: Not applicable.

Data Availability Statement: The data presented in this study are available on request from the corresponding author.

Acknowledgments: The authors are grateful to the reviewers for their valuable comments for revision.

Conflicts of Interest: The authors declare no conflict of interest.

References

1. Shen, N.; Wang, T.F.; Gan, Q.; Liu, S.; Wang, L.; Jin, B. Plant flavonoids: Classification, distribution, biosynthesis, and antioxidant activity. *Food Chem.* **2022**, *383*, 132531. [CrossRef] [PubMed]
2. Liu, W.X.; Feng, Y.; Yu, S.H.; Fan, Z.Q.; Li, X.L.; Li, J.Y.; Yin, H.F. The flavonoid biosynthesis network in plants. *Int. J. Mol. Sci.* **2021**, *22*, 12824. [CrossRef] [PubMed]
3. Li, H.; Yang, Z.; Zeng, Q.W.; Wang, S.B.; Luo, Y.W.; Huang, Y.; Xin, Y.C.; He, N.J. Abnormal expression of *bHLH3* disrupts a flavonoid homeostasis network, causing differences in pigment composition among mulberry fruits. *Hortic. Res.* **2020**, *7*, 83. [CrossRef]
4. Mizuno, T.; Yabuya, T.; Kitajima, J.; Iwashina, T. Identification of novel Cglycosylflavones and their contribution to flower colour of the Dutch iris cultivars. *Plant Physiol. Biochem.* **2013**, *72*, 116–124. [CrossRef] [PubMed]
5. Shi, J.; Simal-Gandara, J.; Mei, J.F.; Ma, W.J.; Peng, Q.H.; Shi, Y.L.; Xu, Q.; Lin, Z.; Lv, H.P. Insight into the pigmented anthocyanins and the major potential co-pigmented flavonoids in purple-coloured leaf teas. *Food Chem.* **2021**, *363*, 130278. [CrossRef] [PubMed]
6. Zhou, X.H.; Liu, S.Y.; Liu, Y.P.; Liu, J.; Yang, Y.; Wang, D.; Zhuang, Y. Extensive transcriptome changes underlying the fruit skin colour intensity variation in purple eggplant. *Not. Bot. Horti. Agrobo.* **2021**, *49*, 12434. [CrossRef]
7. Xu, W.; Dubos, C.; Lepiniec, L. Transcriptional control of flavonoid biosynthesis by MYB-bHLH-WDR complexes. *Trends Plant Sci.* **2015**, *20*, 176–185. [CrossRef]
8. An, J.P.; Zhang, X.W.; You, C.X.; Bi, S.Q.; Wang, X.F.; Hao, Y.J. MdWRKY40 promotes wounding-induced anthocyanin biosynthesis in association with MdMYB1 and undergoes MdBT2-mediated degradation. *New Phytol.* **2019**, *224*, 380–395. [CrossRef]
9. Liu, H.N.; Su, J.; Zhu, Y.F.; Yao, G.F.; Allan, A.C.; Ampomah-Dwamena, A.; Shu, Q.; Lin-Wang, K.; Zhang, S.L.; Wu, J. The involvement of PybZIPa in light-induced anthocyanin accumulation via the activation of PyUFGT through binding to tandem G-boxes in its promoter. *Hortic. Res.* **2019**, *6*, 134. [CrossRef]
10. Ni, J.B.; Premathilake, A.T.; Gao, Y.H.; Yu, W.J.; Tao, R.Y.; Teng, Y.W.; Bai, S.L. Ethylene-activated PpERF105 induces the expression of the repressor-type R2R3-MYB gene PpMYB140 to inhibit anthocyanin biosynthesis in red pear fruit. *Plant J.* **2021**, *105*, 167–181. [CrossRef]
11. Toppino, L.; Barchi, L.; Mercati, F.; Acciarri, N.; Perrone, D.; Martina, M.; Gattolin, S.; Sala, T.; Fadda, S.; Mauceri, A.; et al. A new intra-specific and high-resolution genetic map of eggplant based on a RIL population, and location of QTLs related to plant anthocyanin pigmentation and seed vigour. *Genes* **2020**, *11*, 745. [CrossRef] [PubMed]
12. Gürbüz, N.; Uluişik, S.; Frary, A.; Frary, A.; Doğanlar, S. Health benefits and bioactive compounds of eggplant. *Food Chem.* **2018**, *268*, 602–610. [CrossRef] [PubMed]
13. Stommel, J.R.; Dumm, J.M. Coordinated regulation of biosynthetic and regulatory genes coincides with anthocyanin accumulation in developing eggplant fruit. *J. Am. Soc. Hortic. Sci.* **2015**, *140*, 129–135. [CrossRef]
14. Jiang, M.M.; Ren, L.; Lian, H.L.; Liu, Y.; Chen, H.Y. Novel insight into the mechanism underlying light-controlled anthocyanin accumulation in eggplant (*Solanum melongena* L.). *Plant Sci.* **2016**, *249*, 46–58. [CrossRef] [PubMed]
15. Li, J.; Ren, L.; Gao, Z.; Jiang, M.M.; Liu, Y.; Zhou, L.; He, Y.J.; Chen, H.Y. Combined transcriptomic and proteomic analysis constructs a new model for light-induced anthocyanin biosynthesis in eggplant (*Solanum melongena* L.). *Plant Cell Environ.* **2017**, *40*, 3069–3087. [CrossRef] [PubMed]
16. He, Y.; Chen, H.; Zhou, L.; Liu, Y.; Chen, H. Comparative transcription analysis of photosensitive and non-photosensitive eggplants to identify genes involved in dark regulated anthocyanin synthesis. *BMC Genom.* **2019**, *20*, 678. [CrossRef]
17. Li, Y.; Chen, Y.; Zhou, L.; You, S.; Deng, H.; Chen, Y.; Alseekh, S.; Yuan, Y.; Fu, R.; Zhang, Z.; et al. MicroTom metabolic network: Rewiring tomato metabolic regulatory network throughout the growth cycle. *Mol. Plant* **2020**, *13*, 1–16. [CrossRef]
18. Wang, S.; Li, Y.; He, L.; Yang, J.; Fernie, A.R.; Luo, J. Natural variance at the interface of plant primary and specialized metabolism. *Curr. Opin. Plant Biol.* **2022**, *67*, 102201. [CrossRef]
19. Sadilova, E.; Stintzing, F.C.; Carle, R. Anthocyanins, colour and antioxidant properties of eggplant (*Solanum melongena* L.) and violet pepper (*Capsicum annuum* L.) peel extracts. *Z. Für Nat. C* **2006**, *61c*, 527–535. [CrossRef]

20. Azuma, K.; Ohyama, A.; Ippoushi, K.; Ichiyanagi, T.; Takeuchi, A.; Saito, T.; Fukuoka, H. Structures and antioxidant activity of anthocyanins in many accessions of eggplant and its related species. *J. Agric. Food Chem.* **2008**, *56*, 10154–10159. [CrossRef]
21. Zhao, C.L.; Yu, Y.Q.; Chen, Z.J.; Wen, G.S.; Wei, F.G.; Zheng, Q.; Wang, C.D.; Xiao, X.L. Stability-increasing effects of anthocyanin glycosyl acylation. *Food Chem.* **2017**, *214*, 119–128. [CrossRef] [PubMed]
22. Li, W.F.; Mao, J.; Yang, S.J.; Guo, Z.G.; Ma, Z.H.; Dawuda, M.M.; Zuo, C.W.; Chu, M.Y.; Chen, B.H. Anthocyanin accumulation correlates with hormones in the fruit skin of 'Red Delicious' and its four generation bud sport mutants. *BMC Plant Biol.* **2018**, *18*, 363. [CrossRef] [PubMed]
23. Feng, X.; Gao, G.; Yu, C.; Zhu, A.; Chen, J.; Chen, K.; Wang, X.; Abubakar, A.S.; Chen, P. Transcriptome and metabolome analysis reveals anthocyanin biosynthesis pathway associated with ramie (*Boehmeria nivea* (L.) Gaud.) leaf color formation. *BMC Genom.* **2021**, *22*, 684. [CrossRef] [PubMed]
24. Yang, G.B.; Li, L.J.; Wei, M.; Li, J.; Yang, F.J. SmMYB113 is a key transcription factor responsible for compositional variation of anthocyanin and color diversity among eggplant peels. *Front. Plant Sci.* **2022**, *13*, 843996. [CrossRef] [PubMed]
25. Zhang, Q.; Wang, L.; Liu, Z.; Zhao, Z.; Zhao, J.; Wang, Z.; Zhou, G.; Liu, P.; Liu, M. Transcriptome and metabolome profiling unveil the mechanisms of *Ziziphus jujube* Mill. *peel coloration. Food Chem.* **2020**, *312*, 125903. [CrossRef]
26. Shi, Q.; Du, J.; Zhu, D.; Li, X.; Li, X. Metabolomic and transcriptomic analyses of anthocyanin biosynthesis mechanisms in the color mutant *Ziziphus jujube* cv. Tailihong. *J. Agric. Food Chem.* **2020**, *68*, 15186–15198. [CrossRef]
27. Meng, R.; Qu, D.; Zhao, Z.; Gao, Z.; Liu, Y.; Yang, H.; Shi, X.; Zhao, Z. Anthocyanin accumulation and related gene family expression in the skin of dark-grown red and non-red apples (*Malus domestica* Borkh.) in response to sunlight. *Sci. Hortic.* **2015**, *189*, 66–73. [CrossRef]
28. Li, L.Z.; He, Y.J.; Ge, H.Y.; Liu, Y.; Chen, H.Y. Functional characterization of SmMYB86, a negative regulator of anthocyanin biosynthesis in eggplant (*Solanum melongena* L.). *Plant Sci.* **2021**, *302*, 110696. [CrossRef]
29. Shi, S.L.; Liu, Y.; He, Y.J.; Li, L.Z.; Li, D.L.; Chen, H.Y. R2R3-MYB transcription factor SmMYB75 promotes anthocyanin biosynthesis in eggplant (*Solanum melongena* L.). *Sci. Hortic.* **2021**, *282*, 110020. [CrossRef]
30. Li, L.Z.; Li, S.H.; Ge, H.Y.; Shi, S.L.; Li, D.L.; Liu, Y.; Chen, H.Y. A light-responsive transcription factor SmMYB35 enhances anthocyanin biosynthesis in eggplant (*Solanum melongena* L.). *Planta* **2022**, *255*, 12. [CrossRef]
31. Duan, Z.F.; Tian, S.Y.; Yang, G.B.; Wei, M.; Li, J.; Yang, F.J. The basic helix-loop-helix transcription factor SmbHLH1 represses anthocyanin biosynthesis in eggplant. *Front. Plant Sci.* **2021**, *12*, 757936. [CrossRef] [PubMed]
32. Xi, H.C.; He, Y.J.; Chen, H.Y. Functional characterization of SmbHLH13 in anthocyanin biosynthesis and flowering in eggplant. *Hortic. Plant J.* **2021**, *7*, 73–80. [CrossRef]
33. Zhou, L.; He, Y.; Li, J.; Liu, Y.; Chen, H.Y. CBFs function in anthocyanin biosynthesis by interacting with MYB113 in eggplant (*Solanum melongena* L.). *Plant Cell Physiol.* **2020**, *61*, 416–426. [CrossRef] [PubMed]
34. Kobayashi, S.; Goto-Yamamoto, N.; Hirochika, H. Retrotransposon-induced mutations in grape skin color. *Science* **2004**, *304*, 982. [CrossRef]
35. Sun, Q.; Jiang, S.; Zhang, T.; Xu, H.; Fang, H.; Zhang, J.; Su, M.; Wang, Y.; Zhang, Z.; Wang, N.; et al. Apple NAC transcription factor MdNAC52 regulates biosynthesis of anthocyanin and proanthocyanidin through MdMYB9 and MdMYB11. *Plant Sci.* **2019**, *289*, 110286. [CrossRef]
36. LaFountain, A.M.; Yuan, Y.W. Repressors of anthocyanin biosynthesis. *New Phytol.* **2021**, *31*, 933–949. [CrossRef]
37. Wang, X.; Wang, X.; Hu, Q.; Dai, X.; Tian, H.; Zheng, K.; Wang, X.; Mao, T.; Chen, J.G.; Wang, S. Characterization of an activation-tagged mutant uncovers a role of GLABRA2 in anthocyanin biosynthesis in Arabidopsis. *Plant J.* **2015**, *83*, 300–311. [CrossRef]
38. He, Y.J.; Wang, Z.W.; Ge, H.Y.; Liu, Y.; Chen, H.Y. Weighted gene co-expression network analysis identifies genes related to anthocyanin biosynthesis and functional verification of hub gene SmWRKY44. *Plant Sci.* **2021**, *309*, 110935. [CrossRef]
39. Das, P.K.; Shin, D.H.; Choi, S.B.; Park, Y. Sugar-hormone cross-talk in anthocyanin biosynthesis. *Mol. Cells* **2012**, *34*, 501–507. [CrossRef]
40. Qi, T.; Song, S.; Ren, Q.; Wu, D.; Huang, H.; Chen, Y.; Fan, M.; Peng, W.; Ren, C.; Xie, D. The jasmonate-ZIM-domain proteins interact with the WD-Repeat/bHLH/MYB complexes to regulate jasmonate-mediated anthocyanin accumulation and trichome initiation in Arabidopsis thaliana. *Plant Cell* **2011**, *23*, 1795–1814. [CrossRef]
41. An, J.; Xu, R.; Liu, X.; Zhang, J.; Wang, X.; You, C.; Hao, Y. Jasmonate induces biosynthesis of anthocyanin and proanthocyanidin in apple by mediating the JAZ1-TRB1-MYB9 complex. *Plant J.* **2021**, *106*, 1414–1430. [CrossRef] [PubMed]
42. Zhang, J.; Xu, H.; Wang, N.; Jiang, S.; Fang, H.; Zhang, Z.; Yang, G.; Wang, Y.; Su, M.; Xu, L.; et al. The ethylene response factor MdERF1B regulates anthocyanin and proanthocyanin biosynthesis in apple. *Plant Mol. Biol.* **2018**, *98*, 205–218. [CrossRef] [PubMed]
43. An, J.P.; Zhang, X.W.; Liu, Y.J.; Wang, X.F.; You, C.X.; Hao, Y.J. ABI5 regulates ABA-induced anthocyanin biosynthesis by modulating the MYB1-bHLH3 complex in apple. *J. Exp. Bot.* **2021**, *72*, 1460–1472. [CrossRef] [PubMed]
44. Zhang, Y.; Hu, Z.; Chu, G.; Huang, C.; Tian, S.; Zhao, Z.; Chen, G. Anthocyanin accumulation and molecular analysis of anthocyanin biosynthesis-associated genes in eggplant (*Solanum melongena* L.). *J. Agric. Food Chem.* **2014**, *62*, 2906–2912. [CrossRef]
45. Florea, L.; Salzberg, S.L. Thousands of exon skipping events differentiate among splicing patterns in sixteen human tissues. *F1000Research* **2013**, *2*, 188. [CrossRef]

46. Love, M.I.; Huber, W.; Anders, S. Moderated estimation of fold change and dispersion for RNA-Seq data with DESeq2. *Genome Biol.* **2014**, *15*, 550. [CrossRef]
47. Xie, C.; Mao, X.; Huang, J.; Ding, Y.; Wu, J.; Dong, S.; Kong, L.; Gao, G.; Li, C.; Wei, L. KOBAS 2.0: A web server for annotation and identification of enriched pathways and diseases. *Nucleic Acids Res.* **2011**, *39*, 316–322. [CrossRef]
48. Langfelder, P.; Horvath, S. WGCNA: An R package for weighted correlation network analysis. *BMC Bioinf.* **2008**, *9*, 559. [CrossRef]
49. Livak, K.J.; Schmittgen, T.D. Analysis of relative gene expression data using real-time quantitative PCR and 2(-Delta Delta C(T)) method. *Methods* **2001**, *25*, 402–408. [CrossRef]

International Journal of
Molecular Sciences

MDPI

Article

Putative Transcription Factor Genes Associated with Regulation of Carotenoid Biosynthesis in Chili Pepper Fruits Revealed by RNA-Seq Coexpression Analysis

Maria Guadalupe Villa-Rivera [1], Octavio Martínez [2] and Neftalí Ochoa-Alejo [1,*]

[1] Departamento de Ingeniería Genética, Unidad Irapuato, Centro de Investigación y de Estudios Avanzados del Instituto Politécnico Nacional, Irapuato 36824, Mexico
[2] Unidad de Genómica Avanzada, Unidad Irapuato, Centro de Investigación y de Estudios Avanzados del Instituto Politécnico Nacional, Irapuato 36824, Mexico
* Correspondence: neftali.ochoa@cinvestav.mx; Tel.: +52-(462)-6239654

Abstract: During the ripening process, the pericarp of chili pepper (*Capsicum* spp.) fruits accumulates large amounts of carotenoids. Although the carotenoid biosynthesis pathway in the *Capsicum* genus has been widely studied from different perspectives, the transcriptional regulation of genes encoding carotenoid biosynthetic enzymes has not been elucidated in this fruit. We analyzed RNA-Seq transcriptomic data from the fruits of 12 accessions of *Capsicum annuum* during the growth, development, and ripening processes using the R package named *Salsa*. We performed coexpression analyses between the standardized expression of genes encoding carotenoid biosynthetic enzymes (target genes (TGs)) and the genes of all expressed transcription factors (TFs). Additionally, we analyzed the promoter region of each biosynthetic gene to identify putative binding sequences for each selected TF candidate. We selected 83 TFs as putative regulators of the carotenogenic structural genes. From them, putative binding sites in the promoters of the carotenoid-biosynthesis-related structural genes were found for only 54 TFs. These results could guide the search for transcription factors involved in the regulation of the carotenogenic pathway in chili pepper fruits and might facilitate the collection of corresponding experimental evidence to corroborate their participation in the regulation of this biosynthetic pathway in *Capsicum* spp.

Keywords: chili pepper; carotenoids; transcriptional regulation; RNA-Seq; transcription factors

1. Introduction

Chili pepper (*Capsicum* spp.) is a globally cultivated horticultural crop. Chili pepper fruits are popular because of their great variety of forms, sizes, flavors, and aromas, especially for their unique and characteristic pungency. Chili pepper pods produce a series of compounds, such as anthocyanins, carotenoids, vitamin C, flavonoids, which are powerful antioxidants, and hot capsaicinoid analogs, among others. Chili pepper represents an important model for studying the carotenogenesis process as well as investigating and establishing the carotenoid biosynthetic pathway (Figure 1) [1]. Carotenoids are the second most abundant pigments in nature behind chlorophylls. These pigments are responsible for the yellow, orange, and red colors of flowers and fruits and have important functions as attractors of pollinators, complementary pigments during photosynthesis, and efficient protectors of photosystems against the damage caused by the free radicals generated during the photosynthetic process. Nutraceutic benefits of chili pepper carotenoids for human health have been reported recently [2]. Although the carotenoid biosynthetic pathway in chili pepper fruits has been fully described, the regulation mechanisms at the transcriptional level of all carotenogenic genes are currently poorly understood [3]. Regarding this aspect, a tomato fruit gene regulatory network was generated by Pan et al. (2013) using artificial neural network inference analysis and transcription gene expression patterns derived from fruits sampled at various points during development and ripening, and one of the transcription factor gene expression profiles, with a sequence related to the Arabidopsis (*Arabidopsis thaliana*) *ARABIDOPSIS PSEUDO RESPONSE REGULATOR2-LIKE* gene (*APRR2-Like*), was upregulated at the breaker stage in wild-type tomato fruits and, when overexpressed in transgenic lines, increased the plastid number, area, and pigment content, enhancing the levels of chlorophyll in immature unripe fruits and carotenoids in red ripe fruits. Additionally, a putative ortholog of the tomato *APPR2-Like* gene in sweet pepper (*Capsicum annuum*) was associated with pigment accumulation in fruit tissues [4]. Song et al. (2020) reported an analysis of the *ERF* transcription factor gene family in the *Capsicum* genome to identify transcription factor genes possibly involved in the regulation of species-specific metabolites, including carotenoids, and they observed that the expression patterns of *CaERF82, CaERF97, CaERF66, CaERF107,* and *CaERF101* were consistent with the expression profiles of carotenoid-biosynthesis-related pathway genes and the accumulation of carotenoids (β-carotene, zeaxanthin, and capsorubin) in the pericarp of chili pepper fruits [5]. In a genome-wide identification and analysis of the *MYB* transcription factor gene family in chili pepper (*Capsicum* spp.), Arce-Rodríguez et al. (2021) sought to identify *CaMYB* genes that were positively correlated with the expression patterns of the carotenoid-related genes *PSY (phytoene synthase), β-carotene hydroxylase 1 (BCH1),* and *capsanthin-capsorubin synthase* (*CCS*) during the development and ripening of Serrano 'Tampiqueño 74' chili pepper fruits (0, 10, 20, 30, 40, 50, and 60 days postanthesis) as an approach to predict transcription factors possibly involved in the regulation of this pathway, and they found that the transcriptomic expression patterns of those genes clustered with the *MYB*-related genes *CaDIV1, CaDIV3, CaMYBR13, CaTRF2, CaMYBC1,* and *CaPHR9* and an atypical *MYB* (*CaMYB5R*) [6]. Li et al. (2021) carried out a transcriptome analysis and carotenoid content determination in orange and red chili pepper fruits at three developmental stages (25, 40, and 55 days after flowering) and found that the expression patterns of the *BCH1* and *CCS* genes were significantly different during fruit development. Moreover, through a weighted gene coexpression network analysis (WGCNA), the transcription factor genes F-box protein *SKIP23, GATA transcription factor 26, U-box domain-containing protein 52,* zinc finger family *FYVE/PHD-type, RING/FYVE/PHD-type,* and *CONSTANS-LIKE 9* were proposed to be involved in the regulation of the *CCS* gene [7]. More recently, Ma et al. (2022) analyzed the expression patterns of the transcription factor gene *CaMYB306* in chili pepper fruits of contrasting colors ('XHB' (red fruits) and mutant H0809 pepper (yellow fruits)) at three developmental stages (25, 40, and 55 days after flowering). *CaMYB306* expression was significantly higher in 'XHB' fruit than in mutant H0809 fruit. A knockdown of *CaMYB306* by gene silencing inhibited carotenoid biosynthesis and markedly decreased the expression of carotenoid-related genes (*CaPSY, CaBCH,* and *CaZEP*) in chili pepper fruit. Furthermore, *CaMYB306*-overexpressing (*CaMYB306*-OE) 'Micro-Tom' tomato plants showed earlier coloration with increased carotenoid accumulation. The expression levels of key carotenoid-biosynthesis-related genes (*SlPSY* and *SlPDS*) were increased in *CaMYB306*-OE fruit [8].

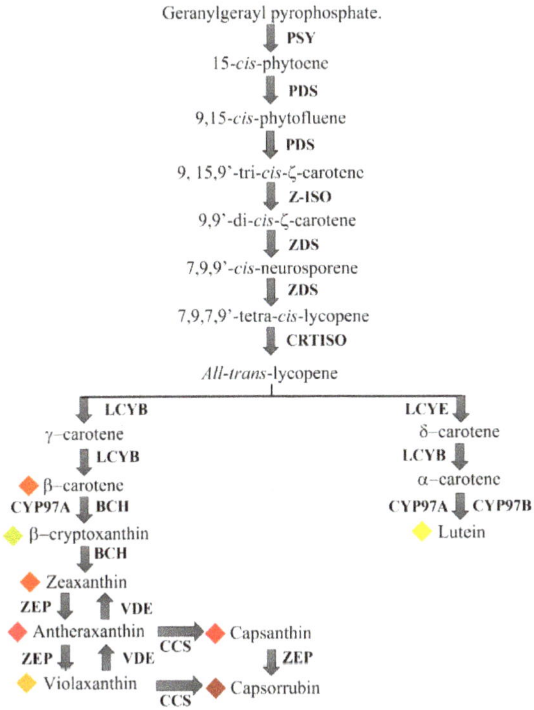

Figure 1. Chili pepper fruit carotenoid biosynthetic pathway. PSY (phytoene synthase), PDS (phytoene desaturase), Z-ISO (ζ-carotene isomerase), ZDS (ζ-carotene desaturase), CRTISO (carotene isomerase), LCYB (β-lycopene cyclase), LCYE (ε-lycopene cyclase), BCH (β-carotene hydroxylase), CYP (β-carotene hydroxylase cytochrome 450 types A and B), ZEP (zeaxanthin epoxidase), VDE (violaxanthin epoxidase), and CCS (capsanthin-capsorubin synthase). Modified from [1]. Colored rhombuses represent the color conferred by each carotenoid.

However, in addition to these findings in chili pepper fruits, transcriptional regulation of carotenoid biosynthesis seems to be a very complex process since diverse transcription factors have been proposed as regulators of this pathway in different plant species using different experimental approaches; for example, Welsch et al. (2007) demonstrated that AtRAP2.2, a member of the APETALA2 (AP2)/ethylene-responsive element-binding protein transcription family, showed positive binding to the ATCTA cis-element present in the promoter of the *phytoene desaturase* (*PDS*) gene in *Arabidopsis* [9]. Similar interactions between transcription factors and promoters of carotenoid biosynthetic genes have been reported as the mechanism regulating this pathway, and the tomato MADS-box transcription factor RIPENING INHIBITOR (RIN) was found to interact with the promoter of the *PSY1* gene, which was enriched following a chromatin immunoprecipitation assay (ChIP); moreover, the *PSY1* gene showed an increase in expression at the onset of ripening in wild-type but not in *rin* mutant fruits [10]. According to Mohanty et al. (2016), the promoter analysis of genes upregulated in response to blue and red lights in rice leaves revealed the involvement of key transcription factors, such as bHLH, bZIP, MYB, WRKY, ZnF, and ERF (jasmonic-acid-inducible), in the regulation of a higher accumulation of carotenoids and phenolic compounds under blue light [11]. In papaya (*Carica papaya*), Fu et al. (2017) indicated that CpEIN3a regulated the carotenoid-biosynthesis-related genes *CpPDS4* and *CpBCH* during fruit ripening, as demonstrated by electrophoretic mobility shift assays (EMSA) and transient expression analyses. Moreover, CpEIN3a physically interacted with the transcription factor CpNAC2, which acted as a transcriptional activator of *CpPDS2/4*, *CpZDS*, *CpLCYE*, and *CpBCH* by directly binding to their promoters [12]. According to Lu et al. (2018), the promoter sequences of the *lycopene β-cyclase* genes *LCYB1* and *LCYB2* were used as bait in a yeast one-hybrid screen cDNA library for promoter-binding proteins from sweet orange (*Citrus sinensis*) to identify TFs that might regulate carotenoid metabolism, and they isolated the transcription factor gene *CsMADS6*, which was demonstrated to modulate carotenoid metabolism since its overexpression increased the expression of the *PSY*, *PDS*, *carotenoid cleavage dioxygenase 1* (CCD1), and *LCYB1* genes in transformed overexpressing transgenic citrus calli and tomato (*Solanum lycopersicum*) plants. Furthermore, CsMADS6 upregulated the expression of the *PSY*, *PDS*, and *CCD1* genes by directly binding to their promoters [13]. Ampomah-Dwamena et al. (2019) identified and characterized the kiwifruit TF *MYB7* gene, whose protein product was found to activate the promoter of the kiwifruit *AdLCYb* gene [14]. Recently, Y. Liu et al. (2021) demonstrated the participation of the SlGRAS4 transcription factor as a regulator of carotenoid biosynthesis by overexpressing this gene in tomato. The fruits from overexpressing tomato plants showed an increase in carotenoids and a higher expression of *PSY1* than those from nontransformed plants, and the expression of the negative regulator of tomato fruit ripening, *SlMADS1*, was repressed [15]. Another approach to search for transcription factors possibly involved in the transcriptional regulation of carotenoid-biosynthesis-related genes has been the comparative expression pattern analysis of structural biosynthetic genes and transcription factor genes in color-contrasting fruits or organs and by coexpression analyses during fruit development and ripening; for example, Ríos et al. (2010) identified a GCC transcription factor that belongs to a subgroup of the GARP (GOLDEN2, ARR-B, and Psr1) subfamily of MYB-related proteins containing a coiled-coil domain, designated GCC (GARP and coiled-coil), in *Citrus clementina* through differential transcriptomic analyses (a *Citrus* microarray containing approximately 20,000 cDNA fragments) associated with fruit color changes in two mutants (*39B3* and *39E7*) differing in delayed color break (carotenoid accumulation and chlorophyll degradation), and they observed that *CcGCC1* expression was strongly induced at the onset of color change in the flavedo and that the expression of *CcGCC1* was associated with peel ripening [16]. Gómez-Gómez et al. (2012) isolated a putative *CsMYB1* transcription factor gene from *Crocus* stigmas (saffron) by comparing the expression patterns during development in *C. sativus* (from yellow to the dark red stage) and two species with branched stigmas and yellow color (*C. cancellatus* and *C. pulchellus*), and they observed that the *CsMYB1* transcription factor gene was highly expressed in *C. sativus* stigmas in the dark red stage [17]. Grassi et al. (2013) generated the expression patterns of 19,324 genes from a transcriptomic analysis of ripening watermelon (*Citrus lanatus*) fruits by Illumina RNA sequencing, and RNA-Seq expression profiling yielded a set of transcription factor genes showing a positive correlation with ripening and carotenoid accumulation. Among them, 19 were homologous to tomato carotenoid-related genes (2 *RIN-MADS* (MADS-box transcription factor; 2 *TAGL1* (TAGL1 transcription factor); 5 *CNR* (COLORLESS NON RIPENING squamosa promoter binding-like protein); 1 *NAC-NOR* (NAC domain protein); 3 *SlAP2a* (AP2 transcription factor); 2 *SlERF6* (ERF6 mRNA), 1 *DET1* (*deetiolated1* homolog), 1 *DDB1* (UV-damaged DNA binding protein 1); 1 *CUL4* (cullin 4); and 1 *HB-1* (homeodomain leucine zipper protein)), all homologous to *S. lycopersicum*. Among these, six (three *CNR*, one *SlAP2a*, one *SlERF6*, and one *HB-1*) were differentially expressed in the flesh during fruit development and ripening [18]. Ye et al. (2015) performed a coexpression analysis of transcription factors and biosynthetic genes of the ascorbic acid, carotenoid, and flavonoid biosynthesis pathways to identify putative transcription factors that might regulate these biosynthetic processes; for this goal, the transcriptomes of tomato 'Ailsa Craig' and 'HG6-61' at seven developmental stages were evaluated using the Illumina sequencing platform, and the coexpression analysis revealed transcription factor genes with expression profiles that correlated with the patterns of genes related to each biosynthetic pathway during fruit development and ripening. In the case of carotenoid biosynthesis, 37 transcription factor genes, including a *MADS-box* gene and 4 *CCT* domain transcription factors, showed positive coexpression with carotenoid-biosynthesis-related genes. Other transcription factors, such as the *AUX/IAA*, *bHLH*, *MYB*, *SBP-box*, transcription factor *B3*, and *zinc finger protein* genes, also exhibited a high correlation with carotenoid metabolism, suggesting a complex underlying regulatory network [19]. Jiang et al. (2019) used the RNA-Seq technique to investigate candidate genes of carotenoid metabolism in the flesh of pummelo (*Citrus maxima*) 'GXMY' and its mutants 'HRMY' and 'HJMY' in three fruit developmental stages and found 12 differentially expressed genes related to carotenoid biosynthesis; a coexpression analysis revealed that the expression patterns of several transcription factors, such as *bHLH*, *MYB*, *ERF*, *NAC*, and *WRKY*, were highly correlated with differentially expressed genes involved in carotenoid biosynthesis [20]. Lu et al. (2019) identified six transcription factors (*CmMYB305*, *CmMYB29*, *CmRAD3*, *CmbZIP61*, *CmAGL24*, and *CmNAC1*) in *Chrysanthemum* X *morifolium* (a pink chrysanthemum cultivar, 'Jianliuxiang Pink', and its three bud sport mutants (including the white, yellow, and red color mutants 'Jianliuxiang White', 'Jianliuxiang Yellow', and 'Jianliuxiang Red', respectively)), using a weighted gene coexpression correlation network analysis (WGCNA) combined with a correlative analysis to determine whether they played an important role in carotenoid accumulation by regulating structural genes related to the carotenoid metabolism pathway and plastid development [21].

In the present work, we applied a coexpression analysis method developed by our research team [22] to identify transcription factor candidate genes possibly involved in the regulation of each carotenoid-biosynthesis-related structural gene in chili pepper fruit transcriptomes at different developmental stages from domesticated and wild accessions. This approach was complemented with the confirmation of cis-sequences in the promoters of the carotenoid-biosynthesis-related structural genes for each potentially selected transcription factor as a second criterion of selection.

2. Results

2.1. Expression Profiles of Carotenoid-Biosynthesis-Related Genes

Transcriptomic data were obtained from 12 accessions of *Capsicum annuum*: 4 wild, 6 domesticated, and 2 reciprocal crosses between a domesticated and a wild-type accession (Table 1). Fruits were collected at seven time points (0, 10, 20, 30, 40, 50, and 60 days after anthesis (DAA)) during the growth, development, and ripening of chili peppers. Data from all RNA-Seq libraries were analyzed as standardized expression profiles (SEPs). SEPs are seven dimensional numeric vectors in which each one of the numbers summarizes the standardized mean expression (mean = 0 and standard deviation = 1) at the corresponding time point based on ternary models in scale-free measurements [22,23].

Table 1. Chili pepper accessions used for the RNA-Seq analyses.

Accession Type	Accession Name	Abbreviation
D	Ancho San Luis	AS
D	Criollo de Morelos 334 (CM334)	CM
D	California Wonder	CW
D	Jalapeño Espinalteco	JE
D	Serrano Tampiqueño 74	ST
D	Zunla-1	ZU
W	Piquín Coahuila	CO
W	Piquín Querétaro	QU
W	Piquín Sonora Red	SR
W	Piquín Sonora Yellow	SY
C	F1: CM female x QU male	CQ
C	F1: QU female x CM male	QC

D, domesticated; W, wild; C, crosses.

The carotenoid structural biosynthetic genes used as target genes (TGs) (Figure 1) for coexpression analysis are summarized in Table 2. One SEP of *PDS*, *Z-ISO*, *LCYE*, *CYP97A*, *VDE*, *ZEP*, and *CCS* was identified in *Salsa*, but for *ZDS*, *CRTISO*, and *LCYB*, two different transcription profiles were identified for each gene, and both SEPs were analyzed for TF candidate prediction. Moreover, three transcript types encoding the *PSY* gene, and two types of *BCH* were identified in the *Salsa* database (Table 2). Nevertheless, only *PSY1* and *BCH1* presented a similar expression pattern as previously reported [24]. The expression profiles of carotenoid structural genes were grouped according to their pattern similarity.

Table 2. Description of the analyzed target genes involved in carotenoid biosynthesis.

Salsa Id	Description	Abbreviation	Protein Id
20260	Bifunctional 15-*cis*-phytoene synthase, chromoplastic	*PSY1*	XP_016570422.2
10326	Phytoene synthase 2, chloroplastic	*PSY2*	XP_016560212.1
9117	Phytoene synthase 2, chloroplastic-like	*PSY3*	XP_016576229.2
31467	15-*cis*-phytoene desaturase, chloroplastic/chromoplastic	*PDS*	XP_016562405.2
22592	15-*cis*-zeta-carotene isomerase, chloroplastic	*Z-ISO*	XP_016550148.1
5981	Zeta-carotene desaturase, chloroplastic/chromoplastic	*ZDS1*	NP_001311497.1
5979	Zeta-carotene desaturase, chloroplastic/chromoplastic	*ZDS2*	NP_001311497.1
11712	Prolycopene isomerase, chloroplastic	*CRTISO1*	XP_016548432.1
28615	Prolycopene isomerase, chloroplastic	*CRTISO2*	XP_016555023.1
2529	Lycopene epsilon cyclase, chloroplastic isoform X2	*LCYE*	XP_016540361.1
4235	Lycopene beta cyclase, chloroplastic/chromoplastic	*LCYB1*	XP_016571836.1
27689	Lycopene beta cyclase, chloroplastic	*LCYB2*	XP_016543793.2
20137	Protein LUTEIN DEFICIENT 5, chloroplastic	*CYP97A*	XP_016551303.1
4572	Beta-carotene hydroxylase 1, chloroplastic	*BCH1*	NP_001311784.1
13055	Beta-carotene hydroxylase 2, chloroplastic isoform X1	*BCH2*	NP_001385279.1
28396	Zeaxanthin epoxidase, chloroplastic	*ZEP*	XP_016561102.1
16272	Violaxanthin de-epoxidase, chloroplastic isoform X1	*VDE*	XP_016550436.1
34360	Capsanthin/capsorubin synthase, chromoplastic	*CCS*	NP_001311998.1

Id, identifier.

The averages of the gene expression profiles of 12 accessions of *Capsicum annuum*, instead of individual SEPs, are presented in Figure 2. *PSY1*, *PDS*, and *ZDS2* showed transcript accumulation in the flower (0 DAA), although the transcripts of the *PSY* and *PDS* genes decreased during the growth of chili pepper fruit, while the transcripts of *ZDS2* exhibited a slight increase at 20 DAA and then decreased. At close to 50 DAA, a notable increase in the expression levels was observed for these three genes. In addition, the *BCH1* and *CCS* genes showed little or no accumulation of transcripts in the flower and growing fruit (10–40 DAA), although the expression increased at 50 and 60 DAA. *PSY1*, *BCH1*, and *CCS* reached their highest levels of expression at 60 DAA, whereas the transcript levels of *PDS* and *ZDS2* at 60 DAA were only slightly higher than at 50 DAA (Figure 2a). A similar pattern of expression was observed for the *ZDS1* gene, with an accumulation of transcripts in flowers, an increase at 40 DAA, a subsequent slight decrease at 50 DAA, and then maximum levels reached at 60 DAA (Figure 2a).

The *Z-ISO* and *LCYB1* genes displayed a low accumulation of transcripts at 0 DAA and a slight decrease at 10 DAA; nevertheless, a slight increase in the levels of expression was detected from 10 to 40 DAA, followed by a drastic increase at 50 DAA and a subsequent decrease at 60 DAA (Figure 2b).

In contrast, *CRTISO1*, *LCYE*, *LCYB2*, and *ZEP* reached their maximum levels of expression at 0 DAA (flower), with an abrupt decrease in the accumulation of transcripts at 10 DAA (Figure 2c). The expression levels of *LCYE* and *ZEP* remained low from 10 to 40 DAA, with a slight decrease at 50 and 60 DAA. Transcripts of *CRTISO1* gradually increased from 30 to 60 DAA but never reached the levels of expression detected in flowers. Similarly, an increase in the accumulation of *LCYB2* transcripts at 20 DAA was observed, which then decreased at 50 and 60 DAA (Figure 2c).

Finally, different types of transcription profiles were observed for *CRTISO2* and *CYP97A*. In the case of *CRTISO2*, a high accumulation of transcripts was detected from 0 to 40 DAA, and a subsequent decrease was observed at 50 and 60 DAA. The expression profile of *CYP97A* showed a sharp increase from 0 to 10 DAA, when the highest level of transcripts was reached, followed by an abrupt decrease at 20 DAA and then a slow decrease from 30 to 60 DAA (Figure 2d).

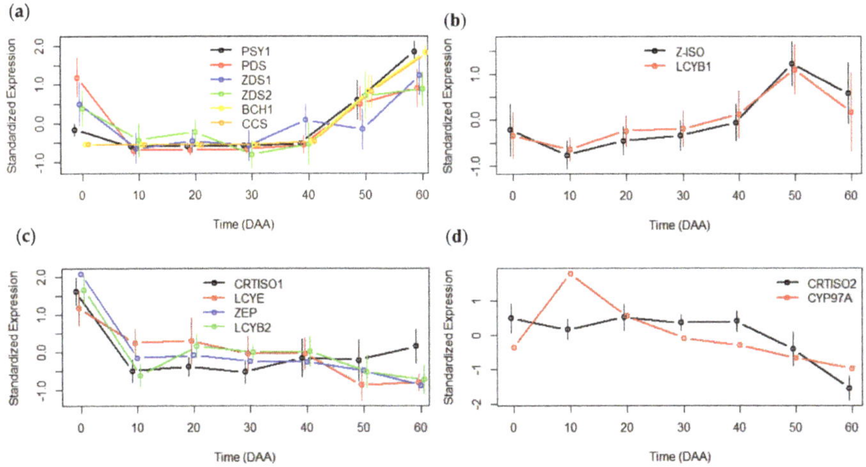

Figure 2. Groups of TGs of carotenoid biosynthesis with similar expression patterns. (**a**) Phytoene synthase (*PSY1*), phytoene desaturase (*PDS*), Z-carotene desaturase 1 (*ZDS1*), Z-carotene desaturase 2 (*ZDS2*), beta-carotene hydroxylase (*BCH1*), and capsanthin/capsorubin synthase (*CCS*) (**b**) Z-carotene isomerase (*Z-ISO*) and lycopene beta cyclase 1 (*LCYB1*) (**c**) lycopene isomerase 1 (*CTRTISO1*), lycopene epsilon cyclase (*LCYE*), zeaxanthin epoxidase (*ZEP*), and lycopene beta cyclase 2 (*LCYB2*) (**d**) lycopene isomerase 2 (*CRTISO2*) and cytochrome P450 97A (*CYP97A*).

2.2. Coexpression Analysis of Carotenoid-Biosynthesis-Related Genes and TF Genes

The prediction of transcription factor (TF) gene candidates as possible participants in the regulation of carotenoid-biosynthesis-related genes was carried out by a coexpression analysis using the function g2g.TFcandidates, which is available in the R package *Salsa* v 1.0 [25]. Hypothesizing that TGs and TF genes should show highly similar expression profiles, strong and significant positive correlations between the SEPs of TFs and TGs were expected and evaluated. The first search for TF gene candidates was performed using the minimum $r^2 = 0.7$, implying that all relations should have $r^2 > 0.7$ to be considered "winners" and should appear in the output. Robustness was evaluated by the number of times that a correlation between the expression profiles of TG and TF genes was independently detected in different accessions at the fixed false discovery rate (FDR) selected by the researcher, and a minimum number of accessions (n.min.acc) = 12. For all analyzed TGs, the list of TF "winners" and the statistical data of coexpression analyses are compiled in Table S1 (Supplementary Materials).

As a second criterion for the selection of TF candidates, we searched for putative binding sites for TF "winners" in the regulatory regions of TGs. The localization of TGs in the *Capsicum* reference genome, the prediction of consensus sequences of the promoter region (the TATA box, initiator element (INR), and downstream promoter element (DPE)), and the putative transcription start site (TSS) are shown in Table S2 (Supplementary Materials). On the other side, TF "winners" with putative binding sites located in TG promoters and selected as good candidates to regulate the expression of structural genes of carotenoid biosynthesis and their *Arabidopsis* locus names and *S. lycopersicum* orthologs are included in Table S3 (Supplementary Materials). Finally, Figure 3 shows the average SEPs of each analyzed TG and the four best candidates selected in each case.

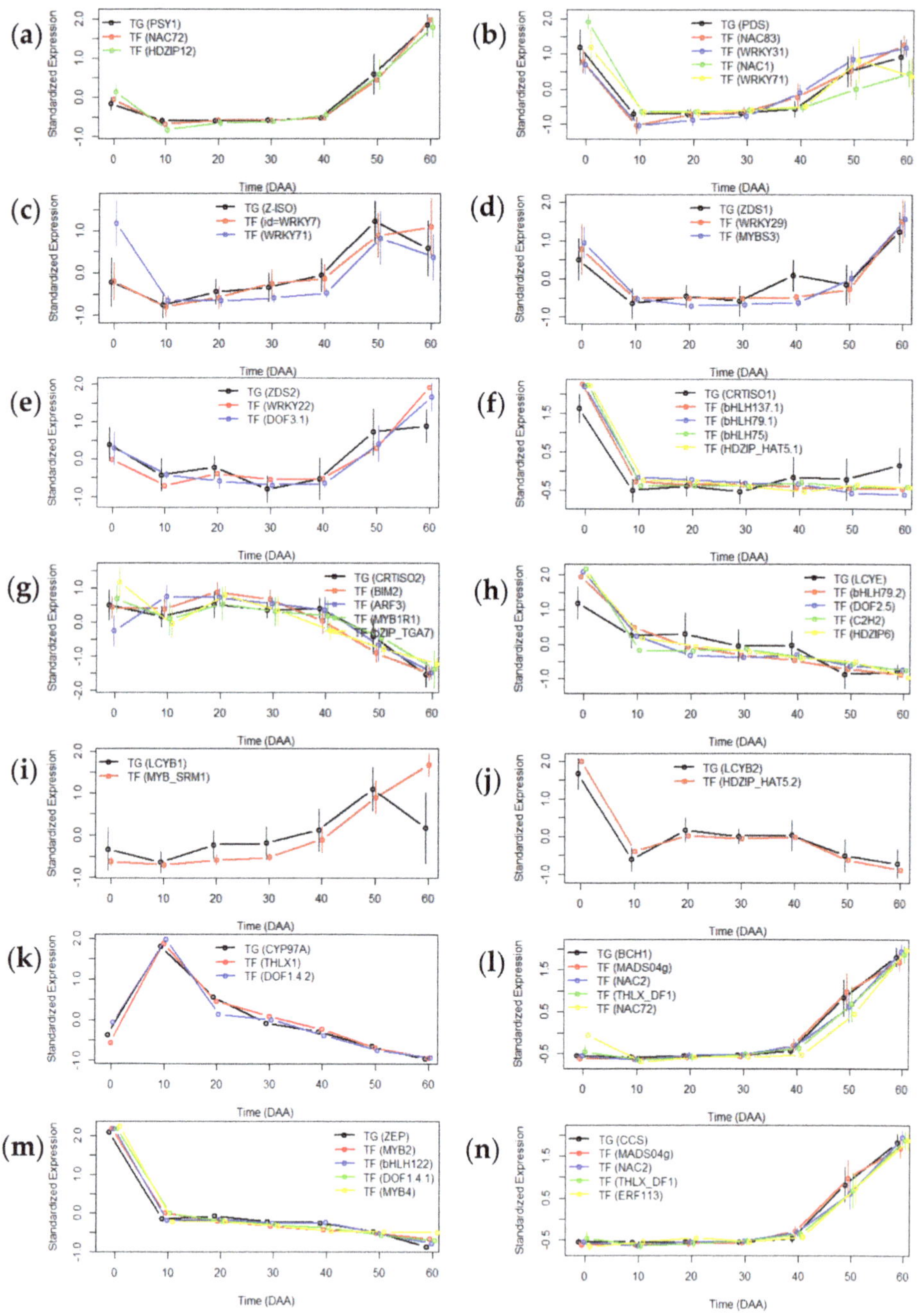

Figure 3. Average SEPs for TGs and the "TF winners". (**a**) Phytoene synthase (*PSY1*), (**b**) phytoene desaturase (*PDS*), (**c**) Z-carotene isomerase (*Z-ISO*), (**d**) Z-carotene desaturase 1 (*ZDS1*), (**e**) Z-carotene desaturase 2 (*ZDS2*), (**f**) lycopene isomerase 1 (*CTRTISO1*), (**g**) lycopene isomerase 2 (*CRTISO2*), (**h**) lycopene epsilon cyclase (*LCYE*), (**i**) lycopene beta cyclase 1 (*LCYB1*), (**j**) lycopene beta cyclase 2 (*LCYB2*), (**k**) cytochrome P450 97A (*CYP97A*), (**l**) beta-carotene hydroxylase (*BCH1*), (**m**) zeaxanthin epoxidase (*ZEP*), and (**n**) capsanthin/capsorubin synthase (*CCS*).

As a result of the analysis of RNA-Seq, transcripts of 1859 transcription factors were identified during the growth and ripening process of *Capsicum* spp. After the coexpression analysis, we selected 83 TFs as possible regulators of the different carotenogenic structural genes (Table S1, Supplementary Materials). From them, putative binding sites in the promoters of those structural genes were identified for only 54 TFs (Table S3, Supplementary Materials).

By applying stringent parameters (min.fdr = 0.1, min.r^2 = 0.6, and n.min.acc = 12), we obtained six good candidates for the transcriptional regulation of the *PSY1* gene (Table S1, Supplementary Materials). All of them presented a mean value for r (r.Med) < 0.94; nevertheless, the analysis of the promoter region of the *PSY* gene showed eight putative motif union sites just for PREDICTED NAC domain-containing protein 72-like (NAC72) and one putative binding site for Homeobox-leucine zipper protein ATHB-12-like (HDZIP12), (Figure 3a, Table S3, Supplementary Materials).

In the case of *PDS*, a coexpression analysis using stringent parameters did not identify appropriate TFs for its regulation; therefore, additional stringent parameters were tested (min.fdr = 0.1, min.r^2 = 0.6, and n.min.acc = 8). The algorithm displayed four TF candidate genes with similar expression profiles as *PDS*: NAC domain-containing protein 83 (NAC83), a putative WRKY transcription factor 31 (WRKY31), NAC domain-containing protein 1 (NAC1), and WRKY transcription factor 71 (WRKY71) (Table S1, Supplementary Materials). Putative binding sites were identified for all these TFs (Figure 3b, Table S3, Supplementary Materials); accordingly, these TFs might be good candidates for *PDS* regulation for only 8 of the 12 analyzed accessions (Table S1, Supplementary Materials).

Less strict parameters were used to search for TF gene candidates possibly involved in the transcriptional regulation of *Z-ISO*: min.fdr = 0.1, min.r^2 = 0.6, and n.min.acc = 6. The transcriptional profile of three TF genes showed a positive correlation (r.Med < 0.87) with the *Z-ISO* expression profile in seven accessions of *Capsicum annuum* (Table S1, Supplementary Materials). A regulatory region analysis in the *Z-ISO* promoter showed 12 and 11 putative binding motifs for WRKY transcription factor 7 (WRKY7) and WRKY71, respectively (Table S3, Supplementary Materials), and both were considered good candidates to regulate the expression of this carotenogenic gene (Figure 3c).

As mentioned above, two different expression profiles for *ZDS* were identified in the *Salsa* database with the same protein Id (*ZDS1* and *ZDS 2*) (Table 2 and Figure 2a); thus, a coexpression analysis was carried out independently for both. Using min.fdr = 0.1, min.r^2 = 0.6, and n.min.acc = 6 as parameters, five TF candidate genes with a positive correlation (r.Med > 0.85) with that of *ZDS1* were predicted by g2gTF.candidates (Table S1, Supplementary Materials). Among them, an analysis of the *ZDS* promoter showed binding sites only for a putative WRKY transcription factor 29 (WRKY29) and a transcription factor MYBS3 (MYBS3) (Figure 3d and Table S3, Supplementary Materials); therefore, both were selected as candidates to regulate the transcription of this gene in seven and six accessions, respectively (Table S1, Supplementary Materials). By using the same search parameters, the *Salsa* algorithm selected two TFs with a positive correlation (r.Med > 0.85) with *ZDS2*: WRKY transcription factor 22 (WRKY22) and Dof zinc finger protein DOF3.1 (DOF3.1) (Figure 3e and Table S1, Supplementary Materials). For these TFs, 3 and 36 putative binding sites in the *ZDS* promoter were identified, respectively (Table S3, Supplementary Materials). Both TF candidate genes showed a positive correlation with the *ZDS* expression profile in six accessions; however, although *WRKY22* was not selected in CW and the *DOF3.1* was not identified in ZU, both were ultimately considered as good TF candidates for the six mentioned accessions of *Capsicum annuum* (Table S1, Supplementary Materials).

In the same way, two distinct SEPs with contrasting expression patterns for *CRTISO* were identified in *Salsa* (*CRTISO1* and *CRTISO2*) (Table 2, Figure 2c,d). They showed different patterns of expression during the growth and ripening processes of *Capsicum* fruits, and both were subjected to coexpression testing. A correlation analysis of *CRTISO1* was performed using min.fdr = 0.05, min.r^2 = 0.7, and n.min.acc = 6 as selection parameters. Approximately 80 candidates with a positive correlation were displayed by the algorithm.

The five best options (r.Med > 0.9) and the number of putative binding sites in the *CRTISO1* promoter are summarized in Tables S1 and S3 (Supplementary Materials). Moreover, a coexpression analysis of *CRTISO2* using less strict parameters (min.fdr = 0.1, min.r^2 = 0.6, and n.min.acc = 7) showed seven candidates with r.Med $\geq$ 0.85 (Table S1, Supplementary Materials). The *CRTISO2* promoter analysis showed binding motifs for five of them: transcription factor BIM2-like isoform X4 (BIM2), auxin response factor 3 (ARF3), transcription factor MYB1R1 (MYB1R1), transcription factor TGA7 (bZIP_TGA7), and a putative WRKY transcription factor 20 isoform X1 (WRKY20) (Table S3). In the case of *CRTISO1*, *CRTISO2*, and other analyzed TGs, only the first four TF "winners" with putative binding sites in their promoter region were graphed in Figure 3f,g; data of the other selected TF candidates can be consulted in Table S3 (Supplementary Materials).

A coexpression analysis for prediction of TF candidates to regulate *LCYE* rendered results in five accessions (CO, CQ, SR, ST, and SY) using less strict parameters: min.fdr = 0.1, min.r^2 = 0.6, and n.min.acc = 5. For these accessions, the g2g.TFcandidates algorithm identified 11 options with a positive correlation (r.Med > 0.9) between TG and TF SEPs (Table S1, Supplementary Materials). However, the *LCYE* promoter analysis showed binding sites for only eight of them: PREDICTED transcription factor bHLH79 (bHLH79.2), Dof zinc finger protein DOF2.5-like isoform X2 (DOF2.5), zinc finger protein NUTCRACKER (C2H2), homeobox-leucine zipper protein ATHB-6 (HDZIP6), PREDICTED homeobox-leucine zipper protein HAT5 (HDZIP_HAT5.2), Dof zinc finger protein DOF2.1 (DOF2.1), Dof zinc finger protein DOF1.4 (DOF1.4.1), and Dof zinc finger protein DOF3.6 (DOF3.6) (Table S3, Supplementary Materials). Figure 3h presents the coexpression patterns of *LCYE* and the first four selected TF gene candidates.

Two SEPs with different expression patterns were identified for *LCYB* (*LCYB1* and *LCYB2*) (Table 2, Figure 2b,c). Therefore, both SEPs were subjected to a coexpression analysis using nonstringent parameters: min.fdr = 0.01, min.r^2 = 0.6, and n.min.acc = 6 (Figure 3i,j). For *LCYB1*, three TF "winners" with r.Med > 0.8 were identified; however, the analysis of the regulatory region showed only cis sequences for the transcription factor SRM1 isoform X1 (MYB_SRM1) in only six accessions (Figure 3i and Table S3, Supplementary Materials). Moreover, the *Salsa* algorithm identified four SEPs that showed a positive correlation (r.Med > 0.85) with *LCYB2* SEPs (Table S1, Supplementary Materials). Among them, binding motifs were located only for PREDICTED homeobox-leucine zipper protein HAT5 (HDZIP_HAT5.2) (Table S3, Supplementary Materials); therefore, this TF represented a good candidate to regulate *LCYB2* in nine accessions of *Capsicum annuum*.

Using the parameters min.fdr = 0.1, min.r^2 = 0.6, and n.min.acc= 12, four TF "winners" were suggested as candidates for *CYP97A* regulation by a coexpression analysis with a correlation coefficient of r.Med < 0.9 (Table S1, Supplementary Materials). Interestingly, the analysis of cis sequences in the promoter of *CYP97A* revealed up to 56 binding motifs for Dof zinc finger protein DOF1,4 (DOF1.4.2) and 13 putative binding motifs for trihelix transcription factor GTL1-like (THLX1) (Figure 3k and Table S3, Supplementary Materials).

On the other hand, very stringent parameters (min.fdr = 0.05, min.r^2 = 0.7, and n.min.acc = 12) were applied to the coexpression analysis between *BCH1* and the TF gene candidates. Eight TF "winners" were displayed by the algorithm, and remarkably, at least six TFs presented a positive correlation of r.Med > 0.95 (Table S1, Supplementary Materials). A minimum of 2 and a maximum of 40 binding sites were localized in the *BCH* promoter region for six TFs: MADS-box protein 04g005320 (MADS04g), NAC domain-containing protein 2 (NAC2), trihelix transcription factor DF1 (THLX_DF1), PREDICTED NAC domain-containing protein 72-like (NAC72), TGACG-sequence-specific DNA-binding protein TGA-2.1 (bZIP_2.1), and homeobox-leucine zipper protein HAT5 (HDZIP_HAT5.3) (Table S3, Supplementary Materials). Although only the four candidates with the highest r.Med are shown in Figure 3l, the six selected TFs can be considered good candidates for *BCH* transcriptional regulation.

A positive correlation (r.Med $\geq$ 0.92) between the SEPs of 13 TFs and *ZEP* was identified based on a coexpression analysis using very strict parameters (min.fdr = 0.05, min.r^2 = 0.7, and n.min.acc = 12) (Table S1, Supplementary Materials). An analysis of the *ZEP* promoter region allowed us to find putative binding sites for nine candidates proposed by the g2g.TFcandidates algorithm: MYB-related protein 2 isoform X1 (MYB2), PREDICTED transcription factor bHLH122-like (bHLH122), Dof zinc finger protein DOF1.4 (DOF1.4.1), MYB-like transcription factor 4 (MYB4), homeobox protein 13-like isoform X2 (HDZIP13), PREDICTED transcription factor HBI1 (bHLH_HBI1), Dof zinc finger protein DOF2.4-like (DOF2.4), transcription factor TCP4 (TCP4), and PREDICTED transcription factor DIVARICATA-like isoform X1 (MYB_DIV) (Table S3, Supplementary Materials), and the expression pattern was apparently similar in all TF SEPs (Figure 3m).

Less strict parameters (min.fdr = 0.1, min.r^2 = 0.6, and n.min.acc = 8) were employed to search for TF candidates for *VDE*. Only two TFs with low correlation coefficients (r.Med > 0.8) were displayed by the *Salsa* algorithm: an ethylene-responsive transcription factor-like protein, At4g13040, and a chloroplastic zinc finger protein, VAR3 (Table S1, Supplementary Materials). Unfortunately, no putative binding sites for these TFs were localized in the *VDE* promoter region. For these reasons, these results are not shown.

A coexpression analysis was carried out between TF SEPs and *CCS* transcripts using very strict parameters (min.fdr = 0.05, min.r^2 = 0.6, and n.min.acc = 12), and it showed at least six TFs with a positive correlation (r.Med $\geq$ 0.94) (Table S1). The search for putative binding sites in the promoter of *CCS* detected several binding motifs for the TF candidates selected by the *Salsa* algorithm: MADS-box protein 04g005320 (MADS04g), ethylene-responsive transcription factor ERF113 (ERF113), trihelix transcription factor DF1 (THLX_DF1), NAC domain-containing protein 2-like (NAC2), and PREDICTED TGACG-sequence-specific DNA-binding protein TGA-2.1 (bZIP_2.1) (Table S3, Supplementary Materials). The expression patterns of *CCS* and the TF candidates with r.Med >95 are shown in Figure 3n.

3. Discussion

Carotenoids are important pigments in plants because they are involved in essential processes, such as photosynthesis, photoprotection, and photomorphogenesis, and they are also precursors for the biosynthesis of the growth regulators abscisic acid (ABA) and strigolactones [26]. The biosynthesis of carotenoids is regulated at different levels. In the first instance, it depends on the availability of isoprenoid substrates; moreover, they are points in the pathway that controls the flux of metabolites through the pathway [27]. Environmental conditions, such as light, are important regulators of the carotenogenic pathway because they stimulate the biosynthesis of carotenoids and lead plastid structures to accumulate pigments [28]. Additionally, the biosynthesis of phytoene represents a rate-limiting step of the carotenoid biosynthesis pathway; the differential expression of *PSY* is regulated by abiotic factors such as temperature, photoperiod, salt, drought, high light, and ABA [29]. Finally, at the genetic level, carotenoid biosynthesis is regulated at the transcriptional, post-transcriptional, and epigenetic levels [27].

In *Capsicum* spp., the content of carotenoids in fruits is variable and changes from green or yellow in immature stages to red, orange, and yellow in mature fruits [1]. It has been reported that the *PSY*, *LCYB*, *BCH*, and *CCS* genes are necessary for the synthesis of capsanthin, and variations in their expression patterns result in the variety of colors of chili pepper fruits [30]. In this sense, the accumulation of transcripts of some biosynthetic genes is congruent with the change in color in the fruits and the accumulation of carotenoids [24].

The transcriptional regulation of the carotenogenic pathway in chili pepper fruits is poorly understood, as mentioned previously. Few transcription factors have been identified as regulators of the expression of the genes encoding the carotenoid biosynthetic enzymes. In our present study, we identified TF candidates that might be involved in the transcriptional regulation of these carotenogenic enzymes by using coexpression analysis and searching for putative binding sites for each TF candidate in the promoter of structural genes of the carotenoid biosynthesis pathway. We consider that this approach represents a good option for selecting TF candidates for subsequent studies of function and characterization through different experimental approaches. In the present work, an analysis of transcriptomic data identified 1859 TFs that are expressed during the growth, development, and ripening of *Capsicum* spp. fruits. After the coexpression analyses, we obtained a list of 83 TF "winners". After the analysis of the binding sites of these TFs in the promoter regions of the carotenoid-biosynthesis-related structural genes, it was possible to compile a list of 54 TF candidates susceptible to experimental assessment. A correlation analysis was previously used with a similar purpose in *Capsicum* spp. A positive correlation between the expression patterns of *CaERF82*, *CaERF97*, *CaERF66*, *CaERF107*, and *CaERF101* and the accumulation patterns of carotene, zeaxanthin, and capsorubin suggested their possible participation in the transcriptional regulation of carotenoid biosynthesis [5]. Additionally, the carotenoid biosynthesis profile of lutein, zeaxanthin, and capsorubin in the pericarp showed a positive correlation with the expression profile of the *CabHLH009*, *CabHLH032*, *CabHLH048*, *CabHLH095*, and *CabHLH100* genes [31]. A coexpression analysis was also used to identify TF candidates for the regulation of the carotenoid biosynthesis pathway in *Capsicum* spp. A positive correlation between the expression pattern of the *CCS* gene and the TF genes *CaDIV1* and *CaMYB3R-5* was previously reported [6]. Finally, by using a coexpression analysis, U-box domain-containing protein 52, GATA transcription factor 26, RING/FYVE/PHD-type, F-box protein SKIP23, CONSTANS-LIKE 9, and zinc finger family FYVE/PHD-type were suggested as possible transcriptional regulators of *CCS* in *Capsicum* [7]. Interestingly, our coexpression approach combined with an analysis of the promoter regions of structural carotenoid biosynthetic genes did not recover any of the candidates mentioned above (Table S3, Supplementary Materials).

The described functions for the families of the TF candidates compiled in Table S3 (Supplementary Materials) are associated with the regulation of responses to different biotic and abiotic factors, such as pathogens, cold, drought, and salt stress, among others. This was also the case for the TF CaMADS, which was found to play an important role in the responses to cold, salt, and osmotic stress in chili pepper [32]; in our analysis, only one TF candidate belonging to the MADS family (MADS04g) was selected by the g2g.TFcandidates algorithm, and it was proposed as a candidate to regulate the transcription of *BCH* and *CCS*. Interestingly, 9 and up to 56 putative binding sites for this TF were identified in the promoters of *BCH* and *CCS*, respectively (Table S3, Supplementary Materials). MADS04g (previously reported as CaMADS and CaMADS-RIN) is an important TF associated with fruit development/ripening in chili pepper [33–35]. In addition, its ortholog in tomato, SlCBM1 (MADS-box protein 04g005320, NCBI reference sequence: NP_001362848.1) was reported to be a positive regulator of tomato fruit ripening, in particular a regulator of ethylene biosynthesis and a transcriptional regulator of *PSY*, *PDS*, *LCYB*, and *LCYE* during carotenoid accumulation [36,37]. Thus, it is plausible to hypothesize that some TFs described as regulators of stress responses could also have a role during the growth, development, and ripening of fruits. Several functions have been assigned to MYB TFs in *Capsicum* spp. fruits; for instance, they have been reported to be involved in the accumulation of anthocyanins [38] and in the regulation of capsaicinoid biosynthesis [39,40] and cold stress [41]. In this work, members of the MYB family were selected as putative regulator candidates of carotenoid biosynthetic genes (Figure 3d,g,i,m; Table S3, Supplementary Materials); previously, Arce-Rodríguez et al. (2021), using a coexpression analysis, proposed *CaDIV1* (XP_016551180.1) as a candidate to regulate *CCS* only in the ST accession, which corresponds to MYB_SRM1 (XP_016551180.1), which was suggested as a regulator of *LCYB1* in the present study (Figure 3l) in six accessions of *Capsicum annuum* (Table S1, Supplementary Materials). Recently, a dual function of an MYB TF was described: the knockdown of *CaMYB306* in chili pepper resulted in decreased expression of carotenogenic structural genes in fruits. Additionally, the overexpression of this gene in *S. lycopersicum* plants increased the accumulation of carotenoids and showed earlier coloration of fruits. On the other hand, CaMYB306 was described as a negative regulator of cold resistance in chili pepper [8]. These results suggest that TFs that have been characterized as regulators of stress responses could also be involved in the regulation of carotenoid biosynthesis.

Two members of the Trihelix TF family (THLX1 and THLX_DF1) were identified as candidates for *CYP97A*, *BCH*, and *CCS* regulation. THLX_DF1 belongs to the GT-2 group and is involved in responses to pathogens; salt stress; flower, trichome, and stoma development; seed abscission; and embryogenesis. Moreover, the orthologs of THLX1 and THLX_DF1 in tomato (*S. lycopersicum*; *SlGT*-30) were demonstrated to be ubiquitously expressed in buds, flowers, leaves, and fruits [42]. Interestingly, Trihelix DF1 has been reported to bind to promoters of genes that are downregulated by light and dark-inducible [43,44].

At least six TFs belonging to the WRKY TF family (WRKY7, WRKY20, WRKY22, WRKY29, WRKY31, and WRKY71) were identified as candidates to regulate some points in the first steps of the carotenogenic pathway in this work. These TFs have been typically associated with environmental stress responses and defense, although important levels of the transcripts of several *CaWRKYs* have been detected during fruit ripening [45]. For example, recent evidence has demonstrated that WRKY9 could be associated with pungency and capsaicinoid content in chili pepper fruits [46]. Among the TF candidates proposed in this work, CaWRKY22 has been described as a positive regulator of innate immunity against *Ralstonia solanacearum*. This process is regulated by a signaling pathway mediated by jasmonic acid, salicylic acid, and ethylene and involves a network of WRKY TFs [47].

Regarding the NAC TF family, we propose four candidates as regulators of carotenoid biosynthesis: NAC1, NAC2, NAC72, and NAC83. Previous reports of NAC TFs in *Capsicum* spp. have associated them with biotic and abiotic stress; for instance, CaNAC1 was found to be induced during incompatible interactions between chili pepper and pathogens (viruses and bacteria) [48], CaNAC2 was found to be involved in the response to abiotic stress tolerance [49], CaNAC064 was found to be associated with cold stress tolerance [50], and CaNAC1 was proposed to regulate low-temperature-induced phospholipid degradation in green bell pepper [51]. In the same way, the orthologs of these four NAC candidates in *S. lycopersicum* have been grouped phylogenetically in the same cluster (Va) and associated with stress responses [52]. However, in tomato, NAC TFs have been associated with fruit ripening and the accumulation of carotenoids, particularly *SlNAC1* and *SlNAC4* (ortholog of *NAC2* selected as candidates in this work) [53,54]. Moreover, NAC TF (NOR-like1) has been identified as a positive regulator of fruit ripening because it is involved in the formation of color [55]. In particular, SlNAC4 is involved in the transcriptional regulation of the *PSY*, *LCYB*, and *LCYE* structural genes of the carotenoid biosynthesis pathway [37].

Additionally, seven different transcription factors of the DOF family were identified as TF candidates in the present study (Tables S1 and S3, Supplementary Materials), and they have been associated with seed development, dormancy, germination, and light-mediated processes [56,57].

Six candidates belonging to the HD-ZIP TF family were selected as candidates to regulate the biosynthesis of carotenoids in *C. annuum*. HDZIP (CaCBF) was reported as a cold-inducible TF [58], and CaHDZ27 was found to be a positive regulator of resistance to *R. solanacearum* [59]. In tomato, HD-ZIP homeobox genes have been associated with the control of ethylene, the regulation of floral organogenesis, and early fruit development [60]. In the present study, we propose two members of the bZIP TF family (bZIP_TGA7 and bZIP_TGA2.1) as TF candidates to regulate *CRTISO2* and *CCS*, respectively. In chili pepper, bZIPs have been involved in the modulation of ABA signaling and drought tolerance [61] and in environmental stress responses [62], while CabZIP2 has been proposed to participate in bacterial disease resistance [63,64].

Regarding the coexpression analyses in the accessions of *Capsicum* spp., these could only be performed for *PSY*, *BCH*, *CYP97A*, *CCS*, and *ZEP* in the 12 accessions, and they were not affected by the late ripening time of AS, CW, and JE, which mature at 80, 70, and 70 DAA, respectively [22]. A correlation analysis was carried out with *PDS*, *Z-ISO*, *ZDS*, *CRTISO*, *LCYE*, *LCYB*, and *VDE* in a minimum of five and a maximum of nine accessions; in these cases, the selected TFs predicted to be good candidates by the correlation analysis only apply for those accessions that were highly correlated.

4. Materials and Methods

4.1. Biological Materials, Sequencing, and RNA-Seq Data Analysis

Transcriptomic data were obtained from 12 accessions of *Capsicum annuum*: 4 wild, 6 domesticated, and 2 reciprocal crosses between domesticated and wild-type accessions (Table 1). Fruits were collected at seven time points (0, 10, 20, 30, 40, 50, and 60 days after anthesis (DAA)) during the growth and ripening process of chili pepper. The procedures of seed germination and growth, RNA-Seq library generation, and processing were as previously described [22]. A total of 84 sampling points (12 accessions $\times$ 7 time points) and two RNA-Seq libraries (biological replicates), 84 $\times$ 2 = 168 RNA-Seq libraries, were analyzed for time expression profile estimation. Data from 179 RNA-Seq libraries were obtained, and more than 3 billion raw reads were mapped to the reference chili pepper genome. These data were deposited in the NCBI gene expression omnibus (GEO) 25 and are accessible through GEO Series accession number GSE165448 (https://www.ncbi.nlm.nih.gov/geo/query/acc.cgi?Acc=GSE165448, accessed on 30 August 2022).

Data from all RNA-seq libraries were analyzed as standardized expression profiles (SEPs). SEPs are seven dimensional numeric vectors in which each of one of the numbers summarizes the standardized mean expression at the corresponding time point (0, 10, 20, 30, 40, 50, and 60 DAA) obtained from RNA-Seq in the 12 accessions of *Capsicum annuum*, based on ternary models in scale-free measurements [22,23].

4.2. Selection of TF Candidates

The prediction of transcription factor (TF) gene candidates as possible participants in the regulation of the carotenoid-biosynthesis-related genes was carried out by a coexpression analysis using the function g2g.TFcandidates, which is available in the R package *Salsa* v 1.0. [25]. Hypothesizing that target genes (TGs) and TF genes should show highly similar expression profiles, strong and significant positive correlations between the standardized expression profiles (SEPs) of TFs and TGs were expected and evaluated. This method implements a search for TF gene candidates by evaluating Pearson's correlations (r^2) between the expression profile of the TG and the expression profiles of each one on the TF genes annotated in the *Capsicum* reference genome (also recorded in the transcriptome *Salsa* database) in turn with each of the accessions that we wanted to consider preferentially with the 12 accessions of *Capsicum annuum*. By evaluating pairwise correlations between the TG and each TF gene profile in each accession at all stages of growth, development, and ripening, the algorithm filtered the "winners" (potential candidates using the set of thresholds input). The first search of TF gene candidates was performed using the minimum r^2 for the correlations, which was set to min.r^2 = 0.7, implying that all relations should have $r^2 > 0.7$ to be considered "winners" and should appear in the output. Note that this threshold was equivalent to asking that $\hat{r} > \sqrt{0.7} = 0.84$. Robustness was evaluated by the number of times that a correlation between the expression profiles of the TG and TF genes was independently detected in different accessions at the fixed false discovery rate (FDR) selected by the researcher. Thus, the stringent parameters employed for predicting TF gene candidates were min.fdr = 0.05, min.r^2 = 0.7, and n.min.acc = 12. If the algorithm did not display options using these strict parameters, less stringent values were tested (min.fdr = 0.1, min.r^2 = 0.6, and n.min.acc = 12). Finally, the number of accessions in which we wanted to identify TF candidates could be modified and tested once more. The statistical analyses were complemented by the maximum of the absolute differences (m.a) between the SEPs of the TG and TF genes. Values of m.a.Max > 1.5 were considered to contain extreme outliers. An m.a.Max < 0.5 was acceptable. Good TF gene candidates should have a "large" value of r^2 and a "small" value of m.a. Complete statistical values of each of the TF "winners" are compiled in Table S1 (Supplementary Materials). Finally, the TGs used for the coexpression analyses are summarized in Figure 2 and Table 2.

For the list of candidates displayed by g2g.TFcandidates, we analyzed the promoter region of each biosynthetic target gene to identify putative binding sequences for each TF candidate. Consensus DNA-binding sequences, including the TATA box, initiator element

(INR), downstream promoter element (DPE), and putative transcriptional starting site (TSS) of each promoter region, were located using the YAPP Eukaryotic Core Promoter program (http://www.bioinformatics.org/yapp/cgi-bin/yapp.cgi, accessed on 24 May 2022), and the results of the promoter region location of each gene are summarized in Table S2 (Supplementary materials). The bp of the TSS column indicated the bp number from ATG. Then, 2500 bp from the TSS was used for searching cis-sequences. Putative binding sites for TF candidates were determined using PlantPan software and *Arabidopsis thaliana* as references [65]. The UniProt [66] and Plant Transcription Factors databases were used for the identification of orthologous genes of *Capsicum* spp. in *A. thaliana* [67]. Only the transcription factor candidates with putative binding sites located in the promoters of target genes are summarized in Table S3 (Supplementary Materials).

5. Conclusions

In this work, we proposed a list of putative TF candidates associated with the regulation of the carotenoid biosynthesis pathway by RNA-Seq coexpression analysis, and further verification assays (ChIP, yeast one-hybrid, and virus-induced gene silencing, among others) are necessary to confirm or discard the participation of them in the transcriptional regulation of carotenoid biosynthesis. Due to the limited information about the transcriptional regulation of the carotenoid biosynthesis pathway in chili pepper and other nonclimacteric fruits, we propose a list of putative candidates that could guide future experimental approaches.

Some TF candidates, such as MADS04g and NAC2, have previously been reported as regulators of the ripening process in chili pepper and tomato, respectively. Thus, our RNA-Seq coexpression analysis represents a valuable tool for the analysis and selection of novel candidates to develop further functional TF gene studies and gene characterization. Gene function analysis using virus-induced gene silencing and the yeast one-hybrid assay of some of the selected candidates (ERF113, HDZIP_HAT5.3, MADS04g, NAC2, NAC72, and THLX_DF1) are currently underway.

It is important to mention that the TF "winners" listed in Table S1 (Supplementary Materials) for which it was not possible to locate cis-binding sequences in the promoters of carotenogenic genes cannot be ruled out as transcriptional regulators of the pathway, but they could act as regulators of other transcription factors and are susceptible to being evaluated by yeast two-hybrid assays.

Although most TFs of plants have been associated with the regulation of different processes other than the biosynthesis of carotenoids, there are several cases (addressed in the discussion section) where different members of families of TFs have been found to act as regulators of this biosynthetic pathway.

Supplementary Materials: The following supporting information can be downloaded at: https://www.mdpi.com/article/10.3390/ijms231911774/s1.

Author Contributions: Conceptualization, M.G.V.-R., O.M. and N.O.-A.; methodology, O.M. and M.G.V.-R.; formal analysis, investigation and data curation, M.G.V.-R.; resources, N.O.-A. and O.M.; supervision, N.O.-A. and O.M.; writing original draft preparation, M.G.V.-R. and N.O.-A.; review and editing, N.O.-A., M.G.V.-R. and O.M. All authors have read and agreed to the published version of the manuscript.

Funding: This work was supported by the Consejo Nacional de Ciencia y Tecnología (Conacyt; Mexico) project CB280755.

Institutional Review Board Statement: Not applicable.

Informed Consent Statement: Not applicable.

Data Availability Statement: Data is available at the Gene Expression Omnibus (GEO) of the NCBI (https://www.ncbi.nlm.nih.gov/geo/query/acc.cgi?Acc=GSE165448, accessed on 30 August 2022).

Conflicts of Interest: The authors declare no conflict of interest.

References

1. Gómez-García, M.d.R.; Ochoa-Alejo, N. Biochemistry and molecular biology of carotenoid biosynthesis in chili peppers (*Capsicum* spp.). *Int. J. Mol. Sci.* **2013**, *14*, 19025. [CrossRef] [PubMed]
2. Villa-Rivera, M.G.; Ochoa-Alejo, N. Chili pepper carotenoids: Nutraceutical properties and mechanisms of action. *Molecules* **2020**, *25*, 5573. [CrossRef] [PubMed]
3. Villa-Rivera, M.G.; Ochoa-Alejo, N. Transcriptional regulation of ripening in chili pepper fruits (*Capsicum* spp.). *Int. J. Mol. Sci.* **2021**, *22*, 12151. [CrossRef] [PubMed]
4. Pan, Y.; Bradley, G.; Pyke, K.; Ball, G.; Lu, C.; Fray, R.; Marshall, A.; Jayasuta, S.; Baxter, C.; van Wijk, R.; et al. Network inference analysis identifies an APRR2-like gene linked to pigment accumulation in tomato and pepper fruits. *Plant Physiol.* **2013**, *161*, 1476–1485. [CrossRef]
5. Song, J.; Chen, C.; Zhang, S.; Wang, J.; Huang, Z.; Chen, M.; Cao, B.; Zhu, Z.; Lei, J. Systematic analysis of the *Capsicum* ERF transcription factor family: Identification of regulatory factors involved in the regulation of species-specific metabolites. *BMC Genom.* **2020**, *21*, 573. [CrossRef]
6. Arce-Rodríguez, M.L.; Martinez, O.; Ochoa-Alejo, N. Genome-wide identification and analysis of the MYB transcription factor gene family in chili pepper (*Capsicum* spp.). *Int. J. Mol. Sci.* **2021**, *22*, 2229. [CrossRef] [PubMed]
7. Li, Q.H.; Yang, S.P.; Yu, Y.N.; Khan, A.; Feng, P.L.; Ali, M.; Shao, D.K.; Wang, Y.Y.; Zhang, R.X.; Gai, W.X.; et al. Comprehensive transcriptome-based characterization of differentially expressed genes involved in carotenoid biosynthesis of different ripening stages of *Capsicum*. *Sci. Hortic.* **2021**, *288*, 110311. [CrossRef]
8. Ma, X.; Yu, Y.-N.; Jia, J.-H.; Li, Q.-H.; Gong, Z.-H. The pepper MYB transcription factor CaMYB306 accelerates fruit coloration and negatively regulates cold resistance. *Sci. Hortic.* **2022**, *295*, 110892. [CrossRef]
9. Welsch, R.; Maass, D.; Voegel, T.; Dellapenna, D.; Beyer, P. Transcription factor RAP2.2 and its interacting partner SINAT2: Stable elements in the carotenogenesis of *Arabidopsis* leaves. *Plant Physiol.* **2007**, *145*, 1073–1085. [CrossRef]
10. Martel, C.; Vrebalov, J.; Tafelmeyer, P.; Giovannoni, J.J. The tomato MADS-Box transcription factor ripening inhibitor interacts with promoters involved in numerous ripening processes in a colorless nonripening-dependent manner. *Plant Physiol.* **2011**, *157*, 1568–1579. [CrossRef]
11. Mohanty, B.; Lakshmanan, M.; Lim, S.-H.; Kim, J.K.; Ha, S.-H.; Lee, D.-Y. Light-specific transcriptional regulation of the accumulation of carotenoids and phenolic compounds in rice leaves. *Plant Signal. Behav.* **2016**, *11*, e1184808. [CrossRef] [PubMed]
12. Fu, C.C.; Han, Y.C.; Kuang, J.F.; Chen, J.Y.; Lu, W.J. Papaya CpEIN3a and CpNAC2 co-operatively regulate carotenoid biosynthesis-related genes CpPDS2/4, CpLCY-e and CpCHY-b during fruit ripening. *Plant Cell Physiol.* **2017**, *58*, 2155–2165. [CrossRef] [PubMed]
13. Lu, S.; Zhang, Y.; Zhu, K.; Yang, W.; Ye, J.; Chai, L.; Xu, Q.; Deng, X. The citrus transcription factor CsMADS6 modulates carotenoid metabolism by directly regulating carotenogenic genes. *Plant Physiol.* **2018**, *176*, 2657–2676. [CrossRef]
14. Ampomah-Dwamena, C.; Thrimawithana, A.H.; Dejnoprat, S.; Lewis, D.; Espley, R.V.; Allan, A.C. A kiwifruit (*Actinidia deliciosa*) R2R3-MYB transcription factor modulates chlorophyll and carotenoid accumulation. *New Phytol.* **2019**, *221*, 309–325. [CrossRef] [PubMed]
15. Liu, Y.; Shi, Y.; Su, D.; Lu, W.; Li, Z. SlGRAS4 accelerates fruit ripening by regulating ethylene biosynthesis genes and SlMADS1 in tomato. *Hortic. Res.* **2021**, *8*, 3. [CrossRef] [PubMed]
16. Ríos, G.; Naranjo, M.A.; Rodrigo, M.-J.; Alós, E.; Zacarías, L.; Cercós, M.; Talón, M. Identification of a GCC transcription factor responding to fruit colour change events in citrus through the transcriptomic analyses of two mutants. *BMC Plant Biol.* **2010**, *10*, 276. [CrossRef]
17. Gómez-Gómez, L.; Trapero-Mozos, A.; Gómez, M.D.; Rubio-Moraga, A.; Ahrazem, O. Identification and possible role of a MYB transcription factor from saffron (*Crocus sativus*). *J. Plant Physiol.* **2012**, *169*, 509–515. [CrossRef]
18. Grassi, S.; Piro, G.; Lee, J.M.; Zheng, Y.; Fei, Z.; Dalessandro, G.; Giovannoni, J.J.; Lenucci, M.S. Comparative genomics reveals candidate carotenoid pathway regulators of ripening watermelon fruit. *BMC Genom.* **2013**, *14*, 781. [CrossRef]
19. Ye, J.; Hu, T.; Yang, C.; Li, H.; Yang, M.; Ijaz, R.; Ye, Z.; Zhang, Y. Transcriptome profiling of tomato fruit development reveals transcription factors associated with ascorbic acid, carotenoid and flavonoid biosynthesis. *PLoS ONE* **2015**, *10*, e0130885. [CrossRef]
20. Jiang, C.C.; Zhang, Y.F.; Lin, Y.J.; Chen, Y.; Lu, X.K. Illumina® sequencing reveals candidate genes of carotenoid metabolism in three pummelo cultivars (*Citrus maxima*) with different pulp color. *Int. J. Mol. Sci.* **2019**, *20*, 2246. [CrossRef]
21. Lu, C.; Pu, Y.; Liu, Y.; Li, Y.; Qu, J.; Huang, H.; Dai, S. Comparative transcriptomics and weighted gene co-expression correlation network analysis (WGCNA) reveal potential regulation mechanism of carotenoid accumulation in *Chrysanthemum × morifolium*. *Plant Physiol. Biochem.* **2019**, *142*, 415–428. [CrossRef] [PubMed]
22. Martínez, O.; Arce-Rodríguez, M.L.; Hernández-Godinez, F.; Escoto-Sandoval, C.; Cervantes-Hernández, F.; Hayano-Kanashiro, C.; Ordaz-Ortíz, J.J.; Reyes-Valdes, M.H.; Razo-Mendivil, F.G.; Garces-Claver, A.; et al. Transcriptome analyses throughout chili pepper fruit development reveal novel insights into the domestication process. *Plants* **2021**, *10*, 585. [CrossRef] [PubMed]
23. Escoto-Sandoval, C.; Flores-Diaz, A.; Reyes-Valdes, M.H.; Ochoa-Alejo, N.; Martínez, O. A method to analyze time expression profiles demonstrated in a database of chili pepper fruit development. *Sci. Rep.* **2021**, *11*, 13181. [CrossRef] [PubMed]
24. Martínez-López, L.A.; Ochoa-Alejo, N.; Martínez, O. Dynamics of the chili pepper transcriptome during fruit development. *BMC Genom.* **2014**, *15*, 143. [CrossRef]

25. Escoto-Sandoval, C.; Flores-Díaz, A.; Reyes-Valdés, M.H.; Ochoa-Alejo, N.; Martínez, O. *An R Package for Data Mining Chili Pepper Fruit Transcriptomes*; Research Square: Durham, CA, USA, 2020.
26. Maoka, T. Carotenoids as natural functional pigments. *J. Nat. Med.* **2020**, *74*, 1–16. [CrossRef]
27. Cazzonelli, C.I.; Pogson, B.J. Source to sink: Regulation of carotenoid biosynthesis in plants. *Trends Plant Sci.* **2010**, *15*, 266–274. [CrossRef]
28. Llorente, B.; Martinez-Garcia, J.F.; Stange, C.; Rodriguez-Concepcion, M. Illuminating colors: Regulation of carotenoid biosynthesis and accumulation by light. *Curr. Opin. Plant Biol.* **2017**, *37*, 49–55. [CrossRef]
29. Sathasivam, R.; Radhakrishnan, R.; Kim, J.K.; Park, S.U. An update on biosynthesis and regulation of carotenoids in plants. *S. Af. J. Bot.* **2021**, *140*, 290–302. [CrossRef]
30. Tian, S.L.; Li, L.; Shah, S.N.M.; Gong, Z.H. The relationship between red fruit colour formation and key genes of capsanthin biosynthesis pathway in *Capsicum annuum*. *Biol. Plant.* **2015**, *59*, 507–513. [CrossRef]
31. Liu, R.; Song, J.; Liu, S.; Chen, C.; Zhang, S.; Wang, J.; Xiao, Y.; Cao, B.; Lei, J.; Zhu, Z. Genome-wide identification of the *Capsicum* bHLH transcription factor family: Discovery of a candidate regulator involved in the regulation of species-specific bioactive metabolites. *BMC Plant Biol.* **2021**, *21*, 262. [CrossRef]
32. Chen, R.; Ma, J.; Luo, D.; Hou, X.; Ma, F.; Zhang, Y.; Meng, Y.; Zhang, H.; Guo, W. CaMADS, a MADS-box transcription factor from pepper, plays an important role in the response to cold, salt, and osmotic stress. *Plant Sci.* **2019**, *280*, 164–174. [CrossRef] [PubMed]
33. Soon-Kee, S.; Yong-Hwan, M.; Jae-Eun, C.; Sook-Yi, L.; HyoGuen, P.; Gynheung, A. Characterization of MADS box genes from hot pepper. *Mol. Cells* **2001**, *11*, 352–359.
34. Dong, T.; Chen, G.; Tian, S.; Xie, Q.; Yin, W.; Zhang, Y.; Hu, Z. A non-climacteric fruit gene CaMADS-RIN regulates fruit ripening and ethylene biosynthesis in climacteric fruit. *PLoS ONE* **2014**, *9*, e95559. [CrossRef] [PubMed]
35. Dubey, M.; Jaiswal, V.; Rawoof, A.; Kumar, A.; Nitin, M.; Chhapekar, S.S.; Kumar, N.; Ahmad, I.; Islam, K.; Brahma, V.; et al. Identification of genes involved in fruit development/ripening in *Capsicum* and development of functional markers. *Genomics* **2019**, *111*, 1913–1922. [CrossRef] [PubMed]
36. Zhang, J.; Hu, Z.; Yao, Q.; Guo, X.; Nguyen, V.; Li, F.; Chen, G. A tomato MADS-box protein, SlCMB1, regulates ethylene biosynthesis and carotenoid accumulation during fruit ripening. *Sci. Rep.* **2018**, *8*, 3413. [CrossRef]
37. Li, C.X.; Hou, X.M.; Qi, N.A.; Liu, H.W.; Li, Y.H.; Huang, D.J.; Wang, C.L.; Liao, W.B. Insight into ripening-associated transcription factors in tomato: A review. *Sci. Hortic.* **2021**, *288*, 110363. [CrossRef]
38. Li, J.G.; Li, H.L.; Peng, S.Q. Three R2R3 MYB transcription factor genes from *Capsicum annuum* showing differential expression during fruit ripening. *Afr. J. Biotechnol.* **2011**, *10*, 8267–8274. [CrossRef]
39. Arce-Rodríguez, M.L.; Ochoa-Alejo, N. An R2R3-MYB transcription factor regulates capsaicinoid biosynthesis. *Plant Physiol.* **2017**, *174*, 1359–1370. [CrossRef]
40. Sun, B.; Zhu, Z.; Chen, C.; Chen, G.; Cao, B.; Chen, C.; Lei, J. Jasmonate-inducible R2R3-MYB transcription factor regulates capsaicinoid biosynthesis and stamen development in *Capsicum*. *J. Agric. Food Chem.* **2019**, *67*, 10891–10903. [CrossRef]
41. Zhang, R.-X.; Cheng, G.-X.; Liu, G.-T.; Chen, S.-Y.; Haq, S.u.; Khan, A.; Li, Q.-H.; Gong, Z.-H. Assessing the functional role of color-related CaMYB gene under cold stress using virus-induced gene silencing in the fruit of pepper (*Capsicum annuum* L.). *Sci. Hortic.* **2020**, *272*, 109504. [CrossRef]
42. Yu, C.; Cai, X.; Ye, Z.; Li, H. Genome-wide identification and expression profiling analysis of trihelix gene family in tomato. *Biochem. Biophys. Res. Commun.* **2015**, *468*, 653–659. [CrossRef] [PubMed]
43. Kaplan-Levy, R.N.; Brewer, P.B.; Quon, T.; Smyth, D.R. The trihelix family of transcription factors—Light, stress and development. *Trends Plant Sci.* **2012**, *17*, 163–171. [CrossRef] [PubMed]
44. Nagano, Y.; Inaba, T.; Furuhashi, H.; Sasaki, Y. Trihelix DNA-binding protein with specificities for two distinct cis-elements—Both important for light down-regulated and dark-inducible gene expression in higher plants. *J. Biol. Chem.* **2001**, *276*, 22238–22243. [CrossRef]
45. Cheng, Y.; Ahammed, G.J.; Yu, J.; Yao, Z.; Ruan, M.; Ye, Q.; Li, Z.; Wang, R.; Feng, K.; Zhou, G.; et al. Putative WRKYs associated with regulation of fruit ripening revealed by detailed expression analysis of the WRKY gene family in pepper. *Sci. Rep.* **2016**, *6*, 39000. [CrossRef]
46. Kondo, F.; Hatakeyama, K.; Sakai, A.; Minami, M.; Nemoto, K.; Matsushima, K. The pungent-variable sweet chili pepper 'Shishito' (*Capsicum annuum*) provides insights regarding the relationship between pungency, the number of seeds, and gene expression involving capsaicinoid biosynthesis. *Mol. Genet. Genom.* **2021**, *296*, 591–603. [CrossRef] [PubMed]
47. Hussain, A.; Li, X.; Weng, Y.; Liu, Z.; Ashraf, M.F.; Noman, A.; Yang, S.; Ifnan, M.; Qiu, S.; Yang, Y.; et al. CaWRKY22 acts as a positive regulator in pepper response to *Ralstonia solanacearum* by constituting networks with CaWRKY6, CaWRKY27, CaWRKY40, and CaWRKY58. *Int. J. Mol. Sci.* **2018**, *19*, 1426. [CrossRef] [PubMed]
48. Oh, S.-K.; Lee, S.; Yu, S.H.; Choi, D. Expression of a novel NAC domain-containing transcription factor (CaNAC1) is preferentially associated with incompatible interactions between chili pepper and pathogens. *Planta* **2005**, *222*, 876–887. [CrossRef]
49. Guo, W.L.; Wang, S.B.; Chen, R.G.; Chen, B.H.; Du, X.H.; Yin, Y.X.; Gong, Z.H.; Zhang, Y.Y. Characterization and expression profile of CaNAC2 pepper gene. *Front. Plant Sci.* **2015**, *6*, 755. [CrossRef]
50. Hou, X.M.; Zhang, H.F.; Liu, S.Y.; Wang, X.K.; Zhang, Y.M.; Meng, Y.C.; Luo, D.; Chen, R.G. The NAC transcription factor CaNAC064 is a regulator of cold stress tolerance in peppers. *Plant Sci.* **2020**, *291*, 110346. [CrossRef]

51. Kong, X.M.; Zhou, Q.; Zhou, X.; Wei, B.D.; Ji, S.J. Transcription factor CaNAC1 regulates low-temperature-induced phospholipid degradation in green bell pepper. *J. Exp. Bot.* **2020**, *71*, 1078–1091. [CrossRef]
52. Su, H.; Zhang, S.; Yin, Y.; Zhu, D.; Han, L. Genome-wide analysis of NAM-ATAF1,2-CUC2 transcription factor family in *Solanum lycopersicum*. *J. Plant Biochem. Biotechnol.* **2015**, *24*, 176–183. [CrossRef]
53. Ma, N.; Feng, H.; Meng, X.; Li, D.; Yang, D.; Wu, C.; Meng, Q. Overexpression of tomato SlNAC1 transcription factor alters fruit pigmentation and softening. *BMC Plant Biol.* **2014**, *14*, 351. [CrossRef] [PubMed]
54. Zhu, M.; Chen, G.; Zhou, S.; Tu, Y.; Wang, Y.; Dong, T.; Hu, Z. A new tomato NAC (NAM/ATAF1/2/CUC2) transcription factor, SlNAC4, functions as a positive regulator of fruit ripening and carotenoid accumulation. *Plant Cell Physiol.* **2014**, *55*, 119–135. [CrossRef] [PubMed]
55. Gao, Y.; Wei, W.; Zhao, X.; Tan, X.; Fan, Z.; Zhang, Y.; Jing, Y.; Meng, L.; Zhu, B.; Zhu, H.; et al. A NAC transcription factor, NOR-like1, is a new positive regulator of tomato fruit ripening. *Hortic. Res.* **2018**, *5*, 75. [CrossRef]
56. Noguero, M.; Atif, R.M.; Ochatt, S.; Thompson, R.D. The role of the DNA-binding One Zinc Finger (DOF) transcription factor family in plants. *Plant Sci.* **2013**, *209*, 32–45. [CrossRef]
57. Ruta, V.; Longo, C.; Lepri, A.; De Angelis, V.; Occhigrossi, S.; Costantino, P.; Vittorioso, P. The DOF transcription factors in seed and seedling development. *Plants* **2020**, *9*, 218. [CrossRef]
58. Kim, S.; An, C.S.; Hong, Y.N.; Lee, K.W. Cold-inducible transcription factor, CaCBF, is associated with a homeodomain leucine zipper protein in hot pepper (*Capsicum annuum* L.). *Mol. Cells* **2004**, *18*, 300–308.
59. Mou, S.; Liu, Z.; Gao, F.; Yang, S.; Su, M.; Shen, L.; Wu, Y.; He, S. CaHDZ27, a Homeodomain-Leucine Zipper I Protein, positively regulates the resistance to *Ralstonia solanacearum* infection in pepper. *Mol. Plant Microbe Interact.* **2017**, *30*, 960–973. [CrossRef]
60. Lin, Z.; Hong, Y.; Yin, M.; Li, C.; Zhang, K.; Grierson, D. A tomato HD-Zip homeobox protein, LeHB-1, plays an important role in floral organogenesis and ripening. *Plant J.* **2008**, *55*, 301–310. [CrossRef]
61. Joo, H.; Lim, C.W.; Lee, S.C. Roles of pepper bZIP transcription factor CaATBZ1 and its interacting partner RING-type E3 ligase CaASRF1 in modulation of ABA signalling and drought tolerance. *Plant J.* **2019**, *100*, 399–410. [CrossRef]
62. Lee, B.J.; Park, C.J.; Kim, S.K.; Kim, K.J.; Paek, K.H. In vivo binding of hot pepper bZIP transcription factor CabZIP1 to the G-box region of pathogenesis-related protein 1 promoter. *Biochem. Biophys. Res. Commun.* **2006**, *344*, 55–62. [CrossRef] [PubMed]
63. Lim, C.W.; Baek, W.; Lim, S.; Han, S.W.; Lee, S.C. Expression and functional roles of the pepper pathogen-induced bZIP transcription factor CabZIP2 in enhanced disease resistance to bacterial pathogen infection. *Mol. Plant Microbe Interact.* **2015**, *28*, 825–833. [CrossRef] [PubMed]
64. Shen, L.; Liu, Z.; Yang, S.; Yang, T.; Liang, J.; Wen, J.; Liu, Y.; Li, J.; Shi, L.; Tang, Q.; et al. Pepper CabZIP63 acts as a positive regulator during Ralstonia solanacearum or high temperature-high humidity challenge in a positive feedback loop with CaWRKY40. *J. Exp. Bot.* **2016**, *67*, 2439–2451. [CrossRef] [PubMed]
65. Chow, C.-N.; Lee, T.-Y.; Hung, Y.-C.; Li, G.-Z.; Tseng, K.-C.; Liu, Y.-H.; Kuo, P.-L.; Zheng, H.-Q.; Chang, W.-C. PlantPAN3.0: A new and updated resource for reconstructing transcriptional regulatory networks from ChIP-seq experiments in plants. *Nucleic Acids Res.* **2019**, *47*, D1155–D1163. [CrossRef]
66. Consortium, T.U. UniProt: The universal protein knowledgebase in 2021. *Nucleic Acids Res.* **2020**, *49*, D480–D489. [CrossRef] [PubMed]
67. Jin, J.; Tian, F.; Yang, D.-C.; Meng, Y.-Q.; Kong, L.; Luo, J.; Gao, G. PlantTFDB 4.0: Toward a central hub for transcription factors and regulatory interactions in plants. *Nucleic Acids Res.* **2016**, *45*, D1040–D1045. [CrossRef]

International Journal of *Molecular Sciences*

MDPI

Article

Genome-Wide Analysis of the MYB-Related Transcription Factor Family in Pepper and Functional Studies of *CaMYB37* Involvement in Capsaicin Biosynthesis

Yi Liu [1,2,†], Zhishuo Zhang [1,2,†], Ke Fang [2], Qingyun Shan [2], Lun He [2], Xiongze Dai [2], Xuexiao Zou [1,2,*] and Feng Liu [1,2,*]

1 Longping Branch, Graduate School of Hunan University, Changsha 410125, China
2 Key Laboratory for Vegetable Biology of Hunan Province, Engineering Research Center for Germplasm Innovation and New Varieties Breeding of Horticultural Crops, College of Horticulture, Hunan Agricultural University, Changsha 410125, China
* Correspondence: zouxuexiao@hunau.edu.cn (X.Z.); jwszjx@hunau.edu.cn (F.L.)
† These authors contributed equally to the work.

Abstract: Chili pepper is an important economic vegetable worldwide. *MYB* family gene members play an important role in the metabolic processes in plant growth and development. In this study, 103 pepper *MYB*-related members were identified and grouped into nine subfamilies according to phylogenetic relationships. Additionally, a total of 80, 20, and 37 collinear gene pairs were identified between pepper and tomato, pepper and *Arabidopsis*, and tomato and *Arabidopsis*, respectively. We performed promoter *cis*-element analysis and showed that *CaMYB*-related members may be involved in multiple biological processes such as growth and development, secondary metabolism, and circadian rhythm regulation. Expression pattern analysis indicated that *CaMYB37* is significantly more enriched in fruit placenta, suggesting that this gene may be involved in capsaicin biosynthesis. Through VIGS, we confirmed that *CaMYB37* is critical for the biosynthesis of capsaicin in placenta. Our subcellular localization studies revealed that CaMYB37 localized in the nucleus. On the basis of yeast one-hybrid and dual-luciferase reporter assays, we found that CaMYB37 directly binds to the promoter of capsaicin biosynthesis gene *AT3* and activates its transcription, thereby regulating capsaicin biosynthesis. In summary, we systematically identified members of the *CaMYB*-related family, predicted their possible biological functions, and revealed that *CaMYB37* is critical for the transcriptional regulation of capsaicin biosynthesis. This work provides a foundation for further studies of the CaMYB-related family in pepper growth and development.

Keywords: genome-wide; pepper; *CaMYB*-related transcription factor family; phylogenetic relationships; capsaicin biosynthesis

1. Introduction

Transcription factors play an important role in regulating plant growth and development [1]. *MYB* transcription factors are ubiquitous in eukaryotes and comprise one of the largest transcription factor families [2,3]. Members of the *MYB* family are involved in the regulation of plant growth, development, and metabolism. For example, overexpression of *EgMYB2* in tobacco leads to increased lignin content and secondary cell-wall thickening [4]. *MYB* transcription factors in tea plants may regulate theanine synthesis in different organs and tissues [5]. *Arabidopsis MYB12* closely regulates the content of flavonols [6]. In apple, *MdMYB1* interacts with transcription factor *MdGSTF6* to regulate anthocyanin synthesis and transport [7]. In apple, *MdMYB6* also inhibits the biosynthesis of anthocyanins by reducing the contents of UDP-glucose and UDP-galactose [8]. In R2R3-*MYB*-type transcriptional activator *WP1*, the C-terminal acidic activation motif can directly regulate the expression of carotenoid biosynthesis genes, such as *MtLYCb* [9]. *MYB* transcription factors usually interact with transcription factors such as *WD40* and *bHLH* to regulate the biosynthesis of anthocyanins. PacMYBA binds to the promoters of *DFR*, *ANS*, and *UFGT* after interacting with several *bHLH* family members in sweet cherry [10]. Silencing *BvMYB1* in sugar beet can downregulate the expression of genes related to betaine biosynthesis and reduce pigmentation [11].

A typical MYB transcription factor contains one or more MYB domains. The MYB domain is characterized by three helices [12], in which the two C-terminal helices form a helix–turn–helix structure (HTH) [13]. The HTH structure recognizes the target motif C/TAACG/TG and provides specificity of MYB [14]. The N-terminal first helix contains a DNA-binding region and is usually composed of several tandem nonrepetitive R structures, with each R structure containing about 51 amino acids. The transcriptional regulatory region at the C-terminus is highly diverse. This may explain the diversified functions of *MYB* transcription factors [15]. According to the number of tandem structures of DNA-binding domains, the MYB transcription factor family can be divided into four subfamilies: 1R-MYB (MYB-Related), R2R3-MYB, 3R-MYB, and 4R-MYB. The R2R3 subfamily has the largest number of members, followed by the 1R subfamily [3].

Capsicum, a crop that originates from tropical regions of Latin America, is an important member of the nightshade family along with tomato, tobacco, potato, and petunia [16]. Pepper has important economic value and is grown worldwide [17]. Pepper has five domesticated species, which are highly different in their spiciness [18]. The spiciness mainly comes from capsaicinoids, economically important alkaloids [19,20]. Capsaicinoids are specifically synthesized in the glandular cells of the fruit placenta and then transported to other organs [21]. About 24 species of capsaicinoids have been identified, of which capsaicin (Cap) and dihydrocapsaicin (DhCap) account for over 90% of the total content and are the main determinants of spiciness [22]. The spicy taste produced by eating chili is essentially the burning sensation caused by the binding of capsaicin to the capsaicin receptor TRPV1 in the human neurosensory system [23]. Purified capsaicinoids have an important pharmaceutical, military, and food processing role [24]. Capsaicinoids can be synthesized through two intermate pathways. Phenylalanine is catalyzed by a series of enzymes to derive vanillylamine, and the key enzymes involved in this pathway are phenylalanine ammonialyase (PAL), cinnamic acid hydrolase (C4H), *p*-coumaric acid hydrolase (C3H), caffeic acid transmethoxylase (CoMT), aminotransferase (AMT), β-ketoacyl-ACP synthase (Kas), etc. Valine is catalyzed by a series of enzymes to derive branched-chain fatty acids of 8-methyl-6-nonene-CoA. The key enzymes involved in this pathway are branched-chain amino-acid transferase (BCAT) and isovalerate dehydrogenase (IVD), acyl transportin (ACL), acyl-ACP-thioesterase (FatA), and acyl-CoA synthase (ACS). 8-methyl-6-nonenoyl-CoA and vanillylamine form capsaicin catalyzed by capsaicin synthase (CS) [21,22,25].

Previous studies showed that *R2R3-MYB* genes are involved in the regulation of biocapsaicin synthesis. For instance, *MYB31* upregulates *CBGs* functions to increase capsaicin accumulation by promoting the expression level [18]. Silencing *MYB48* reduced the expression of key capsaicin biosynthetic enzymes, including *AT3* and *KAS*, and affected the biosynthesis of capsaicin [26]. However, little is known about how *CaMYB*-related genes regulate capsaicin biosynthesis.

In this study, we identified 103 *CaMYB*-related members in the pepper genome and revealed their chromosomal locations, phylogenetic relationships, promoter *cis*-elements, and expression patterns. Transcriptome analysis showed that *CaMYB37* was specifically expressed in the fruit placenta and flower. Silencing of *CaMYB37* resulted in a significant decrease in the capsaicin content. Y1H and dual-luciferase reporter assays proved that CaMYB37 can directly bind to the *AT3* promoter and activate its transcription, thereby promoting capsaicin biosynthesis. This study provides a foundation for further research on the functional mechanism of *CaMYB*-related family members involved in pepper growth, development, and metabolism.

2. Results

2.1. Identification and Protein Characterization of the CaMYB-Related Family

A total of 103 *CaMYB*-related genes and 132 *SlMYB*-related genes were obtained for further analysis (Supplementary Table S1). CaMYB-related proteins range from 101 amino acids (Capana04g000348) to 1260 amino acids (Capana09g002226), with the isoelectric point PI ranging from 4.53 (Capana05g000377) to 11.39 (Capana05g001626), and the MW molecular weight ranging from 11.3 kDa (Capana04g000348) to 140.9 kDa (Capana09g002226) (Supplementary Table S2).

2.2. Chromosomal Localization of the CaMYB-Related Family

All 103 *CaMYB*-related transcription factors are named according to their chromosomal location (*CaMYB1–CaMYB103*) (Figure 1). According to published genome sequences, 94 *CaMYB*-related genes are distributed on 12 chromosomes (91%), whereas 9 *CaMYB*-related genes stayed on as yet unmapped scaffolds. Most *CaMYB*-related genes are clustered at the two ends of the chromosome, while fewer genes are distributed in the middle of the chromosome.

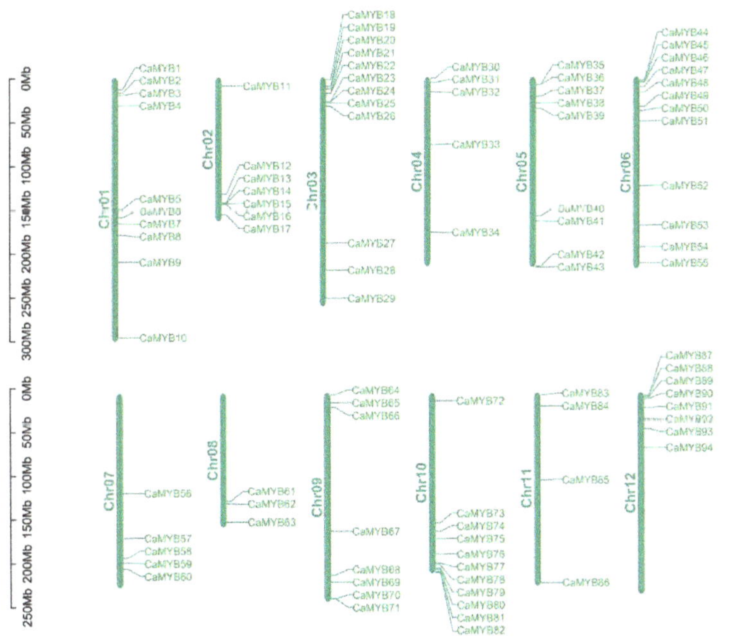

Figure 1. Chromosomal locations of *CaMYB*-related genes.

2.3. Duplication Analysis of Members of the CaMYB-Related Family

To further explore the evolutionary relationship between *MYB*-related genes in different species, we analyzed the gene pairs within pepper and those with tomato and *Arabidopsis* (Figure 2). The intraspecific collinearity of pepper showed that six gene pairs may have undergone segmental duplication in evolution: *CaMYB48/CaMYB21*, *CaMYB46/CaMYB35*, *CaMYB54/CaMYB46*, *CaMYB82/CaMYB79*, *CaMYB63/CaMYB17*, and *CaMYB81/CaMYB80*. The number of *MYB*-related gene pairs between pepper and tomato (80) was higher than that between pepper and *Arabidopsis* (20) and between tomato and *Arabidopsis* (37). This finding indicates that tomato and pepper are more closely related. We found that 64 *CaMYB*-related genes were collinear with 63 *SlMYB*-related genes, most of which were one-to-one, such as *CaMYB44/SlMYB71*; *CaMYB34/SlMYB47*. We also found one-to-many cases: *CaMYB21/(SlMYB29, SlMYB67)* and *CaMYB89/(SlMYB92, SlMYB131)*.

Many-to-one cases also existed such as *(CaMYB63, CaMYB17)/SlMYB11* and *(CaMYB79, CaMYB82)/SlMYB101*. These results indicate that MYB-related genes may have evolved from the same ancestor; accordingly, they possess similar functions and, thus, are evolutionarily conserved (Supplementary Table S3).

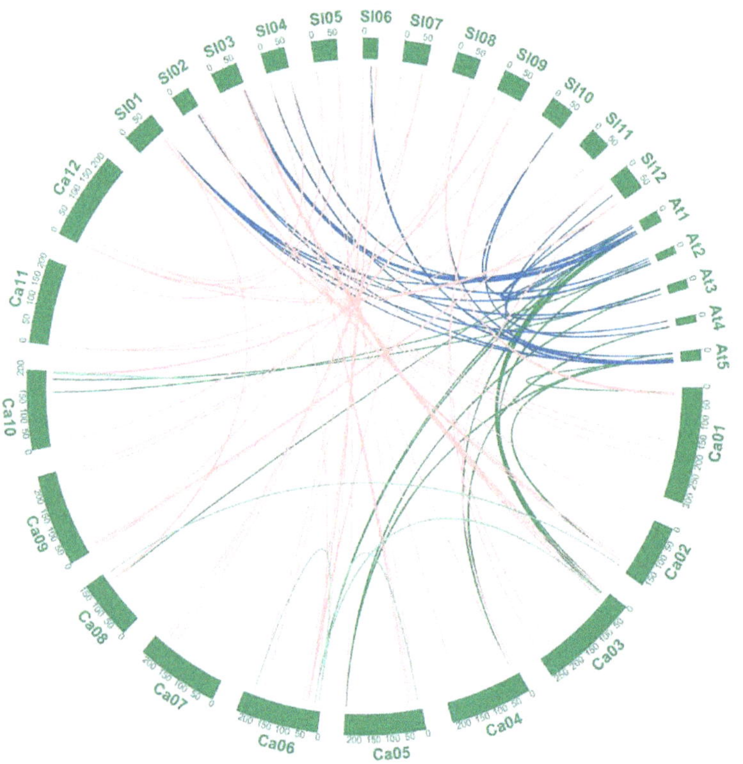

Figure 2. Collinear analysis of *CaMYB*-related genes among pepper (Ca), *Arabidopsis* (At), and tomato (Sl). Blue, green, pink, and blue-green lines represent the collinear gene pairs between chromosomes of tomato and *Arabidopsis*, *Arabidopsis* and pepper, tomato and pepper, and pepper and pepper, respectively. Chromosome numbers are located above the chromosomes.

2.4. Structure and Motif Analysis of CaMYB-Related Family

A total of 103 *CaMYB*-related genes were divided into nine subfamilies according to the phylogenetic relationship and with reference to the classification of the *Arabidopsis* MYB-related family (Figure 3a). The gene structure analysis showed that the *CaMYB*-related genes have 0–22 introns, among which *CaMYB71* has the most introns (22) (Figure 3b). Ten conserved motifs (motifs 1–10) were identified in *CaMYB*-related members (Supplementary Figure S1). Each family member contained 1–6 motifs (Supplementary Figure S2). Members within the same subfamily had similar motifs. Except for individual members in subgroup VII, the remaining members contained motif 2, which is an important part of the conserved domain of the CaMYB-related family. A few subfamily members had unique motifs. For example, all genes in subfamily I contained motif 7, which was not found in other subfamily members. These genes may have unique functions.

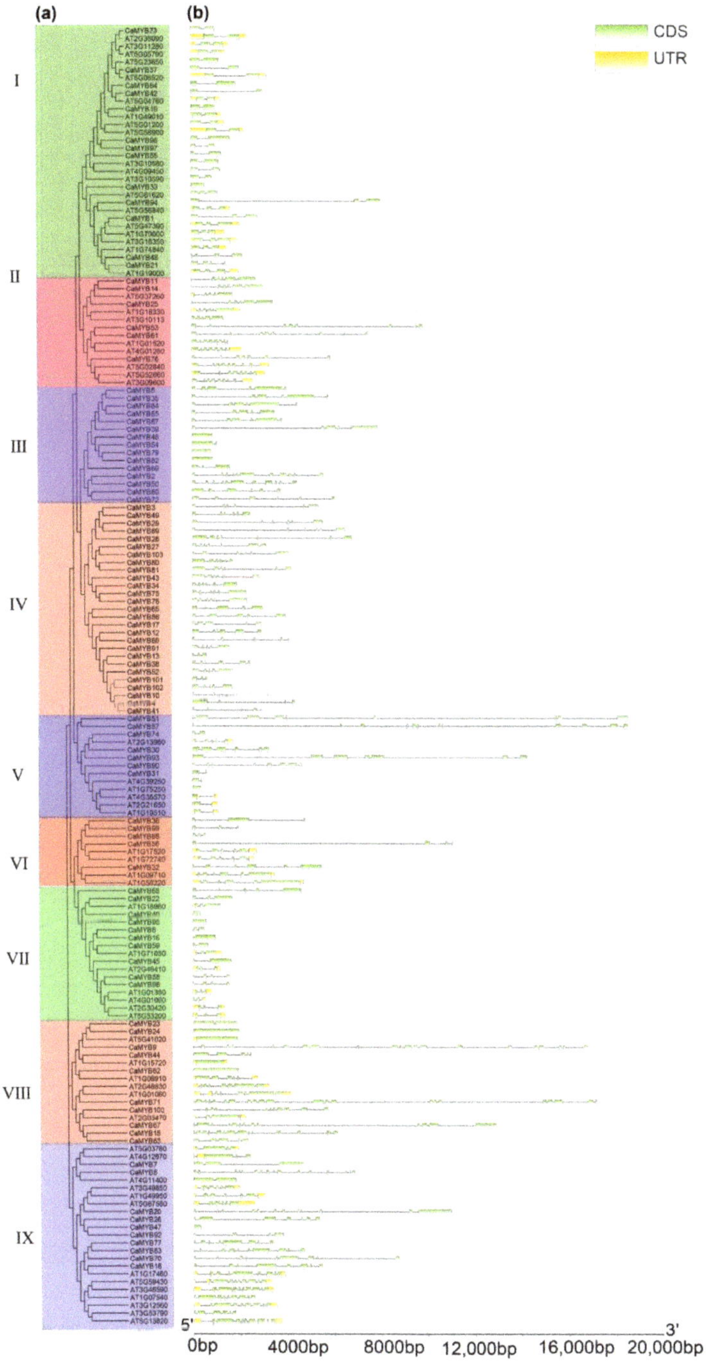

Figure 3. Phylogenetic relationship and gene structure of CaMYB-related and AtMYB-related families. (**a**) Unrooted phylogenetic tree of CaMYB-related and AtMYB-related proteins constructed using neighbor-joining method (NJ). (**b**) Gene structure. Green squares represent introns of genes, yellow squares represent uncoded UTR regions, and black lines represent exons.

2.5. Cis-Element Analysis of CaMYB-Related Family

The 2000 bp promoter sequences of the *CaMYB*-related genes were extracted, and the *cis*-regulatory elements (REs) in the promoters were predicted using the PLANTCARE online website (http://bioinformatics.psb.ugent.be/webtools/plantcare/html/, accessed on 5 May 2022). We identified 51 REs in the promoters of *CaMYB*-related genes. Among these, 20 REs are involved in light responses, such as G-box, CARE, and DRE1. Sixteen REs are involved in the regulation of abscisic acid (ABA), jasmonic acid (JA), and salicylic acid (SA) signaling pathways, including ACE, MYC, and ERE. The remaining REs are involved in biological processes such as stress responses, cell development, and metabolic regulation (Figure 4).

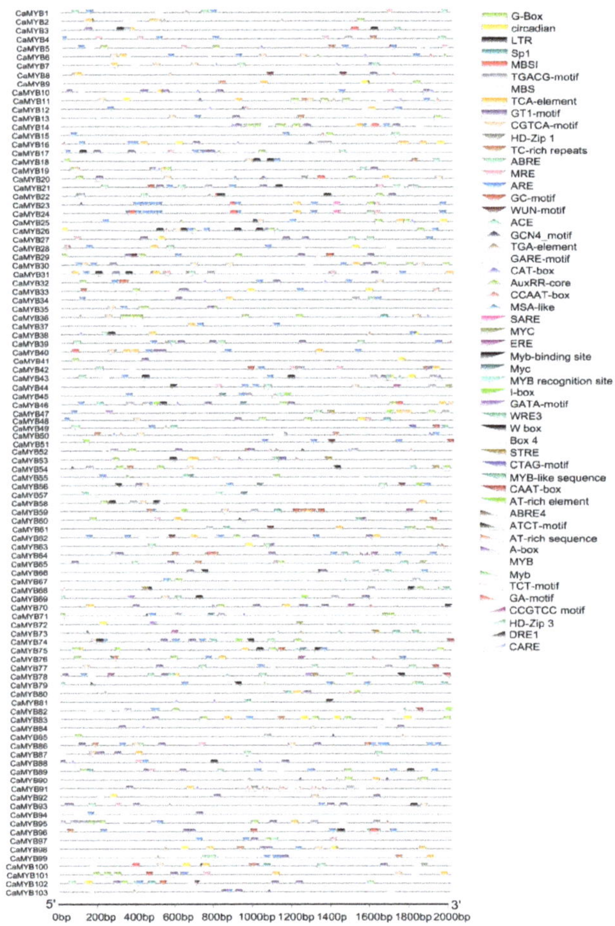

Figure 4. *cis*-Element analysis of CaMYB-related family. Different shapes and colors represent different elements.

In parallel, we closely examined the promoters of the 103 *CaMYB*-related genes. Fifty-eight promoters contain elements involved in the regulation of flavonoid biosynthesis (TGACG), and 95 promoters contain palisade mesophyll cell differentiation-related elements (AREs). Elements involved in circadian rhythm control (LTR) in plants were identified in 25 promoters; 13 promoters contain WRE/STRE, a necessary element for anaerobic induction regulation, whereas 10 promoters contain hypoxia-specific induction element CGTCA motif. A total of 55, 60, and 23 promoters contain the CGTCA motif, GT1 motif, and TGA element, respectively, which are involved in the drought stress process (Supplementary Table S4). The above results indicate that the diversification of *CaMYB*-related family gene functions plays an important role in plant growth and development.

2.6. Expression Patterns of CaMYB-Related Genes in Various Tissues

We obtained transcriptome data of *CaMYB*-related genes in different tissues from previous studies and plotted the expression heatmap (Figure 5, Supplementary Table S5). Some *CaMYB*-related genes were expressed throughout all developmental stages and tissues, such as *CaMYB20/CaMYB27/CaMYB30/CaMYB63/CaMYB100*, while some displayed unique expression patterns. For example, *CaMYB52/CaMYB60/CaMYB99* were highly expressed in leaves, while *CaMYB14/CaMYB23/CaMYB24* were enriched throughout floral development, and *CaMYB22* was restricted to the early stage of floral development. Similarly, while *CaMYB13* was highly expressed throughout seed development, *CaMYB15* was restricted to late developmental stages. These unique expression patterns reflect their functions in different biological and developmental processes. In this study, we focused on *CaMYB37*, which is specifically and highly expressed in developing placenta. We speculated that *CaMYB37* may play an important role in regulating capsaicin biosynthesis.

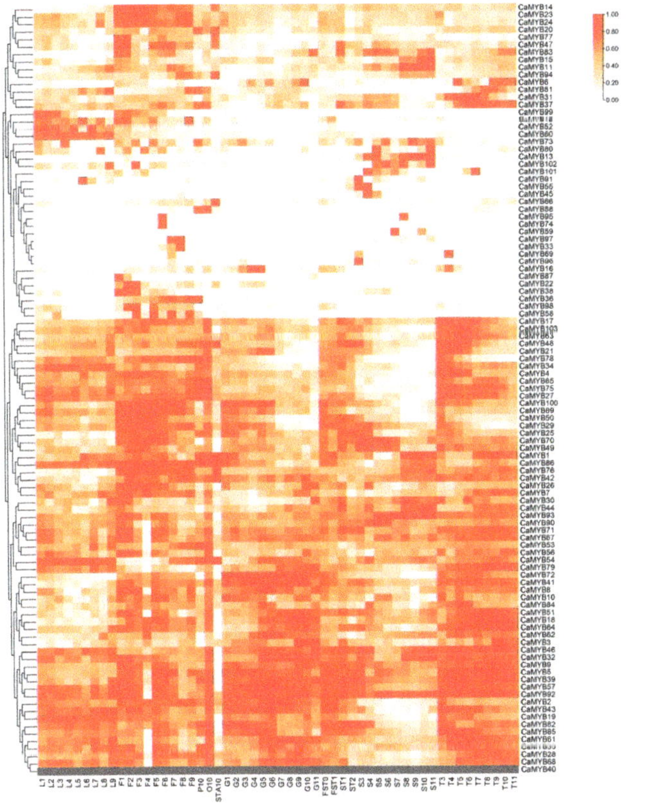

Figure 5. Expression patterns of *CaMYB*-related genes in different tissues and organs of pepper based on transcriptome data. L, F, O, G, S, and T represent the leaf, flower, ovary, pulp, seed, and placenta, respectively. ST denotes early seed and placenta, FST denotes early whole fruit, P10 denotes petal, and STA denotes stamen. Dark colors indicate high expression levels, and light colors indicate low expression levels.

2.7. Homology and Subcellular Localization of CaMYB37

A phylogenetic tree was constructed by comparing the protein sequence of CaMYB37 with homologous protein sequences from other species (such as tomato and *Arabidopsis* 15 species). We found that CaMYB37 is closely related to SlMYB37 and StMYB37 (Figure 6a). However, there are few studies on the function of *MYB37* in Solanaceae. Subcellular localization analysis of CaMYB37 was conducted in *N. benthamiana* leaves. As shown in Figure 6b, 35S:CaMYB37-GFP colocalizes with HY5, which is used as a marker for nuclear proteins. Therefore, CaMYB37 is localized in the nucleus.

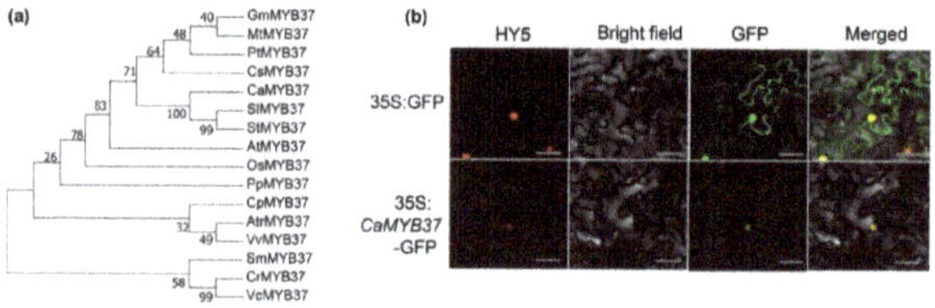

Figure 6. Phylogenetic analysis and subcellular localization of CaMYB37. (**a**) Phylogenetic analysis of amino-acid sequences of CaMYB37 in 16 plants. (**b**) Subcellular localization of the CaMYB37 protein in *N. benthamiana* leaves. Green fluorescent protein (GFP) signals indicated that CaMYB37 was localized in the nucleus. HY5 was used as a nuclear marker. Scale bar, 15 μm.

2.8. Silencing CaMYB37 Reduces the Expression of CBGs and Capsaicin Content

Gene silencing of *CaMYB37* in pepper cultivar 'S8′ was performed using virus-induced gene silencing (VIGS). After about 1 month, the pepper plants injected with the *PDS* solution turned white, indicating that the VIGS test was successful (Figure 7a). The fruit placentas at 16 days after flowering were selected for qRT-PCR verification (Figure 7b). Compared with control plants, the expression levels of *CaMYB37* and capsaicin synthesis genes including *AMT*, *4CL*, and *AT3* were significantly decreased. At the same time, the contents of Cap and DhCap in fruit placentas at 45 days after flowering were determined. The contents of Cap and DhCap were significantly reduced in the placentas of pTRV2-*CaMYB37* plants compared with those in control pTRV2 plants (Figure 7c). Therefore, we can conclude that *CaMYB37* activates the expression of key *CBGs* and leads to increased capsaicin content in pepper.

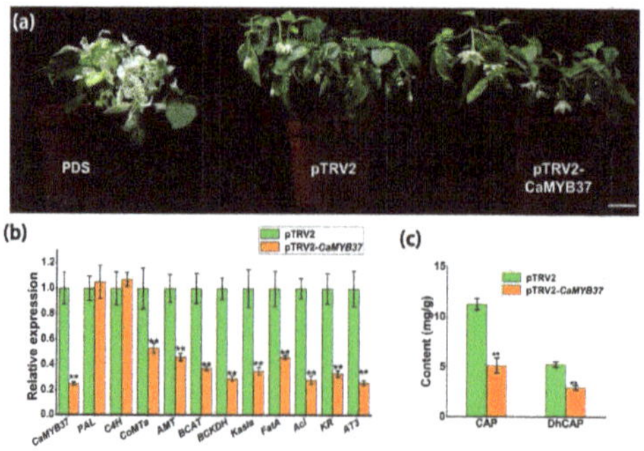

Figure 7. Expression level of *CBGs/CaMYB37* and contents of CaP/DhCaP in the control (empty vector) and *CaMYB37*-silenced fruit placenta. (**a**) Phenotypes of control, pTRV2-*PDS*, and pTRV2-*CaMYB37* plants. Scale bars, 10 cm. (**b**) Relative expression of *CaMYB37* and *CBGs* in the placenta at developmental stage 16 DPA in the *CaMYB37*-silenced fruit compared to levels in the control. The relative expression levels of the *CBGs* and *CaMYB37* in the control were set to 1, and those in the *CaMYB37*-silenced fruit placenta were measured relative to the control. (**c**) Contents of CaP and DhCaP in the control and *CaMYB37*-silenced fruit placenta. Error bars indicate means ± SEs ($n = 3$). ** means extremely significant difference ($p < 0.01$).

2.9. CaMYB37 Binds to Capsaicin Synthesis Genes and Activates Their Transcription

The *AT3* gene encodes capsaicin synthase, which catalyzes the synthesis of capsaicin from vanillylamine and 8-methyl-6-nonenoyl-CoA [27]. According to the PLANTCARE online tool, MYB, MBS, and other elements may be bound by the MYB proteins in the 2000 bp promoter sequence upstream of *AT3*. Therefore, a yeast one-hybrid experiment (Y1H) was performed to verify whether CaMYB37 binds to the *AT3* promoter. Yeast strain Y1HGOLD (pGADT7-CaMYB37) was co-transformed with pAbAi-*AT3*, and the transformants survived on selective medium lacking leucine and containing AbA, suggesting that CaMYB37 binds to the promoter of *AT3* (Figure 8a). To validate this observation, we performed a dual-luciferase assay (Figure 8c). The ratio of LUC to REN in tobacco leaves injected with 62SK-MYB37 and pro *AT3*-LUC increased 2.5-fold relative to the negative control (62SK and proAT3-LUC) (Figure 8d). The results indicate that CaMYB37 directly binds to the promoter of *AT3* and activates its expression in the pepper placenta.

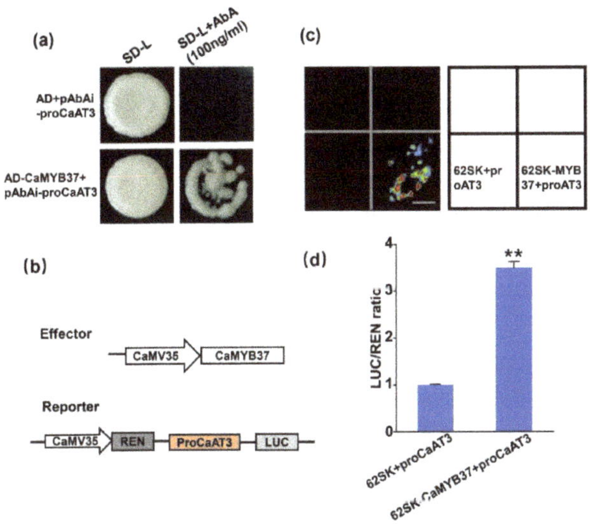

Figure 8. CaMYB37 binds to the promoter of *AT3* and promotes its transcription. (**a**) DNA-binding potential of CaMYB37 with the promoter of *AT3* in yeast. SD/−L, medium lacking leucine. (**b**) Vectors for dual luciferase experiments. (**c**) LUC images of tobacco leaves after transient infiltration. Scale bars, 1 cm. (**d**) Enzyme activity detections in dual luciferase assays. Error bars indicate means ± SEs ($n = 6$). ** means extremely significant difference ($p < 0.01$).

3. Discussion

The MYB-related family has been reported to be involved in a wide range of biological processes such as circadian rhythm, hypocotyl elongation, meristem formation, and secondary metabolism [28–32]. The MYB-related family has been identified in *Arabidopsis* [15], tomato [33], rice [34], wheat [35], watermelon [36], eggplant [37], and other species. Previously, 235 *CaMYB* genes were identified in pepper [38]. Here, we used bioinformatics to identify members of the MYB-related family in pepper and analyze their evolutionary relationships, chromosomal locations, promoter elements, and expression patterns in various pepper tissues. A family member, *CaMYB37*, which is specifically expressed in pepper fruit placentas and flowers, was identified on the basis of transcriptome data. Gene silencing, yeast one-hybrid, and dual-luciferase reporter assays showed that *CaMYB37* regulates capsaicin biosynthesis by binding to the *AT3* promoter and activating its expression. In this study, the systematic analysis and functional verification of *MYB*-related family members can provide a foundation for the further exploration of *CaMYB*-related transcription factors in regulating pepper growth and development.

We identified 103 *CaMYB*-related members in the pepper genome. According to the taxonomy and phylogenetic relationship of *Arabidopsis* members, pepper *MYB*-related genes were divided into nine subgroups. Subfamilies III and IV consist only of pepper members. This may indicate the unique functions of these genes evolved in pepper [39]. In terms of gene structure, obvious differences were found in the number of exons among members of different *CaMYB*-related subfamily members, which may have led to functional differentiation during evolution. However, the number of exons within members of the same subfamily was not very different, indicating that the functions of these genes are conserved to a certain extent. The promoter sequences of the *CaMYB*-related genes contain various types of elements, such as light response, drought stress, and circadian rhythm regulation. Various types of elements have indicated that the *CaMYB*-related genes may play an important role in pepper growth, development, and stress response [40].

Homologous genes in the same branch of the evolutionary tree usually have certain similarities in plant functions. Many studies have been conducted on *Arabidopsis* MYB-related genes, allowing us to predict the function of the corresponding *CaMYB*-related homologous genes according to the known function of *Arabidopsis MYB*-related genes. *At-MYB*-related members of subfamily II are involved in the regulation of circadian rhythms in *Arabidopsis thaliana*, dormant germination of plant seeds, and flower bud development. For example, AT5G37260, AT1G18330, and AT3G10113 contain motifs 2, 1, and 5, respectively, and they are arranged in the same order. Previous studies have shown that *AT1G18330* (*EPR1*) encodes a nuclear-localized protein regulated by phytochromes A and B. Overexpression of *EPR1* increased far-red light-induced cotyledon opening and delayed flowering. *AT5G37260* (*RVE2*), *RVE7*, and *gibberellin 3-oxidase 2* (*GA3ox2*) act downstream of *FHY3*, and the transcriptional cascade composed of phyB-FHY3-RVE2/RVE7/SPT-GA3ox2 can transmit ambient light signals to regulate seed dormancy and germination [41]. *AT2G46410* (*CPC*), *AT1G01380* (*ETC1*), *AT2G30420* (*ETC2*), *AT4G01060* (*CPL3*), and *AT5G53200* (*TRY*) in subfamily VII mediate the differentiation of root hair cells, affecting the density of root hairs, development of trichomes in plants, and development of the waxy layer of the mustard epidermis [42]. The same subgroup of genes *CaMYB53*, *CaMYB61*, and *CaMYB76* may also be involved in the regulation of pepper root hair density and fruit epidermal cell development. *AT5G47390* (*MYBH*) and *AT1G70000* (*MYBD*) affect the accumulation of the secondary metabolite anthocyanin [43]. CaMYB1 of the same subfamily has similar gene domain structures; hence, *CaMYB1* may regulate anthocyanin synthesis in pepper.

Capsaicin is an important alkaloid that exists specifically in pepper, and the placenta is the main tissue producing capsaicin. Several studies have shown that *R2R3-MYB* family members regulate capsaicin synthesis in the placenta. For example, the R2R3-MYB-type transcription factor *MYB31* encoding Cap1 is specifically expressed in pepper placenta. However, no placental-specific expression has been found in other Solanaceae members such as tomato or potato, indicating that the tissue specificity of *MYB31* may be a consequence of capsaicin production unique to pepper placenta [18]. *MYB31* can directly upregulate the expression of CBGs in the placenta to increase capsaicin content. In another study, the transcriptome analysis showed that the expression of the *R2R3-MYB* transcription factor *CaMYB108* was higher in pepper fruits and flowers than in other tissues. Gene silencing (VIGS) and temporal overexpression assays of *CaMYB108* showed that this transcription factor was positively correlated with the content of capsaicin [44]. In the present study, through bioinformatic predictions, we identified *CaMYB37*, which was specifically expressed in the placenta. Our functional and molecular studies showed that *CaMYB37* promotes capsaicin biosynthesis by positively regulating the expression of several key capsaicin biosynthetic genes. Therefore, whether the identified seed-specific expression genes *CaMYB15* and *CaMYB13*, flower-specific expression genes *CaMYB23/CaMYB24*, and leaf-specific expression genes *CaMYB52/CaMYB60/CaMYB99* play an important role in the development of corresponding tissues deserves further study.

4. Materials and Methods

4.1. Identification of the CaMYB-Related Gene Family

Pepper and tomato gene and protein sequences were downloaded from the SGN website (https://solgenomics.net/, accessed on 5 May 2022). The *Arabidopsis* MYB-related protein sequences were downloaded from the TAIR website (https://www.Arabidopsis.org/, accessed on 1 May 2022). HMMER 3.0 [45] was used to screen tomato and pepper protein sequences containing the complete MYB domain (PF00249), with thresholds set at *E*-values of $<1 \times 10^{-5}$. Candidate sequences were further screened using the SMART website (http://smart.embl-heidelberg.de/, accessed on 1 May 2022) and the NCBI database (http://www.ncbi.nlm.nih.gov/, accessed on 1 May 2022), thereby retaining sequences containing the complete MYB domain. Candidate gene sequences were used for further analyses.

4.2. Phylogenetic Analysis and Classification of the CaMYB-Related Gene Family

The protein sequences of *Arabidopsis thaliana* and pepper MYB-related family members were subjected to multiple sequence alignment using ClustalW 2.0 [46]. A phylogenetic tree was constructed using the neighbor-joining (NJ) method in MEGA 7.0 with the following parameter settings: 1000 bootstrap values, Poisson model, and pairwise deletion [47]. The members of the CaMYB-related family were divided into different subfamilies based on AtMYB taxonomy and bootstrapping values in the phylogenetic tree.

4.3. Protein Properties, Conserved Motifs, and Gene Structures of CaMYB-Related Family

CaMYB-related proteins were uploaded to the ExPaSy website (http://web.expasy.org, accessed on 3 May 2022) to calculate the isoelectric point (PI) and molecular weight (MW). At the same time, MEME v5.1.1 (http://meme-suite.org/, accessed on 3 May 2022) was used to identify the conserved motifs of the CaMYB-related proteins. The optimal motif length was 100–200 residues, the number of motifs was 10, and the remaining parameters remained unchanged [48]. The gene structure and intron number of the pepper and tomato genes were determined with reference to the annotation files. Gene structure maps were drawn using TBtools v1.098765 [49].

4.4. Chromosomal Location and Gene Duplication Analysis of CaMYB-Related Family

MapGene2 (http://mg2c.iask.in/mg2c_v2.0/, accessed on 1 May 2022) was used to determine the location of the *CaMYB*-related genes on the chromosomes. MCScanX was used to identify *MYB*-related gene pairs in tomato, *Arabidopsis*, and pepper, as well as intraspecific gene pairs in pepper. The parameters were set as follows: match score of 50, match size of 5, gap score of −3, and *E*-value of 1×10^{-10}. A collinearity graph was drawn using Circos software v0.69 (accessed on 5 May 2022) [50].

4.5. Analysis of cis-Elements in the Promoters of CaMYB-Related Genes

The 2000 bp upstream promoter sequences of the *CaMYB*-related genes were extracted using the SeqKit v0.13.0 software (accessed on 5 May 2022) [51]. After uploading the promoter sequences to PlantCARE [52] for online analysis, predicted promoter *cis*-elements were sorted and classified.

4.6. Expression Pattern Analysis of CaMYB-Related Genes

Previously reported transcriptome data were used to analyze the expression patterns of *CaMYB*-related genes in multiple organs and tissues of pepper [53]. The FPKM value (fragments per kilobase of transcript per million mapped reads) represents the expression level of each gene, which was converted to $\log_2$ (FPKM + 1). A heatmap was drawn after normalizing the data using R v3.6.1 (accessed on 7 May 2022).

4.7. Quantitative Real-Time PCR (qRT-PCR)

The fruit stage used in the experiment was 16 days after pollination [18]. Total RNA from 'S8' placenta was extracted using the TransZol kit (TransGen Biotech, Inc., Beijing,

China). Each group of samples was repeated three times. Extracted RNA was reverse-transcribed using HiScriptIIQ RT SuperMix, and the product from the previous step was used in the qPCR (+gDNA wiper) vazyme kit (Vazyme, Piscataway, NJ, USA). qRT-PCR experiments were performed using a LightCycler® 96 real-time PCR machine. The reaction volume was 20 μL, and the program was as follows: pre-denaturation at 95 °C for 30 s, 95 °C for 10 s, and 60 °C for 30 s, over 40 cycles, followed by 95 °C for 15 s, 60 °C for 60 s, and 95 °C for 15 s. Gene expression level was calculated using the $2^{-\Delta\Delta Ct}$ method [54]. qRT-PCR primers (*PAL/C4H/CoMTa/AMT/Kasla/BCAT/BCKDH/FatA/Acl/AT3/KR/Actin2*) were from the previous study [18]. The qRT-PCR primers of *CaMYB37* were designed by Primer5 software, and the specificity of primers was confirmed using Primer-BLAST (NCBI). The melting curve and standard curve were supplemented to check the amplification efficiency of qRT-PCR primers (*CaMYB37*) (Supplementary Table S6). The internal reference gene for gene expression levels was *Actin2* [18]. *Actin2* encodes an actin expressed during reproductive development. The gene primers used in this study are list in Supplementary Table S6.

4.8. Subcellular Localization

First, a subcellular localization expression vector was constructed. After double-enzyme digestion of the pSuper1300-eGFP plasmid, the obtained fragment was homologously recombined with the gene CDS sequence and then transformed into DH5α competent cells. A single colony was picked and shaken for sequencing, and the extracted plasmid was transferred into *Agrobacterium* competent GV3101. *Agrobacterium* liquid PCR test strips were expanded in LB medium containing kanamycin and rifampicin antibiotics. After 12 h, the OD600 was detected as 0.8–1. The sample was centrifuged at 3000 rpm for 8 min, resuspended in buffer (10 mM MES, 10 mM $MgCl_2$, 100 mM AS), and incubated at 28 °C for 3 h in the dark. pSuper1300-CaMYB37-eGFP and the nuclear marker HY5-mCherry were transiently co-transformed into *Nicotiana benthamiana* leaves. The infected tobacco was cultured in an artificial climate chamber (day/night temperature 18/22 °C, light/dark 8/16 h). After 48 h, the fluorescence was detected using a laser confocal microscope and photographed (LSM800, Zeiss, Jena, Germany).

4.9. Extraction and Detection of Capsaicin and Dihydrocapsaicin

The capsaicin and dihydrocapsaicin contents in 'S8′ placenta were extracted and detected using liquid chromatography tandem mass spectrometry (LC–MS/MS). The fruit stage used in the experiment was 45 days after pollination [18]. Cryopreserved pepper placenta samples were collected and ground into a powder with a grinder. About 0.1 g of the powder sample was weighed into a 2 mL centrifuge tube, before adding 1.0 mL of methanol extract; then, ultrasonic extraction was performed at 4 °C, and the process was repeated three times. The leaching solution was centrifuged at 10,000 rpm for 10 min, concentrated with nitrogen, and dissolved in 200 μL of methanol. All samples were purified by solid-phase extraction (SPE) using an Oasis HLB RP column (30 mg, 1 mL), washed with 1 mL of 100% MeOH and 1 mL of deionized water, and then eluted with 1 mL of methanol to obtain the analytical solution. The analytical solution was then blown dry with constant nitrogen at 40 °C and redissolved in 100 μL of methanol. The diluted samples were analyzed using UPLC–MS/MS. Using the external standard method, the peak areas of all the detected samples were substituted into the standard curve equation for calculation. The absolute contents of capsaicin and dihydrocapsaicin in the actual samples were obtained. All assays were performed in triplicate.

4.10. Virus-Induced Gene Silencing

First, a gene silencing expression vector was constructed. The coding sequence of the *CaMYB37* gene was uploaded to the SGN VIGS website (https://vigs.solgenomics.net/, accessed on 15 May 2022). A specific 300 bp sequence was selected for cloning. *PDS* (*Phytoene desaturase*) was used as an indicative control gene. We digested the pTRV2 vector and performed ligation transformation. Subsequent experiments were performed after the vector was verified to be correct. First, the bacteria were shaken before injection and resuspended when the OD value was approximately 1.0. The resuspended solution was allowed to stand for 4 h in the dark environment at 25 °C and then mixed. The pTRV1 bacterial solution and the bacterial solution containing pTRV2 or pTRV2-*CaMYB37* were mixed at a 1:1 ratio to prepare for inoculation. The pepper seedlings were subjected to drought treatment for 2 days when the cotyledons were fully expanded, and no true leaves were formed. A disposable syringe was used to inject the mixture into the backs of the *capsicum* cotyledons. The injected peppers were returned to an artificial climate chamber for cultivation.

4.11. Yeast One-Hybrid

The Y1HGold yeast strain was used to prepare competent yeast (the Y1HGold strain was purchased from the company). The *CaMYB37* coding sequence was ligated into the PGADT7 (AD) vector using homologous recombination. Specific primers were designed to amplify the promoter sequence of the capsaicin synthesis gene *AT3*. The *AT3* promoter sequence was ligated to the pAbAi vector, linearized, and transformed into competent yeast cells. The correctly sequenced pAbAi-*AT3* vector can screen for the minimum inhibitory concentration of Aureobasidin A (AbA) according to plaque growth. This screening concentration was considered suitable for subsequent one-to-one interaction experiments. The abovementioned Y1HGold (pBait-AbAi) strain was inoculated on SD/−Ura solid plates, and positive plaques with diameters of 2–3 mm were selected and propagated in YPDA liquid medium. The AD-CaMYB37 plasmid and AD empty plasmid were transformed into the Y1HGold strain and spread on SD/Leu plates. Positive yeast plaques were inoculated onto SD/−Leu+AbA plates. After 3–5 days, plaque growth was observed.

4.12. Dual-Luciferase Reporter Gene Assay

Two types of vectors were constructed for the Dual-Luc reporter gene detection experiments. The cloned *AT3* promoter sequence was ligated into a pGreenIIo800 vector. The coding sequence (CDS) region of cloned *CaMYB37* was ligated into the pGreenII62-sk vector. The two vectors were transformed, and the tested bacterial liquid was expanded and cultured in LB medium containing kanamycin and rifampicin. The samples were cultured until the OD600 value reached 1. After centrifugation for 5 min at 5000 rpm, the bacteria were collected. The cells were resuspended in an MES solution (10 mM MES, 10 mM $MgCl_2$, and 100 mM AS). The suspension was then mixed in the dark for 4 h (pGreenII0800:pGreenII62-sk = 1:9). A 1 mL syringe was used to inject the back of the leaves (tobacco seedlings were approximately 1 month old). The injection dose of each leaf in the control and experimental groups remained the same (1 mL). After injection, the tobacco plants were treated in the dark for 48 h. Fluorescence was observed using an in vivo imager. The leaves injected with the mixed solution were sampled using a hole punch, quick-frozen in liquid nitrogen, and ground. Enzyme activity was detected using a dual-luciferase reporter assay system kit. The primer pairs used for the yeast one-hybrid and dual-luciferase reporter assays are listed in Supplementary Table S1.

4.13. Statistical Analysis

Data were analyzed using GraphPad Prism software (version 7; San Diego, CA, USA) and the Student's *t*-test for significance in all assays. Error bars indicate means ± SEs. "*" means significant difference ($p < 0.05$). "**" means extremely significant difference ($p < 0.01$).

5. Conclusions

In conclusion, we systematically identified and analyzed *CaMYB*-related family members. Through bioinformatic prediction, we identified placental-specific *CaMYB37*. We provided functional and molecular evidence that CaMYB37 regulates capsaicin biosynthesis by binding to the *AT3* promoter and activating its expression. The results provide reference for further study on the role of *CaMYB*-related genes in pepper growth and development.

Supplementary Materials: The following supporting information can be downloaded at: https://www.mdpi.com/article/10.3390/ijms231911667/s1.

Author Contributions: Validation, Y.L., Z.Z., F.L., X.D. and X.Z.; data curation, Z.Z., Y.L. and X.Z.; writing—review and editing, Z.Z., X.D., F.L. and X.Z.; investigation, Y.L., Z.Z., K.F., Q.S. and L.H.; funding acquisition, X.Z. and F.L. All authors have read and agreed to the published version of the manuscript.

Funding: This research was funded by the National Natural Science Foundation of China (U21A20230, U19A2028 and 32130097), the Science and Technology Innovation Program of Hunan Province, China (2021NK1006).

Institutional Review Board Statement: Not applicable.

Informed Consent Statement: Not applicable.

Data Availability Statement: The data presented in this study are available in insert article or supplementary material here.

Conflicts of Interest: The authors declare no conflict of interest.

References

1. Riaño-Pachón, D.M.; Ruzicic, S.; Dreyer, I.; Mueller-Roeber, B. PlnTFDB: An integrative plant transcription factor database. *BMC Bioinform.* **2007**, *8*, 42. [CrossRef] [PubMed]
2. Anton, I.A.; Frampton, J. Tryptophans in *myb* proteins. *Nature* **1988**, *336*, 719. [CrossRef] [PubMed]
3. Dubos, C.; Stracke, R.; Grotewold, E.; Weisshaar, B.; Martin, C.; Lepiniec, L. *MYB* transcription factors in *Arabidopsis*. *Trends Plant Sci.* **2010**, *15*, 573–581. [CrossRef] [PubMed]
4. Goicoechea, M.; Lacombe, E.; Legay, S.; Mihaljevic, S.; Rech, P.; Jauneau, A.; Lapierre, C.; Pollet, B.; Verhaegen, D.; Chaubet-Gigot, N.; et al. *EgMYB2*, a new transcriptional activator from *Eucalyptus* xylem, regulates secondary cell wall formation and lignin biosynthesis. *Plant J.* **2005**, *43*, 553–567. [CrossRef]
5. Zhang, S.; Chen, Y.; He, X.; Du, J.; Zhang, R.; Ma, Y.; Hu, X.; Zhang, Z.; Chen, Q.; Wan, X. Identification of *MYB* transcription factors regulating theanine biosynthesis in tea plant using omics-based gene coexpression analysis. *J. Agric. Food Chem.* **2020**, *68*, 918–926. [CrossRef]
6. Mehrtens, F.; Kranz, H.; Bednarek, P.; Weisshaar, B. The *Arabidopsis* transcription factor *MYB12* is a flavonol-specific regulator of phenylpropanoid biosynthesis. *Plant Physiol.* **2005**, *138*, 1083–1096. [CrossRef]
7. Jiang, S.; Chen, M.; He, N.; Chen, X.; Wang, N.; Sun, Q.; Zhang, T.; Xu, H.; Fang, H.; Wang, Y.; et al. *MdGSTF6*, activated by *MdMYB1*, plays an essential role in anthocyanin accumulation in apple. *Hortic. Res.* **2019**, *6*, 40. [CrossRef]
8. Xu, H.; Zou, Q.; Yang, G.; Jiang, S.; Fang, H.; Wang, Y.; Zhang, J.; Zhang, Z.; Wang, N.; Chen, X. *MdMYB6* regulates anthocyanin formation in apple both through direct inhibition of the biosynthesis pathway and through substrate removal. *Hortic. Res.* **2020**, *7*, 72. [CrossRef]
9. Meng, Y.; Wang, Z.; Wang, Y.; Wang, C.; Zhu, B.; Liu, H.; Ji, W.; Wen, J.; Chu, C.; Tadege, M.; et al. The MYB activator *WHITE PETAL1* associates with *MtTT8* and *MtWD40-1* to regulate carotenoid-derived flower pigmentation in *Medicago truncatula*. *Plant Cell* **2019**, *31*, 2751–2767. [CrossRef]
10. Shen, X.; Zhao, K.; Liu, L.; Zhang, K.; Yuan, H.; Liao, X.; Wang, Q.; Guo, X.; Li, F.; Li, T. A role for PacMYBA in ABA-regulated anthocyanin biosynthesis in red-colored sweet cherry cv. Hong Deng (*Prunus avium* L.). *Plant Cell Physiol.* **2014**, *55*, 862–880. [CrossRef]
11. Hatlestad, G.J.; Sunnadeniya, R.M.; Akhavan, N.A.; Gonzalez, A.; Goldman, I.L.; McGrath, J.M.; Lloyd, A.M. The beet R locus encodes a new cytochrome P450 required for red betalain production. *Nat. Genet.* **2012**, *44*, 816–820. [CrossRef] [PubMed]
12. Feller, A.; Machemer, K.; Braun, E.L.; Grotewold, E. Evolutionary and comparative analysis of *MYB* and *bHLH* plant transcription factors. *Plant J.* **2011**, *66*, 94–116. [CrossRef] [PubMed]
13. Stracke, R.; Werber, M.; Weisshaar, B. The R2R3-MYB gene family in *Arabidopsis thaliana*. *Curr. Opin. Plant Biol.* **2001**, *4*, 447–456. [CrossRef]
14. Rosinski, J.A.; Atchley, W.R. Molecular evolution of the *Myb* family of transcription factors: Evidence for polyphyletic origin. *J. Mol. Evol.* **1998**, *46*, 74–83. [CrossRef]
15. Jin, H.; Martin, C. Multifunctionality and diversity within the plant MYB-gene family. *Plant Mol. Biol.* **1999**, *41*, 577–585. [CrossRef]
16. Kim, S.; Park, M.; Yeom, S.I.; Kim, Y.M.; Lee, J.M.; Lee, H.A.; Seo, E.; Choi, J.; Cheong, K.; Kim, K.T.; et al. Genome sequence of the hot pepper provides insights into the evolution of pungency in *Capsicum* species. *Nat. Genet.* **2014**, *46*, 270–278. [CrossRef]

17. Alappat, B.; Alappat, J. Anthocyanin Pigments: Beyond Aesthetics. *Molecules* **2020**, *25*, 5500. [CrossRef]
18. Zhu, Z.; Sun, B.; Cai, W.; Zhou, X.; Mao, Y.; Chen, C.; Wei, J.; Cao, B.; Chen, C.; Chen, G.; et al. Natural variations in the *MYB* transcription factor *MYB31* determine the evolution of extremely pungent peppers. *New Phytol.* **2019**, *223*, 922–938. [CrossRef]
19. Usman, M.G.; Rafii, M.Y.; Ismail, M.R.; Malek, M.A.; Latif, M.A. Capsaicin and dihydrocapsaicin determination in chili pepper genotypes using ultra-fast liquid chromatography. *Molecules* **2014**, *19*, 6474–6488. [CrossRef]
20. Wu, S.; Zeng, J.; Xie, H.; Ng, S.H. Capsaicin determination and chili sauce discrimination using low-cost and portable electrochemical sensors based on all graphite pencil electrodes. *Anal. Methods* **2016**, *8*, 7025–7029. [CrossRef]
21. Suzuki, T.; Fujiwake, H.; Iwai, K. Intracellular localization of capsaicin and its analogues, capsaicinoid, in *Capsicum* fruit 1. Microscopic investigation of the structure of the placenta of *Capsicum annuum* var. *annuum* cv. Karayatsubusa. *Plant Cell Physiol.* **1980**, *21*, 839–853. [CrossRef]
22. Mazourek, M.; Pujar, A.; Borovsky, Y.; Paran, I.; Mueller, L.; Jahn, M.M. A dynamic interface for capsaicinoid systems biology. *Plant Physiol.* **2009**, *150*, 1806–1821. [CrossRef] [PubMed]
23. Wahyuni, Y.; Ballester, A.-R.; Sudarmonowati, E.; Bino, R.J.; Bovy, A.G. Metabolite biodiversity in pepper (*Capsicum*) fruits of thirty-two diverse accessions: Variation in health-related compounds and implications for breeding. *Phytochemistry* **2011**, *72*, 1358–1370. [CrossRef] [PubMed]
24. Meléndez-Martínez, A. An overview of carotenoids, apocarotenoids, and vitamin A in agro-food, nutrition, health, and disease. *Mol. Nutr. Food Res.* **2019**, *63*, 1–43. [CrossRef] [PubMed]
25. Estrada, B.; Bernal, M.A.; Díaz, J.; Pomar, F.; Merino, F. Fruit development in *capsicum annuum*: Changes in capsaicin, lignin, free phenolics, and peroxidase patterns. *J. Agric. Food Chem.* **2000**, *48*, 6234–6239. [CrossRef] [PubMed]
26. Sun, B.; Zhou, X.; Chen, C.; Chen, C.; Chen, K.; Chen, M.; Liu, S.; Chen, G.; Cao, B.; Cao, F.; et al. Coexpression network analysis reveals an *MYB* transcriptional activator involved in capsaicinoid biosynthesis in hot peppers. *Hortic. Res.* **2020**, *7*, 162. [CrossRef]
27. Stewart, C., Jr.; Kang, B.C.; Liu, K.; Mazourek, M.; Moore, S.L.; Yoo, E.Y.; Kim, B.D.; Paran, I.; Jahn, M.M. The *Pun1* gene for pungency in pepper encodes a putative acyltransferase. *Plant J.* **2005**, *42*, 675–688. [CrossRef]
28. Lu, S.X.; Knowles, S.M.; Andronis, C.; Ong, M.S.; Tobin, E.M. *CIRCADIAN CLOCK ASSOCIATED1* and *LATE ELONGATED HYPOCOTYL* function synergistically in the circadian clock of *Arabidopsis*. *Plant Physiol.* **2009**, *150*, 834–843. [CrossRef]
29. Matsui, K.; Umemura, Y.; Ohme-Takagi, M. *AtMYBL2*, a protein with a single MYB domain, acts as a negative regulator of anthocyanin biosynthesis in *Arabidopsis*. *Plant J.* **2008**, *55*, 954–967. [CrossRef]
30. Rawat, R.; Takahashi, N.; Hsu, P.Y.; Jones, M.A.; Schwartz, J.; Salemi, M.R.; Phinney, B.S.; Harmer, S.L. *REVEILLE8* and *PSEUDO-REPONSE REGULATOR5* form a negative feedback loop within the *Arabidopsis* circadian clock. *PLoS Genet.* **2011**, *7*, e1001350. [CrossRef]
31. Kerstetter, R.A.; Bollman, K.; Taylor, R.A.; Bomblies, K.; Poethig, R.S. *KANADI* regulates organ polarity in *Arabidopsis*. *Nature* **2001**, *411*, 706–709. [CrossRef] [PubMed]
32. Izhaki, A.; Bowman, J.L. *KANADI* and class III *HD-Zip* gene families regulate embryo patterning and modulate auxin flow during embryogenesis in *Arabidopsis*. *Plant Cell* **2007**, *19*, 495–508. [CrossRef] [PubMed]
33. Lin, Q.; Hamilton, W.; Merryweather, A. Cloning and initial characterization of 14 *myb*-related cDNAs from tomato (*Lycopersicon esculentum* cv. Ailsa Craig). *Plant Mol. Biol.* **1996**, *30*, 1009–1020. [CrossRef] [PubMed]
34. Yanhui, C.; Xiaoyuan, Y.; Kun, H.; Meihua, L.; Jigang, L.; Zhaofeng, G.; Zhiqiang, L.; Yunfei, Z.; Xiaoxiao, W.; Xiaoming, Q.; et al. The MYB transcription factor superfamily of *Arabidopsis*: Expression analysis and phylogenetic comparison with the rice MYB family. *Plant Mol. Biol.* **2006**, *60*, 107–124. [CrossRef]
35. Zhang, L.; Zhao, G.; Jia, J.; Liu, X.; Kong, X. Molecular characterization of 60 isolated wheat *MYB* genes and analysis of their expression during abiotic stress. *J. Exp. Bot.* **2011**, *63*, 203–214. [CrossRef]
36. Qing, X.U.; Jie, H.E.; Jianhui, D.O.N.G.; Xiaojin, H.O.U.; Zhang, X. Genomic survey and expression profiling of the MYB gene family in watermelon. *Hortic. Plant J.* **2018**, *4*, 1–15.
37. Li, D.; Qian, J.; Li, W.; Yu, N.; Gan, G.; Jiang, Y.; Li, W.; Liang, X.; Chen, R.; Mo, Y.; et al. A high-quality genome assembly of the eggplant provides insights into the molecular basis of disease resistance and chlorogenic acid synthesis. *Mol. Ecol. Resour.* **2021**, *21*, 1274–1286. [CrossRef]
38. Arce-Rodríguez, M.L.; Martínez, O.; Ochoa-Alejo, N. Genome-wide identification and analysis of the *MYB* transcription factor gene family in chili pepper (*Capsicum* spp.). *Int. J. Mol. Sci.* **2021**, *22*, 2229. [CrossRef]
39. Qian, W.F.; Liao, B.Y.; Chang, A.Y.F.; Zhang, J.Z. Maintenance of duplicate genes and their functional redundancy by reduced expression. *Trends Genet.* **2010**, *26*, 425–430. [CrossRef]
40. Islam, S.; Sajib, S.D.; Jui, Z.S.; Arabia, S.; Islam, T.; Ghosh, A. Genome-wide identification of glutathione S-transferase gene family in pepper, its classification, and expression profiling under different anatomical and environmental conditions. *Sci. Rep.* **2019**, *9*, 9101. [CrossRef]
41. Kuno, N.; Møller, S.G.; Shinomura, T.; Xu, X.; Chua, N.H.; Furuya, M. The novel MYB protein *EARLY-PHYTOCHROME-RESPONSIVE1* is a component of a slave circadian oscillator in *Arabidopsis*. *Plant Cell* **2003**, *15*, 2476–2488. [CrossRef] [PubMed]
42. Liu, S.; Yang, L.; Li, J.; Tang, W.; Li, J.; Lin, R. *FHY3* interacts with *phytochrome B* and regulates seed dormancy and germination. *Plant Physiol.* **2021**, *187*, 289–302. [CrossRef]
43. Nguyen, N.H.; Jeong, C.Y.; Kang, G.H.; Yoo, S.D.; Hong, S.W.; Lee, H. *MYBD* employed by *HY5* increases anthocyanin accumulation via repression of *MYBL 2* in *Arabidopsis*. *Plant J.* **2015**, *84*, 1192–1205. [CrossRef]

44. Sun, B.; Zhu, Z.; Chen, C.; Chen, G.; Cao, B.; Chen, C.; Lei, J. Jasmonate-inducible *R2R3-MYB* transcription factor regulates capsaicinoid biosynthesis and stamen development in *Capsicum*. *J. Agric. Food Chem.* **2019**, *67*, 10891–10903. [CrossRef] [PubMed]
45. Eddy, S.R. Profile hidden Markov models. *Bioinformatics* **1998**, *14*, 755–763. [CrossRef]
46. Larkin, M.A.; Blackshields, G.; Brown, N.P.; Chenna, R.; McGettigan, P.A.; McWilliam, H.; Valentin, F.; Wallace, I.M.; Wilm, A.; Lopez, R.; et al. Clustal W and Clustal X version 2.0. *Bioinformatics* **2007**, *23*, 2947–2948. [CrossRef] [PubMed]
47. Tamura, K.; Dudley, J.; Nei, M.; Kumar, S. MEGA4: Molecular Evolutionary Genetics Analysis (MEGA) software version 4.0. *Mol. Biol. Evol.* **2007**, *24*, 1596–1599. [CrossRef] [PubMed]
48. Hu, B.; Jin, J.; Guo, A.-Y.; Zhang, H.; Luo, J.; Gao, G. GSDS 2.0: An upgraded gene feature visualization server. *Bioinformatics* **2015**, *31*, 1296–1297. [CrossRef]
49. Chen, C.; Chen, H.; Zhang, Y.; Thomas, H.R.; Frank, M.H.; He, Y.; Xia, R. TBtools: An integrative toolkit developed for interactive analyses of big biological data. *Mol. Plant* **2020**, *13*, 1194–1202. [CrossRef]
50. Krzywinski, M.; Schein, J.; Birol, I.; Connors, J.; Gascoyne, R.; Horsman, D.; Jones, S.J.; Marra, M.A. Circos: An information aesthetic for comparative genomics. *Genome Res.* **2009**, *19*, 1639–1645. [CrossRef]
51. Shen, W.; Le, S.; Li, Y.; Hu, F. SeqKit: A cross-platform and ultrafast toolkit for FASTA/Q file manipulation. *PLoS ONE* **2016**, *11*, e0163962. [CrossRef] [PubMed]
52. Lescot, M.; Déhais, P.; Thijs, G.; Marchal, K.; Moreau, Y.; Van de Peer, Y.; Rouzé, P.; Rombauts, S. PlantCARE, a database of plant *cis*-acting regulatory elements and a portal to tools for in silico analysis of promoter sequences. *Nucleic Acids Res.* **2002**, *30*, 325–327. [CrossRef] [PubMed]
53. Liu, F.; Yu, H.; Deng, Y.; Zheng, J.; Liu, M.; Ou, L.; Yang, B.; Dai, X.; Ma, Y.; Feng, S.; et al. PepperHub, an informatics hub for the chili pepper research community. *Mol. Plant* **2017**, *10*, 1129–1132. [CrossRef] [PubMed]
54. Schmittgen, T.D.; Livak, K.J. Analyzing real-time PCR data by the comparative C(T) method. *Nat. Protoc.* **2008**, *3*, 1101–1108. [CrossRef] [PubMed]

International Journal of
Molecular Sciences

MDPI

Article

A Large-Scale Genomic Association Analysis Identifies the Candidate Genes Regulating Salt Tolerance in Cucumber (*Cucumis sativus* L.) Seedlings

Dongrang Liu †, Shaoyun Dong †, Han Miao, Xiaoping Liu, Caixia Li, Jianan Han, Shengping Zhang * and Xingfang Gu *

Institute of Vegetables and Flowers, Chinese Academy of Agricultural Sciences, Beijing 100081, China; 82101192231@caas.cn (D.L.); dongshaoyun@caas.cn (S.D.); miaohan@caas.cn (H.M.); liuxiaoping@caas.cn (X.L.); 82101211129@caas.cn (C.L.); 82101191124@caas.cn (J.H.)
* Correspondence: zhangshengping@caas.cn (S.Z.); guxingfang@caas.cn (X.G.)
† These authors contributed equally to this work.

Abstract: Salt stress seriously restricts plant growth and development, affects yield and quality, and thus becomes an urgent problem to be solved in cucumber stress resistance breeding. Mining salt tolerance genes and exploring the molecular mechanism of salt tolerance could accelerate the breeding of cucumber germplasm with excellent salt stress tolerance. In this study, 220 cucumber core accessions were used for Genome-Wide Association Studies (GWAS) and the identification of salt tolerance genes. The salinity injury index that was collected in two years showed significant differences among the core germplasm. A total of seven loci that were associated with salt tolerance in cucumber seedlings were repeatedly detected, which were located on Chr.2 (*gST2.1*), Chr.3 (*gST3.1* and *gST3.2*), Chr.4 (*gST4.1* and *gST4.2*), Chr.5 (*gST5.1*), and Chr.6 (*gST6.1*). Within these loci, 62 genes were analyzed, and 5 candidate genes (*CsaV3_2G035120*, *CsaV3_3G023710*, *CsaV3_4G033150*, *CsaV3_5G023530*, and *CsaV3_6G009810*) were predicted via the functional annotation of *Arabidopsis* homologous genes, haplotype of extreme salt-tolerant accessions, and qRT-PCR. These results provide a guide for further research on salt tolerance genes and molecular mechanisms of cucumber seedlings.

Keywords: cucumber seedlings; salt stress; GWAS; candidate genes

1. Introduction

Cucumber (*Cucumis sativus* L.) is one of the most important vegetables worldwide, and according to the FAOSTAT (https://www.fao.org/faostat/zh/#home (accessed on 17 January 2022)), the global total grown area and production of cucumber was 2.25 million hectares and 90.35 million tons, respectively, in 2020. However, cucumber is very sensitive to salt stress, which is one of the important limiting factors in cucumber cultivation [1,2].

In general, salt stress restricts plant physiological activity through osmotic and toxic effects, depending on the salt concentration [3]. Previous studies have shown that plants have evolved multiple complex mechanisms to adapt to salt stress in environment, including antioxidant systems, osmotic regulation, transcriptional regulation, and signal transduction, etc. [4,5]. Salt stress will cause significant damage to cucumber seedlings, which is mainly manifested in obvious symptoms of chlorosis and the decline of various growth indicators of the plant and root system, and the damage is significantly increased with the increase of salt concentration and treatment time [6,7]. Therefore, mining genes that are related to plant salt tolerance is an important and complex process. Researches have shown that cucumber salt tolerance is a quantitative trait that is controlled by multiple genes and is easily affected by environmental factors [7,8]. Liu, et al. [7] evaluated the salt tolerance of a RILs population in cucumber seedlings and detected a major QTL on chromosome 6 using salt injury index. As this is currently the only reported QTL controlling the salt tolerance of cucumber seedlings, it is urgent to identify more loci and genes to elucidate the mechanism of cucumber salt tolerance.

Recently, linkage mapping and genome-wide association studies have been widely used in mining genes that are related to vegetable phenotypic variation [9]. Hou, et al. [10] conducted GWAS analysis on 212 diverse melon (*Cucumis melo* L.) accession seedlings using chilling injury index and identified 11 low-temperature tolerance genes. Khanzada, et al. [11] identified the phenotypes of 228 *Brassica* accession seedlings under drought stress and 85 drought tolerance genes were identified by GWAS analysis in rapeseed (*Brassica napus*). Zhang, et al. [12] used GWAS analysis of 83 non-heading Chinese cabbage accessions to identify 17 loci that were associated with turnip mosaic virus (TuMV) resistance. In cucumber, a large number of loci and genes that are related to cucumber growth phenotypes and resistance traits have been unearthed through GWAS analysis. Wei, et al. [13] used the heat injury index of 86 cucumber core accessions to conduct GWAS analysis and detected seven GWAS loci (*gHII4.1*, *gHII4.2*, *gHII5.1*, *gHII5.2*, *gHII6.1*, *gHII6.2*, and *gHII7.1*). Wang, et al. [14] repeatedly detected four GWAS loci (*gLTS1.1*, *gLTS3.1*, *gLTS4.1*, and *gLTS5.1*) by using the chilling injury index of 172 cucumber core accessions seedlings. Lee, et al. [15] conducted GWAS analysis of cucumber powdery mildew resistance using 264 accessions from all over the world and predicted two novel candidate genes (*Csa5G453160* and *Csa5G471070*). Liu, et al. [16,17] used the core germplasm to predict five downy mildew-related candidate genes (*Csa1G575030*, *Csa2G060360*, *Csa4G064680*, *Csa5G606470*, and *Csa7G004020*) and eight candidate genes that were associated with powdery mildew (*Csa2G030010*, *Csa2G030020*, *Csa2G030030*, *Csa3G414050*, *Csa4G022350*, *Csa5G48880*, *Csa5G603950*, and *Csa5G603960*), respectively. However, no study on the investigation of cucumber salt tolerance via GWAS has been reported.

Here, GWAS was performed using the salt injury index of 220 cucumber core accessions, the loci that were repeatedly detected in two years, that were further investigated, and candidate genes were mined through haplotype and qRT-PCR analyses. These genes could lay the foundation for the elaboration of cucumber salt stress response mechanism and the molecular improvement of excellent salt-tolerant cucumber varieties.

2. Results

2.1. Genetic Diversity of Salt Tolerance in Cucumber Germplasm

The 220 accessions were evaluated for salt tolerance two times, in June 2020 (Salt_2020) and May 2021 (Salt_2021). A total of three replicates were set for each experiment, each of which had 5 plants. At 14 days old, the seedlings were exposed to salt stress by 200 mmol·L^{-1} NaCl. After 7 days of salt treatment, the salt injury was divided into six scales, based on the chlorosis and wilting degree of the first true leaf and cotyledons (Figure 1a). For each accession, the salt injury index of each experiment was calculated by the formula on the basis of the six scales and shown in Table S2. The phenotypic data of the two treatments showed that the salt injury index of different accessions was significantly different. The average salt injury index was 31.18 and 28.80, with the coefficient of variation of 41.35% and 47.84%, respectively (Table 1). As shown in Table 1 and Figure 1c, the normal distribution test ($p > 0.05$) and the violin plot showed that the salt injury index follow the normal distribution pattern, and the salt injury index correlation among the two experiments was significant (Figure 1c,d).

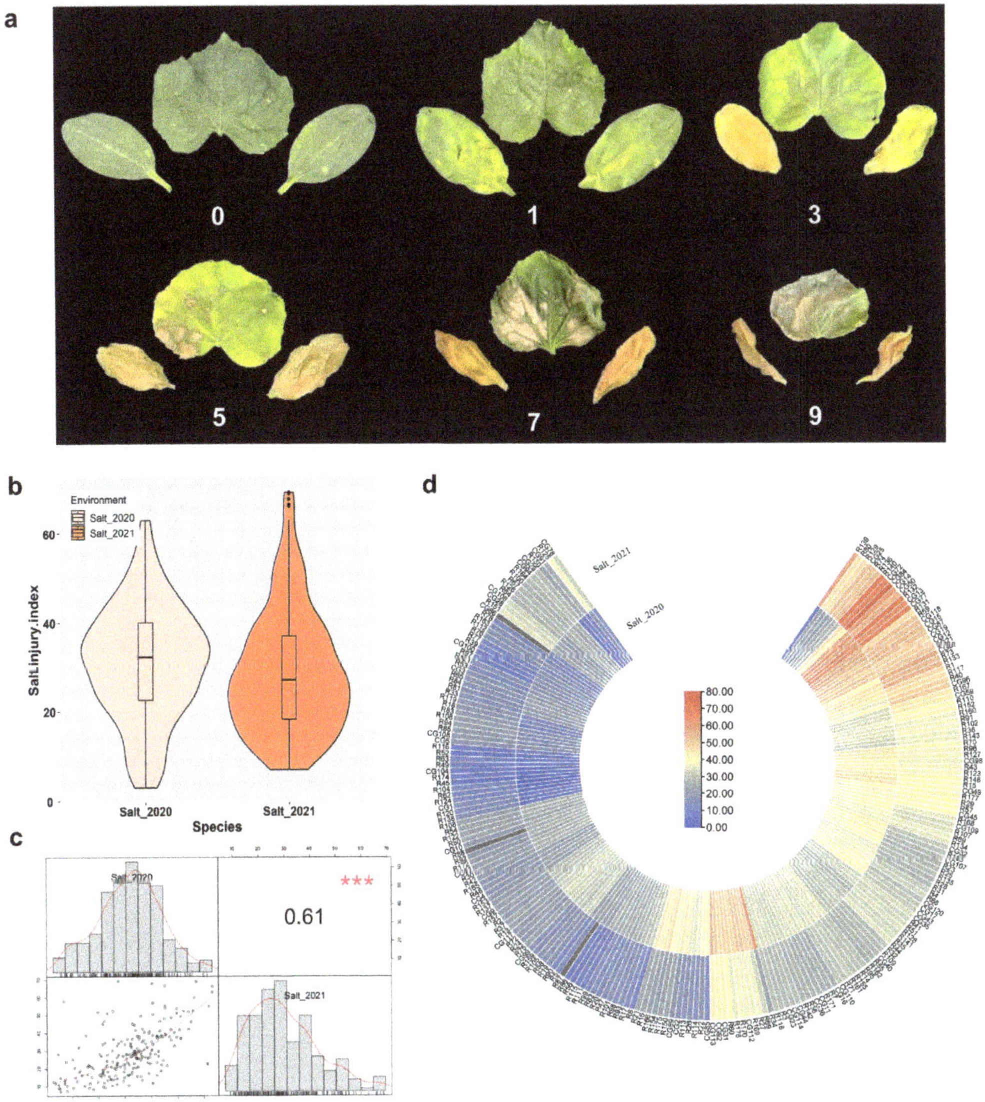

Figure 1. Phenotypic characterization of salt tolerance in 14 day old cucumber seedlings that were treated with 200 mmol·L^{-1} NaCl for seven days. A total of three replicates were set for each experiment, with a total of 15 replicates per accession. (**a**) Salt injury of cucumber seedlings was divided into 6 grades. 0, the cotyledons and the first true leaf show no symptoms of salt injury; 1, the cotyledons are slightly yellowed; 3, the cotyledons and the first true leaf are significantly yellowed; 5, the cotyledons are yellowed and wilted, and the first true leaf is significantly yellowed and wilted; 7, the cotyledons are dehydrated, the large areas of first true leaf are yellowed and wilted; 9, both the cotyledons and the first true leaf are yellowed and dehydrated. (**b**) Violin plot depicting the phenotypic distribution of salt injury index in the two treatments. (**c**) Frequency distribution and Pearson rank correlations of the mean salt injury index in two experiments. *** indicate significance at $p < 0.001$. (**d**) Heatmap depicting the salt injury index in two treatments.

Table 1. The salt injury indexes of the germplasm in two years.

Treatment	Max. Score	Min. Score	Mean	STD	Skewness	Kurtosis	C.V	Normal Distribution Test (p Value)
Salt_2020	63.00	3.00	31.18	12.89	−0.05	−0.25	41.35%	0.83
Salt_2021	69.30	6.95	28.80	13.78	0.77	0.22	47.84%	0.06

According to difference in the salt injury index, 220 accessions were divided into 5 groups (I, II, III, IV, and V) by the Ward method (Figure 2). Group I had 58 highly salt-tolerant accessions, Group II had 45 salt-tolerant accessions, Group III had 66 intermediate salt-tolerant accessions, Group IV had 31 salt-sensitive accessions, and Group V had 20 highly salt-sensitive accessions (Table S3). The results of cluster analysis were further analyzed in combination with the ecotype of the germplasm. The 220 core accessions included 1 European-Asia ecotype, 5 American slicing ecotypes, 6 Xishuangbanna ecotypes, 12 Japan ecotypes, 16 American processing ecotypes, 17 India ecotypes, 18 North China-European hybrid ecotypes, 30 South China ecotypes, 31 European ecotypes, and 84 North China ecotypes. It was difficult to conclude the salt tolerance of the Xishuangbanna ecotype, American slicing ecotype, and the European-Asia ecotype, since only several lines were included. As such, we analyzed the ecotypes which contained more than 10 accessions. The results showed that most of the Japan ecotypes showed high salt-tolerance; the European, North China-European hybrid and North China ecotypes exhibited intermediate salt-tolerance; while the Indian ecotypes were more salt-sensitive (Figure 3).

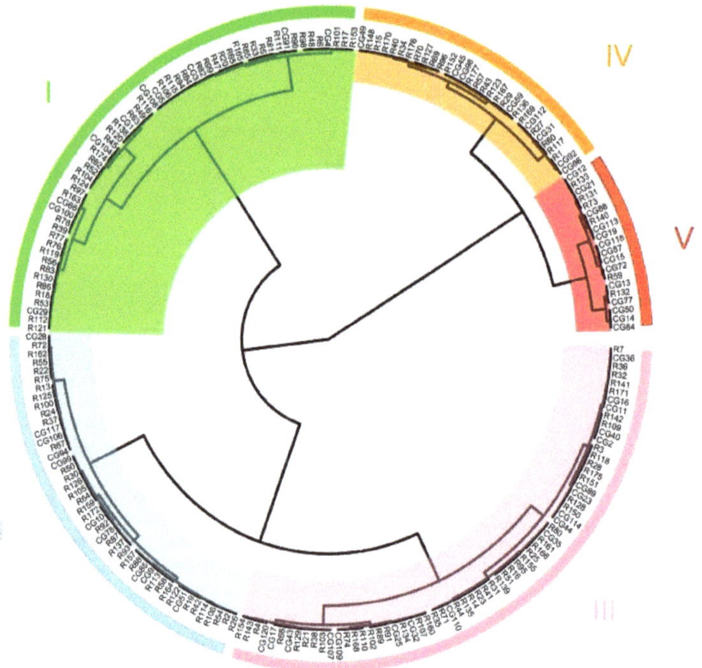

Figure 2. Cluster analysis of salt tolerance of core accession seedlings based on salt injury index. The I, II, III, IV, and V indicates Group I, Group II, Group III, Group IV, and Group V, respectively.

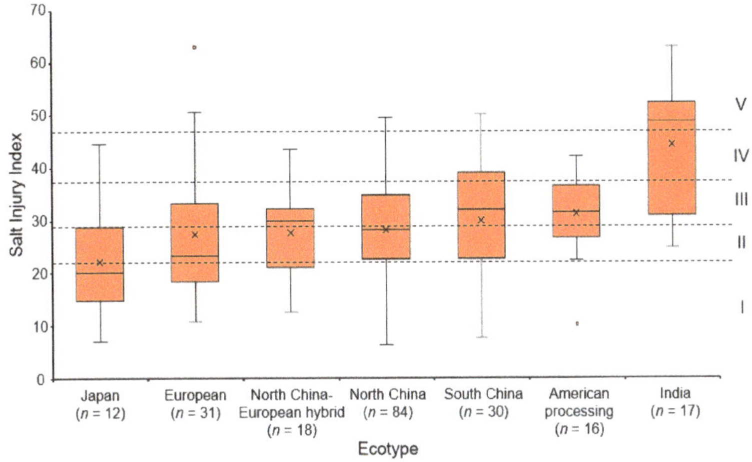

Figure 3. The box plot depicting the salt injury index of different ecotypes. The I, II, III, IV, and V indicates Group I, Group II, Group III, Group IV, and Group V, respectively. The dot represents the outlier. The cross represents the mean.

2.2. Genome-Associated Analysis of Salt Tolerance in Seedlings

Using the salt injury index data of the two experiments in 2020 and 2021, genome-wide association analysis (GWAS) was performed and 7 genetic loci (*gST2.1*, *gST3.1*, *gST3.2*, *gST4.1*, *gST4.2*, *gST5.1*, and *gST6.1*) that were repeatedly detected. These 7 loci were distributed over five chromosomes (Figure 4), which were considered as novel loci for salt tolerance in cucumber seedlings.

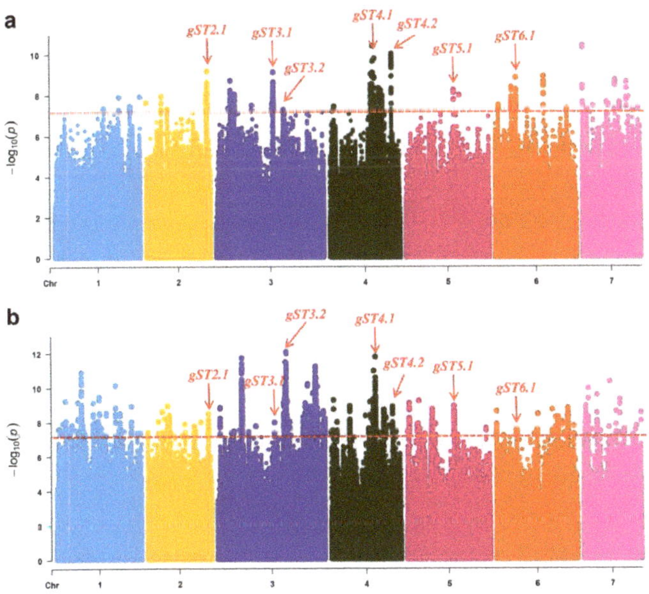

Figure 4. Genome-wide association analysis (GWAS) Manhattan plots of salt tolerance in two experiments: (**a**) Salt_2020 and (**b**) Salt_2021. The different colors represent the different chromosomes. The red dashed line represents the significance threshold. The red arrow indicates the position of the strong peaks that were repeatedly detected in the two experiments.

2.3. Candidate Gene Analysis for the Novel Loci

In order to identify the potential candidate genes in these loci, the 100-kb region around each peak of SNPs was used for further analysis, and genes in each locus are shown in Table S4. To identify the candidate genes, functional annotation of the orthologs in *Arabidopsis*, haplotype of extreme salt-tolerant accessions, and qRT-PCR were analyzed. According to the results of the clustering analysis, Group I (58 highly salt-tolerant accessions) and Group V (20 highly salt-sensitive accessions) were used for haplotype analysis. Based on the phenotypic data and ecotype of accessions, salt-tolerant accessions (R45, R63, R78, and CG104) and salt-sensitive accessions (R59, R132, CG13, and CG14) were used for expression analysis. At 14 days old, the seedlings were exposed to salt stress by 200 mmol·L^{-1} NaCl, with a total of 4 time points, namely 0 h, 10 h, 48 h, and 96 h post-salt treatment. The first true leaf of each seedling was harvested as the sample for qRT-PCR. Through the above analysis, we preliminarily predicted five candidate genes that were related to salt tolerance in *gST2.1*, *gST3.1*, *gST4.2*, *gST5.1*, and *gST6.1* (Table S5).

For *gST2.1*, the 100 kb (23,409,202~23,509,202 bp) region on Chr.2 was analyzed using pairwise LD correlations (Figure 5a). According to CuGenDB, 11 annotated genes are located in this region, and there was one gene (*CsaV3_2G035120*) that was related to salt stress (Figure 5b and Table S5). *CsaV3_2G035120* (encoding an isocitrate lyase) had a SNP in the second exon region that led to the mutation of glycine to alanine (Figure 5c). In the highly salt-tolerant and highly salt-sensitive accessions, 49 out of the 58 highly salt-tolerant accessions (84.48%) carried the G haplotype, while 16 out of the 20 highly salt-sensitive (80%) accessions had the alternate C haplotype (Figure 5d). qRT-PCR analysis revealed that the relative expression level of this gene was significantly up-regulated in highly salt-tolerant accessions (R45 and R63) and highly salt-sensitive accessions (R132, CG13, and CG14) after salt treatment (Figure 5e).

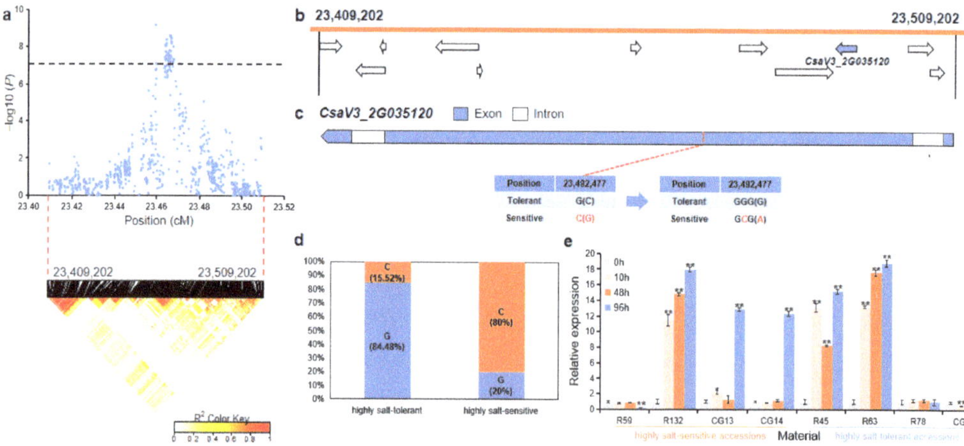

Figure 5. Identification of the candidate gene that was related to salt tolerance for the locus *gST2.1*. (**a**) Manhattan plot and LD heatmap of *gST2.1*. The black dashed line represents the significance threshold and the red dashed lines represent the candidate region (100 kb). (**b**) A total of 11 functional annotated genes were predicted in the *gST2.1* candidate region. The blue arrow indicates that the functional annotation of the gene which is related to salt tolerance. (**c**) Gene structure of *CsaV3_2G035120*. Blue rectangles and white rectangles indicate exons and introns, respectively. (**d**) Orange and blue blocks represent the proportion of C and G haplotype in highly salt-tolerant or highly salt-sensitive accessions, respectively. (**e**) qRT-PCR was used to detect the expression changes of *CsaV3_2G035120* in highly salt-sensitive accessions (R59, R132, CG13 and CG14) and highly salt-tolerant accessions (R45, R63, R78 and CG104) after salt treatment (0 h, 10 h, 48 h and 96 h). Significant difference (* $p < 0.05$, ** $p < 0.01$).

For *gST3.1*, the 100 kb (21,496,729~21,596,729 bp) region on Chr.3 was analyzed using pairwise LD correlations (Figure 6a). According to CuGenDB, six annotated genes are located in this region, and there was one gene (*CsaV3_3G023710*) that was related to salt stress (Figure 6b and Table S5). *CsaV3_3G023710* (encoding an ankyrin repeat family protein) had two SNPs in the first exon region that led to the mutation of lysine to glutamate and isoleucine to methionine (Figure 6c). In the highly salt-tolerant and highly salt-sensitive accessions, 57 out of the 58 highly salt-tolerant accessions (98.28%) carried the AA haplotype, while 15 out of the 20 highly salt-sensitive (75%) accessions had the alternate GG haplotype (Figure 6d). qRT-PCR analysis revealed that the relative expression level of this gene was significantly up-regulated in the highly salt-tolerant accessions (R45 and R63) and the highly salt-sensitive accessions (R132, CG13, and CG14) after salt treatment (Figure 6e).

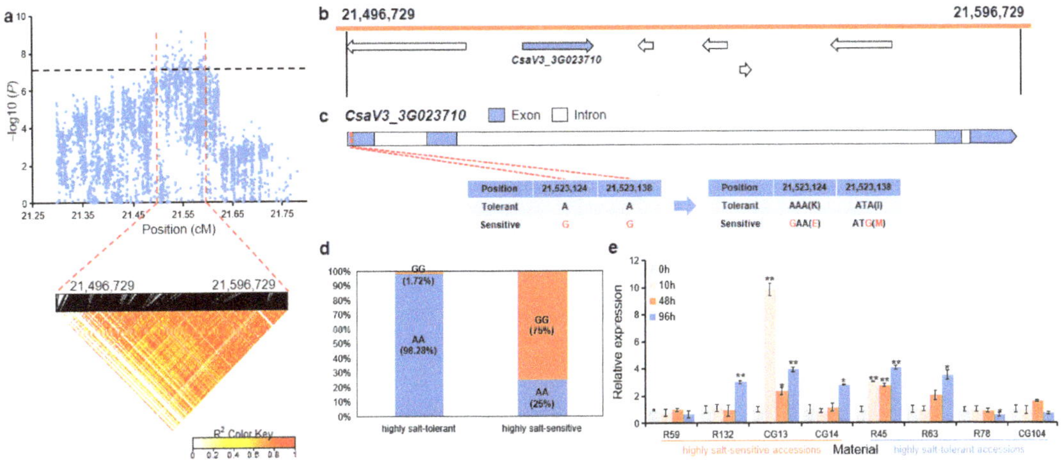

Figure 6. Identification of the candidate gene that was related to salt tolerance for the locus *gST3.1*. (**a**) Manhattan plot and LD heatmap of *gST3.1*. The black dashed line represents the significance threshold and the red dashed lines represent the candidate region (100 kb). (**b**) A total of six functional annotated genes were predicted in the *gST3.1* candidate region. The blue arrow indicates that the functional annotation of the gene which is related to salt tolerance. (**c**) Gene structure of *CsaV3_3G023710*. Blue rectangles and white rectangles indicate exons and introns, respectively. (**d**) Orange and blue blocks represent the proportion of GG and AA haplotype in the highly salt-tolerant or highly salt-sensitive accessions, respectively. (**e**) qRT-PCR was used to detect the expression changes of *CsaV3_3G023710* in the highly salt-sensitive accessions (R59, R132, CG13, and CG14) and highly salt-tolerant accessions (R45, R63, R78, and CG104) after salt treatment (0 h, 10 h, 48 h, and 96 h). Significant difference (* $p < 0.05$, ** $p < 0.01$).

For *gST3.2* and *gST4.1*, we found two genes (*CsaV3_3G030250* and *CsaV3_4G027680*) that were related to salt stress according to CuGenDB. There were nine SNPs on *CsaV3_3G030250* and one SNP on *CsaV3_4G027680*, which were located in the CDS (coding sequence) region that led to amino acid variation (Figures S1c and S2c). But according to the haplotype analysis, all of these SNPs did not have any differential change between the highly salt tolerant and highly salt-sensitive accessions (Figures S1d and S2d). As such, the candidate genes in loci *gST3.2* and *gST4.1* had not been predicted.

For *gST4.2*, the 100 kb (23,423,770~23,523,770 bp) region on Chr.4 was analyzed using pairwise LD correlations (Figure 7a). According to CuGenDB, 13 annotated genes are located in this region, and there was one gene (*CsaV3_4G033150*) that was related to salt stress (Figure 7b and Table S5). *CsaV3_4G033150* (encoding a phosphoenolpyruvate carboxylase protein) had one SNP in the fourth exon region that led to the mutation of leucine to valine (Figure 7c). In the highly salt-tolerant and highly salt-sensitive accessions,

51 out of the 58 highly salt-tolerant accessions (87.93%) carried the C haplotype, while 10 out of the 20 highly salt-sensitive (50%) accessions had the alternate G haplotype (Figure 7d). qRT-PCR analysis revealed that after salt treatment, the relative expression level of this gene was significantly down-regulated in the highly salt-tolerant accessions (R45, R63, R78, and CG104), while was significantly up-regulated in the highly salt-sensitive accessions (R132 and CG14) (Figure 7e).

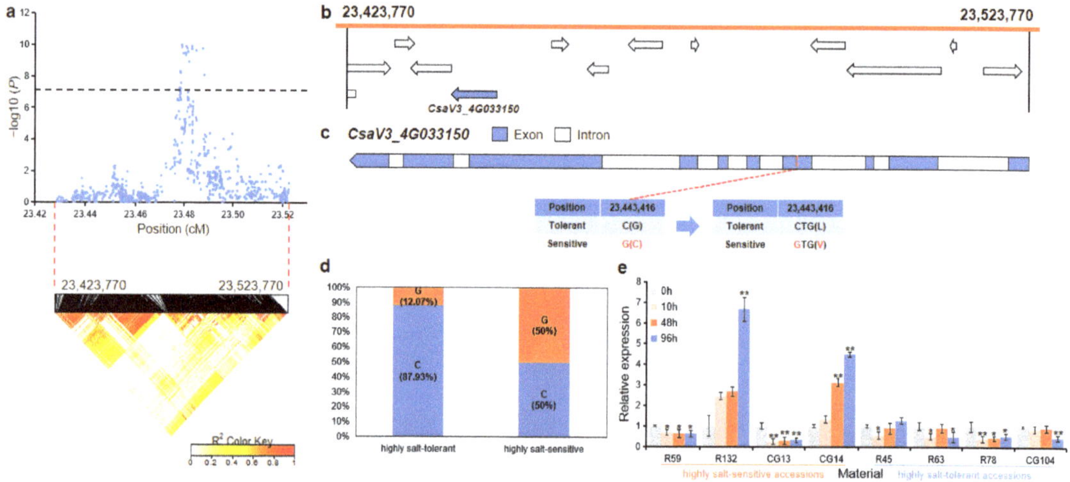

Figure 7. Identification of the candidate gene that was related to salt tolerance for the locus *gST4.2*. (**a**) Manhattan plot and LD heatmap of *gST4.2*. The black dashed line represents the significance threshold and the red dashed lines represent the candidate region (100 kb). (**b**) A total of 13 functional annotated genes were predicted in the *gST4.2* candidate region. The blue arrow indicates that the functional annotation of the gene which is related to salt tolerance. (**c**) Gene structure of *CsaV3_4G033150*. Blue rectangles and white rectangles indicate exons and introns, respectively. (**d**) Orange and blue blocks represent the proportion of G and C haplotypes in the highly salt-tolerant or highly salt-sensitive accessions, respectively. (**e**) qRT-PCR was used to detect the expression changes of *CsaV3_4G033150* in the highly salt-sensitive accessions (R59, R132, CG13, and CG14) and highly salt-tolerant accessions (R45, R63, R78, and CG104) after salt treatment (0 h, 10 h, 48 h, and 96 h). Significant difference (* $p < 0.05$, ** $p < 0.01$).

For *gST5.1*, the 100 kb (17,864,273~17,964,273 bp) region on Chr.5 was analyzed using pairwise LD correlations (Figure 8a). According to CuGenDB, eight annotated genes are located in this region, and there was one gene (*CsaV3_5G023530*) that was related to salt stress (Figure 8b and Table S5). *CsaV3_5G023530* (encoding a PHD finger protein) had one SNP in the 19th exon region that led to the mutation of glutamate to aspartate (Figure 8c). In the highly salt-tolerant and highly salt-sensitive accessions, 24 out of the 58 highly salt-tolerant accessions (58.62%) carried the G haplotype, while all highly salt-sensitive (100%) accessions had the alternate C haplotype (Figure 8d). qRT-PCR analysis revealed that after salt treatment, the relative expression level of this gene was significantly up-regulated in the highly salt-tolerant accessions (R45 and R63), while was significantly down-regulated in the highly salt-sensitive accessions (R59 and CG14) (Figure 8e).

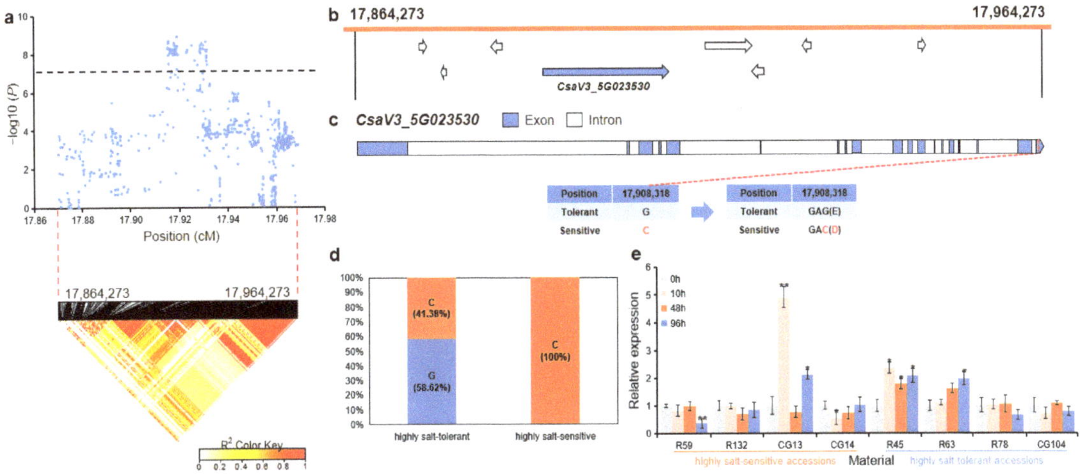

Figure 8. Identification of the candidate gene that was related to salt tolerance for the locus *gST5.1*. (**a**) Manhattan plot and LD heatmap of *gST5.1*. The black dashed line represents the significance threshold and the red dashed lines represent the candidate region (100 kb). (**b**) A total of eight functional annotated genes were predicted in the *gST5.1* candidate region. The blue arrow indicates that the functional annotation of the gene which is related to salt tolerance. (**c**) The gene structure of *CsaV3_5G023530*. Blue rectangles and white rectangles indicate exons and introns, respectively. (**d**) Orange and blue blocks represent the proportion of C and G haplotypes in the highly salt-tolerant or highly salt-sensitive accessions, respectively. (**e**) qRT-PCR was used to detect the expression changes of *CsaV3_5G023530* in the highly salt-sensitive accessions (R59, R132, CG13, and CG14) and highly salt-tolerant accessions (R45, R63, R78, and CG104) after salt treatment (0 h, 10 h, 48 h, and 96 h). Significant difference (* $p < 0.05$, ** $p < 0.01$).

For *gST6.1*, the 100 kb (7,916,377~8,016,377 bp) region on Chr.6 was analyzed using pairwise LD correlations (Figure 9a). According to CuGenDB, 12 annotated genes are located in this region, and there was one gene (*CsaV3_6G009810*) that was related to salt stress (Figure 9b and Table S5). *CsaV3_6G009810* (encoding a pesticidal crystal cry8Ba protein) had three SNPs in the CDS (coding sequence) region that led to the mutation of aspartate to glutamate, histidine to glutamine, and methionine to isoleucine (Figure 9c). In the highly salt-tolerant and highly salt-sensitive accessions, all the highly salt-tolerant accessions (100%) carried the GTT haplotype, while 13 out of the 20 highly salt-sensitive (65%) accessions had the alternate AGG haplotype (Figure 9d). qRT-PCR analysis revealed that after salt treatment, the relative expression level of this gene was significantly up-regulated in the highly salt-tolerant accessions (R45 and R63) and highly salt-sensitive accessions (R132, CG13, and CG14), while was significantly down-regulated in the highly salt-tolerant accessions (R78 and CG104) (Figure 9e).

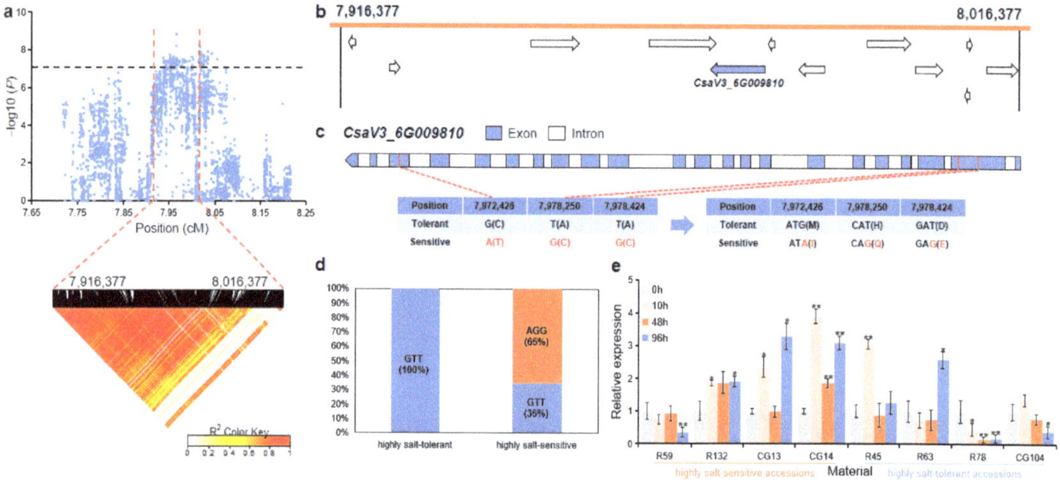

Figure 9. Identification of the candidate gene that as related to salt tolerance for the locus *gST6.1*. (**a**) Manhattan plot and LD heatmap of *gST5.1*. The black dashed line represents the significance threshold and the red dashed lines represent the candidate region (100 kb). (**b**) A total of 12 functional annotated genes were predicted in the *gST6.1* candidate region. The blue arrow indicates that the functional annotation of the gene which is related to salt tolerance. (**c**) Gene structure of *CsaV3_6G009810*. Blue rectangles and white rectangles indicate exons and introns, respectively. (**d**) Orange and blue blocks represent the proportion of AGG and GTT haplotype in the highly salt-tolerant or highly salt-sensitive accessions, respectively. (**e**) qRT-PCR was used to detect the expression changes of *CsaV3_6G009810* in the highly salt-sensitive accessions (R59, R132, CG13, and CG14) and highly salt-tolerant accessions (R45, R63, R78, and CG104) after salt treatment (0 h, 10 h, 48 h, and 96 h). Significant difference (* $p < 0.05$, ** $p < 0.01$).

3. Discussion

The acquisition of phenotypic data is the basis of GWAS. At present, there are many studies on the identification of salt tolerance phenotypic data of cucumber, involving morphological indicators, physiological and biochemical indicators, and cytological indicators [18]. We obtained effective phenotypic data by using the salt injury index classification survey [7]. It can efficiently obtain germplasm salt injury index data that can be used for subsequent analysis. As such, we identified the salt tolerance of different accessions by referring to this method (Figure 1a). Meanwhile, we conducted the experiments in two different environments (Shouguang in 2020 and Beijing in 2021) to ensure the accuracy of the results. The results showed that the two phenotypic data in this study have significant correlations, demonstrating the reliability of the data and the applicability of this method (Table 1 and Figure 1b).

The stress resistance of accessions in different ecotype groups was different. Cao, et al. [19] believed that the salt tolerance of South China ecotype cucumber germplasm was generally better than that of North China cucumber germplasm at the germination stage. Wei, et al. [13] identified the heat injury index of 86 cucumber core accession seedlings and found that the Indian ecotype and Xishuangbanna ecotype accessions were mostly highly heat-tolerant. Wang, et al. [14] used the chilling injury index of 87 cucumber core accession seedlings as an indicator and found that most of the North China ecotype and Japan ecotype accessions were tolerant to low temperature, while most of the European ecotype and India ecotype accessions were sensitive to low temperature. In this study, 220 core accessions were used to identify the salt tolerance of seedlings and we found that Japan ecotype and Indian ecotype accessions were generally highly salt-tolerant and highly salt-sensitive, respectively (Figure 3).

In recent years, great progress has been made in revealing the key genes of plant salt tolerance through GWAS in a variety of crops. Baxter, et al. [20] used 337 *Arabidopsis* accessions to conduct GWAS analysis of Na^+ accumulation in leaves under salt stress and identified the key locus of *Arabidopsis* salt tolerance *AtHKT1;1* via combined genetic complementation and gene expression studies. Patishtan, et al. [21] used 306 rice accessions as materials and used Na^+ and K^+ content in the tissues as indictors for GWAS analysis and predicted 70 candidate genes. Do, et al. [22] detected a GWAS locus that was associated with four traits, including leaf scorch degree, chlorophyll content ratio, leaf sodium content, and leaf chlorine content using 305 soybean accessions. Oyiga, et al. [23] used 150 wheat accessions to conduct GWAS analysis with Na^+ and K^+ contents in leaves and predicted 22 candidate genes for salt tolerance with combined gene expression analysis. Sun, et al. [24] measured the relative survival rate and salt tolerance of 713 cotton accessions under salt stress and predicted six candidate genes by combining GWAS analysis and expression analysis. In this study, *gST2.1*, *gST3.1*, *gST3.2*, *gST4.1*, *gST4.2*, *gST5.1*, and *gST6.1* were new salt tolerance loci in cucumber through the identification of salt tolerance of core germplasm and GWAS analysis, which was also the first application of GWAS analysis in mining salt tolerance genes in cucumber (Figure 4 and Table S4).

The mechanism of cucumber response to salt stress is a complex and dynamic process. Through homologous gene cloning and expression analysis, it was found that the main mechanisms of salt stress regulation in *Arabidopsis* also play an important role in cucumber, such as SOS (salt-overly-sensitive) and ABA [18]. Currently, a QTL locus that is associated with salt tolerance in cucumber has been reported [7]. In addition, a large number of gene families that might be related to salt tolerance of cucumber have been identified in recent years, such as CDPK [25], SOD [26], AOP [27], bHLH [28], WRKY [29], GRX [30], and CAMTA [31]. In this study, five candidate genes that were associated with salt stress were predicted from these seven loci (Table S5). *CsaV3_2G035120* encodes a glyoxylate cycle enzyme isocitrate lyase (ICL) that is involved in salt tolerance. Transcriptome analysis showed that rice isocitrate lyase gene (*OsICL*) was induced by salt stress, and the overexpression of rice calmodulin gene (*OsCAM1-1*) enhanced *OsICL* expression by salt stress [32]. The overexpression of *OsICL* in *Arabidopsis AtICL* knockout mutant by Yuenyong, et al. [33] further supports the role of ICL in plant salt tolerance through the glyoxylate cycle. Our study found the expression of *CsaV3_2G035120* was also significantly up-regulated by salt stress in cucumber core accessions (R45, R63, R132, CG13, and CG14) (Figure 5e).

CsaV3_3G023710 encodes an ankyrin repeat family protein and the homologous gene in *Arabidopsis* is *ITN1* (*AT3G12360*). *ITN1* encodes a protein with an ankyrin motif and transmembrane domains that is involved in salt tolerance and may act through abscisic acid (ABA) signaling pathways and promote reactive oxygen species (ROS) production in *Arabidopsis* [34]. The expression of *CsaV3_3G023710* was also significantly up-regulated by salt stress in highly salt-sensitive cucumber core accessions (R132, CG13, and CG14) (Figure 6e).

CsaV3_4G033150 encodes *Arabidopsis* phosphoenolpyruvate carboxylase (PEPC) proteins. Sanchez, et al. [35] found that *PEPC* genes show differential expression in *Arabidopsis* organs in response to salt and drought. Zhong, et al. [36] found the expression of the *PEPC* gene in cucumber leaves increased under NaCl stress. We also found that *CsaV3_4G033150* expression was significantly up-regulated in two highly salt-sensitive accessions (R132 and CG14), while significantly down-regulated in all the highly salt-tolerant accessions (Figure 7e).

CsaV3_5G023530 encodes a plant homologous domain (PHD) finger family protein. PHD finger proteins are a family of important zinc finger transcription factors that are responsible for regulating the transcription and chromatin state and responding to various stresses, and this family of genes has been reported to be associated with salt tolerance in many plants, such as *Arabidopsis thaliana* [37], *Moso bamboo* [38], *Brassica rapa* [39], and *Solanum tuberosum* [40]. Zheng, et al. [41] found that the expression of *CsaV3_5G023530* homologous gene *PRE2* (*AT1G77800*) in *Arabidopsis* was significantly increased under salt treatment, which may be involved in ABA and salt responses in *Arabidopsis*. We also found that the expression level of *CsaV3_5G023530* changed significantly in several cucumber accessions under salt stress, especially in the highly salt-sensitive accession CG13, which showed an extremely significant increase (Figure 8e).

CsaV3_6G009810 is predicted to encode a pesticidal crystal cry8Ba protein in CuGenDB, but more information in NCBI shows that it encodes a peptide of unknown function with no similarity to any known sequence to date. *AtRASD1* (*AT5G48310*), a homologue gene of *CsaV3_6G009810* in *Arabidopsis*, has been shown to activate abiotic stress-related responses such as salt and drought in *Arabidopsis* by encoding an unknown protein that is involved in ABA-dependent signal transduction pathways [42]. In this study, there were significant changes in the expression level of *CsaV3_6G009810* in all the highly salt-tolerant and -sensitive cucumber accessions (Figure 9e).

In fact, the change trend of gene expression in different accessions may be variable. We found that the changes in the expression of five genes in these accessions were diverse, and the expression of the same gene was up-regulated in some of the highly salt-tolerant or -sensitive accessions, while down-regulated in others. For four highly salt-tolerant accessions, the expression of *CsaV3_2G035120* was up-regulated in two accessions, while down-regulated in one accession; the expression of *CsaV3_3G023710* was up-regulated in two accessions, while down-regulated in one accession; the expression of *CsaV3_6G009810* was up-regulated in one accession, while down-regulated in two accessions. For four highly salt-sensitive accessions, the expression of *CsaV3_2G035120* was significantly up-regulated by salt stress in three accessions, while significantly down-regulated in one accession; the expression of *CsaV3_4G033150* was up-regulated in two accessions, while down-regulated in two accessions; the expression of *CsaV3_5G023530* was up-regulated in one accession, while down-regulated in two accessions; the expression of *CsaV3_6G009810* was up-regulated in three accessions, while down-regulated in one accession. In addition, we also found that the expression of *CsaV3_5G023530* was down-regulated in two cucumber accessions under salt stress, while its homologous gene in *Arabidopsis* was significantly up-regulated. We think that the possible reason for this result is not only that the gene expression is variable, but also that the regulation mechanism of this gene in cucumber is different from that in *Arabidopsis*. In any case, the up-regulation or down-regulation of gene expression is strong evidence that these genes respond to salt stress in cucumber.

To sum up, *CsaV3_2G035120*, *CsaV3_3G023710*, *CsaV3_4G033150*, *CsaV3_5G023530*, and *CsaV3_6G009810* are predicted to be involved in the salt stress response in cucumber. The next research will focus on the functional validation of these candidate genes via overexpression and CRISPR in cucumber. This work could provide some references for elucidating the mechanism of cucumber salt stress response and breeding salt-tolerant cucumber cultivars. In addition, the highly salt-tolerant and highly salt-sensitive accessions that were identified in this study can be used as experimental materials for the follow-up gene cloning and salt-tolerant cucumber cultivars breeding, such as selecting accessions to construct RILs populations for QTL mapping.

4. Materials and Methods

4.1. Plant Materials

The plant materials with a total of 220 accessions that were used in this study are from the cucumber research group of the Institute of Vegetables and Flowers, Chinese Academy of Agricultural Sciences, Beijing, China. The code and ecotype information of 220 accessions can be found in Table S1.

4.2. Evaluation of Salt Tolerance Ability and Data Analysis

The salt treatment experiments were carried out in two different environments, including Shouguang (36°51′ N, 118°50′ E) in June 2020 and Beijing (39°9′ N, 116°3′ E) in May 2021. A total of three replicates were set for each experiment, each of which had 5 plants. The cucumbers were cultivated in 54 cm × 28 cm × 9 cm soil pots which had sufficient nursery substrate, with trays for water collection at the bottom. The nursery substrate that was used in each experiment was the same, ideal for cucumber growth and did not affect the salt stress responses. The average ambient temperature in the greenhouse during the experiments was 26.65 °C (Shouguang) and 25.05 °C (Beijing), respectively. At 14 days old, the seedlings were exposed to salt stress by adding 2 L of 200 mmol·L^{-1} NaCl into the trays regularly to ensure that each seedling was saturated with the same amount of salt solution. The first true leaf and cotyledon showed different degrees of chlorosis and wilting after 7 days of salt treatment. The salt treatment and the calculation of the salt injury index were carried as described by Liu, et al. [7]. The salt injury index was calculated as below: salt injury index = S × $\sum(s \times n)/N$. Where "S" represents the highest salt injury rating scale, "s" represents the salt injury grade, "n" represents the number of plants in the salt injury grade, and "N" represents the total number of plants. For each of the cucumber seedlings that were grown under 200 mmol·L^{-1} NaCl, the phenotypic data were collected for the degree of salt injury. The salt symptoms within the 220 cucumber accessions were divided into six scales. WPS Office v.11.1 (Kingsoft, Beijing, China) was used to calculate salt injury index.

Variance analysis and cluster analysis of the phenotypic data were implemented using SPSS v.19.0 (International Business Machines Corporation, Armonk, NY, USA). The cluster analysis used the Ward method and the Euclidean distance method for measurement.

4.3. Genome-Wide Association Study (GWAS) and Linkage-Disequilibrium (LD) Analysis

The re-sequenced genome data and SNPs of these 220 core accessions are available in the NCBI Database (PRJNA171718 and PRJNA831637). A general linear model (GLM) was used for the association tests, with an estimated relatedness matrix as a covariate [43]. GWAS was conducted, and the genome-wide lowest p-value was recorded. The 5% lowest tail was taken from the 200 recorded minimal p-values as the threshold for genome-wide significance.

The software Plink [44] was used to calculate the LD coefficient (r^2) between the pairwise high-quality SNPs and the results were used to estimate LD decay [16]. Manhattan plots and the LD heatmaps were drawn by qqman package in R environment [45].

4.4. Identification of QTLs and Candidate Gene Analysis

The regions of the significant trait-associated SNPs were identified as candidate regions of QTLs by repeated loci of two sets of the salt injury index. The cucumber genome database website CuGenDB (http://cucurbitgenomics.org/ (accessed on 2 August 2021)) was used to retrieve candidate resistance genes that were located within a 100-kb interval, from 50 kb upstream and downstream of each peak SNPs from the candidate regions [46]. Gene function was analyzed and predicted by BLAST in the CuGenDB, Swiss-Prot (https://www.uniprot.org/ (accessed on 2 August 2021)), TAIR (https://www.arabidopsis.org/index.jsp (accessed on 2 August 2021)) and Gene Ontology databases (http://amigo.geneontology.org/amigo/landing (accessed on 2 August 2021)) [47]. Furthermore, highly salt-tolerant and highly salt-sensitive core accessions that were obtained by cluster analysis were used as materials, and the genes were analyzed by combination of germplasm haplotype analysis and qRT-PCR to determine the candidate genes that were related to salt tolerance.

4.5. RNA Extraction and qRT-PCR Verification

Using four highly salt-tolerant accessions (R45, R63, R78, and CG104) and four highly salt-sensitive accessions (R59, R132, CG13, and CG14) as experimental materials, a total of 4 time points were set, namely 0 h, 10 h, 48 h, and 96 h after salt treatment. There were three replicates for each time point. The first true leaf of each seedling was harvested and immediately put into liquid nitrogen, and stored at −80 °C. The TaKaRa MiniBEST Universal RNA Extraction Kit (TakaRa, Kusatsu, Japan) was used to extract the total RNA of each sample that was collected at the different time points. Reverse transcription of the extracted RNA into cDNA was performed with the UEIris II RT-PCR System (Biodee, Beijing, China). According to the CDS sequence of the candidate gene on the reference genome, primer 6.0 software was used to design the primers (Table S6). qRT-PCR was performed with ChamQ Universal SYBR qPCR Master Mix (Vazyme, Nanjing, China), using the CFX96 Real-Time System (Bio-Rad, Hercules, CA, USA). Analysis of the relative expression data of the candidate genes was performed using the $2^{-\Delta\Delta Ct}$ method [48].

5. Conclusions

We reported 58 highly salt-tolerant accessions and 20 highly salt-sensitive accessions from 220 cucumber core accessions. A total of seven loci, namely *gST2.1*, *gST3.1*, *gST3.2*, *gST4.1*, *gST4.2*, *gST5.1*, and *gST6.1*, that regulate salt tolerance in cucumber seedlings, were repeatedly detected. Furthermore, five candidate genes (*CsaV3_2G035120*, *CsaV3_3G023710*, *CsaV3_4G033150*, *CsaV3_5G023530*, and *CsaV3_6G009810*) within these loci that are involved in the salt stress response were predicted through gene analysis, haplotype, and qRT-PCR analyses. This study not only provides a reference for the subsequent excavation of candidate genes and the molecular mechanism of salt tolerance in cucumber, but also provides a theoretical basis for cultivating new high-yield and high-quality cucumber varieties with highly salt-tolerant, which will accelerate the cultivation of salt-tolerant cucumber varieties.

Supplementary Materials: The following supporting information can be downloaded at: https://www.mdpi.com/article/10.3390/ijms23158260/s1.

Author Contributions: Conceptualization, X.G., S.Z. and H.M; methodology, D.L., C.L. and J.H.; investigation, D.L. and S.D.; data curation, X.L.; writing—original draft preparation, D.L.; writing—review and editing, D.L. and S.D.; visualization, D.L.; supervision, X.G., S.Z. and H.M. All authors have read and agreed to the published version of the manuscript.

Funding: This research was funded by Key-Area Research and Development Program of Shandong Province (2021LZGC016), Key-Area Research and Development Program of Guangdong Province (2020B020220001), the Earmarked Fund for Modern Agro-industry Technology Research System (CARS-23), Science and Technology Innovation Program of the Chinese Academy of Agricultural Science (CAAS-ASTIP-IVFCAAS), Central Public-Interest Scientific Institution Basal Research Fund (No.Y2017PT52), and the key Laboratory of Biology and Genetic Improvement of Horticultural Crops, Ministry of Agriculture and Rural Affairs, P.R. China.

Institutional Review Board Statement: Not applicable.

Informed Consent Statement: Not applicable.

Data Availability Statement: Not applicable.

Acknowledgments: The authors express their appreciation to the reviewers for their valuable comments on this manuscript.

Conflicts of Interest: The authors declare no conflict of interest.

References

1. Baysal, G.; Tipirdamaz, R. The effect of salinity on lipid peroxidation and some antioxidant enzyme activities in two cucumber cultivars. *Acta Hortic.* **2007**, *729*, 199–203. [CrossRef]
2. Zhang, T.; Shi, Z.; Zhang, X.; Zheng, S.; Wang, J.; Mo, J. Alleviating effects of exogenous melatonin on salt stress in cucumber. *Sci. Hortic.* **2020**, *262*, 109070. [CrossRef]
3. Shu, S.; Chen, L.; Lu, W.; Sun, J.; Guo, S.; Yuan, Y.; Li, J. Effects of exogenous spermidine on photosynthetic capacity and expression of Calvin cycle genes in salt-stressed cucumber seedlings. *J. Plant Res.* **2014**, *127*, 763–773. [CrossRef]
4. Liang, W.; Ma, X.; Wan, P.; Liu, L. Plant salt-tolerance mechanism: A review. *Biochem. Biophis. Res. Commun.* **2018**, *495*, 286–291. [CrossRef]
5. Gupta, B.; Huang, B.R. Mechanism of Salinity Tolerance in Plants: Physiological, Biochemical, and Molecular Characterization. *Int. J. Genom.* **2014**, *2014*, 701596. [CrossRef] [PubMed]
6. Shen, J.; Jiang, J.; Tian, S.; Liu, X. Effects of NaCl Stress on Growth and Cell Membrane Permeability of Cucumber Seedlings. *Guizhou Agric. Sci.* **2016**, *8*, 19–24. [CrossRef]
7. Liu, D.; Dong, S.; Bo, K.; Miao, H.; Li, C.; Zhang, Y.; Zhang, S.; Gu, X. Identification of QTLs Controlling Salt Tolerance in Cucumber (*Cucumis sativus* L.) Seedlings. *Plants* **2021**, *10*, 85. [CrossRef]
8. Li, Y.; Si, L. Genetic analysis of salt-tolerance of cucumber at seedling stage. *J. Northwest A F Univ.* **2011**, *39*, 127–131. [CrossRef]

9. Simko, I.; Jia, M.; Venkatesh, J.; Kang, B.-C.; Weng, Y.; Barcaccia, G.; Lanteri, S.; Bhattarai, G.; Foolad, M.R. Genomics and Marker-Assisted Improvement of Vegetable Crops. *Crit. Rev. Plant Sci.* **2021**, *40*, 303–365. [CrossRef]
10. Hou, J.; Zhou, Y.; Gao, L.; Wang, Y.; Yang, L.; Zhu, H.; Wang, J.; Zhao, S.; Ma, C.; Sun, S.; et al. Dissecting the Genetic Architecture of Melon Chilling Tolerance at the Seedling Stage by Association Mapping and Identification of the Elite Alleles. *Front. Plant Sci.* **2018**, *9*, 1577. [CrossRef]
11. Khanzada, H.; Wassan, G.M.; He, H.; Mason, A.S.; Keerio, A.A.; Khanzada, S.; Faheem, M.; Solangi, A.M.; Zhou, Q.; Fu, D.; et al. Differentially evolved drought stress indices determine the genetic variation of *Brassica napus* at seedling traits by genome-wide association mapping. *J. Adv. Res.* **2020**, *24*, 447–461. [CrossRef] [PubMed]
12. Zhang, R.; Liu, C.; Song, X.; Sun, F.; Xiao, D.; Wei, Y.; Hou, X.; Zhang, C. Genome-wide association study of turnip mosaic virus resistance in non-heading Chinese cabbage. *3 Biotech* **2020**, *10*, 363. [CrossRef] [PubMed]
13. Wei, S.; Zhang, S.; Bo, K.; Wang, W.; Miao, H.; Dong, S.; Gu, X.; Zhang, S. Evaluation and Genome-wide Association Study (GWAS) of Seedling Thermotolerance in Cucumber Core Germplasm. *J. Plant Genet. Resour.* **2019**, *20*, 1223–1231. [CrossRef]
14. Wang, W.; Song, Z.; Bo, K.; Dong, S.; Wei, S.; Miao, H.; Li, J.; Zhang, S.; Gu, X. Evaluation and Genome-wide Association Study (GWAS) of Low-temperature Tolerance at Seedling Stage in Cucumber Core Germplasm. *J. Plant Genet. Resour.* **2019**, *20*, 1606–1612. [CrossRef]
15. Lee, H.Y.; Kim, J.G.; Kang, B.C.; Song, K. Assessment of the Genetic Diversity of the Breeding Lines and a Genome Wide Association Study of Three Horticultural Traits Using Worldwide Cucumber (*Cucumis* spp.) Germplasm Collection. *Agronomy* **2020**, *10*, 1736. [CrossRef]
16. Liu, X.; Lu, H.; Liu, P.; Miao, H.; Bai, Y.; Gu, X.; Zhang, S. Identification of Novel Loci and Candidate Genes for Cucumber Downy Mildew Resistance Using GWAS. *Plants* **2020**, *9*, 1659. [CrossRef]
17. Liu, X.; Gu, X.; Lu, H.; Liu, P.; Miao, H.; Bai, Y.; Zhang, S. Identification of Novel Loci and Candidate Genes for Resistance to Powdery Mildew in a Resequenced Cucumber Germplasm. *Genes* **2021**, *12*, 584. [CrossRef]
18. Liu, D.; Dong, S.; Miao, H.; Bo, K.; Zhang, S.; Gu, X. Research Progress on Genetic Breeding of Cucumber Tolerance for Salt Stress. *China Veg.* **2021**, *7*, 14–23. [CrossRef]
19. Cao, Q.; Li, L.; Kong, S.; Qiu, A.; Zhang, Y.; Sun, X. Evaluation of Salt Tolerance of Newly-bred Greenhouse Cucumber Cultivars at Germination Stage. *J. Plant Genet. Resour.* **2014**, *15*, 546–553. [CrossRef]
20. Baxter, I.; Brazelton, J.N.; Yu, D.; Huang, Y.S.; Lahner, B.; Yakubova, E.; Li, Y.; Bergelson, J.; Borevitz, J.O.; Nordborg, M.; et al. A coastal cline in sodium accumulation in *Arabidopsis thaliana* is driven by natural variation of the sodium transporter AtHKT1;1. *PLoS Genet.* **2010**, *6*, e1001193. [CrossRef]
21. Patishtan, J.; Hartley, T.N.; Fonseca de Carvalho, R.; Maathuis, F.J.M. Genome-wide association studies to identify rice salt-tolerance markers. *Plant Cell Environ.* **2018**, *41*, 970–982. [CrossRef] [PubMed]
22. Do, T.D.; Vuong, T.D.; Dunn, D.; Clubb, M.; Valliyodan, B.; Patil, G.; Chen, P.; Xu, D.; Nguyen, H.T.; Shannon, J.G. Identification of new loci for salt tolerance in soybean by high-resolution genome-wide association mapping. *BMC Genom.* **2019**, *20*, 318. [CrossRef] [PubMed]
23. Oyiga, B.C.; Sharma, R.C.; Baum, M.; Ogbonnaya, F.C.; Leon, J.; Ballvora, A. Allelic variations and differential expressions detected at quantitative trait loci for salt stress tolerance in wheat. *Plant Cell Environ.* **2018**, *41*, 919–935. [CrossRef] [PubMed]
24. Sun, Z.; Li, H.; Zhang, Y.; Li, Z.; Ke, H.; Wu, L.; Zhang, G.; Wang, X.; Ma, Z. Identification of SNPs and Candidate Genes Associated With Salt Tolerance at the Seedling Stage in Cotton (*Gossypium hirsutum* L.). *Front. Plant Sci.* **2018**, *9*, 1011. [CrossRef]
25. Xu, X.; Liu, M.; Lu, L.; He, M.; Qu, W.; Xu, Q.; Qi, X.; Chen, X. Genome-wide analysis and expression of the calcium-dependent protein kinase gene family in cucumber. *Mol. Genet. Genom.* **2015**, *290*, 1403–1414. [CrossRef] [PubMed]
26. Zhou, Y.; Hu, L.; Wu, H.; Jiang, L.; Liu, S. Genome-Wide Identification and Transcriptional Expression Analysis of Cucumber Superoxide Dismutase (SOD) Family in Response to Various Abiotic Stresses. *Int. J. Genom.* **2017**, *2017*, 7243973. [CrossRef]
27. Zhu, Y.; Yang, L.; Liu, N.; Yang, J.; Zhou, X.; Xia, Y.; He, Y.; He, Y.; Gong, H.; Ma, D.; et al. Genome-wide identification, structure characterization, and expression pattern profiling of aquaporin gene family in cucumber. *BMC Plant Biol.* **2019**, *19*, 345. [CrossRef]
28. Li, J.; Wang, T.; Han, J.; Ren, Z. Genome-wide identification and characterization of cucumber *bHLH* family genes and the functional characterization of *CsbHLH041* in NaCl and ABA tolerance in *Arabidopsis* and cucumber. *BMC Plant Biol.* **2020**, *20*, 272. [CrossRef]
29. Chen, C.; Chen, X.; Han, J.; Lu, W.; Ren, Z. Genome-wide analysis of the *WRKY* gene family in the cucumber genome and transcriptome-wide identification of WRKY transcription factors that respond to biotic and abiotic stresses. *BMC Plant Biol.* **2020**, *20*, 443. [CrossRef]
30. Yang, Y.; Xue, W.; Chen, P.; Yuan, X.; Li, X.; Zhang, T.; Chen, S. Identification and expression analyzes of CC-type glutaredoxin in cucumber (*Cucumis sativus* L.) under abiotic stress. *Sci. Hortic.* **2021**, *289*, 110417. [CrossRef]
31. Gao, R.; Luo, Y.; Yun, F.; Wu, X.; Wang, P.; Liao, W. Genome-Wide Identification, Expression Profile, and Alternative Splicing Analysis of *CAMTA* Family Genes in Cucumber (*Cucumis sativus* L.). *Agronomy* **2021**, *11*, 1827. [CrossRef]
32. Yuenyong, W.; Chinpongpanich, A.; Comai, L.; Chadchawan, S.; Buaboocha, T. Downstream components of the calmodulin signaling pathway in the rice salt stress response revealed by transcriptome profiling and target identification. *BMC Plant Biol.* **2018**, *18*, 335. [CrossRef] [PubMed]
33. Yuenyong, W.; Sirikantaramas, S.; Qu, L.; Buaboocha, T. Isocitrate lyase plays important roles in plant salt tolerance. *BMC Plant Biol.* **2019**, *19*, 472. [CrossRef]

34. Sakamoto, H.; Matsuda, O.; Iba, K. *ITN1*, a novel gene encoding an ankyrin-repeat protein that affects the ABA-mediated production of reactive oxygen species and is involved in salt-stress tolerance in *Arabidopsis thaliana*. *Plant J.* **2008**, *56*, 411–422. [CrossRef] [PubMed]
35. Sanchez, R.; Flores, A.; Cejudo, F. *Arabidopsis* phosphoenolpyruvate carboxylase genes encode immunologically unrelated polypeptides and are differentially expressed in response to drought and salt stress. *Planta* **2006**, *223*, 901–909. [CrossRef] [PubMed]
36. Zhong, M.; Yuan, Y.; Shu, S.; Sun, J.; Guo, S.; Yuan, R.; Tang, Y. Effects of exogenous putrescine on glycolysis and Krebs cycle metabolism in cucumber leaves subjected to salt stress. *Plant Growth Regul.* **2016**, *79*, 319–330. [CrossRef]
37. Zang, D.; Li, H.; Xu, H.; Zhang, W.; Zhang, Y.; Shi, X.; Wang, Y. An *Arabidopsis* Zinc Finger Protein Increases Abiotic Stress Tolerance by Regulating Sodium and Potassium Homeostasis, Reactive Oxygen Species Scavenging and Osmotic Potential. *Front. Plant Sci.* **2016**, *7*, 1272. [CrossRef]
38. Gao, Y.; Liu, H.; Wang, Y.; Li, F.; Xiang, Y. Genome-wide identification of PHD-finger genes and expression pattern analysis under various treatments in moso bamboo (*Phyllostachys edulis*). *Plant Physiol. Biochem.* **2018**, *123*, 378–391. [CrossRef]
39. Alam, I.; Liu, C.; Ge, H.; Batool, K.; Yang, Y.; Lu, Y. Genome wide survey, evolution and expression analysis of PHD finger genes reveal their diverse roles during the development and abiotic stress responses in *Brassica rapa* L. *BMC Genom.* **2019**, *20*, 773. [CrossRef]
40. Qin, M.; Luo, W.; Zheng, Y.; Guan, H.; Xie, X. Genome-wide identification and expression analysis of the PHD-finger gene family in *Solanum tuberosum*. *PLoS ONE* **2019**, *14*, e0226964. [CrossRef]
41. Zheng, K.; Wang, Y.; Wang, S. The non-DNA binding bHLH transcription factor Paclobutrazol Resistances are involved in the regulation of ABA and salt responses in *Arabidopsis*. *Plant Physiol. Biochem.* **2019**, *139*, 239–245. [CrossRef] [PubMed]
42. Podia, V.; Milioni, D.; Martzikou, M.; Haralampidis, K. The role of *Arabidopsis thaliana RASD1* gene in ABA-dependent abiotic stress response. *Plant Biol.* **2018**, *20*, 307–317. [CrossRef] [PubMed]
43. Bo, K.; Wei, S.; Wang, W.; Miao, H.; Dong, S.; Zhang, S.; Gu, X. QTL mapping and genome-wide association study reveal two novel loci associated with green flesh color in cucumber. *BMC Plant Biol.* **2019**, *19*, 243. [CrossRef] [PubMed]
44. Purcell, S.; Neale, B.; Todd-Brown, K.; Thomas, L.; Ferreira, M.A.; Bender, D.; Maller, J.; Sklar, P.; de Bakker, P.I.; Daly, M.J.; et al. PLINK: A tool set for whole-genome association and population-based linkage analyses. *Am. J. Hum. Genet.* **2007**, *81*, 559–575. [CrossRef]
45. Turner, S.D. qqman: An R package for visualizing GWAS results using Q-Q and manhattan plots. *bioRxiv* **2014**. [CrossRef]
46. Yao, L.; Li, Y.; Ma, C.; Tong, L.; Du, F.; Xu, M. Combined genome-wide association study and transcriptome analysis reveal candidate genes for resistance to Fusarium ear rot in maize. *J. Integr. Plant Biol.* **2020**, *62*, 1535–1551. [CrossRef]
47. Shi, L.; Yang, Y.; Xie, Q.; Miao, H.; Bo, K.; Song, Z.; Wang, Y.; Xie, B.; Zhang, S.; Gu, X. Inheritance and QTL mapping of cucumber mosaic virus resistance in cucumber (*Cucumis Sativus* L.). *PLoS ONE* **2018**, *13*, e0200571. [CrossRef]
48. Livak, K.J.; Schmittgen, T.D. Analysis of relative gene expression data using real-time quantitative PCR and the $2^{-\Delta\Delta CT}$ Method. *Methods* **2001**, *25*, 402–408. [CrossRef]

International Journal of
Molecular Sciences

MDPI

Article

Fine Mapping and Gene Analysis of *restorer-of-fertility* Gene *$CaRf_{HZ}$* in Pepper (*Capsicum annuum* L.)

Zhixing Nie [1,2,3,†], Yunpeng Song [1,†], Hong Wang [2], Jianying Chen [2], Qingliang Niu [3,*] and Weimin Zhu [1,*]

1 Shanghai Key Laboratory of Protected Horticultural Technology, Horticulture Research Institute, Shanghai Academy of Agricultural Sciences, Shanghai 201403, China; niezhixing@gmail.com (Z.N.); wupolalala@gmail.com (Y.S.)
2 Vegetable Research Institute, Hangzhou Academy of Agricultural Sciences, Hangzhou 310024, China; zqianglee@gmail.com (H.W.); luoyeguigen2@yeah.net (J.C.)
3 School of Agriculture and Biology, Shanghai Jiao Tong University, Shanghai 200240, China
* Correspondence: qlniu@sjtu.edu.cn (Q.N.); yy17@saas.sh.cn (W.Z.)
† These authors contributed equally to this work.

Abstract: Cytoplasmic male sterility (CMS) is a common biological phenomenon used in hybrid production of peppers (*Capsicum annuum* L.). Although several *restorer-of-fertility* (*Rf*) genes of pepper CMS lines have been mapped, there is no report that the *Rf* gene with clear gene function has been isolated. Here, pepper CMS line HZ1A and its restorer line HZ1C were used to construct (HZ1A × HZ1C) F_2 populations and map the *Rf* gene. A single dominant gene *CaRfHZ* conferred male fertility according to inheritance analysis. Using sterile plants from (HZ1A × HZ1C) F_2 populations and bulked segregant analysis (BSA), the *$CaRf_{HZ}$* gene was mapped between P06gInDel-66 and P06gInDel-89 on chromosome 6. This region spans 533.81 kb, where four genes are annotated according to Zunla-1 V2.0 gene models. Based on the analysis of genomic DNA sequences, gene expressions, and protein structures, *Capana06g002968* was proposed as the strongest candidate for the *$CaRf_{HZ}$* gene. Our results may help with hybrid pepper breeding and to elucidate the mechanism of male fertility restoration in peppers.

Keywords: pepper; cytoplasmic male sterility; *restorer-of-fertility* gene; fine mapping

1. Introduction

Cytoplasmic male sterility (CMS) is a common biological phenomenon observed in plants, and it is widely used for seed production of hybrid crops. Fertility restoration of CMS is critical in hybrid production, and *restorer-of-fertility* (*Rf*) gene can restore and affect the fertility of hybrids. More than ten *Rf* genes have been isolated and studied in major crops [1] such as rice, maize, rapeseed, wheat, and soybean. Most of the *Rf* genes in crops contain multiple pentatricopeptide repeat (PPR) domains, such as *Rf1* [2], *Rf4* [3], *Rf5* [4], and *Rf6* [5] in rice; *Rf1* [6] and *Rf3* [7] in maize; *Rfp* [8] in rapeseed; *Rfo* [9] in radish; *RFL79* and *RFL29a* [10] in wheat; and *GmPPR576* [11] in soybean. Additionally, certain *Rf* genes do not contain PPR domains, such as *Rf2* [6] and *Rf4* [12] in maize. *Rf2* is similar to the mammalian aldehyde dehydrogenase, and *Rf4* is a transcription factor that contains basic helix–loop–helix (bHLH) domains. Most *Rf* genes restore the fertility of sterile lines by affecting the transcription of sterile genes in mitochondria. For example, the rice gene *Rf4* restores fertility by reducing the transcripts of the sterility gene *orf352* in the cytoplasm of wild-type rice [4,13]; the rice gene *Rf6* reduces the accumulation of the transcript *atp6-orfH79*, thereby restoring fertility [5]; the maize gene *Rf3* inhibits mitochondrial gene *orf77* editing and degradation, and accelerates *orf355* degradation, leading to CMS-S fertility restoration [14]; the rapeseed gene *Rfp* restores fertility of the sterile line by processing the transcript of the mitochondrial sterility gene *atp6-orf224* [8].

Pepper is one of the largest vegetable crops globally. Peterson (1958) [15] was the first to report cytoplasmic–nucleus interaction type male sterility in pepper, following which, breeders bred numerous sterile pepper lines using CMS lines [16]. Due to the different fertility segregating populations used for genetic analysis in different studies, the inheritance pattern of pepper *Rf* gene also has different conclusions. Wang et al. [17] used the male sterile line 77013A to cross 114 doubled haploid lines to identify the fertility of F_1 offspring, and considered that fertility restoration of pepper CMS was controlled by one major and four minor genes. Wei et al. [18] analyzed the segregation of fertility using an F_2 population. The fertility restoration of pepper CMS may be conferred by two pairs of additive-dominant epistatic major genes and additive-dominant polygenes. Gulyas et al. [19] and Ye et al. [20] investigated the fertility separation of F_1–F_4 and F_2 populations constructed using male sterile and restorer lines, respectively. The results showed that a pair of dominant nuclear genes controls fertility restoration in CMS.

In early studies, owing to the limitations of molecular biology techniques and the absence of a reference genome, the *Rf* genes of pepper were only initially mapped. Zhang et al. [21] used the F_2 population to locate the major *Rf* gene between two randomly amplified polymorphic DNA (RAPD) markers using bulked segregant analysis (BSA), and the genetic distance was 0.37 centimorgans (cM) and 8.12 cM. Wang et al. [17] mapped a major quantitative trait locus (QTL) of fertility restoration to chromosome 6 and four minor QTLs on chromosomes 1, 2, 3, and 5. Kim et al. [22] used BSA–amplified fragment length polymorphism (AFLP) to segregate populations by fertility, constructed a linkage map of the *Rf* gene, and obtained a cleaved amplified polymorphic sequence (CAPS) marker AFRF8CAPS with a genetic distance of 1.8 cM from the *Rf* gene. Lee et al. [23] used 205 F_2 individual plants and obtained three CAPS and one sequence characterized amplified region (SCAR) markers that were closely linked to the *Rf* site of pepper CMS. Yang et al. [24] used the F_2 population to obtain the simple sequence repeat (SSR) molecular marker AF208834 linked to the pepper *Rf* gene, with a genetic distance of 20.8 cM. Jo et al. [25] screened the pepper BAC library with the petunia *Rf* gene, cloned the *Rf* candidate gene *PePPR1*, and found three markers closely linked to the *Rf* gene.

Along with the in-depth research and publication of the pepper reference genome [26,27], the pepper *Rf* gene has been mapped to an adjacent region of chromosome 6 in different reports (Table 1). Jo et al. [28] used BSA–AFLP and comparative genomics to locate the *Rf* gene within the 821 kb segment on chromosome 6 and identified the candidate gene *CaPPR6*. Zhang et al. [29] narrowed down the region of *CaPPR6* to 128.96 Kb using kompetitive allele-specific PCR (KASP) markers. Kang et al. [30] mapped an unstable *Rf* gene, *Rfu*, using the molecular markers from Jo et al. [28] and newly developed CAPS markers. Ye et al. [20] located the *Rf* gene between SSR markers pep43 and pep20 on chromosome 6, with a physical distance of 498.6 kb. Wu et al. [31] performed association analysis on 287 pepper lines using specific-locus amplified fragment sequencing (SLAF-seq) and genome-wide association study (GWAS) and identified two candidate *Rf* genes, *Capana06g002967* and *Capana06g002969*. Cheng et al. [32] used high-density genetic mapping and collinearity analysis to map the *Rf* gene *CaRf* to a 270.10 kb region on chromosome 6, and analyzed the candidate gene *Capana06g003028*. Zhang et al. [33] used high-throughput sequencing combined with BSA to initially map the *Rf* gene in the F_2 population, then mapped the *Rf* gene to a 148.05 kb segment on chromosome 6, and identified the candidate gene *CaRf032* (*CA00g82510*), which has a base variation in the maintainer line leading to premature termination of transcription of the gene. Wei et al. [34] performed RNA sequencing on fertile and sterile pools constructed from the F_2 population and found a candidate gene, *NEDD8* (*Capana06g002866*), on chromosome 6. However, to date, there is no report that the *Rf* gene with a clear gene function of the pepper CMS line has been cloned.

Table 1. Reported *Rf* candidate genes on chromosome 6 in pepper.

Gene Name	Mapping Method	Range	Flanking Markers	Mapping Interval (Zunla-1 V2.0)	Candidate Genes	Locations of Candidate Genes (Zunla-1 V2.0)	References
Rf	CAPS, SSR, SCAR markers and BSA	498.60 kb	pep20~pep43	Chr06: 214,671,854.. 215,170,454	*Capana06g002962* *Capana06g002963* *Capana06g002965*	Chr06: 214,791,005..214,791,707 Chr06: 214,928,883..214,932,088 Chr06: 215,102,194..215,102,754	[20]
CaPPR6	BSA–AFLP and comparative mapping	128.96 kb	S423~S424	Nearby Chr06: 216,988,988	*CaPPR6*	Chr06: 214,033,734..214,035,560	[28,29]
Rfu	CAPS, SCAR markers	398.00 kb	4162-SCAR ~G16-CAPS	Nearby Chr06: 214,076,189	*CA00g30080*	-	[30]
Rf	SLAF-seq, and GWAS	858.26 kb	P06-247 ~PW6-126	Chr06: 214,868,888.. 215,727,145	*Capana06g002967* *Capana06g002969*	Chr06: 215,172,352..215,173,332 Chr06: 215,340,394..215,340,521	[31]
CaRf	Conjoint analysis of recombinants and collinearity	270.10 kb	S3~S1	Nearby Chr06: 217,065,420	*Capana06g003028*	Chr06: 217,273,835..217,280,134	[32]
CaRf032	Genome resequencing and recombination analysis	148.05 kb	S1402~S1354	Chr06: 213,923,525.. 214,071,576	*CA00g82510*	Chr06: 214,033,838..214,035,589	[33]
NEDD8	Bulked segregant RNA sequencing	5.1 Mb	KS18~KS22	Chr06: 210,576,870.. 215,685,280	*Capana06g002866*	Chr06: 210,452,333..210,452,887	[34]

"-" indicates there is no corresponding region. In some studies, CM334 v1.55 was selected as the reference genome, therefore, the flanking markers of the mapping interval were compared on the Zunla-1 V2.0 genome to obtain the corresponding positions. For the markers without corresponding positions, according to the corresponding positions of the annotated genes in the mapping interval on the Zunla-1 V2.0 genome, the approximate locations of the mapping intervals, such as *CaPPR6* [28,29], *Rfu* [30], and *CaRf* [32] were speculated. The mapping interval of *NEDD8* [34] in the original text is inconsistent with the candidate gene position.

In the present study, we constructed an F_2 segregating population and fine-mapped the *Rf* gene in the pepper line HZ1C. The *Rf* gene was mapped to a 533.81 kb region between two insertion or deletion (InDel) markers on chromosome 6 of the Zunla-1 V2.0 reference genome [26]. According to DNA sequence alignment, gene expression, and protein structure analysis, *Capana06g002968* was proposed as the strongest candidate gene for the *Rf* gene in pepper line HZ1C.

2. Results

2.1. The Inheritance Analysis of Fertility Restoration in CMS Line HZ1A

HZ1A shows the typical characteristics of male sterility, including no pollen dissemination around its anthers and fewer and shriveled pollen grains [35]. The fertility phenotypes were identified through visible pollen on anthers and Alexander's staining of pollen. All (HZ1A × HZ1C) F_1 plants were completely male fertile. The segregation of male-fertile to male-sterile plants in 11 F_2 populations of different sizes, which had been planted in the autumn and spring of 2020 and 2021, respectively, conformed to a 3:1 segregation ratio under the χ^2 criterion ($p > 0.05$). Overall, 1658 F_2 plants were classified as 1277 male-fertile and 381 male-sterile plants, which also conformed to a 3:1 segregation ratio ($\chi^2 = 3.50$, $p = 0.06$) (Table 2). These results demonstrated that a single dominant gene conferred male fertility in HZ1A, which we designated as $CaRf_{HZ}$.

Table 2. Segregation patterns of male fertility and sterility in the pepper F_2 population derived from the cross HZ1A × HZ1C.

Year and Season	Line Name	No. of Plants			Expected Ratio	χ^2	Probability
		Total	Fertile	Sterile			
2020 Autumn	P20210	66	53	13	3:1	0.73	0.39
	P20217	63	53	10		2.33	0.13
	P20218	97	65	22		0.00	0.95
2021 Spring	P211014	134	102	32	3:1	0.04	0.84
	P211017	370	279	91		0.01	0.90
	P211091	129	102	27		0.93	0.33
	P211103	168	134	34		1.79	0.18
	P211115	160	128	32		1.88	0.17
	P211129	161	121	40		0.00	0.96
	P211157	218	165	53		0.02	0.88
	P211162	102	75	27		0.05	0.82
Total		1658	1277	381	3:1	3.50	0.06

2.2. BSA and Genetic Linkage Mapping of the $CaRf_{HZ}$ Gene

In total, 226 plants of the (HZ1A × HZ1C) F_2 population, which had been planted in the autumn of 2020, were used to map the *CaRfHZ* gene according to BSA. DNA from 10 male-fertile plants was mixed to composite the male-fertile DNA bulk, and DNA from 10 male-sterile plants was mixed to composite the male-sterile DNA bulk. In total, 449 pairs of SSR markers [36] evenly distributed on pepper chromosomes were selected to detect the parents: CMS line HZ1A and restorer line HZ1C. Polymorphic markers with clear bands were detected in male-fertile and male-sterile DNA bulks. Finally, the SSR marker P06g8490 was found to be polymorphic among the DNA bulks. This implied that the $CaRf_{HZ}$ gene is close to P06g8490 and is located on chromosome 6. In the 12 Mb region near P06g8490, 11 polymorphic SSR markers among the parents were selected. The above 11 SSR markers and P06g8490 were amplified in 45 male-sterile plants from the (HZ1A × HZ1C) F_2 population. According to the recombinant number and genetic distance between the $CaRf_{HZ}$ and its linked markers (Table 3), $CaRf_{HZ}$ was initially located between the markers P06G8229 and P06G8560 (Figure 1a). In this region, 6 SSR markers, P06g8264, P06g8490, P06g8494, P06g8497, P06g8536, and P06g8527, were co-segregated with $CaRf_{HZ}$.

Table 3. The recombinant number and genetic distance between the $CaRf_{HZ}$ gene and its linked markers.

	Marker	Position (bp)	Recombinant Individuals	Genetic Distance (cM)
Primary mapping	P06g8405	203,656,182	3 HO	6.71
	P06g8077	212,732,221	2 HO, 1 HE	5.58
	P06g8089	204,569,609	2 HO	4.46
	P06g8229	208,899,271	2 HE	2.22
	$CaRf_{HZ}$			
	P06g8264	209,767,667	0	0.00
	P06g8490	214,353,887	0	0.00
	P06g8494	214,428,701	0	0.00
	P06g8497	214,494,826	0	0.00
	P06g8536	215,672,797	0	0.00
	P06g8527	215,336,423	0	0.00
	P06g8560	215,995,316	1 HE	1.11
	P06g8618	216,556,879	2 HO	4.46

Table 3. *Cont.*

	Marker	Position (bp)	Recombinant Individuals	Genetic Distance (cM)
Fine mapping	P06g8229	208,899,271	43 HE	6.43
	P06g8335	210,959,629	35 HE	5.23
	P06g8405	212,732,234	24 HE	3.58
	P06g8434	213,257,869	23 HE	3.43
	P06g8219	208,736,376	22 HE	3.28
	P06gInDel-15	213,906,630	20 HE	2.98
	P06g8528	215,338,579	6 HE	0.89
	P06gInDel-38	214,431,109	4 HE	0.60
	P06g8490	214,353,887	4 HE	0.60
	P06g8499	214,498,896	4 HE	0.60
	P06g8508	214,872,305	3 HE	0.45
	P06gInDel-46	214,885,975	2 HE	0.30
	P06gInDel-48	214,903,798	2 HE	0.30
	P06gInDel-56	214,950,959	2 HE	0.30
	P06gInDel-66	215,097,259	1 HE	0.15
	CaRf$_{HZ}$			
	P06g8527	215,336,423	0	0.00
	P06gInDel-79	215,419,831	0	0.00
	P06gInDel-81	215,498,386	0	0.00
	P06gInDel-89	215,631,069	1 HE	0.15
	P06gInDel-90	215,633,948	1 HE	0.15
	P06gInDel-91	215,636,889	2 HE	0.30
	P06gInDel-92	215,669,032	2 HE	0.30
	P06gInDel-94	215,669,384	2 HE	0.30
	P06gInDel-95	215,674,072	2 HE	0.30
	P06g8543	215,710,422	3 HE	0.45
	P06gInDel-99	215,741,617	3 HE	0.45
	P06g8549	215,773,438	3 HE	0.45
	P06g8560	215,995,331	6 HE	0.89

'HO' indicates a recombinant individual with a homozygous band from the fertile parent HZ1C, and 'HE' indicates a recombinant individual with heterozygous bands from the two parents.

2.3. Fine Mapping of the *CaRf*$_{HZ}$ Gene

A larger population composed of 336 male-sterile plants out of the (HZ1A × HZ1C) F_2 populations, which had been planted in the spring of 2021, was used to narrow down the region of the *CaRfHZ* gene. More SSR markers and newly developed InDel markers (Table 3) in the primary mapping region were selected for fine mapping. Twenty-eight markers were polymorphic in HZ1A and HZ1C, and were used to map the *CaRf*$_{HZ}$ gene. The recombinant numbers and genetic distances between the *CaRf*$_{HZ}$ gene and its linked markers are shown in Table 3. Some of these markers identified the same number of recombinant individuals, such as P06gInDel-46, P06gInDel-48, and P06gInDel-56, and these markers had the same genetic distance from the *CaRf*$_{HZ}$ gene. However, owing to image size limitations, the markers are not listed in the mapping linkage (Figure 1a,b). The figure shows five F_2 recombinant individuals between the P06G8508 and P06gInDel-95 markers (Figure 1c). Finally, the *CaRf*$_{HZ}$ gene was mapped between the markers P06gInDel-66 and P06gInDel-89 with the same genetic distance of 0.15 cM (Figure 1b,c). Three markers, P06G8527, P06gInDel-79, and P06gInDel-81, were co-segregated with the *CaRf*$_{HZ}$ gene.

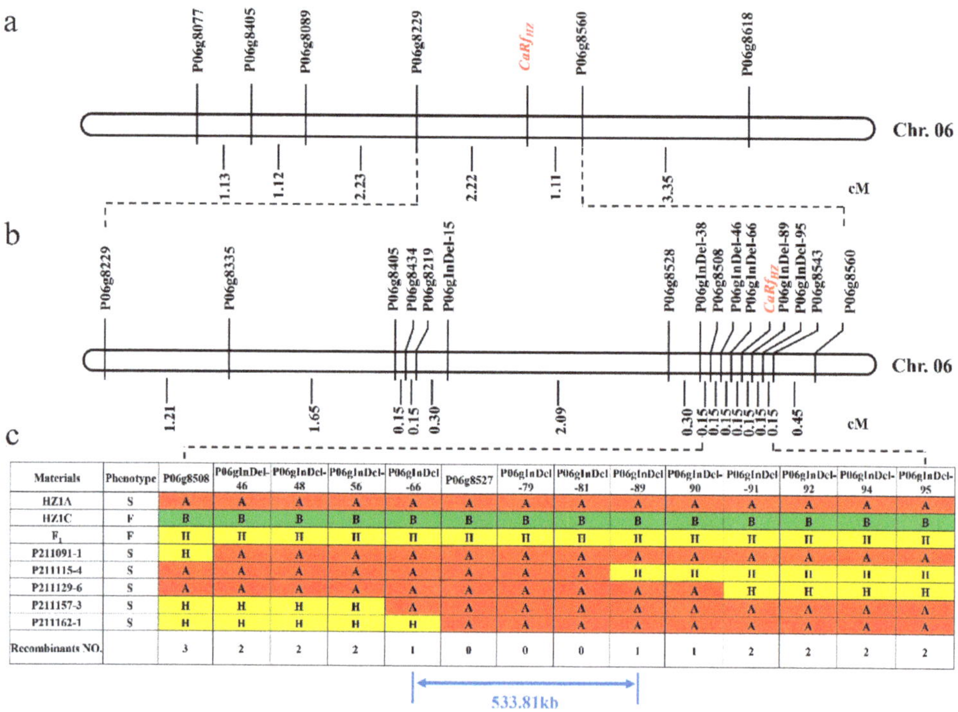

Figure 1. Mapping of the pepper $CaRf_{HZ}$ gene. (**a**) Primary mapping of $CaRf_{HZ}$ using a small population of (HZ1A × HZ1C) F_2. (**b**) Fine mapping of $CaRf_{HZ}$ using a large population of (HZ1A × HZ1C) F_2. (**c**) Genotyping results from 5 F_2 recombinants by 2 SSR markers and 12 newly developed InDel markers: 'A' = 'HZ1A' genotype; 'B' = 'HZ1C' genotype; 'H' = (HZ1A × HZ1C) F_1 genotype. Distances are shown in cM. Due to image size limitations, the markers that co-segregated with the fertility phenotype are not all listed in the mapping linkage (**a**,**b**).

2.4. Analysis of the Annotation Genes

According to the pepper reference genome sequence (Zunla-1 V2.0) [26], the physical distance between the two markers, P06gInDel-66 and P06gInDel-89, is 533.81kb on Chr. 06: 215,097,259..215,631,069. In this region, there are four annotated genes (Table 4): *Capana06g002965*, *Capana06g002967*, *Capana06g002968*, and *Capana06g002969* according to Zunla-1 V2.0 gene models.

Table 4. List of four annotation genes in the mapping region of $CaRf_{HZ}$.

Gene ID	Position (bp)	Annotation Description
Capana06g002965	Chr06:215,102,194..215,102,754	Unknown protein
Capana06g002967	Chr06:215,172,352..215,173,332	CW-type zinc finger protein
Capana06g002968	Chr06:215,328,827..215,334,970	Tetratricopeptide repeat protein
Capana06g002969	Chr06:215,340,394..215,341,013	Unknown protein

Full-length genomic DNA sequences of the four annotated genes were sequenced in HZ1A and HZ1C, and the coding sequences were analyzed (Table 5). There were two missense variations at 129 bp and 504 bp and a synonymous variation at 436 bp in *Capana06g002965*. There was a missense variation of 935 bp in *Capana06g002967*. There were two missense variations at 20 bp and 467 bp in *Capana06g002968*. There were two missense variations at 196 bp and 318 bp and a synonymous variation at 144 bp in *Capana06g002969*. Differences in the above sites lead to differences in the amino acid sequence (Figure S1), but only the protein structure of *Capana06g002969* was different. Due to the missense variation at 318 bp, the stop codon TGA of *Capana06g002969* in HZ1C was changed to TGG in HZ1A, which resulted in 31 more amino acids in HZ1A.

Table 5. The variations of four annotated genes in the CMS, maintainer, and restorer lines.

Line name	Type	*Capana06g002965*			*Capana06g002967*	*Capana06g002968*		*Capana06g002969*		
		129 bp	436 bp	504 bp	935 bp	20 bp	467 bp	144 bp	196 bp	318 bp
HZ1A	CMS line	A	T	G	G	G	T	G	T	G
HZ1C	Restorer line	C	C	A	A	C	G	A	G	A
P21204	Maintainer line	C	T	G	G	G	G	G	G	G
P21238	Maintainer line	A	T	G	A	G	T	A	T	G
P21239	Maintainer line	A	T	G	A	G	T	A	T	G
P21241	Maintainer line	A	T	G	A	G	T	A	T	G
P21243	Maintainer line	A	T	G	A	G	T	A	T	G
P21244	Maintainer line	A	T	G	A	G	T	A	T	G
P21246	Maintainer line	A	T	G	A	G	T	A	T	G
P21270	Maintainer line	A	T	G	A	G	T	A	T	G
P21273	Maintainer line	A	T	G	A	G	T	A	T	G
P21215	Restorer line	A	T	G	A	C	G	A	G	A
P21240	Restorer line	A	T	G	A	C	G	A	G	A
P21247	Restorer line	A	T	G	A	C	G	A	G	A
P21274	Restorer line	A	T	G	A	G	T	A	T	G

Shading indicates the nucleobases are consistent with the restored traits.

The expression of the four annotated genes in young leaves and three developmental stages (stage 1 (S1), sporogenous tissue to meiotic stage; stage 2 (S2), tetrad to mononuclear stage; stage 3 (S3), mature pollen stage) of anthers was analyzed in HZ1A, HZ1C, and (HZ1A × HZ1C) F_1 using quantitative real-time PCR (qRT-PCR) (Figure 2). The expression of *Capana06g002965* in (HZ1A × HZ1C) F_1 was low in young leaves but was not detected in HZ1A and HZ1C. At the S1 stage, the expression level of *Capana06g002965* in HZ1A was significantly lower than that in HZ1C and (HZ1A × HZ1C) F_1. Conversely, the expression level of *Capana06g002965* in HZ1A was significantly higher than that in HZ1C and (HZ1A × HZ1C) F_1 at the S2 stage. The expression of *Capana06g002965* was not detected in HZ1A, HZ1C, or (HZ1A × HZ1C) F_1. The expression of *Capana06g002967* in HZ1A was not significantly different from that in HZ1C, but was higher than that in (HZ1A × HZ1C) F_1 in young leaves and the S3 stage. The expression trends of *Capana06g002967* were similar to *Capana06g002965* at the S1 and S2 stages. The expression level of *Capana06g002968* in HZ1A was significantly lower than that in HZ1C and (HZ1A × HZ1C) F_1 in young leaves and the S1 stage. At the S2 and S3 stages, the expression levels of *Capana06g002968* in HZ1A were significantly higher than those in HZ1C. The expression of *Capana06g002969* was not detected in the young leaves of HZ1A, HZ1C, and (HZ1A × HZ1C) F_1. At the S1, S2, and S3 stages, the expression level of *Capana06g002969* in HZ1A was significantly lower than that in HZ1C. In summary, in HZ1A and HZ1C, four genes were differentially expressed at the S1, S2, and S3 developmental stages of anther, with the exception of *Capana06g002967* at the S3 stage. The gene expression trends in HZ1C and (HZ1A × HZ1C) F_1 were mostly the same.

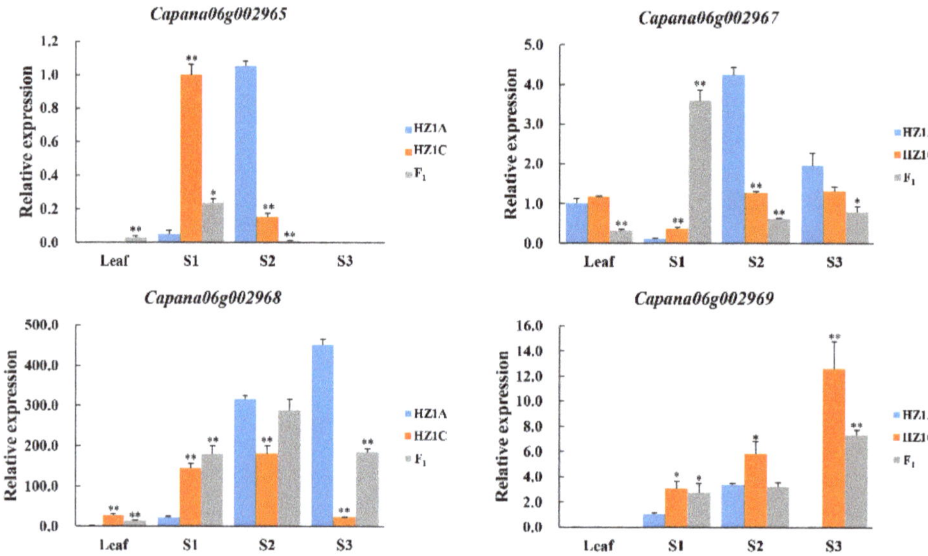

Figure 2. Expression levels of four annotated genes in young leaves and in anthers. * and ** indicate significant difference at $p < 0.05$ and $p < 0.01$, respectively.

Although there were differences in the amino acid sequences and gene expression levels of the four annotated genes, it was not possible to determine which gene was the candidate gene for *CaRf*$_{HZ}$ based on these differences.

3. Discussion

3.1. Comparison of CaRf$_{HZ}$ Mapping Interval and the Published Rf Gene Position in Pepper

Thus far, the seven pepper *Rf* genes have been fine mapped using a forward genetics strategy (Table 1). According to the gene mapping intervals, these seven genes are located in an adjacent region of approximately 7 Mb from 210–217 Mb on chromosome 6, and the mapping results overlapped in some studies [20,31]. The situation in which the mapping results are adjacent is presumed to be related to the similarity of the materials used for mapping or to the possible clustering of pepper *Rf* genes on chromosome 6. In the present study, the *CaRf*$_{HZ}$ gene was fine-mapped within a 533.81 kb segment in the adjacent region. The mapping interval of the *CaRf*$_{HZ}$ gene partially overlaps the mapping region in *Rf* [20] and is located within the mapping range of *Rf* (858.26 kb) [31] and *NEDD8* (5.11 Mb) [34], which are related to the larger mapping range of the above research results. Similar mapping results proved the accuracy of our results in this study but were not conducive to the discovery of new pepper *Rf* genes. As pepper *Rf* genes may exist in clusters on chromosome 6, there may be multiple genes controlling fertility restoration simultaneously. This leads to inconsistent genotypes for the molecular marker *Rf*, making it difficult to locate the *Rf* gene within a shorter interval.

3.2. Prediction and Characteristics of the Candidate Gene

According to the annotation information of the Zunla-1 V2.0 genome [26], there are four annotated genes in the localization interval of the *CaRfHZ* gene. All four annotated genes showed sense variations and differences in expression between HZ1A and HZ1C. The candidate gene for *CaRf*$_{HZ}$ could not be determined based on the above information.

Therefore, variations between HZ1A and HZ1C were detected in nine maintainer and four restorer lines of HZ1A (Table 5), and missense variations were analyzed. In *Capana06g002965*, missense variations were the same as HZ1A in eight maintainer lines, but they were different from HZ1C in all four restorer lines, and the correlation between missense variations and restored traits was 61.54% in 13 lines. In *Capana06g002967*, missense variations were the same as HZ1C in all four restorer lines, but they were different from HZ1A in eight maintainer lines, and the correlation between missense variations and restored traits was 38.46% in 13 lines. The detection results of missense variations in *Capana06g002968* and *Capana06g002969* were similar; the missense variations were the same as HZ1A in eight maintainer lines and were the same as HZ1C in three restorer lines, and the correlation between missense variations and restored traits was 84.62% in 13 lines. Conclusively, the missense variations in *Capana06g002965* and *Capana06g002967* were not highly correlated with restored traits, while the missense variations in *Capana06g002968* and *Capana06g002969* were highly correlated with restored traits and may be candidate genes for $CaRf_{HZ}$. *Capana06g002968* was not reported as a candidate gene for pepper *Rf* in previous studies, but *Capana06g002969* was identified as a candidate gene. There was no difference in *Capana06g002969* sequence between the CMS and restorer lines in a study by Wu et al. [31], which is different from the findings in the present study.

Using the simple modular architecture research tool (SMART) (https://smart.embl.de/, accessed on 15 March 2022) [37] to analyze protein structures, it was found that *Capana06g002968* in HZ1A and HZ1C both contained tetratricopeptide repeat (TPR) domains, and *Capana06g002969* in HZ1A, but not HZ1C, contained an IBR domain. No reports related to plant fertility were found for the IBR domain, but there have been reports on the TPR domain. In rice, silencing the gene *ORF3* containing the TPR domain can restore the spikelet fertility of heterozygous plants [38]. The C-terminal TPR and CaMbd domains of the *wFKBP73* gene were critical for male fertility in transgenic rice [39]. Overexpression of *AtTRP1*, which contains a TRP domain, in wild-type *Arabidopsis* results in dwarf plants and reduced fertility [40]. High expression of TPR may affect normal plant fertility. In our study, anther development of HZ1A started abnormally from the S2 stage, and the expression of *Capana06g002968* in HZ1C was lower than that in HZ1A in the S2 and S3 stages. This suggests that lower expression of *Capana06g002968* may be involved in the restoration of fertility of HZ1A. Therefore, *Capana06g002968* was proposed as the strongest candidate gene for $CaRf_{HZ}$, but this needs to be validated in the future.

3.3. Application of the Related Markers of $CaRf_{HZ}$ Gene in Pepper Breeding

Since the discovery of the first pepper CMS by Peterson [15], CMS sterile lines have been used for pepper breeding. The 'three-line' hybrid seed production system consisting of CMS, maintainer, and restorer lines is used in pepper seed production. This method can reduce the link of manual emasculation, thereby reducing the production cost and improving the purity of the hybrid. Several excellent pepper hybrid varieties were obtained by this method in China [41,42]. Fertility stability of the hybrid is an important guarantee for the large-scale promotion of the 'three-line' hybrid, and the *Rf* gene in the restorer line plays a key role.

Based on the differential variations in *Capana06g002968* (467 bp) and *Capana06g002969* (318 bp) in HZ1A and HZ1C, we developed two CAPS markers (results not shown). Amplification and enzyme digestion were performed on 13 lines (Table 5), and the results were consistent with the sequencing results. The accuracy of detecting both maintainer and restorer lines was 84.62% (*Capana06g002968*) and 92.31% (*Capana06g002969*), but not 100%, which may be related to the existence of other fertility restorer genes. The accuracies of 84.62% and 92.31% were similar to those of previous studies, such as Co1Mod1-CAPS (88.0% and 92.2%, respectively) [28], and CRF3S1S, CRF-SCAR, and Co1Mod1-CAPS markers (77.2%, 79.2%, and 70.3%, respectively) [33]. The newly developed CAPS markers in this study can be used for sterile and restorer line screening, but they need validation in more maintainer and restorer lines.

4. Materials and Methods

4.1. Plant Materials

The CMS line HZ1A is a pepper line with complete sterility that is obtained by backcrossing for more than 15 generations. The F_2 segregating populations (Table 2) were derived from 11 F_1 hybrid plants, with HZ1A and HZIC as the female and male parents, respectively. In total, 226 F_2 plants and 1442 F_2 plants were planted in the autumn of 2020 and spring of 2021, respectively. Plants of F_2 populations were used for inheritance analysis and mapping of the *Rf* gene $CaRf_{HZ}$. All plant materials, consisting of HZ1A, HZ1C, maintainer lines, restorer lines, F_1 plants, and F_2 plants, were grown in a plastic-covered greenhouse at the Zhuanghang Experimental Station of Shanghai Academy of Agricultural Sciences (Shanghai, China).

4.2. Pollen Fertility Evaluation

Pollen fertility was evaluated visually with at least three blooming flowers and further discriminated using Alexander's solution staining [43] under microscope, as described previously [35]. Plants that produced abundant pollen grains were considered male-fertile, whereas plants with no pollen dissemination were considered male-sterile. After Alexander's solution staining, the pollen grains that stained magenta red were classified as male-fertile, and those stained blue–green were classified as male-sterile.

4.3. Inheritance Analysis of Fertility Restoration

Segregation of male fertility/sterility was identified based on the fertility of the plants from the (HZ1A × HZ1C) F_2 population. Chi-square goodness-of-fit test was performed using the following equation: $\chi^2 = \Sigma\ (O - E)^2/E$, where O is the observed frequency, and E is the expected frequency, with $p = 0.05$ as the threshold.

4.4. Nucleic Acid Extraction

Genomic DNA was extracted from the young leaves using a DNA extraction kit (Tiangen, Beijing, China). The anthers of HZ1A, HZ1C, and (HZ1A × HZ1C) F_1 at three developmental stages: stage 1 (S1), sporogenous tissue to meiotic stage; stage 2 (S2), tetrad to mononuclear stage; and stage 3 (S3), mature pollen stage, based on an earlier study [35] were isolated from the flower buds and immediately placed into an RNA stabilization solution (Qiagen, Hilden, Germany). Total RNA was isolated from anthers using a Biospin Plant Total RNA Extraction Kit (Bioer Technology, Hangzhou, China). Reverse transcriptase was used to enrich full-length cDNA using a HiScript II One Step RT-PCR Kit (Vazyme, Nanjing, China). According to the manufacturer's instructions, DNA and RNA were extracted, and reverse transcription was performed.

4.5. SSR and InDel Markers

SSR markers used in the present study were derived from Cheng et al. [36]. SSR markers are numbered according to their position on the chromosome, and chromosome information is added, such as P01g0001. The markers used in the $CaRf_{HZ}$ linkage map are shown in Table S1.

Chromosome 6 sequences of two published pepper genomes, Zunla-1 V2.0 and Chiltepin V2.0, were obtained from the China National GeneBank Sequence Archive (CNSA) (https://db.cngb.org/search/project/CNPhis0000547/, accessed on 25 October 2020). According to the initial and fine mapping of the $CaRf_{HZ}$ gene, 4 Mb (Chr06:195,000,000..196,000,000 and Chr06: 213,000,000..216,000,000) sequences were selected to search for InDel. The InDel variations between 'Zunla-1' and 'Chiltepin' were detected using MUMmer 3.0 software [44] (Hamburg, Germany), and the parameter of variation length $\geq$ 5 bp was used for screening. The 500 bp sequences before and after the InDel variation were selected, and Primer-BLAST software (Bethesda, MD, USA) was used for primer sequence design. In total, 102 InDel markers were developed, and the information is listed in Table S2. Sangon Biotech (Shanghai, China) synthesized the primers. The design and synthesis of all primers used in the present study were the same as those described above.

4.6. BSA and Linkage Analysis of $CaRf_{HZ}$ Gene

The $CaRf_{HZ}$ gene was mapped using BSA. DNA from 10 plants each from male-fertile and male-sterile plants of the (HZ1A × HZ1C) F_2 population were randomly selected and mixed to composite the male-fertile and male-sterile DNA bulks. Then, 449 SSR markers [36], which covered the entire pepper genome, were tested for polymorphism in the parent HZ1A and HZ1C. Thereafter, the polymorphic markers between parents were detected in male-fertile and male-sterile DNA bulks. The polymorphic marker between DNA bulks was recognized as linked to the $CaRf_{HZ}$ gene.

In the present study, the male-sterile plants from the (HZ1A × HZ1C) F_2 populations were used to map the $CaRf_{HZ}$ gene (Table 2) according to Zhang et al. [45]. The polymorphic markers between DNA bulks and the SSR and InDel markers near this marker were amplified and analyzed in male-sterile plants from the (HZ1A × HZ1C) F_2 population. The recombination rate (c) between the marker and the $CaRf_{HZ}$ gene was calculated using the formula $c = (N1 + N2/2)/N$, where N is the number of all sterile plants, $N1$ is the number of plants with homozygous bands from the fertile parent HZ1C, and $N2$ is the number of plants with heterozygous bands from the two parents [45,46]. Kosambi's formula: $d = 1/4 \log((1 + 2c)/(1 - 2c))$ was used to calculate the genetic distance (d) [47]. The unit of d is centimorgans (cM). MapDraw v2.1 was used to construct the genetic map [48].

PCR amplification was performed using a reaction mixture containing the following: 5 μL of 2×Hieff® PCR Master Mix (YEASEN, Shanghai, China), 1 μL of 10 μM PCR upstream primer, 1 μL of 10 μM PCR downstream primer, 20 ng template DNA, and ddH_2O up to 10 μL. The PCR cycling program was as follows: pre-denaturation at 94 °C for 2 min; followed by 32 cycles of 94 °C for 30 s, primer annealing temperature for 30 s, and 72 °C for 30 s; final extension at 72 °C for 5 min; cooling to 12 °C. Amplification products were electrophoresed using 8% non-denaturing polyacrylamide gels at 150 V for 1 h. The gels were visualized using 0.1% $AgNO_3$ solution [46].

4.7. Candidate Gene Amplification and qRT-PCR

Amplification of the candidate genes of $CaRf_{HZ}$ in the HZ1A and HZ1B parental lines was performed following the instructions using GoldenStar® T6 Super PCR Mix Ver.2 (Tsingke Biotechnology, Beijing, China) in a total reaction volume of 50 μL containing 45 μL of 1.1× GoldenStar® Mix Ver.2, 2 μL of 10 μM upstream primer, 2 μL of 10 μM downstream primer, and 200 ng template DNA. The PCR cycling conditions were as follows: pre-denaturation at 98 °C for 2 min; followed by 32 cycles of 98 °C for 10 s, primer annealing temperature for 15 s, and 72 °C for 40 s; final extension at 72 °C for 5 min, and cooling to 12 °C. The primers used for cloning the candidate genes of $CaRf_{HZ}$ are listed in Table S3. PCR products were sequenced by Sangon Biotech (Shanghai, China). The amino acid sequences of the candidate genes of $CaRf_{HZ}$ were aligned using BioXM 2.7.120 (Nanjing, China).

The qRT-PCR was performed using 2x Hieff UNICON® Universal Blue qPCR SYBR Green Master Mix (YEASEN, Shanghai, China) following the instructions. The PCR reactions were analyzed using the Quant Studio 5 real-time PCR system (Waltham, MA, USA). The *β-actin* gene from Lv et al. [49] was used as an internal control to normalize expression data. The $2^{-\Delta\Delta Ct}$ method was used to calculate the relative expression levels. Three biological and three technical replicates were used for each experiment. The primers used for qRT-PCR are listed in Table S4.

4.8. Analysis of Candidate Genes

The physical positions of the molecular markers were determined with reference to the pepper reference Zunla-1 V2.0 genome sequence as described above. The annotated genes in the identified target region were obtained according to the Zunla-1 V2.0 gene models [26]. The protein structures of annotated genes were predicted by SMART (https://smart.embl.de/, accessed on 15 March 2022) [37].

5. Conclusions

To summarize, the male fertility of the pepper CMS line HZ1A could be restored by a single dominant gene, $CaRf_{HZ}$. The *Rf* gene, $CaRf_{HZ}$, was fine-mapped to a 533.81 kb region on chromosome 6 using BSA. Four annotated genes were analyzed according to Zunla-1 V2.0 gene models. Based on the comparison of genomic DNA sequences, qRT-PCR, and protein structures in HZ1A and its maintainer and restorer lines, *Capana06g002968* was proposed as the strongest candidate gene for the $CaRf_{HZ}$ gene. Functional verification of the *CaRfHZ* gene needs to be performed in the future.

Supplementary Materials: The following supporting information can be downloaded at: https://www.mdpi.com/article/10.3390/ijms23147633/s1.

Author Contributions: Conceptualization, Z.N., W.Z. and Q.N.; methodology, Y.S., H.W. and J.C.; software, Z.N. and Y.S.; investigation, Z.N. and J.C.; resources, J.C.; data curation, Z.N. and Y.S.; writing—original draft preparation, Z.N.; writing—review and editing, W.Z. and Q.N.; supervision, W.Z. and Q.N.; funding acquisition, Z.N. and W.Z. All authors have read and agreed to the published version of the manuscript.

Funding: This research was funded by the Independent Declaration Project of Agricultural and Social Research in Hangzhou, grant number 20191203B58; the Excellent Team Building Program of Shanghai Academy of Agricultural Sciences, grant number 2022019; the National Major Vegetable Industry Technology System Construction Project, grant number CARS-25; and the Science and Technology Project of the Hangzhou Academy of Agricultural Sciences, grant number 2020HNCT-10.

Institutional Review Board Statement: Not applicable.

Informed Consent Statement: Not applicable.

Data Availability Statement: Not applicable.

Conflicts of Interest: The authors declare no conflict of interest.

References

1. Kim, Y.J.; Zhang, D. Molecular control of male fertility for hybrid crop breeding. *Trends Plant Sci.* **2018**, *23*, 53–65. [CrossRef] [PubMed]
2. Wang, Z.; Zou, Y.; Li, X.; Zhang, Q.; Chen, L.; Wu, H.; Su, D.; Chen, Y.; Guo, J.; Luo, D. Cytoplasmic male sterility of rice with Boro II cytoplasm is caused by a cytotoxic peptide and is restored by two related PPR motif genes via distinct modes of mRNA silencing. *Plant Cell* **2006**, *18*, 676–687. [CrossRef] [PubMed]
3. Kazama, T.; Toriyama, K. A fertility restorer gene, *Rf4*, which is widely used for hybrid rice breeding, encodes a pentatricopeptide repeat protein. *Rice* **2014**, *7*, 28. [CrossRef] [PubMed]
4. Hu, J.; Wang, K.; Huang, W.; Liu, G.; Gao, Y.; Wang, J.; Huang, Q.; Ji, Y.; Qin, X.; Wan, L.; et al. The rice pentatricopeptide repeat protein RF5 restores fertility in Hong-Lian cytoplasmic male-sterile lines via a complex with the glycine-rich protein GRP162. *Plant Cell* **2012**, *24*, 109–122. [CrossRef]
5. Huang, W.; Yu, C.; Hu, J.; Wang, L.; Dan, Z.; Zhou, W.; He, C.; Zeng, Y.; Yao, G.; Qi, J. Pentatricopeptide-repeat family protein RF6 functions with hexokinase 6 to rescue cytoplasmic male sterility in rice. *Proc. Natl. Acad. Sci. USA* **2015**, *112*, 14984–14989. [CrossRef]
6. Cui, X.; Wise, R.P.; Schnable, P.S. The *rf2* nuclear restorer gene of male-sterile T-cytoplasm maize. *Science* **1996**, *272*, 1334–1336. [CrossRef] [PubMed]
7. Zabala, G.; Gabay-Laughnan, S.; Laughnan, J.R. The nuclear gene *Rf3* affects the expression of the mitochondrial chimeric sequence R implicated in S-type male sterility in maize. *Genetics* **1997**, *147*, 847–860. [CrossRef]
8. Yang, G.; Fu, T.; Yi, B.; Hong, D. Mitochondria-targeted PPR protein restores pol cytoplasmic male sterility by reducing *orf224* transcript levels in oilseed rape. *Mol. Plant* **2016**, *9*, 1082–1084.
9. Uyttewaal, M.; Arnal, N.; Quadrado, M.; Martin-Canadell, A.; Vrielynck, N.; Hiard, S.; Gherbi, H.; Bendahmane, A.; Budar, F.; Mireau, H. Characterization of *Raphanus sativus* pentatricopeptide repeat proteins encoded by the fertility restorer locus for Ogura cytoplasmic male sterility. *Plant Cell* **2008**, *20*, 3331–3345. [CrossRef]
10. Melonek, J.; Duarte, J.; Martin, J.; Beuf, L.; Murigneux, A.; Varenne, P.; Comadran, J.; Specel, S.; Levadoux, S.; Bernath-Levin, K.; et al. The genetic basis of cytoplasmic male sterility and fertility restoration in wheat. *Nat. Commun.* **2021**, *12*, 1036. [CrossRef]
11. Wang, T.L.; He, T.T.; Ding, X.L.; Zhang, Q.Q.; Yang, L.S.; Nie, Z.X.; Zhao, T.J.; Gai, J.Y.; Yang, S.P. Confirmation of *GmPPR576* as a fertility restorative gene of cytoplasmic male sterility in soybean. *J. Exp. Bot.* **2021**, *72*, 7729–7742. [CrossRef] [PubMed]

12. Jaqueth, J.S.; Hou, Z.; Zheng, P.; Ren, R.; Nagel, B.A.; Cutter, G.; Niu, X.; Vollbrecht, E.; Greene, T.W.; Kumpatla, S.P. Fertility restoration of maize CMS-C altered by a single amino acid substitution within the *Rf4* bHLH transcription factor. *Plant J.* **2020**, *101*, 101–111. [CrossRef] [PubMed]
13. Tang, H.; Luo, D.; Zhou, D.; Zhang, Q.; Tian, D.; Zheng, X.; Chen, L.; Liu, Y.G. The rice restorer *Rf4* for wild-abortive cytoplasmic male sterility encodes a mitochondrial-localized PPR protein that functions in reduction of *WA352* transcripts. *Mol. Plant* **2014**, *7*, 1497–1500. [CrossRef] [PubMed]
14. Qin, X.; Tian, S.; Zhang, W.; Zheng, Q.; Wang, H.; Feng, Y.; Lin, Y.; Tang, J.; Wang, Y.; Yan, J.; et al. The main restorer *Rf3* of maize S-type cytoplasmic male sterility encodes a PPR protein that functions in the reduction of the transcripts of *orf355*. *Mol. Plant* **2021**, *14*, 1961–1964. [CrossRef] [PubMed]
15. Peterson, P. Cytoplasmically inherited male sterility in capsicum. *Am. Nat.* **1958**, *92*, 111–119. [CrossRef]
16. Xu, F.; Yang, X.; Zhao, N.; Hu, Z.; Mackenzie, S.A.; Zhang, M.; Yang, J. Exploiting sterility and fertility variation in cytoplasmic male sterile vegetable crops. *Hortic. Res.* **2022**, *9*, uhab039. [CrossRef]
17. Wang, L.H.; Zhang, B.X.; Lefebvre, V.; Huang, S.W.; Daubeze, A.M.; Palloix, A. QTL analysis of fertility restoration in cytoplasmic male sterile pepper. *Theor. Appl. Genet.* **2004**, *109*, 1058–1063. [CrossRef]
18. Wei, B.; Wang, L.; Chen, L.; Zhang, R. Genetic analysis on the restoration of cytoplasmic male sterility with mixed model of major gene plus polygene in pepper. *Acta Hortic. Sin.* **2013**, *40*, 2263–2268.
19. Gulyas, G.; Pakozdi, K.; Lee, J.S.; Hirata, Y. Analysis of fertility restoration using cytoplasmic male-sterile red pepper (*Capsicum annuum* L.) lines. *Breed. Sci.* **2006**, *56*, 331–334. [CrossRef]
20. Ye, Q.; Ruan, M.; Wang, R.; Yao, Z.; Li, Z.; Wan, H.; Cheng, Y.; Yang, Y.; Zhou, G. Genetic analysis and molecular mapping of the restorer gene for cytoplasmic male sterility in pepper. *Mol. Plant Breed.* **2017**, *15*, 4985–4991.
21. Zhang, B.; Huang, S.; Yang, G.; Guo, J. Two RAPD markers linked to a major fertility restoration gene in pepper. *Euphytica* **2000**, *113*, 155–161.
22. Kim, D.S.; Kim, D.H.; Yoo, J.H.; Kim, B.D. Cleaved amplified polymorphic sequence and amplified fragment length polymorphism markers linked to the fertility restorer gene in chili pepper (*Capsicum annuum* L.). *Mol. Cells* **2006**, *21*, 135–140.
23. Lee, J.; Yoon, J.B.; Park, H.G. A CAPS marker associated with the partial restoration of cytoplasmic male sterility in chili pepper (*Capsicum annuum* L.). *Mol. Breed.* **2008**, *21*, 95–104. [CrossRef]
24. Yang, J.; Wang, W.; Shen, H. SSR marker mapping and marker-assisted selection of *Rf* gene in pepper. *Chin. Cucurbits Veg.* **2010**, *23*, 1–5.
25. Jo, Y.D.; Kim, Y.M.; Park, M.N.; Yoo, J.H.; Park, M.; Kim, B.D.; Kang, B.C. Development and evaluation of broadly applicable markers for restorer-of-fertility in pepper. *Mol. Breed.* **2010**, *25*, 187–201. [CrossRef]
26. Qin, C.; Yu, C.; Shen, Y.; Fang, X.; Chen, L.; Min, J.; Cheng, J.; Zhao, S.; Xu, M.; Luo, Y.; et al. Whole-genome sequencing of cultivated and wild peppers provides insights into *Capsicum* domestication and specialization. *Proc. Natl. Acad. Sci. USA* **2014**, *111*, 5135–5140. [CrossRef]
27. Kim, S.; Park, M.; Yeom, S.I.; Kim, Y.M.; Lee, J.M.; Lee, H.A.; Seo, E.; Choi, J.; Cheong, K.; Kim, K.T.; et al. Genome sequence of the hot pepper provides insights into the evolution of pungency in *Capsicum* species. *Nat. Genet.* **2014**, *46*, 270–278. [CrossRef]
28. Jo, Y.D.; Ha, Y.; Lee, J.H.; Park, M.; Bergsma, A.C.; Choi, H.I.; Goritschnig, S.; Kloosterman, B.; van Dijk, P.J.; Choi, D.; et al. Fine mapping of *Restorer-of-fertility* in pepper (*Capsicum annuum* L.) identified a candidate gene encoding a pentatricopeptide repeat (PPR)-containing protein. *Theor. Appl. Genet.* **2016**, *129*, 2003–2017. [CrossRef]
29. Zhang, Z.H.; An, D.L.; Cao, Y.C.; Yu, H.L.; Zhu, Y.S.; Mei, Y.J.; Zhang, B.X.; Wang, L.H. Development and application of KASP markers associated with *Restorer-of-fertility* gene in *Capsicum annuum* L. *Physiol. Mol. Biol. Plant* **2021**, *27*, 2757–2765. [CrossRef]
30. Kang, M.C.; Kang, H.J.; Jung, S.Y.; Lee, H.Y.; Kang, M.Y.; Jo, Y.D.; Kang, B.C. The *Unstable Restorer-of-fertility* locus in pepper (*Capsicum annuum* L.) is delimited to a genomic region containing PPR genes. *Theor. Appl. Genet.* **2022**, *135*, 1923–1937. [CrossRef]
31. Wu, L.; Wang, P.; Wang, Y.; Cheng, Q.; Lu, Q.; Liu, J.; Li, T.; Ai, Y.; Yang, W.; Sun, L.; et al. Genome-wide correlation of 36 agronomic traits in the 287 pepper (*Capsicum*) accessions obtained from the SLAF-seq-Based GWAS. *Int. J. Mol. Sci.* **2019**, *20*, 5675. [CrossRef]
32. Cheng, J.; Chen, Y.; Hu, Y.; Zhou, Z.; Hu, F.; Dong, J.; Chen, W.; Cui, J.; Wu, Z.; Hu, K. Fine mapping of *restorer-of-fertility* gene based on high-density genetic mapping and collinearity analysis in pepper (*Capsicum annuum* L.). *Theor. Appl. Genet.* **2020**, *133*, 889–902. [CrossRef]
33. Zhang, Z.; Zhu, Y.; Cao, Y.; Yu, H.; Bai, R.; Zhao, H.; Zhang, B.; Wang, L. Fine mapping of the male fertility restoration gene *CaRf032* in *Capsicum annuum* L. *Theor. Appl. Genet.* **2020**, *133*, 1177–1187. [CrossRef] [PubMed]
34. Wei, B.; Bosland, P.W.; Zhang, Z.; Wang, Y.; Zhang, G.; Wang, L.; Yu, J. A predicted *NEDD8* conjugating enzyme gene identified as a *Capsicum* candidate *Rf* gene using bulk segregant RNA sequencing. *Hortic. Res.* **2020**, *7*, 210. [CrossRef] [PubMed]
35. Nie, Z.; Chen, J.; Song, Y.; Fu, H.; Wang, H.; Niu, Q.; Zhu, W. Comparative transcriptome analysis of the anthers from the cytoplasmic male-sterile pepper line HZ1A and its maintainer line HZ1B. *Horticulturae* **2021**, *7*, 580. [CrossRef]
36. Cheng, J.; Zhao, Z.; Li, B.; Qin, C.; Wu, Z.; Trejo-Saavedra, D.L.; Luo, X.; Cui, J.; Rivera-Bustamante, R.F.; Li, S.; et al. A comprehensive characterization of simple sequence repeats in pepper genomes provides valuable resources for marker development in *Capsicum*. *Sci. Rep.* **2016**, *6*, 18919. [CrossRef] [PubMed]
37. Letunic, I.; Khedkar, S.; Bork, P. SMART: Recent updates, new developments and status in 2020. *Nucleic Acids Res.* **2021**, *49*, D458–D460. [CrossRef]

38. Yu, Y.; Zhao, Z.; Shi, Y.; Tian, H.; Liu, L.; Bian, X.; Xu, Y.; Zheng, X.; Gan, L.; Shen, Y.; et al. Hybrid sterility in rice (*Oryza sativa* L.) involves the tetratricopeptide repeat domain containing protein. *Genetics* **2016**, *203*, 1439–1451. [CrossRef]
39. Kurek, I.; Dulberger, R.; Azem, A.; Tzvi, B.B.; Sudhakar, D.; Christou, P.; Breiman, A. Deletion of the C-terminal 138 amino acids of the wheat FKBP73 abrogates calmodulin binding, dimerization and male fertility in transgenic rice. *Plant Mol. Biol.* **2002**, *48*, 369–381. [CrossRef]
40. Lin, Z.; Ho, C.W.; Grierson, D. *AtTRP1* encodes a novel TPR protein that interacts with the ethylene receptor ERS1 and modulates development in *Arabidopsis*. *J. Exp. Bot.* **2009**, *60*, 3697–3714. [CrossRef] [PubMed]
41. Wang, L.; Zhang, B.; Zhang, Z.; Cao, Y.; Yu, H.; Feng, X. Status in breeding and production of *Capsicum* spp. in China during 'The Thirteenth Five-year Plan' period and future prospect. *China Veg.* **2021**, *2*, 21–29.
42. Zhou, X.; Ma, Y.; Dai, X.; Li, X.; Yang, S. Spread and industry development of pepper in China. *Acta Hortic. Sinica* **2020**, *47*, 1715–1726.
43. Peterson, R.; Slovin, J.P.; Chen, C. A simplified method for differential staining of aborted and non-aborted pollen grains. *J. Plant Biol.* **2010**, *1*, e13. [CrossRef]
44. Kurtz, S.; Phillippy, A.; Delcher, A.L.; Smoot, M.; Shumway, M.; Antonescu, C.; Salzberg, S.L. Versatile and open software for comparing large genomes. *Genome Biol.* **2004**, *5*, R12. [CrossRef]
45. Zhang, Q.; Shen, B.Z.; Dai, X.K.; Mei, M.H.; Maroof, M.A.S.; Li, Z.B. Using bulked extremes and recessive class to map genes for photoperiod-sensitive genic male sterility in rice. *Proc. Natl. Acad. Sci. USA* **1994**, *91*, 8675–8679. [CrossRef]
46. Nie, Z.; Zhao, T.; Liu, M.; Dai, J.; He, T.; Lyu, D.; Zhao, J.; Yang, S.; Gai, J. Molecular mapping of a novel male-sterile gene ms_{NJ} in soybean [*Glycine max* (L.) Merr.]. *Plant Reprod.* **2019**, *32*, 371–380. [CrossRef]
47. Kosambi, D.D. The estimation of map distance from recombination values. *Ann. Eugen.* **1943**, *12*, 172–175. [CrossRef]
48. Liu, R.H.; Meng, J.L. MapDraw: A Microsoft excel macro for drawing genetic linkage maps based on given genetic linkage data. *Hereditas* **2003**, *25*, 317–321.
49. Lv, J.H.; Liu, Z.B.; Liu, Y.H.; Ou, L.J.; Deng, M.H.; Wang, J.; Song, J.S.; Ma, Y.Q.; Chen, W.C.; Zhang, Z.Q.; et al. Comparative transcriptome analysis between cytoplasmic male-sterile line and its maintainer during the floral bud development of pepper. *Hortic. Plant J.* **2020**, *6*, 89–98. [CrossRef]

International Journal of
Molecular Sciences

MDPI

Article

Biosynthesis and Metabolism of Garlic Odor Compounds in Cultivated Chinese Chives (*Allium tuberosum*) and Wild Chinese Chives (*Allium hookeri*)

Shi-Wei Xia [1,†], Lin-Feng Hang [1,†], Siyad Ali [1], Xiao-Yu Xu [1], Yan-Jun Liu [1], Qian-Qian Yan [1], Qiu-Yu Luo [1], Yu Li [1], Li-Jing Lin [1], Huan-Xiu Li [1], Xiao-Ai Zhang [1], Lin-Kai Huang [2], Xiao Ma [2] and Yun-Song Lai [1,*]

1 College of Horticulture, Sichuan Agricultural University, Chengdu 611130, China; xia-sw@stu.sicau.edu.cn (S.-W.X.); 15756269241@163.com (L.-F.H.); siyad.abdali1994@gmail.com (S.A.); 2020305038@stu.sicau.edu.cn (X.-Y.X.); liuyanjun@stu.sicau.edu.cn (Y.-J.L.); yqianqian@stu.sicau.edu.cn (Q.-Q.Y.); 202001479@stu.sicau.edu.cn (Q.-Y.L.); liyu@sicau.edu.cn (Y.L.); 14208@sicau.edu.cn (L.-J.L.); lihuanxiu@sicau.edu.cn (H.-X.L.); 13181@sicau.edu.cn (X.-A.Z.)

2 College of Grassland Science and Technology, Sichuan Agricultural University, Chengdu 611130, China; huanglinkai@sicau.edu.cn (L.-K.H.); maroar@126.com (X.M.)

* Correspondence: laiys@sicau.edu.cn

† These authors contributed equally to this work.

Abstract: Chinese chives is a popular herb vegetable and medicine in Asian countries. Southwest China is one of the centers of origin, and the mountainous areas in this region are rich in wild germplasm. In this study, we collected four samples of germplasm from different altitudes: a land race of cultivated Chinese chives (*Allium tuberosum*), wide-leaf chives and extra-wide-leaf chives (*Allium hookeri*), and ovoid-leaf chives (*Allium funckiaefolium*). Leaf metabolites were detected and compared between *A. tuberosum* and *A. hookeri*. A total of 158 differentially accumulated metabolites (DAM) were identified by Gas Chromatography—Mass Spectrometry (GC-MS) and Liquid Chromatography—Mass Spectrometry (LC-MS), among which there was a wide range of garlic odor compounds, free amino acids, and sugars. *A. hookeri* contains a higher content of fructose, garlic odor compounds, and amino acids than *A. tuberosum*, which is supported by the higher expression level of biosynthetic genes revealed by transcriptome analysis. *A. hookeri* accumulates the same garlic odor compound precursors that *A. tuberosum* does (mainly methiin and alliin). We isolated full-length gene sequences of *phytochelatin synthase* (*PCS*), *γ-glutamyltranspeptidases* (*GGT*), *flavin-containing monooxygenase* (*FMO*), and *alliinase* (*ALN*). These sequences showed closer relations in phylogenetic analysis between *A. hookeri* and *A. tuberosum* (with sequence identities ranging from 86% to 90%) than with *Allium cepa* or *Allium sativum* (which had a lower sequence identity ranging from 76% to 88%). Among these assayed genes, *ALN*, the critical gene controlling the conversion of odorless precursors into odor compounds, was undetected in leaves, bulbs, and roots of *A. tuberosum*, which could account for its weaker garlic smell. Moreover, we identified a distinct *FMO1* gene in extra-wide-leaf *A. hookeri* that is due to a CDS-deletion and frameshift mutation. These results above reveal the molecular and metabolomic basis of impressive strong odor in wild Chinese chives.

Keywords: *Allium hookeri*; wild species; ASCO; allicin; RNA-seq; metabolome

1. Introduction

Cultivated Chinese chives (*Allium tuberosum*) is a popular herb vegetable in Asia that is also known as garlic chives or Chinese leek. Like many other *Allium* plants, *A. tuberosum* is taken as a medicinal food due to its various phytochemicals and physiological effects [1,2]. *A. tuberosum* belongs to the *Allium* genus, which is one of the biggest genera in the family Liliaceae, containing more than 800 species [3,4]. There is a consensus in taxonomic study that the *Allium* genus should be divided into subgenera or species groups, although there is some debate over the specific taxonomic categories [5,6]. In China, the *Allium* genus is divided into the following nine species groups according to their botanic traits: Sect. *Porrum*, Sect. *Schoenoprasum*, Sect. *Cepa*, Sect. *Rhizirdium*, Sect. *Bromatoorrhiza*, Sect. *Haplostemon*, Sect. *Anguium*, Sect. *Molium*, and Sect. *Caloscordum*. *A. tuberosum* belongs to the group Sect. *Rhizirdium* [7]. Wild Chinese chives (e.g., *A. hookeri*) have a wide distribution in mountainous areas of southwest China and are called "wide-leaf chives" by locals. *A. hookeri* is a clump-forming herb that is also native to north India [8]. This species can be further classified into two subspecies according to their habits of withering in the winter, which is also supported by genotype analysis [9]. The subspecies that does not wither in the winter usually has extra-wide leaves.

A. hookeri is consumed as a wild vegetable or an herb medicine but is not commonly cultivated in China even though it has a similar appearance to the cultivated *A. tuberosum*. Commercial cultivars of *A. tuberosum* with wide leaves, which have emerged in recent years, are not *A. hookeri* nor did they originate from *A. hookeri* [10]. Recent studies have shed light on the exploration and utilization of these wild species due to their high nutrient value, stronger odor, and elite agronomic traits. *A. hookeri* has a totally different flavonoid compound and much higher amino acid contents and mineral nutrient elements than cultivated Chinese chives [11]. A methanol extract of *A. hookeri* roots can alleviate inflammation by suppressing inducible nitric oxide synthase (iNOS) and cyclooxygenase-2 (COX-2) expression in mouse cells [12]. A dietary supplementation of *A. hookeri* fresh roots and fermented root food improve antioxidant activities and growth in birds [13]. The long history of the domestication of *A. tuberosum* may have resulted in the loss of some traits that are considered valuable today.

Most *Allium* species, such as garlic (*A. sativum*), chives (*A. fistulosum*), and onion (*A. cepa*)—as well as the subject of this study, *A. tuberosum*—are well-known for their characteristic odor that is due to a sulfide, lipophilic, and volatile component usually referred to as garlic odor compound. The most common flavor precursors are S-allyl cysteine sulphoxide (Alliin, or ACSO), S-methyl cysteine sulphoxide (Methiin, or MCSO), and trans-S-1-propenyl cysteine sulphoxide (Isoalliin, or PeCSO) [14]. The abundance and composition of sulfoxides vary among *Allium* species; *A. tuberosum* contains all three precursors [15]. Flavor precursors themselves are odorless, and their immediate conversion to odor compounds relies on tissue damage that allows the sulfoxide stored in the cytosol to react with vacuolar enzyme alliinase (ALN) [16]. Today, the biosynthetic pathway framework of flavor precursors has become much clearer despite some of its details remaining controversial [17–19]. Several critical enzyme genes have been functionally identified [20]. They are *phytochelatin synthase* (*PCS*), *γ-glutamyltranspeptidases* (*GGT*), and *flavin-containing monooxygenase* (*FMO*) [21]. In contrast, the formation process of volatile compounds from precursors such as diallyl sulfide (DAS), diallyl disulfide (di-2-propenyl disulfide; DADS), and diallyl trisulfide (di-2-propenyltrisulfide; DATS) remains mysterious, which may be partially due to the very fast chemical reaction and subsequent difficulties in the extraction and detection of mid-stage compounds [19]. ALN, lachrymatory factor synthase (LFS), and spontaneous chemical processes are involved in the formation of downstream violate compounds [22].

Species in the genus *Allium* have an impressively large genome size with a range of 6.86 gigabase (Gb) to 30.87 Gb [23,24]. Due to this large genome size, there was no genome sequence release of *Allium* plants for a long time. Until now, only two *Allium* species have had a complete genome sequence reported and deposited into public databases; these are *A. sativum* (garlic) and *A. cepa* (onion) [25]. The assembled genome size of garlic is about 16 Gb, which is nearly the same size detected by flow cytometry. Because of the lack of high-quality reference genomes, omics studies of *Allium* plants lag behind those of many other vegetable crops even though the *Allium* genus contains many popular vegetables. Two transcriptomic studies of garlic chives provide a preliminary information of unigene sequences [26,27].

In this study, we compared the transcriptome and metabolome between cultivated Chinese chives (*A. tuberosum*) and wild Chinese chives (*A. hookeri*). The integrative network analysis explained the difference in garlic odor compounds. We isolated full-length gene sequences of critical enzymes that catalyze the formation of odor compound precursors, and analyzed the correlation of biosynthesis and metabolite in the odor compounds. The findings provide useful information on wild *Allium* species.

2. Results

2.1. Germplasm Curation and Nutrition Evaluation

We collected one cultivated *A. tuberosum* plant and three wild materials from a mountainous area of Ya'an, Sichuan Province, China. Among the three wild species, one was *A. funckiaefolium*, which strictly grows in high-altitude areas (>2500 m) and is called "ovoid-leaf Chinese chives" by locals; and two were subspecies of *A. hookeri*, which are grown in a wide area below 2000 m and are called "wide-leaf Chinese chives" by locals (Figure 1). The two subspecies of *A. hookeri* are characterized by their wide leaves but differ from each other in leaf width. Wide-leaf *A. hookeri* (WL Hookeri) has a leaf width of 1.5 cm, while extra-wide-leaf *A. hookeri* (EWL Hookeri) has a leaf width of 2.4 cm. Both subspecies of *A. hookeri* have high levels of adaptability, whereas *A. funckiaefolium* cannot survive in low altitudes (Figure 1). As shown in Table 1, cultivated Chinese chives had a higher water content implying better edibility; however, *A. hookeri* had a stronger odor, and a higher content in soluble sugars and proteins. This difference might be a result of domestication and artificial selection.

Figure 1. Pictures of Chinese chives and wild Chinese chives. The white vertical bar in the pictures indicate a scale of 5 cm. WL Hookeri, Wide-leaf *A. hookeri*; EWL Hookeri, extra-wide-leaf *A. hookeri*.

Table 1. Physiological comparison between Chinese chives and wild Chinese chives.

	Tuberosum	WL Hookeri	EWL Hookeri	Funckiaefolium
Dry matter (mg/gFW)	67.09 ± 2.09 c	95.01 ± 4.44 b	96.75 ± 3.74 b	105.52 ± 3.86 a
Soluble sugar (mg/gFW)	10.87 ± 0.95 c	15.49 ± 0.29 a	14.79 ± 0.36 b	10.49 ± 0.75 c
Vitamin C (mg/gFW)	0.23 ± 0.02 ab	0.21 ± 0.01 b	0.23 ± 0.01 ab	0.24 ± 0.02 a
Soluble protein (mg/gFW)	2.37 ± 0.10 c	2.97 ± 0.05 a	2.94 ± 0.03 a	2.88 ± 0.03 b

The different letters following the numbers indicate significant differences ($p < 0.05$) in Duncan's multiple range test.

2.2. Metabolite Comparison

Metabolites were compared between the leaves of *A. tuberosum* and WL Hookeri. For quality control and quality assurance in the detection of Gas Chromatography—Mass Spectrometry (GC-MS) and Liquid Chromatography—Mass Spectrometry (LC-MS), each sample was assayed six times (Figure S1). Both principal component analysis (PCA) and partial least squares-discriminate analysis (PLS-DA), a supervised PCA algorithm, showed well-clustered samples indicating high quality; the permutations plot indicated there was no overfitting in PLS-DA analysis (Figure S2).

A total of 69 and 245 chemical molecules were detected by GC-MS and LC-MS, respectively (Table S1). After discarding redundancies between the two methods, there were 304 components identified in leaves. Among the identified metabolites, there were 64 organic acids and 46 amino acids (Figure 2A). There were 177 significant differentially accumulated metabolites (DAMs) between *A. tuberosum* and WL Hookeri, whose composition was similar to that of the total metabolites (Figure 2B). However, the amino acids and polyols were exceptions, which took up 18% and 5%, respectively, in the DAMs. There were as many as 26 amino acids showing significant differences in content, most of which were much higher in WL Hookeri (Figure 3A). A Kyoto Encyclopedia of Genes and Genomes (KEGG) enrichment analysis of the DAMs indicated that more than half of the top 10 significant enriched pathways are amino acid metabolisms, e.g., alanine, glycine, pyruvate (Figure 2C). In the amino acid biosynthesis pathway (ko01230), all the DAMs showed higher contents in WL Hookeri, except for an intermediate pyruvate (Figure 3B). Notably, glutathione metabolism pathway as well as cysteine and methionine metabolisms pathway were significantly different; in particular, 19% of detected metabolites in the glutathione metabolism pathway were DAMs. These two pathways are known to be involved in the biosynthesis of garlic odor compounds.

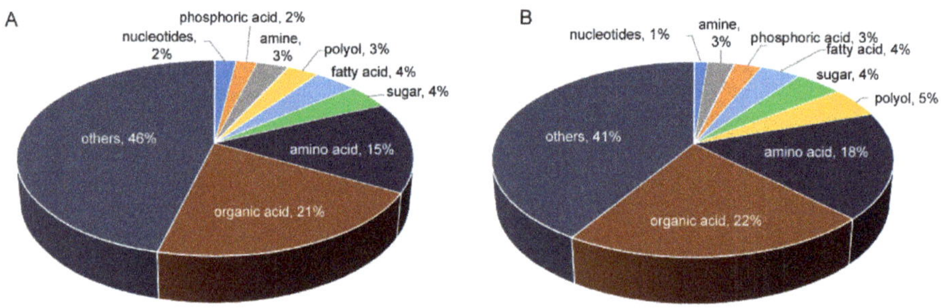

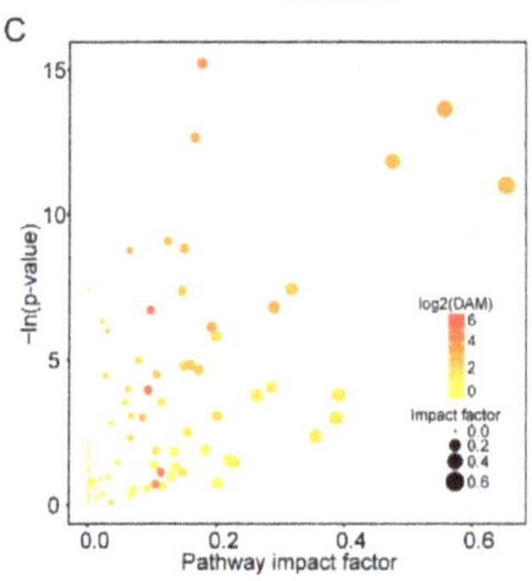

Pathway Name	Pathway ID	Total	Hits	*p*-value	Impact
Alanine, aspartate and glutamate metabolism	ko00250	28	7	0.00	0.65
Glycine; serine and threonine metabolism	ko00260	47	10	0.00	0.56
beta-Alanine metabolism	ko00410	25	7	0.00	0.47
Taurine and hypotaurine metabolism	ko00430	20	3	0.02	0.39
Pyruvate metabolism	ko00620	27	3	0.05	0.39
Starch and sucrose metabolism	ko00500	35	3	0.09	0.35
Arginine biosynthesis	ko00220	23	5	0.00	0.32
Arginine and proline metabolism	ko00330	68	8	0.00	0.29
D-Glutamine and D-glutamate metabolism	ko00471	8	2	0.02	0.26
C5-Branched dibasic acid metabolism	ko00660	20	3	0.02	0.26

Figure 2. Metabolites identified by GC-MS and LC-MS. (**A**) Total metabolites. (**B**) Differentially accumulated metabolites (DAMs). (**C**) KEGG pathway enrichment analysis of DAMs. Left, overview of the enrichment analysis; right, top 10 enriched pathways (with the highest value of impact factor). Total, number of genes annotated in a pathway; Hit, number of genes hits in a pathway.

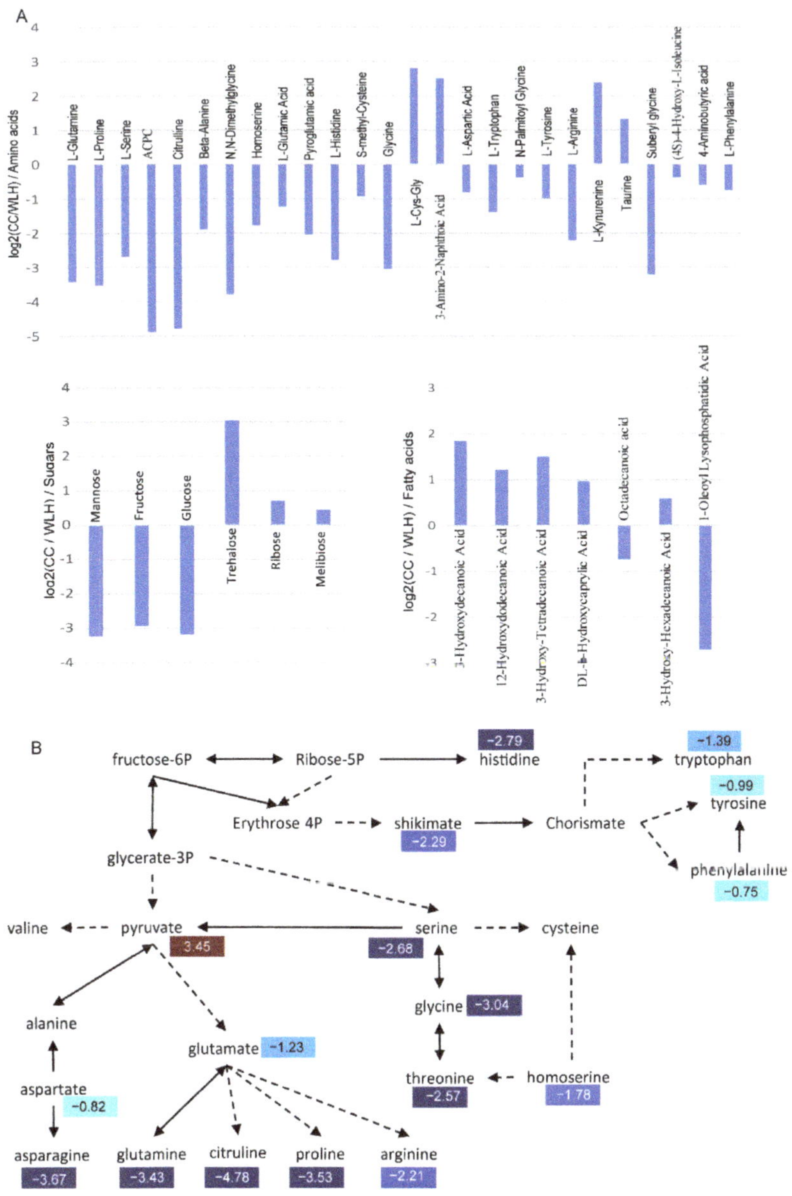

Figure 3. Differentially accumulated metabolites (DAMs) of amino acids, sugars, and fatty acids. (**A**) Change of metabolite indicated by Log2 (Fold change). ACPC, 1-Aminocyclopropanecarboxylic Acid. (**B**) DAMs in amino acids biosynthesis pathway (ko01230). Numbers in color indicate a heatmap of log2 (fold change). Fold change, change in *A. tuberosum* relative to *A. hookeri*.

Interestingly, the much higher accumulation of amino acid in WL Hookeri is accompanied by a much higher level of fructose, which is one of the substrates of amino acid biosynthesis (Figure 3B). Fructose in WL Hookeri is 7.7 times higher than in *A. tuberosum*, which might also account for the sweet taste of WL Hookeri. In addition to fructose, WL Hookeri accumulated more mannose and glucose, despite having less trehalose, ribose, and melibiose (Figure 3A). There are some interesting findings for other metabolites. Organic acids were the most common metabolites in the DAMs. There were seven fatty acids showing significant difference in content, and most of them had a lower abundance in WL Hookeri (Figure 3A).

2.3. *Leaf Transcriptome Comparisons*

The leaf transcriptomes of *A. tuberosum* and WL Hookeri were compared. A total of 39.9 Gb of clean data were obtained, and for each sample, the clean data were over 6.2 Gb. The percentage of high-quality bases ($\geq$Q30) was over 92.5% for each sample (Table S2). We tried to assemble transcripts by first mapping to the garlic reference genome. However, the ratio of mapped reads was too low (<70%). Then, we carried out de novo assemblies without mapping to a reference genome. A total of 146,037 and 124,248 unigenes were obtained for *A. tuberosum* and WL Hookeri, respectively, and the N50 length of total unigenes was 1630 bp. We mapped clean reads against the unigene pool and the mapped ratio ranged from 72–77%. To further reduce low-quality unigenes, we blasted amino acid sequences against both the Swiss-Prot and the garlic genome, and the unigenes were discarded when the blast e-value was $>1 \times 10^{-5}$. As a result, we finally obtained 33,179 and 34,315 unigenes for *A. tuberosum* and WL Hookeri, respectively.

We identified a total of 12,362 differentially expressed genes (DEGs), among which 4946 genes were higher and 7416 genes were lower in *A. tuberosum* (Figure 4). A KEGG enrichment analysis indicated that terpenoid backbone biosynthesis was the most significant pathway. Notably, there was significant difference in fructose and mannose metabolism. A Gene Ontology (GO) enrichment analysis indicated that photosynthesis was the most significant pathway (Table 2).

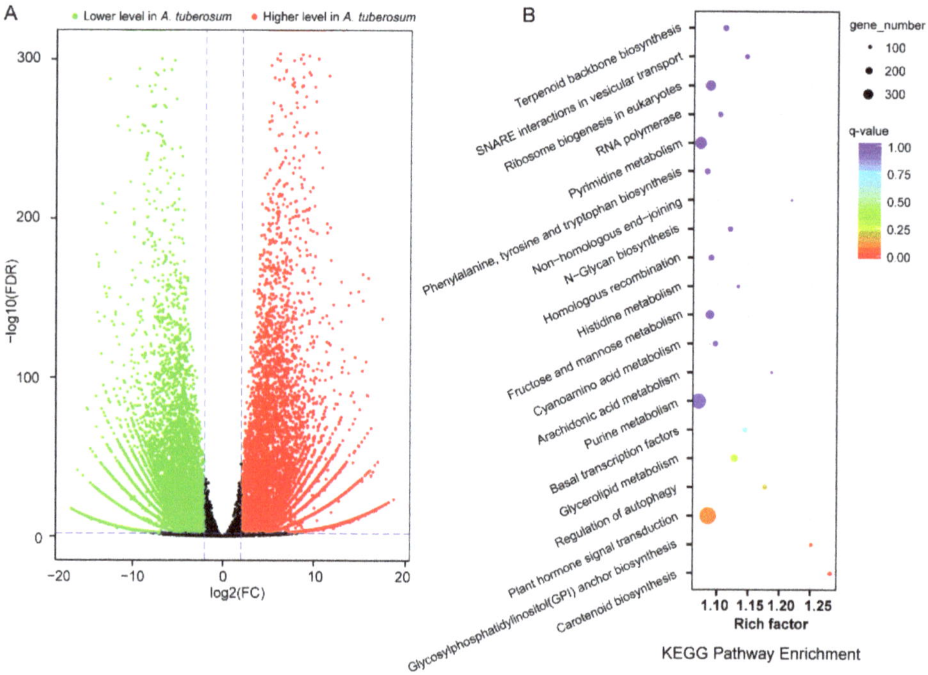

Figure 4. Profile of differentially expressed genes (DEGs) in comparative transcriptome analysis (*A. tuberosum* vs. *A. hookeri*.). (**A**) Volcano plot showing statistical significance of DEGs. (**B**) KEGG enrichment analysis.

Table 2. Top 20 GO terms in GO enrichment analysis of DEGs.

GO ID	GO Term	Annotated	DEG N	Expected N	p-KS
GO:0015979	photosynthesis	240	205	185.14	4.3×10^{-12}
GO:0009658	chloroplast organization	81	76	62.49	1.1×10^{-8}
GO:0018298	protein-chromophore linkage	62	52	47.83	9.9×10^{-8}
GO:0009765	photosynthesis, light harvesting	59	50	45.51	1.5×10^{-7}
GO:0055114	oxidation-reduction process	1986	1489	1532.04	1.5×10^{-5}
GO:0016117	carotenoid biosynthetic process	26	26	20.06	1.6×10^{-5}
GO:0009793	embryo development ending in seed dormancy	153	135	118.03	2.1×10^{-5}
GO:0000413	protein peptidyl-prolyl isomerization	91	82	70.2	2.5×10^{-5}
GO:0006418	tRNA aminoacylation for protein translation	107	88	82.54	3.8×10^{-5}
GO:0010027	thylakoid membrane organization	22	21	16.97	5.4×10^{-5}
GO:0046835	carbohydrate phosphorylation	48	42	37.03	1.4×10^{-4}
GO:0010207	photosystem II assembly	19	19	14.66	2.0×10^{-4}
GO:0006400	tRNA modification	81	74	62.49	4.2×10^{-4}
GO:0006000	fructose metabolic process	6	6	4.63	5.2×10^{-4}
GO:0006003	fructose 2,6-bisphosphate metabolic process	6	6	4.63	5.2×10^{-4}
GO:0048366	leaf development	88	79	67.89	6.0×10^{-4}
GO:1901566	organonitrogen compound biosynthetic process	877	689	676.54	6.6×10^{-4}
GO:0001731	formation of translation preinitiation complex	41	32	31.63	7.2×10^{-4}
GO:0009814	defense response, incompatible interaction	38	35	29.31	7.7×10^{-4}
GO:0009657	plastid organization	111	105	85.63	8.5×10^{-4}

DEG N, number of DEGs enriched; Expected N, expected number enriched; p-KS, p-value of KS test.

2.4. Gene Cloning of Garlic Odor Compounds

Currently, sequence information of garlic odor compound biosynthetic genes is only known in garlic and onion. Here, we identified full-length sequences of genes related to the garlic odor compound by assembled unigenes; these are *GGT1*, *GGT2*, *GGT3*, *FMO1*, *PCS1*, and *ALN*. We manually assembled *AtGGT1* from four unigenes (BMK11240, BMK3806, BMK3807, and BMK94228) and *AhGGT3* from two unigenes (BMK121251 and BMK108624). The rest of the transcripts were determined from a single unigene sequence. To verify the gene sequence based on transcriptome assemblies, we sequenced PCR products of the above genes in *A. tuberosum*, WL Hookeri, and EWL Hookeri by Sanger sequencing. As a result, the final amino acid sequences of cloned genes show a very high level of identity (>99%) to the transcripts (Table 3). Some nucleotides with high divergency in RNA-seq reads were corrected by Sanger sequencing of PCR-products (Figure 5A; Table S4). Orthologue genes in WL Hookeri and EWL Hookeri show identical sequences.

Table 3. Sequence information of biosynthetic genes of garlic odor compounds.

	Length of AA [1]	Identity with Transcript	Identity with NCBI Database [2]
AhGGT1	593	99.1%	90.35%; BAQ21911.1
AhGGT1E	593	99.3%	89.92%; BAQ21911.1
AtGGT1	629	99.8%	92.24%; BAQ21911.1
AhGGT2	622	99.5%	87.24%; BAQ21912.1
AhGGT2E	622	100%	88.27%; BAQ21912.1
AtGGT2	622	99.5%	89.23%; BAQ21912.1
AhGGT3	626	98.9%	86.92%; BAQ21913.1
AhGGT3E	626	98.9%	87.20%; BAQ21913.1
AtGGT3	584	99.3%	86.11%; BAQ21913.1
AhPCS1	505	98.8%	89.35%; AAO13809.1
AhPCS1E	505	97.7%	89.16%; AAO13809.1
AtPCS1	505	98.8%	92.49%; AAO13809.1
AhFMO1	458	99.4%	87.00%; 6WPU_A
AhFMO1E	317	92.4%	88.73%; 6WPU_A
AtFMO1	462	97.5%	91.88%; 6WPU_A
AhALN	471	91.9%	91.15%; AYN25508.1
AhALNE	471	91.7%	90.88%; AYN25510.1
AtALN	471	100%	100%; BAA20358.1

[1] Length of amino acid sequence. [2] Sequences in the database that show the highest identity with the query sequences. AYN25508.1 was derived from *Allium chinense*; AYN25510.1 and BAA20358.1 were derived from *A. tuberosum*; the rest of the sequences were from *A. sativum*.

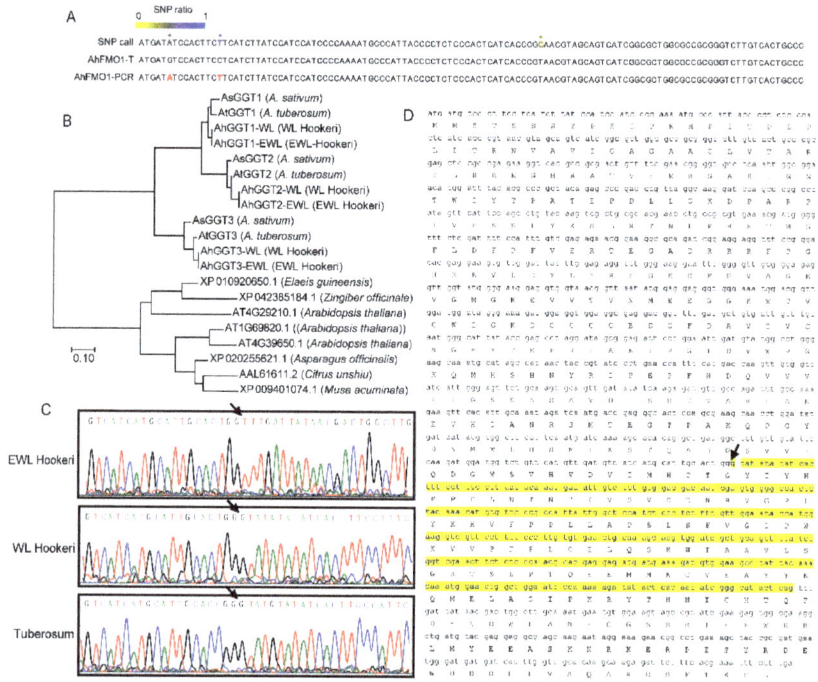

Figure 5. Gene sequential analysis of *GGTs* and *FMO1*. (**A**) A partial illustration of SNP calls that were obtained by mapping RNA-seq reads back to an assembled *AhFMO1-WL* transcript (*AhFMO1-T*). The ratio of SNP was calculated using RNA-seq reads. (**B**) Phylogenetic analysis of *GGTs* gene. (**C,D**) *FMO1* in EWL Hookeri lost a 310-bp CDS region (denoted in yellow color) which led to a frameshift and large deletion. The arrow indicates the start point of the frameshift.

In *Allium* plants, there are three *GGT* genes (*GGT1*, *GGT2*, and *GGT3*) as identified in *A. sativum*; in Arabidopsis, there are four *GGT* genes. Phylogenetic tree analysis indicated the three *GGT* genes in *Allium* are very different to *GGTs* in other plants, indicating the three *GGTs* were independently evolved after the formation of the *Allium* genus (Figure 5B). *GGT2* and *GGT3* were clustered, indicating a closer relation. *AhFMO1-EWL* encoded 317 amino acids in EWL Hookeri, which is much shorter than other sequences (Figure 5B). This is due to a deletion of 310 bp in mRNA, which resulted in 104 deletions of amino acids and a frameshift. This sequence variant is verified by PCR-based Sanger sequencing (Figure 5C,D).

2.5. Integrative Analysis of Garlic Odor Compound Biosynthesis

Precursors of methiin and alliin, two major garlic odor compounds, were identified in both *A. tuberosum* and WL Hookeri. The content of alliin-precursor (*S*-allyl cysteine) and methiin-precursor (*S*-methyl cysteine) in WL Hookeri is 1.39 times and 1.9 times of those in *A. tuberosum*, respectively (Figure 6). No isoalliin- and cycloalliin-related compounds were identified. Nearly all the substrates in the biosynthetic pathway had a higher abundance in WL Hookeri, except for L-cys-gly. Especially of note, the content of glutamine in WL Hookeri is 3.6 times of that in *A. tuberosum*. Cysteine provides the critical sulfur in garlic odor compounds, which has a content 1.9 times the amount in WL Hookeri than in *A. tuberosum*. Lower substrate metabolites and stronger consumption by L-cys-gly together led to less accumulation of garlic odor compounds in *A. tuberosum*.

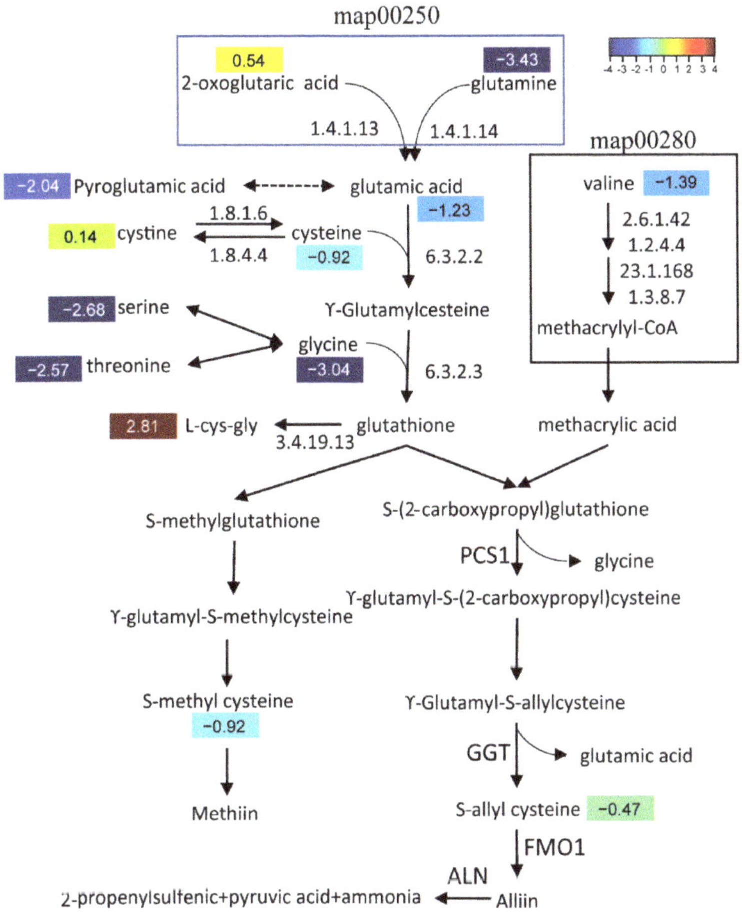

Figure 6. Biosynthetic pathway of a garlic odor compound. Pathways of valine, leucine, and isoleucine degradation (ko00280); alanine, aspartate, and glutamate metabolism (ko00250); cysteine and methionine metabolism (ko00270); and glutathione metabolism (ko00480) were involved in this pathway. Numbers in color indicate a heatmap of log2 (fold change). Fold change, change in *A. tuberosum* relative to *A. hookeri*.

We then compared the transcription level of structural genes in the biosynthetic pathway of the garlic odor compound precursor. Some DEGs were found at steps annotated by KEGG (Figure 6), but there were multiple DEGs for each step and they showed divergent change when compared between *A. tuberosum* and *A. hookeri*. For example, there were six DEGs functionally annotated to encode glutamate dehydrogenase (1.4.1.13) and five of them showed a significantly higher transcription level in Hookeri. For the six newly assembled transcript sequences, *GGT2*, *GGT3*, *PCS1*, and *ALN* all showed a higher transcription level in Hookeri while *GGT1* and *FMO1* did not (Figure 7). The gene transcription level quantified by reads of RNA-seq might be inaccurate in this study, because the unigene sequences were de novo assembled without a reference genome and some of the transcript sequences were manually assembled from multiple unigenes. Thus, we performed quantitative Real-Time PCR (qRT-PCR) to detect the relative transcription level (Figure 7). To select reliable reference genes for qRT-PCR, we validated the expression stability of candidate reference genes in different tissues by using BestKeeper (Table 4). As a result, *elongation factor-1α* (*EF-1α*) is the most stable reference gene, which is a ubiquitous and vastly conserved cytosolic protein in all eukaryotic organisms. The primers designed to quantify the six target genes in qRT-PCR showed high PCR-amplification efficiency (90–110%) (Figure S3). Basically, we get the same conclusion in comparative analysis of gene transcription levels by using qRT-PCR and RNA-seq. However, there are still some exceptions. For example, *GGT2* and *PCS1* had a higher FPKM value in WL-Hookeri which was quantified by reads number in RNA-seq, but a lower relative transcription level quantified in qRT-PCR. Among the three *GGTs*, *GGT2* is the major active gene which had the highest expression level in the three germplasms, which was followed by *GGT3*; *GGT1* seemed inactive in Hookeri. Interestingly, all six genes can be detected in leaves, bulbs, and roots, except for *ALN* in *A. tuberosum*. *ALN* was highly expressed in bulbs of Hookeri but seemed inactive in cultivated *A. tuberosum*, which might account for the weaker garlic odor smell in *A. tuberosum*.

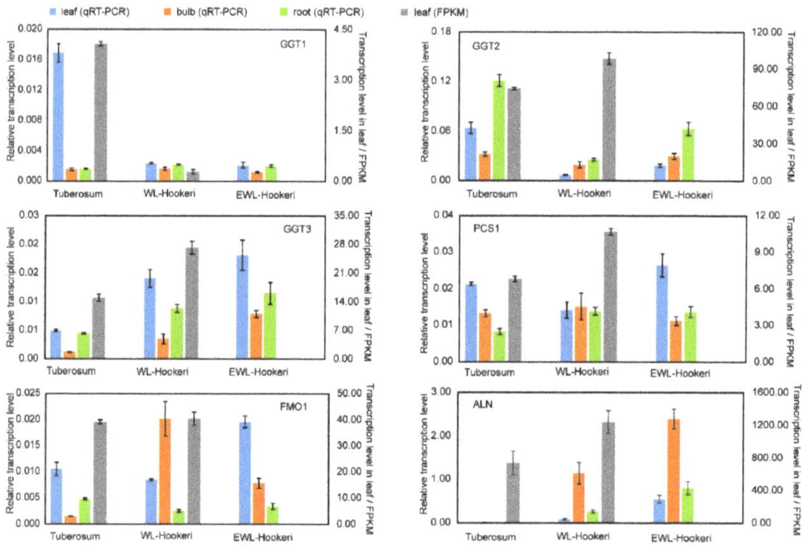

Figure 7. Transcription levels of *GGT1*, *GGT2*, *GGT3*, *PCS1*, *FMO1*, and *ALN* detected by qRT-PCR and RNA-seq counts. Left *y*-axis, relative transcription level modified by Actin in qRT-PCR; right *y*-axis, transcription level quantified by fragments per kilobase of transcript per million mapped fragments (FPKM). The bar represents SE of the mean.

Table 4. Stability comparison of the candidate genes analyzed and ranked by BestKeeper.

Group	Rank	Gene	SD	CV
Leaf	1	*EF-1α*	0.37	2.07
	2	*18SrRNA*	0.57	5.44
	3	*Actin*	1.27	5.92
Bulb	1	*EF-1α*	0.58	3.35
	2	*18SrRNA*	1.12	11.02
	3	*Actin*	1.32	6.16
Root	1	*18SrRNA*	0.74	7.75
	2	*EF-1α*	0.80	4.78
	3	*Actin*	0.89	4.23

3. Discussion

Allium plants are often taken as vegetable and medicine plants due to their rich biochemical components that humans can benefit from. Therefore, many researchers make an effort to dissect metabolite components and profile their response to environmental treatment. However, people usually focus on onion and garlic but not too much on other *Allium* crops, let alone *Allium* wild species, although there are some reports on *Allium hirtifolium* [28–31]. Here, we determined metabolites in *A. hookeri*, which is widely distributed in Asia. *A. hookeri* is also called wild Chinese chives, and therefore, we compared between *A. hookeri* and cultivated Chinese chives (*A. tuberosum*). A total of 341 metabolites were detected by targeted metabolome analysis based on GC-MS and LC-MS assays; these metabolites can be classified into nine groups. A recent extensive study of 30 garlic accessions using a similar LC-MS method reported 472 metabolites [32]. Apparently, there is a technical limitation to identify more metabolites. Higher resolution mass spectrometry and accurate mass tools must be developed to get a more comprehensive profile of *Allium* metabolome [33]. We observed a big difference in amino acid content between *A. tuberosum* and *A. hookeri*. There are as many as 28 DAMs of amino acids, and 86% of these differentially accumulated amino acids show much higher content in *A. hookeri*. The big difference is not only in term of the DAMs number but also the extent of changes. Content of 1-Aminocyclopropanecarboxylic acid in *A. hookeri* is incredibly 29.34 times of that in *A. tuberosum*. Such a tremendous difference is also observed in citrulline (27.40 times), dimethylglycine (13.91 times), asparagine (12.74 times), and proline (11.54 times). The content and composition of amino acids seems to vary in different *Allium* plants, which may be a result of a long history of cultivation or evolution [34]. A higher content of amino acids in wild species probably contributes to their higher tolerance to biotic and abiotic stress [35].

The composition of the flavor precursors and S-Alk (en) yl-L-Cysteine (SAC) derivatives varied among *Allium* species [15,36]. There is a characteristic ratio of different precursor components in particular species. In the onion species *Allium fistulosum* and *Allium schoenoprasum*, isoalliin is the predominant component, which accounts for more than 80% of the total contents; in contrast, the predominant component in *A. tuberosum* is methiin, which accounts for 80% of the total contents [15]. In this study, methiin and alliin, but not isoalliin, were detected by GC-MS and LS-MS, which is probably due to the contents of isoalliin being too low in the materials and its unstable chemical properties [37]. In a previous study, isoalliin accounted for only 7% of the total contents in *A. tuberosum* [15]. The ratio of methiin to alliin is about 6:1 in both *A. tuberosum* and *A. hookeri*, which indicates similar and specific ratios of precursor components in these two species. Together with the high identity of critical structural genes in precursor biosynthetic pathways, *A. tuberosum* and *A. hookeri* seem to have close phylogenetic relationships. However, *A. hookeri* contains more flavor precursors, about 1.5 times the amount in *A. tuberosum*. This indicates the domestication and cultivation of *A. tuberosum* may have caused a decrease in its garlic odor contents.

Allium plants are characterized by their flavors and odors. However, the biosynthetic pathway of these odor compounds is still under debate, and, in particular, the enzyme genes catalyzing each of them are not yet well identified [18]. For precursor alliin biosynthesis, three enzyme genes (*PCS*, *GGT*, and *FMO*) have been identified and characterized. Even so, these three structural genes have only been sequentially reported in garlic and onion. In this study, we identified these sequences in *A. tuberosum* and in two wild *A. hookeri* subspecies (Table S3). Unigene sequences from assemblies of RNA-seq were used as templated sequences. *GGTs* from *Allium* species formed an independent cluster in the phylogenetic analysis, indicating a special evolution of sulfite-containing odor compounds (Figure 5A). We detected a sequence variance of the *FMO1* gene, which catalyzed the last step in forming alliin. In WHL Hookeri, it is highly likely that *FMO1* lost its function due to a fatal deletion of 310 bp, which resulted in a deletion of the amino acid sequence and a frameshift. The definite influence of this sequence variance on *AhFMO1-EWL* function should be determined in the future.

We tried to use the garlic genome as a reference at first, but the mapping ratio was lower than 70%. Due to the large genome size of *Allium* plants, genome sequences have only been released for garlic and onion. Recently, whole genome sequence data for *A. hookeri* (SRR15098630) were released, but the sequence depth was too low and there was no gene annotation. Therefore, we carried out a *de novo* assembly of the unigene sequence without a reference genome. To overcome the disadvantage of not having a reference genome, we restrictively discarded low-quality unigenes by blasting against both the Swiss-Prot database and the garlic genome. We assembled transcripts of garlic odor compound-related genes from RNA-seq reads by a HISAT2 tool and manual operation. To check the reliability, we mapped the RNA-seq reads back to the assembled transcripts and found some nucleotides were divergent. This is probably due to a sequencing error or multiple allele sequences. To make a correct full-length sequence, PCR amplicons were sequenced again by Sanger sequencing. Besides number-limited errors of single nucleotides, only one 18-bp insertion of the *AtFMO1* gene was corrected. This indicates a high level of accuracy of unique assemblies in this study.

4. Conclusions

In this study, we compared metabolome and transcriptome between cultivated Chinese chives and Hookeri to reveal the metabolic and transcriptomic bases accounting for the stronger garlic smell and sweeter taste of wild Chinese chives. Hookeri accumulated amazing much more free amino acids, fructose, and glucose in leaves. In addition, higher level of garlic odor precursors was detected in Hookeri, which is due to the stronger metabolic flux from glutamine, valine, and cystine into the garlic odor biosynthetic pathway. *ALN* gene that controls the conversion from odorless precursors to volatile odor compounds was undetected in cultivated Chinese chives, which probably contribute to its weaker smell in addition. For the first time, we isolated full-length sequences encoding key enzymes of garlic odor biosynthetic pathway from wild *Allium* species. These sequences show closer phylogenetic relationship with those of cultivated Chinese chives than onion and garlic. A big size deletion was observed in the CDS region of *FMO1* in EWL hookeri, which result in serious deletion of the C-terminal region and subsequently possible function loss.5. Plant Materials and Methods.

4.1. Plant Materials

We collected all the garlic germplasm from a mountainous area in Ya'an, Sichuan Province, China at different altitudes. Land race *A. tuberosum* was collected at 980 m in a field, two materials of *A. hookeri* were collected at 1500 m in the wild, and *A. funckiaefolium* was collected at 2700 m in the wild. All the plants were transplanted in Sichuan Agricultural University's Chongzhou planting base at an altitude of 560 m in May 2017. *A. funckiaefolium* failed to survive in the subsequent domestication, and the leaf organs were collected in the wild. *A. tuberosum* and *A. hookeri* leaves were collected in July 2018 for the following assay.

4.2. Physiological Indices

Vitamin C content in the leaves was determined by the 2,6-Dichloroindophenol sodium salt hydrate method in accordance with the national standard GB 5009.86-2016. Dry matters content was determined in accordance with the national standard GB 5009.3-2016. Reducing sugars were measured by the classic Fehling's test in accordance with the national standard GB 5009.7-2016. Soluble protein content was determined by Coomassie Brilliant Blue in accordance with the standard NY/T 1205-2006 (with a modification of initiation mass).

4.3. GC-MS Assay

About 100 mg of leaf tissue for each sample was homogenized in liquid nitrogen by using a high-flux organization grinding apparatus. The powder was then suspended in 1.4 mL of precooled methanol in a tube and 60 μL of ribitol was added (0.2 mg/mL stock in methanol) as an internal quantitative standard. Metabolites were then extracted by ultrasound machine for 30 s. Then, 750 μL of precooled chloroform and 1.4 mL of deionized water were added (at 4 °C). They were then centrifuged for 10 min at 14,000 rpm (at 4 °C), and 1 mL of supernatant was transferred into a new centrifuge tube. Samples were blow-dried by vacuum concentration. Sixty µL of a 15 mg/mL methoxyamine pyridine solution was added and vortexed for 30 s and then reacted for 120 min at 37 °C. Sixty µL of a BSTFA reagent (containing 1% TMCS) was added into the mixture and reacted for 90 min at 37 °C. After a centrifuge at 12,000 rpm (at 4 °C) for 10 min, the supernatant was pipetted and stored for the following steps. For the quality control (QC) samples, 20 µL from each prepared sample extract was taken and mixed. QC samples were used to monitor analytical result deviations from these pool mixtures and compared to the errors caused by the analytical instrument itself. The rest of the samples were used for GC-MS detection.

Gas chromatography was performed on a HP-5MS capillary column (5% phenyl/95% methylpolysiloxane, 30 m × 250 µm i.d., 0.25 µm film thickness, Agilent J & W Scientific, Folsom, CA, USA) to separate the derivatives at a constant flow of 1 mL/min. An amount of 1 µL of the sample was injected in split mode by the autosampler and the split ratio was 20:1. The injection temperature was 280 °C; the interface was set to 150 °C, and the ion source was adjusted to 230 °C. The oven temperature was 60 °C for 2 min, ascending to 230 °C with an increase of 10 °C/min (300 °C for 5 min). The scanning model was 35–750 *m*/*z* (full scanning).

4.4. LC-MS Assay

About 100 mg of leaf tissue for each sample was used. The methods of homogenization, methanol extraction, and chloroform purification were the same for the GC-MS assay. The extract solution was dried by vacuum concentration and then dissolved with a 250 µL methanol aqueous solution (1:1, 4 °C). The sample solution was filtered by a 0.22 µm membrane and then was prepared for LC-MS detection. For the QC samples, 20 µL was taken from each prepared sample extract and mixed. The rest of the samples were used for LC-MS detection.

Chromatographic separation was accomplished in a Thermo Ultimate 3000 system (Thermo Fisher Scientific, Waltham, MA, USA) equipped with an ACQUITY UPLC® HSS T3 (length 150 mm, inner diameter 2.1 mm, particle size 1.8 µm, Waters Co., Milford, MA, USA) column, which was maintained at 40 °C. The temperature of the autosampler was 8 °C. Gradient elution of the analytes was carried out with 0.1% formic acid in water (A) and 0.1% formic acid in acetonitrile (B) or 5 mM ammonium formate in water (C) and acetonitrile (D) at a flow rate of 0.25 mL/min. An amount of 2 µL of each sample was injected after equilibration. An increasing linear gradient of solvent B (*v*/*v*) was used as follows: 0–1 min, 2% B/D; 1–9 min, 2–50% B/D; 9–14 min, 50–98% B/D; 14–15 min, 98% B/D; 15–15.5 min, 98–2% B/D; 15.5–17 min, 2%B/D. For MS, a Thermo Q Exactive Focus mass spectrometer (Thermo Fisher Scientific, Waltham, MA, USA) was set to an ESI mode with sMRM scanning. The spray voltage was 3.8 kV and −2.5 kV in positive and negative modes, respectively. The ion source temperature was set at 400 °C. The analyzer scanned over a mass range of *m*/*z* 81–1000 for a full scan at a mass resolution of 70,000. Data-dependent acquisition (DDA) MS/MS experiments were performed with an HCD scan. The normalized collision energy was 30 eV. Dynamic exclusion was implemented to remove some unnecessary information in the MS/MS spectra.

4.5. RNA Extraction and RNA-Seq

Since the unigenes were assembled without a reference genome, the gene expression level denoted as FPKM is not reliable. The transcript level of biosynthetic genes of the garlic odor compound pathway was validated by qRT-PCR. Total RNA was extracted using the Trizol method, as previously described [38]. A total amount of 1 µg of RNA per sample was used as input material for the RNA sample preparations. The purity, concentration, and integrity of the RNA samples were examined by NanoDrop, Qubit 2.0, Agilent 2100 (Agilent Technologies, de novo, Santa Clara, CA, USA). Qualified RNA was processed for library construction and Illumina sequencing was carried out by Biomarker (Beijing, China). Clean data were obtained after trimming adapter contaminations and removing nucleotides with low-quality scores. Clean reads were mapped to the garlic genome (Garlic V2 assembly, PRJNA606385) by using an alignment tool with high efficiency (HISAT2) [39]. Alternatively, unigenes were assembled without a reference genome by using Trinity [40]. Unigenes were annotated by BLAST against such databases as NR, Swiss-Prot, COG, KOG, and KEGG [41]. Amino acid sequences were searched against Pfam databases by using HMMER [42]. Reads from clean data were then mapped to unigenes, and only the mapped reads were used in the following analysis. Fragments per kilobase of transcript, per million mapped reads (FPKM) was used to measure the expression level of an unigene by using Bowtie and RSEM [43,44].

4.6. Gene Sequence Cloning and qRT-PCR

Total RNA was extracted from leaves and roots by using TaKaRaMiniBEST Plant RNA Extraction Kit (Takara, Japan). cDNA was synthesized by using the PrimeScript™ RT reagent Kit (Perfect Real Time) (Takara, Japan). qRT-PCR was done by using TB Green® *Premix Ex Taq*™ II (TliRNaseH Plus) (Takara, Japan) and CFX96 Real-Time PCR detection systems (Bio-Rad, Hercules, CA, USA). Primers for gene sequence cloning were designed according to unigenes identified from transcriptomes (Table S5); PCR products were purified by a Takara MiniBEST DNA Fragment Purification Kit (Takara, Japan) and sequenced by Sangon Biotech (Shanghai) Co. Primers for qRT-PCR were designed according to the final sequences corrected by PCR-method (Table S5). Sequences of candidate inner reference genes *EF-1α*, *18SrRNA*, and *Actin* were downloaded following a study of selection and validation of garlic reference genes for qRT-PCR [45], but new primers were designed based on our assembled transcript sequence. The stability of these candidate inner reference genes were evaluated by the BestKeeper tool, which is based on the standard deviation (SD) and the coefficient of variation (CV) calculated with the Ct values of all candidate reference genes [46]. Reference genes with the highest SD and CV values are considered as the least stable genes. Amplification efficiency (E, $E = 10[-1/slope] - 1$) of all the newly designed primers in qRT-PCR were calculated by utilizing standard curves with ten-fold serial dilutions (10, 10^2, 10^3, 10^4, and 10^5 dilutions) of each cDNA template. Those primers with an E-value ranging 90–110% were taken as high PCR-amplification efficiency. Following qRT-PCR data collection, Ct values were converted to relative quantities using the formula: $2^{-\Delta Ct}$, where ΔCt is the target gene Ct value minus the reference gene Ct value.

4.7. Statistical Analysis, Data Processing, and Figure Presentation

The experiments of metabolite detection were repeated 6 times for each sample, while other experiments including RNA-seq were repeated 3 times for each sample. The significant difference of physiological indices was analyzed using the IBM SPSS Statistics 22 (IBM Co., New York, NY, USA). Differential accumulated metabolites (DAMs) were identified by Student's *t*-test (p-value ≤ 0.05). The DAMs were subjected to pathway analysis using the MetaboAnalysis 4.0 database (http://www.metaboanalyst.ca/, accessed on 10 May 2019). Pathway impact factor was used to quantify the enrichment of a pathway.

Differential expressed genes (DEGs) were identified by DESeq2 [47], and the criteria for differentially expressed genes was set as Fold Change (FC) ≥ 4 and FDR < 0.01. Gene functional annotation and pathway enrichment analysis of DEGs were analyzed by using the Kyoto Encyclopedia of Genes and Genomes (KEGG) and gene ontology (GO). The GO terms and pathways with q value (adjusted p-value by BH method) < 0.05 were considered to be significantly enriched ones. The enrichment factor = (number of pathway DEGs/total number of all DEGs number)/(number of pathway genes/total number of all genes).

Supplementary Materials: The following supporting information can be downloaded at: https://www.mdpi.com/article/10.3390/ijms23137013/s1.

Author Contributions: Conceptualization, X.M., L.-K.H. and Y.-S.L.; methodology, S.-W.X., L.-F.H. and L.-K.H.; software, Q.-Q.Y. and L.-J.L.; validation, Y.L., X.-A.Z. and Q.-Y.L.; investigation, S.-W.X., S.A., X.-Y.X. and Y.-J.L.; formal analysis, H.-X.L. and Y.-S.L.; writing, S.-W.X. and Y.-S.L. All authors have read and agreed to the published version of the manuscript.

Funding: This research was funded by the Science & Technology Department of Sichuan Province (Grant No. 2018NZ0131 and 2019JDKP0031).

Institutional Review Board Statement: Not applicable.

Informed Consent Statement: Not applicable.

Data Availability Statement: Data are contained within the article or Supplementary Materials. Data of RNA-seq were deposited in NCBI: reference number (PRJNA832428). Gene sequences of garlic odor compound biosynthesis can be found in supplementary tables.

Acknowledgments: Haiping Wang from Institution of Vegetables and Flowers, Chinese Academy of Agricultural Sciences provided valuable advice on *Allium* taxonomy. Simian Agriculture Bureau provided necessary assistance in wild germplasm collection.

Conflicts of Interest: The authors declare no conflict of interest.

References

1. Oh, M.; Kim, S.-Y.; Park, S.; Kim, K.-N.; Kim, S. Phytochemicals in Chinese chive (*Allium tuberosum*) induce the skeletal muscle cell proliferation via PI3K/Akt/mTOR and smad pathways in C2C12 Cells. *Int. J. Mol. Sci.* **2021**, *22*, 2296. [CrossRef] [PubMed]
2. Sharifi-Rad, J.; Mnayer, D.; Tabanelli, G.; Stojanović-Radić, Z.Z.; Sharifi-Rad, M.; Yousaf, Z.; Vallone, L.; Setzer, W.N.; Iriti, M. Plants of the genus *Allium* as antibacterial agents: From tradition to pharmacy. *Cell. Mol. Biol.* **2016**, *62*, 57–68. [PubMed]
3. Fenwick, G.R.; Hanley, A.B.; Whitaker, J.R. The genus *Allium*—Part 3. *Crit. Rev. Food Sci. Nutr.* **1985**, *23*, 1–73. [CrossRef] [PubMed]
4. Fritsch, R.M.; Blattner, F.R.; Gurushidze, M. New classification of *Allium* L. subg. *Melanocrommyum* (Webb &Berthel) Rouy (*Alliaceae*) based on molecular and morphological characters. *Phyton* **2010**, *49*, 145–220.
5. Li, M.J.; Guo, X.L.; Li, J.; Zhou, S.D.; Liu, Q.; He, X.J. Cytotaxonomy of *Allium* (Amaryllidaceae) subgenera *Cyathophora* and *Amerallium* sect. *Bromatorrhiza*. *Phytotaxa* **2017**, *331*, 185–198. [CrossRef]
6. Xie, D.F.; Tan, J.B.; Yu, Y.; Gui, L.J.; Su, D.M.; Zhou, S.D.; He, X.J. Insights into phylogeny, age and evolution of *Allium* (Amaryllidaceae) based on the whole plastome sequences. *Ann. Bot.* **2020**, *125*, 1039–1055. [CrossRef]
7. Wu, Y.; Zhao, Q.; Wang, H. Research progress on the classification of *Allium* plants. *Acta Hortic. Sin.* **2021**, *48*, 1418–1428.
8. Ayam, V.S. Allium hookeri, Thw. Enum. A lesser known terrestrial perennial herb used as food and its ethnobotanical relevance in Manipur. *Afr. J. Food Agric. Nutr. Dev.* **2011**, *11*, 5389–5412. [CrossRef]
9. Ni, N.C.; He, X.J.; Xu, J.M. Random amplified polymorphic DNA analysis on intraspecific differentiation of *Allium hookeri*. *J. Sichuan Univ.* **2002**, *39*, 944–947. (In Chinese)
10. Zhang, G.; Wang, M.; Wang, X.; Huang, Z.; Tian, Q.; Wang, C.; Zhang, Z.; Tian, Y.; Tian, X. Breeding on the new Chinese chive varieties of broad leaf and the high quality of the leeks 'Youkuan No.1'. *North. Hortic.* **2015**, *15*, 153–156. (In Chinese)
11. Singh, K.D.; Chetia, D.; Biplab, D.E. New flavonoid compound from *Allium Hookeri* thwaites as a gastroprotective agent. *Int. J. Pharm. Pharm. Sci.* **2018**, *10*, 24–30. [CrossRef]
12. Jang, J.Y.; Lee, M.J.; You, B.R.; Jin, J.S.; Lee, S.H.; Yun, Y.R.; Kim, H.J. *Allium hookeri* root extract exerts anti-inflammatory effects by nuclear factor-kappaB down-regulation in lipopolysaccharide-induced RAW264.7 cells. *BMC Complement. Altern. Med.* **2017**, *17*, 126.
13. Lee, Y.S.; Lee, S.H.; Gadde, U.D.; Oh, S.T.; Lee, S.J.; Lillehoj, H.S. *Allium hookeri* supplementation improves intestinal immune response against necrotic enteritis in young broiler chickens. *Poult. Sci.* **2018**, *97*, 1899–1908. [CrossRef] [PubMed]
14. Rose, P.; Whiteman, M.; Moore, P.K.; Zhu, Y.Z. Bioactive S-alk (en) yl cysteine sulfoxide metabolites in the genus *Allium*: The chemistry of potential therapeutic agents. *Nat. Prod. Rep.* **2005**, *22*, 351–368. [CrossRef]
15. Yamazaki, Y.; Iwasaki, K.; Mikami, M.; Yagihashi, A. Distribution of eleven flavor precursors, S-Alk(en)yl-L-Cysteine derivatives, in seven *Allium* vegetables. *Food Sci. Technol. Res.* **2010**, *17*, 55–62. [CrossRef]
16. Van Damme, E.M.J.; Smeets, K.; Torrekens, S.; Vanleuven, F.; Peumans, W.J. Isolation and characterisation of alliinase cDNA clones from garlic (*Allium sativum* L.) and related species. *Eur. J. Biochem.* **1992**, *209*, 751–757. [CrossRef]
17. Jones, M.G.; Hughes, J.; Tregova, A.; Milne, J.; Tomsett, A.B.; Collin, H.A. Biosynthesis of the flavour precursors of onion and garlic. *J. Exp. Bot.* **2004**, *55*, 1903–1918. [CrossRef]
18. Yamaguchi, Y.; Kumagai, H. Characteristics, biosynthesis, decomposition, metabolism and functions of the garlic odour precursor, S-allyl-l-cysteine sulfoxide. *Exp. Ther. Med.* **2020**, *19*, 1528–1535. [CrossRef]
19. Yoshimoto, N.; Saito, K. S-Alk(en)ylcysteine sulfoxides in the genus *Allium*: Proposed biosynthesis, chemical conversion, and bioactivities. *J. Exp. Bot.* **2019**, *70*, 4123–4137. [CrossRef]

20. Jones, M.G.; Collin, H.A.; Tregova, A.; Trueman, L.; Brown, L.; Cosstick, R.; Hughes, J.; Milne, J.; Wilkinson, M.C.; Tomsett, A.B.; et al. The biochemical and physiological genesis of alliin in garlic. *Med. Aromat. Plant Sci. Biotechnol.* **2007**, *1*, 21–24.
21. Yoshimoto, N.; Onuma, M.; Mizuno, S.; Sugino, Y.; Nakabayashi, R.; Imai, S.; Tsuneyoshi, T.; Sumi, S.; Saito, K. Identification of a flavin-containing S-oxygenating monooxygenase involved in alliin biosynthesis in garlic. *Plant J.* **2015**, *83*, 941–951. [CrossRef] [PubMed]
22. Imai, S.; Tsuge, N.; Tomotake, M.; Nagatome, Y.; Sawada, H.; Nagata, T.; Kumagai, H. Plant biochemistry: An onion enzyme that makes the eyes water. *Nature* **2002**, *419*, 685. [CrossRef] [PubMed]
23. Ohri, D.; Fritsch, R.M.; Hanelt, P. Evolution of genome size in *Allium* (Alliaceae). *Plant Syst. Evol.* **1998**, *210*, 57–86. [CrossRef]
24. Ricroch, A.; Yockteng, R.; Brown, S.C.; Nadot, S. Evolution of genome size across some cultivated *Allium* species. *Genome* **2005**, *48*, 511–520. [CrossRef] [PubMed]
25. Finkers, R.; Kaauwen, M.; Ament, K.; Burger-Meijer, K.; Egging, R.; Huits, H.; Kodde, L.; Kroon, L.; Shigyo, M.; Sato, S.; et al. Insights form the first genome assembly of onion (*Allium cepa*). *G3* **2021**, *11*, jkab243. [CrossRef]
26. Zhou, S.M.; Chen, L.M.; Liu, S.Q.; Wang, X.F.; Sun, X.D. De novo assembly and annotation of the Chinese chive (*Allium tuberosum* Rottler ex Spr.) transcriptome using the Illumina platform. *PLoS ONE* **2015**, *10*, e0133312. [CrossRef]
27. Liu, N.; Tong, J.; Hu, M.; Ji, Y.; Wang, B.; Liang, H.; Liu, M.; Wu, Z. Transcriptome landscapes of multiple tissues highlight the genes involved in the flavor metabolic pathway in Chinese chive (*Allium tuberosum*). *Genomics* **2021**, *113*, 2145–2157. [CrossRef]
28. Kusano, M.; Kobayashi, M.; Iizuka, Y.; Fukushima, A.; Saito, K. Unbiased profiling of volatile organic compounds in the headspace of Allium plants using an in-tube extraction device. *BMC Res. Notes* **2016**, *9*, 133. [CrossRef]
29. Böttcher, C.; Krähmer, A.; Stürtz, M.; Widder, S.; Schulz, H. Comprehensive metabolite profiling of onion bulbs (*Allium cepa*) using liquid chromatography coupled with electrospray ionization quadrupole time-of-flight mass spectrometry. *Metabolomics* **2017**, *13*, 35. [CrossRef]
30. Basundari, F.R.A.; Sulistyaningsih, E.; Murti, R.H.; Nuringtyas, T.R. Metabolite profile of two Allium cepa L. aggregatum group cultivars by Nuclear Magnetic Resonance. *Biodiversitas J. Biol. Divers.* **2021**, *22*, 3127–3135. [CrossRef]
31. Soorni, A.; Akrami, A.M.; Abolghasemi, R.; Vahedi, M. Transcriptome and phytochemical analyses provide insights into the organic sulfur pathway in Allium hirtifolium. *Sci. Rep.* **2021**, *11*, 768. [CrossRef] [PubMed]
32. Abdelrahman, M.; Hirata, S.; Mukae, T.; Yamada, T.; Sawada, Y.; El-Syaed, M.; Yamada, Y.; Sato, M.; Hirai, M.Y.; Shigyo, M. Comprehensive metabolite profiling in genetic resources of garlic (*Allium sativum* L.) collected from different geographical regions. *Molecules* **2021**, *26*, 1415. [CrossRef] [PubMed]
33. Collins, S.L.; Koo, I.; Peters, J.M.; Smith, P.B.; Patterson, A.D. Current challenges and recent developments in mass spectrometry-based metabolomics. *Annu. Rev. Anal. Chem.* **2021**, *14*, 467–487. [CrossRef] [PubMed]
34. Soininen, T.H.; Jukarainen, N.; Soininen, P.; Auriola, S.O.K.; Julkunen-Tiitto, R.; Oleszek, W.; Stochmal, A.; Karjalainen, R.O.; Vepsäläinen, J.J. Metabolite profiling of leek (*Allium porrum* L) cultivars by 1H NMR and HPLC-MS. *Phytochem. Anal.* **2014**, *25*, 220–228. [CrossRef]
35. Trovato, M.; Funck, D.; Forlani, G.; Okumoto, S.; Amir, R. Editorial: Amino acids in plants: Regulation and functions in development and stress defense. *Front. Plant Sci.* **2021**, *12*, 772810. [CrossRef]
36. Jivishov, E. Investigations on Wild *Allium* Species. Part I: Cysteine Sulfoxides of Flowers. Part 2: Anticancer Activity of Bulb Extracts. Ph.D. Thesis, Philipps University of Marburg, Marburg, Germany, December 2016.
37. Ueda, Y.; Tsubuku, T.; Miyajima, R. Composition of sulfur-containing components in onion and their flavor characters. *Biosci. Biotechnol. Biochem.* **1994**, *58*, 108–110. [CrossRef]
38. Feinbaum, R.L.; Ausubel, F.M. Transcriptional regulation of the *Arabidopsis thaliana* chalcone synthase gene. *Mol. Cell Biol.* **1988**, *8*, 1985–1992.
39. Kim, D.; Langmead, B.; Salzberg, S.L. HISAT: A fast spliced aligner with low memory requirements. *Nat. Methods* **2015**, *12*, 357–360. [CrossRef]
40. Grabherr, M.G.; Haas, B.J.; Yassour, M.; Levin, J.Z.; Thompson, D.A.; Amit, I.; Adiconis, X.; Fan, L.; Raychowdhury, R.; Zeng, Q.; et al. Full-length transcriptome assembly from RNA-Seq data without a reference genome. *Nat. Biotechnol.* **2011**, *29*, 644–652. [CrossRef]
41. Buchfink, B.; Xie, C.; Huson, D.H. Fast and sensitive protein alignment using DIAMOND. *Nat. Methods* **2015**, *12*, 59–60. [CrossRef]
42. Eddy, S.R. Profile hidden Markov models. *Bioinform. Rev.* **1998**, *14*, 755–763. [CrossRef] [PubMed]
43. Langmead, B.; Trapnell, C.; Pop, M.; Salzberg, S.L. Ultrafast and memory-efficient alignment of short DNA sequences to the human genome. *Genome Biol. Ital.* **2009**, *10*, R25. [CrossRef] [PubMed]
44. Li, B.; Colin, N.D. RSEM: Accurate transcript quantification from RNA Seq data with or without a reference genome. *BMC Bioinform. Ital.* **2011**, *12*, 323. [CrossRef]
45. Liu, M.; Wu, Z.; Jiang, F. Selection and validation of garlic reference genes for quantitative real-time PCR normalization. *Plant Cell Tissue Organ Cult.* **2015**, *122*, 435–444. [CrossRef]
46. Pfaffl, M.W.; Tichopad, A.; Prgomet, C.; Neuvians, T.P. Determination of stable housekeeping genes, differentially regulated target genes and sample integrity: BestKeeper—Excel-based tool using pair-wise correlations. *Biotechnol. Lett.* **2004**, *26*, 509–515. [CrossRef]
47. Anders, S.; Huber, W. Differential expression analysis for sequence count data. *Genome Biol.* **2010**, *11*, R106. [CrossRef]

International Journal of
Molecular Sciences

MDPI

Article

Genome-Wide Identification of Binding Sites for SmTCP7a Transcription Factors of Eggplant during Bacterial Wilt Resistance by ChIP-Seq

Xi'ou Xiao [1,2,3,*], Wenqiu Lin [1,2,3], Enyou Feng [4], Caiyu Wu [1,2,3] and Xiongchang Ou [1,2,3]

1 South Subtropical Crop Research Institute Chinese, Academy of Tropical Agricultural Sciences, Zhanjiang 524091, China; linwenqiu1989@163.com (W.L.); wucaiyu1995@163.com (C.W.); oxc9527@163.com (X.O.)
2 Key Laboratory for Post-harvest Physiology and Technology of Tropical Horticultural Products of Hainan, Zhanjiang 524091, China
3 Zhanjiang City Key Laboratory for Tropical Crops Genetic Improvement, Zhanjiang 524091, China
4 Zhanjiang Academy of Agricultural Sciences, Zhanjiang 524091, China; enyou2008@163.com
* Correspondence: xiao-forlearning@163.com

Abstract: Teosinte branched 1/cycloidea/proliferating cell factor (TCP) transcription factors play a key role in the regulation of plant biotic and abiotic stresses. In this study, our results show that SmTCP7a positively regulated bacterial wilt that was caused by *Ralstonia solanacearum*. ChIP-seq was conducted to analyze the transcriptional regulation mechanism of SmTCP7a before (R0 h) and 48 h after infection (R48 h). SmTCP7a regulated a total of 92 and 91 peak-associated genes in R0 h and R48 h, respectively. A KEGG (Kyoto encyclopedia of genes and genomes) pathway analysis showed that phenylpropanoid biosynthesis, MAPK (mitogen activated protein kinas) signaling pathway, plant hormone signal transduction and plant-pathogen interactions were involved. The difference in peaks between R0 h and R48 h showed that there were three peak-associated genes that were modulated by infection. A better understanding of the potential target genes of SmTCP7a in response to *R. solanacearum* will provide a comprehensive understanding of the SmTCP7a regulatory mechanism during the eggplant defense response to bacterial wilt.

Keywords: gene regulation; MAP kinase; *R. solanacearum*; resistance; TCP transcription factors

1. Introduction

TCP transcription factor, which contains a basic helix–loop–helix DNA binding domain, the TCP domain, is a plant-specific transcription factor. Based on TCP domain differences, TCP families can be further divided into class I and class II. The conserved DNA binding site of class I and class II TCPs is GGNCCCAC and GTGGNCCC, respectively [1]. The TCPs play a key role in the regulation of plant growth and development, such as pistil abortion [2], leaf development [3], root growth [4], plant architecture [5,6] and so on. In *Arabidopsis*, AtTCPs can interact with 236 proteins involved in the biosynthesis of phytohormones and their signal transduction [7]. Several results have shown that the TCPs can directly regulate the biosynthesis and signaling pathways of SA(Salicylic acid), JA(Jasmonic acid), CK(Cytokinin), ABA(Abscisic acid) and BR(Brassinoids) [1]. In addition, TCPs can directly bind to promoters and regulate target gene expression, thereby positively or negatively regulating the biosynthesis of phytohormones [1,8]. In Arabidopsis, AtTCP4, AtTCP9 and AtTCP20 can bind to the *LOX2* promoter to regulate JA biosynthesis [9]. AtTCP8 can bind to the *ICS1* promoter to regulate SA biosynthesis [9,10] as well as the key BR transcriptional regulators, *BZR1* and *BZR2/BES1* [11]. However, the molecular mechanism underlying their transcription control remains unclear.

The TCPs not only regulate plant development, they can regulate plant biotic and abiotic stress responses [8,9,12–16]. In addition, TCPs can be recognized by pathogen effectors and can initiate plant defense responses [17]. Previous studies have shown that TCP13, TCP14, TCP15, TCP19 and TCP21 can interact with several pathogen effectors [18,19]. The *Phytophthora capsica* effector CRN12_997 can bindSlTCP14-2 and repress its transcriptional activity to reduce plant resistance [19]. TCPs can interact with other proteins and form a protein-complex, such as AtTCP8, AtTCP14 and AtTCP15 interacting with SRFR1 to regulate the defense response [14]. In addition, AtTCP8, AtTCP14 and AtTCP15 can physically interact with NPR1 to enhance the binding of the *PR5* promoter, which leads to systemic acquired resistance [13]. In cucumber, CsTCP14 can repress the expression of *CsNBS-LRR*, resulting in susceptibility to foliage diseases [20].

Eggplant (*Solanum melongena* L.) is an important vegetable in subtropical and tropical regions. It contains various phenolic compounds that can improve human health. However, bacterial wilt caused by *R. solanacearum* has caused severe yield reductions worldwide. Thus, several studies have been conducted to comprehensively understand the molecular mechanism of eggplant defense against *R. solanacearum*. Thousands of DEGs (differentially expressed genes) were identified by RNA-seq after eggplant inoculation with *R. solanacearum*. KEGG analysis of these DESs showed that the signaling pathways of phytohormones such as SA and JA play a vital role in the defense response [21]. Several transcriptional factors, such as MYB, NAC and WRKY, also regulate the defense response. The MYB transcriptional factor SmMYB44 directly regulates *SmSPDS* expression to increase spermidine content, which increases resistance to bacterial wilt [22]. By contrast, the NAC transcriptional factor SmNAC negatively regulates the expression of *ICS1* (isochorismate synthase 1) and reduces resistance to bacterial wilt [23]. Genome-wide analysis showed that *SiWRKY53* is a source of resistance to bacterial wilt [24]. Our previous study showed that reactive oxygen species are also involved in bacterial wilt resistance in eggplant [25]. Moreover, SmEDS1 (enhanced disease susceptibility 1), which is involved in the SA biosynthetic pathway, positively regulates bacterial wilt resistance [26]. A yeast two-hybrid study showed that SmEDS1 interacts with SmTCP7a [27]. In the present study, our results show that SmTCP7a positively regulates bacterial wilt resistance. Thus, ChIP-seq of SmTCP7a was analyzed to detect the specific genes regulated by SmTCP7a during the eggplant response to *R. solanacearum*. Our results identify several potential candidate genes, and provide a comprehensive understanding of genes regulated by SmTCP7a during eggplant defense against *R. solanacearum*.

2. Results

2.1. Molecular Characterization of SmTCP7a

The CDS of SmTCP7a contains 783 bp, and codes 260 amino acids. The molecular weight of SmTCP7a was 27.09 kDa and the theoretical pI was 8.79. Phylogenetic analysis indicated that SmTCP7a belongs to class I TCPs (Figure 1). In addition, a subcellular localization assay showed that SmTCP7a is localized in the nucleus (Figure 2).

2.2. Analysis of SmTCP7a Transactivation Functions

Compared with 62SK-DB expression, SmTCP7a-DB stimulated the GAL4-response 2.49 times (Figure 3A), demonstrating that SmTCP7a is an activator of transcription. In yeast one-hybrid experiments, SmTCP7a-pGBKT7 and pGBKT7-p53 could grow in SD-T and SD-THA medium, whereas pGBKT7-laminC could grow in SD-T medium but not SD-THA medium (Figure 3B). This result also demonstrates that SmTCP7a is an activator of transcription.

2.3. SmTCP7a Positively Regulates Eggplant Bacterial Wilt Resistance

After inoculation of eggplants with *R. solanacearum*, the SmTCP7a expression level significantly increased in the R genotype, but not in the S genotype (Figure 4A). The result of further VIGS (virus induced gene silencing) showed that SmTCP7a positively regulated bacterial wilt resistance in eggplant (Figure 4B).

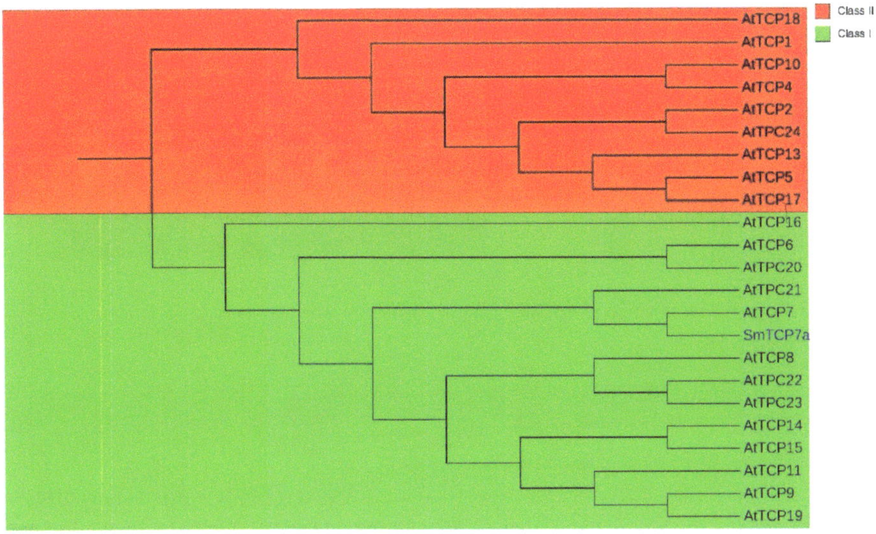

Figure 1. The phylogenetic tree for SmTCP7a. The phylogenetic tree was analyzed by NJ tree with 1000 bootstrap replicates.

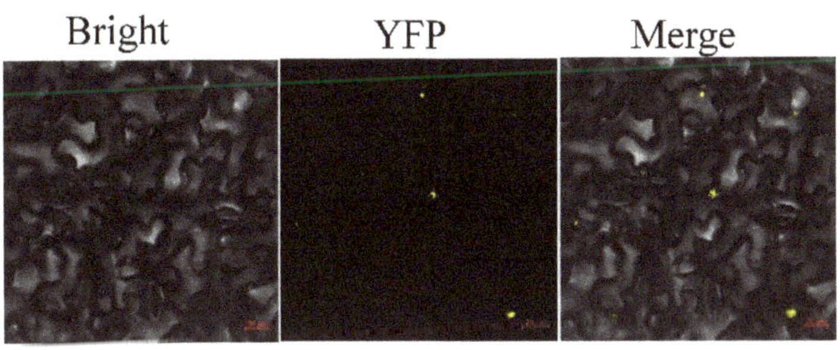

Figure 2. The subcellular location of SmTCP7a. SmTCP7a-YFP was detected using *Nicotiana benthamiana* leaves. Bar = 20 μm.

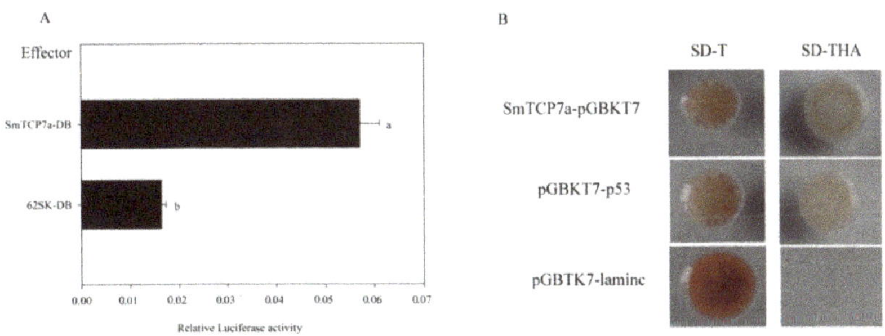

Figure 3. Analysis of SmTCP7a transactivation functions. (**A**) Transactivation of GAL4 response transcription by SmTCP7a fused to the DNA-binding domain of GAL4 in *N. benthamiana* leaves. Different letters indicate statistically significant difference (independent-sample *t*-test (confidence interval = 95%), $n = 3$, $\alpha = 0.05$). The data were analyzed by the SPSS17 software. (**B**) Transactivation by SmTCP7a in a yeast one-hybrid.

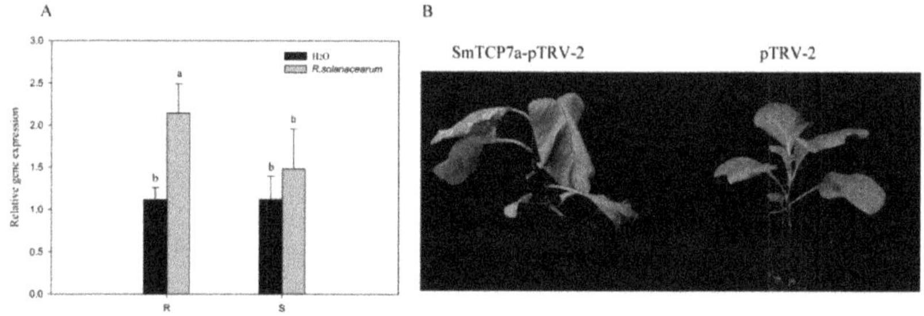

Figure 4. SmTCP7a positively regulates eggplant bacterial wilt resistance. (**A**) Analysis of the expression level of the SmTCP7a after inoculation with *R. solanacearum* at 48 hpi. Different letters indicate statistically significant difference (ANOVA, Duncan post hoc test, $n = 3$, $\alpha = 0.05$). The data were analyzed by the SPSS17 software. (**B**) Eggplant wilted after VIGS of SmTCP7a. The photos were taken 7 days after inoculation with *R. solanacearum*.

2.4. ChIP-Seq Libraries and Detection of Peaks for SmTCP7a

Western-blot analysis with a polyclonal antibody showed the presence of the SmTCP7a target band (Figure S2). Eight ChIP-seq libraries were created for eggplants before (R0 h) and 48 h after infection (R48 h) with *R. solanacearum*. We obtained 82.0 million clean reads for the R0 h antibody-treated library compared to 82.6 million clean reads for the R0 h control sample (pre-immune sera). A total of 94.3 million clean reads was obtained for the R48 h antibody-treated library and 87.4 million clean reads for the R48 h control sample (Table 1).

Table 1. Summary of ChIP-seq reads from eight different libraries.

Treatment	Condition	Clean Reads	Genome Mapped Reads
R0 h-1	Control	37245858	34352758
	Antibody treated	36252688	27346750
R0 h-2	Control	45348560	40775010
	Antibody treated	45783576	27961628
R48 h-1	Control	39735838	30495054
	Antibody treated	44687510	37500882
R48 h-2	Control	47621298	37103632
	Antibody treated	49651010	43919431

After alignment to the eggplant reference genome, the peaks were called by MACS software with a *p*-value < 0.5 [28]. There was a total of 4378 peaks in the R0 h sample and 6555 peaks in the R48 h sample (Tables S1 and S2). The number of peaks in R48 h was higher than R0 h, and the distribution of the peaks in the genome is shown in Figure 5. In R0 h, the E7 and E11 chromosomes contained the most peaks (354) and the E2 chromosome contained the least (180 peaks). In R48 h, the E7 chromosome contained the most peaks (568) and the E2 chromosome contained the least (297 peaks).

2.5. Annotation of Detected Peaks and Discovery of Motifs in Promoter Regions

The genomic location of peaks was analyzed by using the ChIPseeker R package [29]. The result showed that about 94% of peaks were located in the intergenic region (Figure 6). Figure 7 shows that the peaks were located in close proximity to transcription start sites (TSS) in both R0 h and R48 h. In R0 h, 95 peaks were located in promoter regions and 92 genes were identified (Table S3). In R48 h, 93 peaks were located in promoter regions and 91 genes were identified (Table S4). Most of peak-associated genes were annotated as hypothetical proteins or uncharacterized proteins. Genes that may be related to plant disease resistance are listed in Table 2. Interestingly, histone H2A (Smechr0702191) was annotated in both R0 h and R48 h, whereas TCP9-like transcription factor was only detected in R48 h.

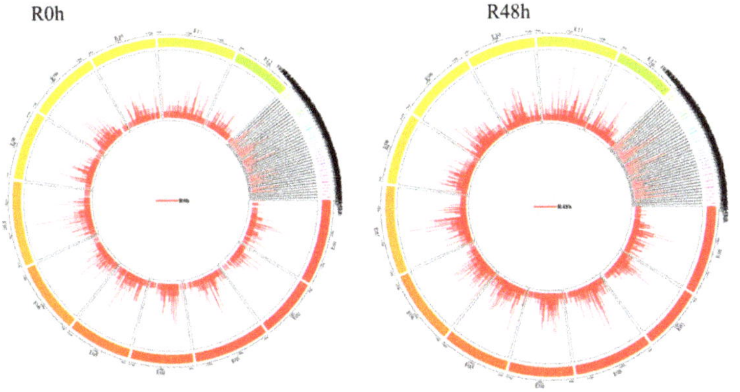

Figure 5. An overview of ChIP-Seq peak distribution in eggplant chromosomes for R0 h and R48 h.

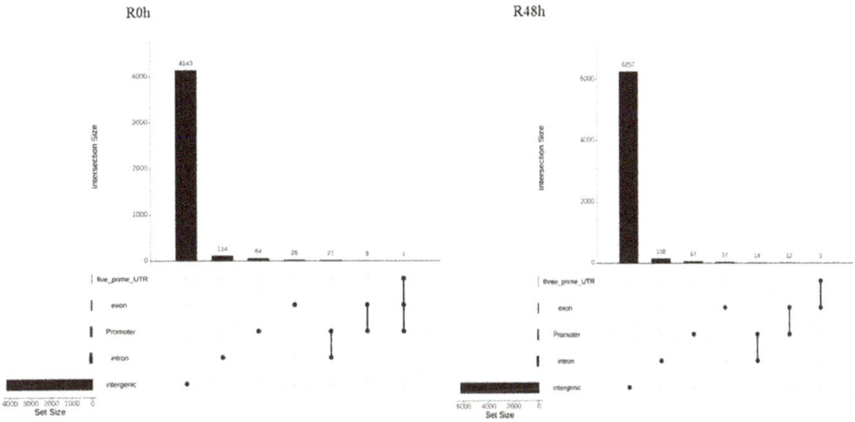

Figure 6. An overview of the genomic distribution of SmTCP7a binding sites for R0 h and R48 h.

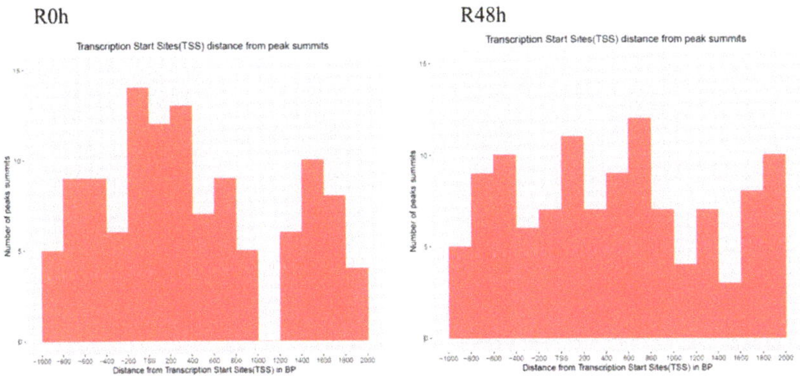

Figure 7. The distance between transcription start sites and MACS peak summits for the gene-associated peaks in R0 h and R48 h.

Table 2. ChIP-seq data reveals peak-associated genes regulated by SmTCP7a.

Gene	NR Annotation	Sample
Smechr0702191	Histone H2A	R0 h
Smechr0601128	protein argonaute 4	
Smechr0501578	WRKY transcription factor protein 1	
Smechr0500083	ubiquitin-conjugating enzyme E2	
Smechr0400288	glutathione S-transferase parC	R48 h
Smechr0702191	Histone H2A	
Smechr0800428	subtilisin-like protease SBT3.3	
Smechr0802294	transcription factor TCP9-like	
Smechr1100895	putative late blight resistance protein homolog R1B-23	

Figure 8 shows that 39 peak-associated genes were common to R0 h and R48 h. In addition, 53 genes and 52 genes were unique to R0 h and R48 h, respectively. The KEGG enrichment results showed that pathways for phenylpropanoid biosynthesis, MAPK signaling, plant hormone signal transduction and plant-pathogen interactions were active in R48 h (Figure 9). However, in R0 h only flavonoid biosynthesis may have been related to the defense response.

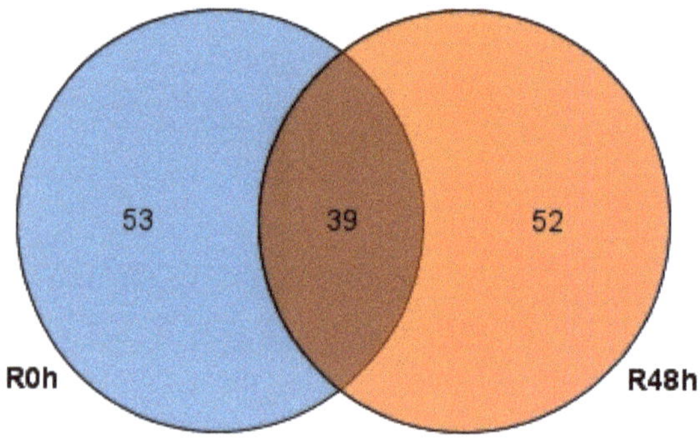

Figure 8. Venn diagram showing the overlap of peak-associated genes for R0 h and R48 h.

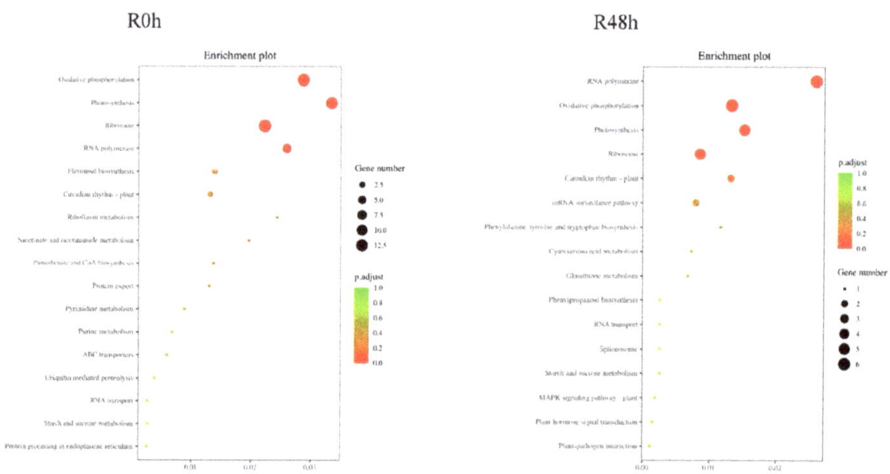

Figure 9. KEGG enrichment pathways for the peak-associated genes in R0 h and R48 h.

The motifs in the promoter region were analyzed using Homer software by de novo methods. There was a total of 17 and 22 predicted motifs in R0 h in R48 h, respectively, and only one motif (Bach2 (bZIP)/OCILy7-Bach2-ChIP-Seq (GSE44420)/Homer (0.746)) was identified in both R0 h and R48 h. All motifs were identified in JASPAR CORE plants or yeast, where two R48 h motifs were identified as TCP binding motifs (Tables S5 and S6).

2.6. Differences in Peaks between R0 h and R48 h

The difference in the peaks between R0 h and R48 h was detected by the Diffbind R package with a p-value < 0.05 and fold > 0 [30]. A total of 89 different peaks were identified, 48 of which were up-regulated and 41 were down-regulated (Table S7). However, 86 peaks were located in the intergenic region and only 3 peaks were located in the promoter region (Figure S2). One of the gene modules (Smechr0702191) was annotated as histone H2A and the others were annotated as uncharacterized proteins (Table 3).

Table 3. The difference in peak-associated genes between R0 h and R48 h.

Gene ID	Annotation	Regulation
Smechr0300595	uncharacterized protein LOC109121249	Up
Smechr0103296	uncharacterized protein LOC107855847 isoform X1	Up
Smechr0702191	Histone H2A	down

3. Discussion

In the present study, our results showed that SmTCP7a, a transcriptional factor, positively regulated bacterial wilt resistance. Furthermore, ChIP-seq results identified several potential genes that were regulated by SmTCP7a during the eggplant response to bacterial wilt.

TCP factors regulate plant resistance through various ways, such as targeting pathogenic effectors, interconnecting hormonal networks, utilization of the circadian clock and transcriptional control [9]. To better understand the defense response of plants, it is important to identify which TCP family members regulate downstream target genes during pathogen infection. Several TCP target genes are involved in plant development, plant hormone biosynthesis, cold tolerance, disease and insect defense [31].To the best of our best knowledge, this study is the first to report genome-wide identification of binding sites for TCP transcription factors during plant disease resistance by ChIP-seq. Our results showed that SmTCP7a positively regulated bacterial wilt resistance in eggplant. In our previous report, SmEDS1 also positively regulated bacterial wilt resistance in eggplant, and SmEDS1 can directly interact with SmTCP7a [26,27]. To discover SmTCP7a target genes in relation to bacterial wilt, highly enriched peaks were detected by ChIP-seq. A total of 4378 and 6555 peaks were identified in R0 h and R48 h samples, respectively, but only 2.62% and 1.59% of these peaks were located in promoter regions with 92 and 91 gene modules, respectively. The percentage of peaks in the promoter region was lower than other results [2,15,30] and this difference may be due to non-specific binding by the polyclonal antibody against SmTCP7a.

DNA binding motifs within the peaks in promoter regions were detected by Homer. The motifs between R0 h and R48 h were different. This result suggests that after inoculation with *R. solanacearum*, SmTCP7a regulated different genes to produce wilt resistance in eggplant. Based on the TCP domain difference, TCP families can be divided into class I and class II. Although both TCP classes can recognize some common cis-elements [32], the conserved DNA binding sites of class I and class II are quite distinct [1]. Several results have shown that class I TCPs, such as TCP8 [8,11], TC9 [10], TCP14 [20], TCP15 [26] and TCP23 [27], play a key role in plant disease resistance. Our polygenetic tree results showed that SmTCP7a is a class I TCP. Motif analysis also showed that the best match motif in R0 h was G(T/C)(G/T)CGGGCCCCC. In R48 h, two motifs (TCP23 and PCF5) were identified as TCP binding motifs in JASPAR CORE plants. TCP23 also belongs to class I [32]. This result indicates that although the polyclonal antibody against SmTCP7a exhibited non-specific binding, the detected peaks are credible. However, the interaction between SmTCP7a and the promoters of these genes should be verified by EMSA or yeast one-hybrid experiments.

Several of the peak-related genes have been reported to be involved in the plant disease response, namely, anthranilate synthase alpha subunit 1 (Smechr0600250), histone H2A (Smechr0702191), TCP9-like (Smechr0802294) transcription factor and R1B-23 (Smechr1100895). Based on previous reports, the anthranilate synthase gene can be induced by wound and pathogenic bacterial infection, and the production of tryptophan is part of the Arabidopsis defense response [33]. The histone H2A variant H2A.Z has been shown to interact with SWR1 to control immunity in Arabidopsis [34,35]. In future studies, the expression level and function of histone H2A during the eggplant response to bacterial wilt will be studied. The KEGG analysis for R48 h also showed that several pathways, such as glutathione biosynthesis, phenylpropanoid biosynthesis, MAPK signaling pathway, plant hormone signal transduction and plant-pathogen interactions were involved in plant biotic resistance. This result indicates that after inoculation with *R. solanacearum*, SmTCP7a regulates defense-related genes to respond to and resist infection.

4. Materials and Methods

4.1. Plant Materials and Growth Conditions

The resistant eggplant inbred line "E35-1" [25,27] was planted in a small pot (4.5 inches). The growing material comprised sterile vermiculite and clay mixed in a 3:1 volume/volume ratio. The seedlings were grown in a temperature of 28 °C/25 °C (day/night) with a 16 h light/8 h dark photoperiod. After 4-week culture, when the seedlings were at the four-leaf stage, the culture was incubated with the *R.solanacearum* P2 strain. The root of the eggplant was cut using a scalpel and then inoculated with 50 mL of *R. solanacearum* (1.0 × 108 cfu). Then the eggplant was grown in a temperature of 32 °C/30 °C (day/night). At 0 and 48 h post-inoculation (hpi), the roots of five eggplants were collected and washed with water, then immediately frozen in liquid nitrogen and stored at −80 °C. All treatments were repeated two times.

4.2. Bioinformation Analysis of SmTCP7a

SmTCP7a (Smechr0103973) MW, theoretical isoelectric point, protein hydrophobicity and the aliphatic amino acid index were determined using the PortParam online tool available from ExPASy (https://expasy.org/proteomics (20 June 2022). Unrooted neighbor-joining phylogenetic trees were constructed using MEGA software with 1000 bootstrap replicates.

4.3. Expression Analysis and VIGS

The expression level of SmTCP7a was analyzed to determine the function of SmTCP7a during the eggplant response to *R. solanacearum*. After inoculation with *R. solanacearum*, SmTCP7a expression level was detected at the root of R and S eggplant. After the pTRV2 vector was digested by *Sma* I, 150 bp SmTCP7a was cloned into the pTRV2 vector by in-fusion methods. The primers are listed in Table S8.

The pTRV2-SmTCP7a, pTRV2, and pTRV1 vectors were then transferred into *Agrobacterium tumefaciens* strain GV3101. After the pTRV1 equivalent was mixed with pTRV2-SmTCP7a, the mixture was infiltrated into four leaves of seedlings using a 1 mL needleless syringe. After injection, the samples were treated in the dark for one day and then inoculated with *R. solanacearum*. Each treatment had at least 10 biological replicates. The plant was grown in a temperature of 32 °C/30 °C (day/night).

4.4. Subcellular Localization and Transactivation Analysis of SmTCP7a Protein

The SmTCP7a CDS without a termination codon was cloned into the pCAMBIA1301-35S-EYFP vector by in-fusion. The recombinant vector was transformed into the *A. tumefaciens* strain GV3101. Transient expression assays were conducted in *N. benthamiana* leaves. Transactivation by SmTCP7a was examined in a strain of *Saccharomyces cerevisiae*. The effector vector was pGreenII 62SK-GAL4BD and the reporter vector was pGreeII-0800-5*GAL4-TATA-Luc (Figure S3). The effector/reporter rotation was 1:9 and the transient expression assays were conducted in *N. benthamiana* leaves. Luciferase assays were performed with the Dual-Luciferase® Reporter Assay System (Promega, E1960) using a TECAN spark microplate reader.

4.5. Production of a Polyclonal Antibody against SmTCP7a and Western Blot Analysis

The SmTCP7a CDS without a termination codon was cloned into the pET-28a(+) vector. After the SmTCP7a protein was purified, it was injected into a healthy New Zealand white rabbit. The polyclonal antibody against SmTCP7a was purified in accordance with the methods of Iqbal [2]. The purity of the antibody was then evaluated using SDS-PAGE and stored at −20 °C. Western blot analysis was carried out in accordance with the method of Li to examine the quality and specificity of the antibody for the ChIP-Seq assay [36].

4.6. ChIP-Seq Library Preparation and Data Analysis

Approximately 4g of eggplant root was cross sectioned using scissors and then cross linked with 1% formaldehyde in a vacuum for 15 min. After the samples were ground to powder in liquid nitrogen, the chromatin complexes were isolated following [2]. Later, the chromatin was sonicated to shear the DNA into 100–500 bp fragments using a Branson digital probe sonicator. Subsequently, the sonicated DNA was incubated with a polyclonal antibody developed against the SmTCP7a. The control treatment, which was treated with pre-immune sera, was used for each experiment. DNA-antibody complexes were then precipitated, and the DNA was recovered by dissociating the complexes. ChIP-seq library construction and high-throughput sequencing were carried out by using the Illumina HiSeq2000.

After the reads were filtered by fastqc, the clean reads were aligned to the eggplant reference genome HQ1315 [37] using bwa-mem2. The experiment was conducted with two conditions: the control library and the antibody-treated library. MACS with standard parameters was used for peak calling to locate the enriched regions of binding sites. All the binding peaks in the genomic locations were sorted into intergenic, intron, promoter and exon regions using the ChIPseeker R package [29]. A motif search was performed using Homer. The different peaks in R0 h and R48 h were analyzed by using the DiffBind R package with default parameters.

5. Conclusions

In this study, our results showed that the SmTCP7a transcriptional factor was inducible and up-regulated after *R. solanacearum* treatment. Furthermore, the VIGS result showed that SmTCP7a positively regulated bacterial wilt resistance. The potential target genes were analyzed by ChIP-seq. The identification of DNA motifs and regulatory genes provides a new approach for understanding the molecular mechanism of SmTCP7a during the eggplant response to *R. solanacearum*.

Supplementary Materials: The following supporting information can be downloaded at: https://www.mdpi.com/article/10.3390/ijms23126844/s1.

Author Contributions: X.X. designed the experiment and composed the manuscript. W.L. analyzed SmsTCP7a function and revised the manuscript. E.F. investigated and analyzed the polyclonal antibody. C.W. and X.O. analyzed the ChIP-seq data and revised the manuscript. All authors have read and agreed to the published version of the manuscript.

Funding: This research was supported by National Natural Science Foundation of China (No. 31872117), Central Public-interest Scientific Institution Basal Research Fund (No. 1630062022003), Central Public-interest Scientific Institution Basal Research Program for Scientific Research Innovation Team (No. 1630062017014), and National Natural Science Foundation of Hainan Province (No. 321RC632).

Conflicts of Interest: The authors declare that the research was conducted in the absence of any commercial or financial relationships that could be construed as a potential conflict of interest.

References

1. Nicolas, M.; Cubas, P. TCP factors: New kids on the signaling block. *Curr. Opin. Plant Biol.* **2016**, *33*, 33–41. [CrossRef] [PubMed]
2. Iqbal, S.; Pan, Z.; Wu, X.; Shi, T.; Ni, X.; Bai, Y.; Gao, J.; Khalil-Ur-Rehman, M.; Gao, Z. Genome-wide analysis of PmTCP4 transcription factor binding sites by ChIP-Seq during pistil abortion in Japanese apricot. *Plant Genome* **2020**, *13*, e20052. [CrossRef] [PubMed]
3. Koyama, T.; Sato, F.; Ohme-Takagi, M. Roles of miR319 and TCP Transcription Factors in Leaf Development. *Plant Physiol.* **2017**, *175*, 874–885. [CrossRef] [PubMed]
4. Hao, J.; Lou, P.; Han, Y.; Chen, Z.; Chen, J.; Ni, J.; Yang, Y.; Jiang, Z.; Xu, M. GrTCP11, a Cotton TCP Transcription Factor, Inhibits Root Hair Elongation by Down-Regulating Jasmonic Acid Pathway in Arabidopsis thaliana. *Front. Plant Sci.* **2021**, *12*, 769675. [CrossRef]
5. Zhang, W.; Tan, L.; Sun, H.; Zhao, X.; Liu, F.; Cai, H.; Fu, Y.; Sun, X.; Gu, P.; Zhu, Z.; et al. Natural Variations at TIG1 Encoding a TCP Transcription Factor Contribute to Plant Architecture Domestication in Rice. *Mol. Plant* **2019**, *12*, 1075–1089. [CrossRef]
6. Manassero, N.G.; Viola, I.L.; Welchen, E.; Gonzalez, D.H. TCP transcription factors: Architectures of plant form. *Biomol. Concepts* **2013**, *4*, 111–127. [CrossRef]
7. Arabidopsis Interactome Mapping Consortium. Evidence for network evolution in an Arabidopsis interactome map. *Science* **2011**, *333*, 601–607. [CrossRef]
8. Spears, B.J.; Howton, T.C.; Gao, F.; Garner, C.M.; Mukhtar, M.S.; Gassmann, W. Direct Regulation of the EFR-Dependent Immune Response by Arabidopsis TCP Transcription Factors. *Mol. Plant Microbe. Interact.* **2019**, *32*, 540–549. [CrossRef]
9. Lopez, J.A.; Sun, Y.; Blair, P.B.; Mukhtar, M.S. TCP three-way handshake: Linking developmental processes with plant immunity. *Trends Plant Sci.* **2015**, *20*, 238–245. [CrossRef]
10. Wang, X.; Gao, J.; Zhu, Z.; Dong, X.; Wang, X.; Ren, G.; Zhou, X.; Kuai, B. TCP transcription factors are critical for the coordinated regulation of isochorismate synthase 1 expression in Arabidopsis thaliana. *Plant J.* **2015**, *82*, 151–162. [CrossRef]
11. Spears, B.J.; McInturf, S.A.; Chlebowski, M.; Su, J.; Mendoza-Cózatl, D.G.; Gassmann, W. The class I TCP transcription factor AtTCP8 is a modulator of phytohormone-responsive signaling networks. *bioRxiv* **2021**. [CrossRef]
12. Chaturvedi, S.; Khan, S.; Usharani, T.R.; Tiwari, S. Analysis of TCP Transcription Factors Revealed Potential Roles in Plant Growth and *Fusarium oxysporum* F.Sp. Cubense Tolerance in Banana (cv. Rasthali). *Res. Sq.* **2021**.
13. Li, M.; Chen, H.; Chen, J.; Chang, M.; Palmer, I.A.; Gassmann, W.; Liu, F.; Fu, Z.Q. TCP Transcription Factors Interact with NPR1 and Contribute Redundantly to Systemic Acquired Resistance. *Front. Plant Sci.* **2018**, *9*, 1153. [CrossRef] [PubMed]
14. Kim, S.H.; Son, G.H.; Bhattacharjee, S.; Kim, H.J.; Nam, J.C.; Nguyen, P.D.; Hong, J.C.; Gassmann, W. The Arabidopsis immune adaptor SRFR1 interacts with TCP transcription factors that redundantly contribute to effector-triggered immunity. *Plant J.* **2014**, *78*, 978–989. [CrossRef]
15. Martin, R.C.; Vining, K.; Dombrowski, J.E. Genome-wide (ChIP-seq) identification of target genes regulated by BdbZIP10 during paraquat-induced oxidative stress. *BMC Plant Biol.* **2018**, *18*, 58. [CrossRef]
16. Xu, Y.; Liu, H.; Gao, Y.; Xiong, R.; Wu, M.; Zhang, K.; Xiang, Y. The TCP transcription factor PeTCP10 modulates salt tolerance in transgenic Arabidopsis. *Plant Cell Rep.* **2021**, *40*, 1971–1987. [CrossRef]
17. Pecher, P.; Moro, G.; Canale, M.C.; Capdevielle, S.; Singh, A.; MacLean, A.; Sugio, A.; Kuo, C.H.; Lopes, J.R.S.; Hogenhout, S.A. Phytoplasma SAP11 effector destabilization of TCP transcription factors differentially impact development and defence of Arabidopsis versus maize. *PLoS Pathog.* **2019**, *15*, e1008035. [CrossRef]
18. Wessling, R.; Epple, P.; Altmann, S.; He, Y.; Yang, L.; Henz, S.R.; McDonald, N.; Wiley, K.; Bader, K.C.; Glasser, C.; et al. Convergent targeting of a common host protein-network by pathogen effectors from three kingdoms of life. *Cell Host Microbe* **2014**, *16*, 364–375. [CrossRef]
19. Stam, R.; Motion, G.; Boevink, P.C.; Huitema, E. A conserved oomycete CRN effector targets and modulates tomato TCP14-2 to enhance virulence. *bioRxiv* **2013**. [CrossRef]
20. Zheng, X.; Yang, J.; Lou, T.; Zhang, J.; Yu, W.; Wen, C. Transcriptome Profile Analysis Reveals that CsTCP14 Induces Susceptibility to Foliage Diseases in Cucumber. *Int. J. Mol. Sci.* **2019**, *20*, 2582. [CrossRef]
21. Chen, N.; Yu, B.; Dong, R.; Lei, J.; Chen, C.; Cao, B. RNA-Seq-derived identification of differential transcription in the eggplant (*Solanum melongena*) following inoculation with bacterial wilt. *Gene* **2018**, *644*, 137–147. [CrossRef]
22. Qiu, Z.; Yan, S.; Xia, B.; Jiang, J.; Yu, B.; Lei, J.; Chen, C.; Chen, L.; Yang, Y.; Wang, Y.; et al. The eggplant transcription factor MYB44 enhances resistance to bacterial wilt by activating the expression of spermidine synthase. *J. Exp. Bot.* **2019**, *70*, 5343–5354. [CrossRef]
23. Na, C.; Shuanghua, W.; Jinglong, F.; Bihao, C.; Jianjun, L.; Changming, C.; Jin, J. Overexpression of the Eggplant (*Solanum melongena*) NAC Family Transcription Factor SmNAC Suppresses Resistance to Bacterial Wilt. *Sci. Rep.* **2016**, *6*, 31568. [CrossRef] [PubMed]
24. Mishra, P.; Tripathi, A.N.; Kashyap, S.P.; Aamir, M.; Tiwari, K.N.; Singh, V.K.; Tiwari, S.K. In silico mining of WRKY TFs through *Solanum melongena* L. and *Solanum incanum* L. transcriptomes and identification of SiWRKY53 as a source of resistance to bacterial wilt. *Plant Gene* **2021**, *26*, 100278. [CrossRef]
25. Xiao, X.-o.; Lin, W.; Li, K.; Li, W.; Gao, X.; Lv, L. Early burst of reactive oxygen species positively regulates resistance of eggplant against bacterial wilt. *J. Phytopathol.* **2017**, *165*, 652–661. [CrossRef]

26. Li, K.; Lin, W.Q.; Li, W.; Lv, L.L.; Cao, B.H. Bacterial Wilt Resistance of EDS1 Positive Regulating Eggplant. *Chin. J. Trop. Crops* **2018**, *39*, 332–337.
27. Xiao, X.; Lin, W.; Li, K.; Feng, X.; Li, W.; Zou, H.; Hui, J. Screening and Analysis of Candidate Interactors of Enhanced disease susceptibility 1 (SmEDS1))in Eggplant. *Chin. J. Trop. Crops* **2018**, *39*, 1580–1584.
28. Zhang, Y.; Liu, T.; Meyer, C.A.; Eeckhoute, J.; Johnson, D.S.; Bernstein, B.E.; Nusbaum, C.; Myers, R.M.; Brown, M.; Li, W.; et al. Model-based Analysis of ChIP-Seq (MACS). *Genome Biol.* **2008**, *9*, R137. [CrossRef]
29. Yu, G.; Wang, L.G.; He, Q.Y. ChIPseeker: An R/Bioconductor package for ChIP peak annotation, comparison and visualization. *Bioinformatics* **2015**, *31*, 2382–2383. [CrossRef]
30. Stark, R.B.G. DiffBind: Differential binding analysis of ChIP-Seq peak data. *R Package Version* **2011**, *100*, 3.
31. Fang, Y.; Zheng, Y.; Lu, W.; Li, J.; Duan, Y.; Zhang, S.; Wang, Y. Roles of miR319-regulated TCPs in plant development and response to abiotic stress. *Crop J.* **2021**, *9*, 17–28. [CrossRef]
32. Martin-Trillo, M.C.P. TCP genes: A family snapshot ten years later. *Trends Plant Sci.* **2010**, *15*, 31–39. [CrossRef] [PubMed]
33. Niyogi, K.K.; Fink, G.R. Two anthranilate synthase genes in Arabidopsis: Defense-related regulation of the tryptophan pathway. *Plant Cell* **1992**, *4*, 721–733.
34. March-Diaz, R.; Garcia-Dominguez, M.; Lozano-Juste, J.; Leon, J.; Florencio, F.J.; Reyes, J.C. Histone H2A.Z and homologues of components of the SWR1 complex are required to control immunity in Arabidopsis. *Plant J.* **2008**, *53*, 475–487. [CrossRef] [PubMed]
35. Berriri, S.; Gangappa, S.N.; Kumar, S.V. SWR1 Chromatin-Remodeling Complex Subunits and H2A.Z Have Non-overlapping Functions in Immunity and Gene Regulation in Arabidopsis. *Mol. Plant* **2016**, *9*, 1051–1065. [CrossRef] [PubMed]
36. Lee, S.C.; Knowles, T.J.; Postis, V.L.; Jamshad, M.; Parslow, R.A.; Lin, Y.P.; Goldman, A.; Sridhar, P.; Overduin, M.; Muench, S.P.; et al. A method for detergent-free isolation of membrane proteins in their local lipid environment. *Nat. Protoc.* **2016**, *11*, 1149–1162. [CrossRef]
37. Wei, Q.; Wang, J.; Wang, W.; Hu, T.; Hu, H.; Bao, C. A high-quality chromosome-level genome assembly reveals genetics for important traits in eggplant. *Hortic. Res.* **2020**, *7*, 153. [CrossRef]

International Journal of
Molecular Sciences

MDPI

Article

A Comparative Transcriptome and Metabolome Combined Analysis Reveals the Key Genes and Their Regulatory Model Responsible for Glucoraphasatin Accumulation in Radish Fleshy Taproots

Xiaoman Li [1,2,†], Peng Wang [2,†], Jinglei Wang [2,3], Haiping Wang [2], Tongjin Liu [2,4], Xiaohui Zhang [2], Jiangping Song [2], Wenlong Yang [2], Chunhui Wu [2], Haohui Yang [2], Liwang Liu [1,*] and Xixiang Li [2,*]

1 National Key Laboratory of Crop Genetics and Germplasm Enhancement, Key Laboratory of Horticultural Crop Biology and Genetic Improvement (East China) of Ministry of Agriculture and Rural Affairs, College of Horticulture, Nanjing Agricultural University, Nanjing 210095, China; lxm911001@163.com
2 Key Laboratory of Biology and Genetic Improvement of Horticultural Crops, Institute of Vegetables and Flowers, Chinese Academy of Agricultural Sciences, Ministry of Agriculture and Rural Affairs, Beijing 100081, China; 82101172227@caas.cn (P.W.); syauwjl@163.com (J.W.); wanghaiping@caas.cn (H.W.); tongjinliu@163.com (T.L.); zhangxiaohui01@caas.cn (X.Z.); songjiangping@caas.cn (J.S.); yangwenlong@caas.cn (W.Y.); wchd163@163.com (C.W.); yanghaohui127@163.com (H.Y.)
3 Institute of Vegetables Research, Zhejiang Academy of Agricultural Sciences, Hangzhou 310021, China
4 Jinling Institute of Technology, College of Horticulture, Nanjing 210038, China
* Correspondence: nauliulw@njau.edu.cn (L.L.); lixixiang@caas.cn (X.L.); Tel.: +86-25-8439-5563 (L.L.); +86-10-8210-5947 (X.L.); Fax: +86-25-8439-5266 (L.L.); +86-10-6217-4123 (X.L.)
† These authors contributed equally to this work.

Abstract: Radish (*Raphanus sativus* L.) is rich in specific glucosinolates (GSLs), which benefit human health and special flavor formation. Although the basic GSLs metabolic pathway in Brassicaceae plants is clear, the regulating mechanism for specific glucosinolates content in radish fleshy taproots is not well understood. In this study, we discovered that there was a significant difference in the GSLs profiles and the content of various GSLs components. Glucoraphasatin (GRH) is the most predominant GSL in radish taproots of different genotypes as assessed by HPLC analysis. Further, we compared the taproot transcriptomes of three radish genotypes with high and low GSLs content by employing RNA-seq. Totally, we identified forty-one differentially expressed genes related to GSLs metabolism. Among them, thirteen genes (*RsBCAT4*, *RsIPMDH1*, *RsMAM1a*, *RsMAM1b*, *RsCYP79F1*, *RsGSTF9*, *RsGGP1*, *RsSUR1*, *RsUGT74C1*, *RsST5b*, *RsAPK1*, *RsGSL-OH*, and *RsMYB28*) were significantly higher co-expressed in the high content genotypes than in low content genotype. Notably, correlation analysis indicated that the expression level of *RsMYB28*, as an R2R3 transcription factor directly regulating aliphatic glucosinolate biosynthesis, was positively correlated with the GRH content. Co-expression network showed that *RsMYB28* probably positively regulated the expression of the above genes, particularly *RsSUR1*, and consequently the synthesis of GRH. Moreover, the molecular mechanism of the accumulation of this 4-carbon (4C) GSL in radish taproots was explored. This study provides new perspectives on the GSLs accumulation mechanism and genetic improvements in radish taproots.

Keywords: glucosinolate profiles; RNA-seq; *RsMYB28*; key genes; regulatory network; radish fleshy taproots; genotypes

1. Introduction

Glucosinolates (GSLs) are a large class of important sulfur-rich secondary metabolites of Brassicaceae plants [1,2]. Based on the precursor amino acids and R groups modification, GSLs are divided into aliphatic, aromatic, and indole groups [1–3]. So far, more than 200 types of GSLs have been found in the cruciferous family [4]. Different *Brassicaceae* plants have diverse GSLs, such as 4-methylthiobutyl as the main GSL in *A. thaliana* root [5]; gluconapin is mainly present in *B. rapa* [6] and Chinese kale [7]; glucoraphanin (GRA) is present in broccoli seed [8], and glucoerucin exists in cabbage (*Brassica oleracea*) [9]. GSLs and myrosinases (b-thioglycoside glucose hydrolase, EC 3.2.3.1) are usually stored in different cell compartments. After tissue damage, GSLs are released from the vacuole and rapidly hydrolyzed by myrosinases to glucose and thiohydroxymate O-sulfonate, and the latter is unstable and spontaneously becomes isothiocyanate (ITCs), thiocyanate, nitrile, acetonitrile, or oxazolidine-2-thioketone [6,7]. GSLs and their degradation products have significant biological activities, such as plant disease resistance and inhibition of human cancer cells, and they are also special flavor substances [1,10–12], although some of them have side effects, such as sinigrin [13] and cyanogenic glucosides [14], which may influence the normal function of the thyroid, and Epithionitriles, nitriles, and sulforaphane nitrile, which have less opportunity to be converted into beneficial products [15,16].

The aliphatic GSLs biosynthetic pathway in *Arabidopsis thaliana* has been elucidated and includes three stages: side chain elongation of amino acids, development of the core structure, and secondary side chain modifications [2,17]. The GSLs biosynthetic pathway is involved in many genes and transcription factors (TFs). At the stage of GSLs chain elongation, *BCAT3* and *BCAT4* perform the deamination and transamination of the Met chain elongation cycle in the formation of aliphatic GSLs [18–20]. *MAM1* and *IPMDH1* catalyze the elongation of the side chain length of aliphatic GSLs [21,22]. With regard to the biosynthesis of core GSLs structure, firstly, precursor amino acids are converted to aldoximes by *CYP79F1*/*CYP79F2* [23,24], and aldoximes are then oxidized to activated compounds (either nitrile oxides or aci-nitro compounds) by *CYP83A1* [25]. After conjugation of the activated aldoximes to a sulfur donor, which can occur nonenzymatically, the produced S-alkylthiohydroximates are converted to thiohydroximates by the C-S lyase SUR1 [26]. Thiohydroximates are then S-glucosylated by glucosyltransferases of the *UGT74* family to form desulfo-GSLs [27]. Glucosylation gives rise to desulfo-GSLs, which are ultimately sulfated by sulfotransferases (SOT16, 17, and 18) to form GSLs [28], and then *FMOGS-OX5*, *AOP1*, *GSL-OH*, *CYP81F2*, and *CYP81F4* genes are involved in the next steps of secondary modification. Although this basic GSL pathway is shared by other Brassicaceae plants, there are some modifications in their GSL metabolic pathways responsible for diverse GSLs and catabolites in different plants [29,30].

In *A. thaliana*, *MYB34*, *MYB51*, *MYB122*, *MYB28*, *MYB29*, and *MYB76* are six TFs that have been identified as regulators of GSLs biosynthesis. The former three MYBs regulate the biosynthesis of indole GSLs, whereas the latter three MYBs regulate the aliphatic biosynthesis of GSLs [31,32]. *MYB28* as main regulator is followed by *MYB29*, while *MYB76* plays a minor role in the aliphatic GSLs formation [33]. *BoaMYB28* has the potential to alter the aliphatic GSLs content in *Brassica oleracea* at the genetic level [34]. *MYB28* is also a regulator of Met-derived GSLs biosynthesis in *Brassica juncea* [35,36]. *BoMYB29* enhances the expression of aliphatic GSLs biosynthetic genes and the accumulation of aliphatic GSLs in *Brassica oleracea* [37]. *MYB28* genes have been detected at three QTLs associated with the GSLs content in rapeseed [38]. Besides, *IQD1* integrates intracellular Ca^{2+} signals to fine-tune GSLs accumulation in response to biotic challenge [39]. *OBP2* is also part of the regulatory network that regulates GSLs biosynthesis in *A. thaliana* [40]. It is obvious that different TFshave different roles in different plants.

Radish (*Raphanus sativus* L., 2n = 2x = 18) is a member of the plant family Brassicaceae and is also a historically important cultivated crop all over the word [41]. It was reported that there were different types of main GSLs in *R. sativus* [9,42–46]. Among them, glucoraphasatin and glucoraphenin, which belong to the aliphatic group, are the two predominant types [3,42,47]. Sulforaphene, produced by the breakdown of glucoraphenin, obtained from hydrolysis of glucoraphasatin, strongly influences the taste of *R. sativus* roots [42] and has potential antioxidative and antimutagenic [48], anti-inflammatory [49], cardiovascular protection, and anticarcinogenic activities [50]. Compared with other Brassicaceae plants, the GSLs metabolic pathway in radish is specific in some respects with regard to the synthesis of particular components, although many similarities among the pathways exist [29]. *RsFMOs* are responsible for converting GRH into glucoraphenin (GRE). With regard to GSLs accumulation in radish taproot, one study reported that three large QTL intervals may be related to the content of aliphatic GSLs in radish fleshy roots [51]. The *FMOGS-OX2* gene expressed differentially in radish taproots and siliques produced different aliphatic GSLs components—that is, GRH in radish taproots and GRE in seeds, respectively [43,52]. *RsGRS1* participates in the dehydrogenation reaction to generate unsaturated 4-carbon (4C) GSLs from glucoerucin or glucoraphanin to obtain GRH or GRE, respectively [9]. Kang et al. [53] analyzed the GSLs content of 59 accessions of radish germplasm and selected out 5 radish accessions with a significant difference in total GSL content for transcriptome analysis. They identified 13 differentially expressed genes (DEGs) in the aliphatic GSL biosynthesis pathway and one differentially expressed transcription factor *MYB29* (*RSG00789*) among these five roots. However, it is noteworthy that all the DEGs showed very low expression without any transition in the fleshy taproots of different genotypes with low and moderate GRH concentrations. In addition, they did not elucidate both the regulation relationship among these DEGs and the relationship between GRH content and *MYB29*. Therefore, the key genes and regulation mechanism for the specific GSLs accumulation is still poorly understood in the fleshy taproots of different radish genotypes. In this study, we selected out three radish genotypes with marked differences in total GSLs content from a previous study [54]. We used HPLC to analyze the GSL profiles and their component contents of the fleshy taproots of these three genotypes. At the same time, we employed a transcriptome method to compare the taproot gene expression profiles of these three genotypes, identified the crucial gene-related GSL biosynthesis, and constructed a metabolic network for the accumulation of GSLs in radish fleshy taproots by combining co-expression analysis and qRT–PCR verification of DEGs and correlation analysis of key gene expressions and major metabolites content. These results provide insights into the relationship among the expression of regulatory factors of GSLs biosynthesis and the key genes related in GSLs synthesis in radish fleshy roots. This study will be beneficial for an in-depth understanding of the mechanism regulating GSLs accumulation and for the molecular genetic improvement in the GSLs content in radish.

2. Results

2.1. GSLs Profiles and Content of Various GSL Components in the Radish Taproots of Different Genotypes

We analyzed GSLs components and their content by using the HPLC method (Figure 1, Table S2). Totally, we identified out six GSL components, which included three aliphatic glucosinolates (GRA, GRE, and GRH) and three indolyl glucosinolates (4-hydroxyglucobrassicin: 4HGBS, glucobrassicin: GBS, and 4-Methoxyglucobrassicin: 4MOGBS). The aliphatic GSLs were prominent, and the proportion of aliphatic GSLs accounting for the total GSLs content was 89.67% in LOW1, 98.53% in HIGH1, and 93.31% in HIGH2. The GRH as the most predominant component had an average value of 2.74, 31.04, and 29.14 μmol/g dry weight (DW) in LOW1, HIGH1, and HIGH2, respectively. The highest values of GRH in HIGH1 was almost 13 times that of the lowest value in LOW1. As for GRE, as the component of second highest content, its content in HIGH1 was almost 4-fold higher than that in LOW1.

2.2. Transcriptome Analysis of Fleshy Taproots of Three Different Radish Varieties

To clarify the differences in transcription levels between above three radish varieties, nine cDNA libraries (HIGH1, HIGH2, and LOW1; three replicates of each variety) were constructed using total RNA extracted from the fleshy taproots (Table S3). After removing low-quality reads, the average number of reads per library was higher than 44 million. The Q20 percentages (sequencing error rates lower than 1%) ranged from 97.84% to 98.02%, and the GC percentages ranged from 47.37% to 47.70%. After alignment with the radish reference genome [55] and subsequent analysis using TopHat2, 33,265,682 (70.72%), 33,349,417 (71.10%), 33,048,869 (70.45%), 31,011,460 (66.00%), 31,065,284 (66.04%), 31,005,222 (65.90%),

31,582,144 (70.85%), 33,257,097 (71.20%), and 32,740,986 (70.40%) reads in nine libraries were mapped to the radish reference genome, respectively. Among these, 67.43%, 67.52%, 67.06%, 62.84%, 63.05%, 62.61%, 67.08%, 67.99%, and 67.14% were mapped uniquely to one location. The above-mentioned quality analysis showed that the data were reliable and sufficient.

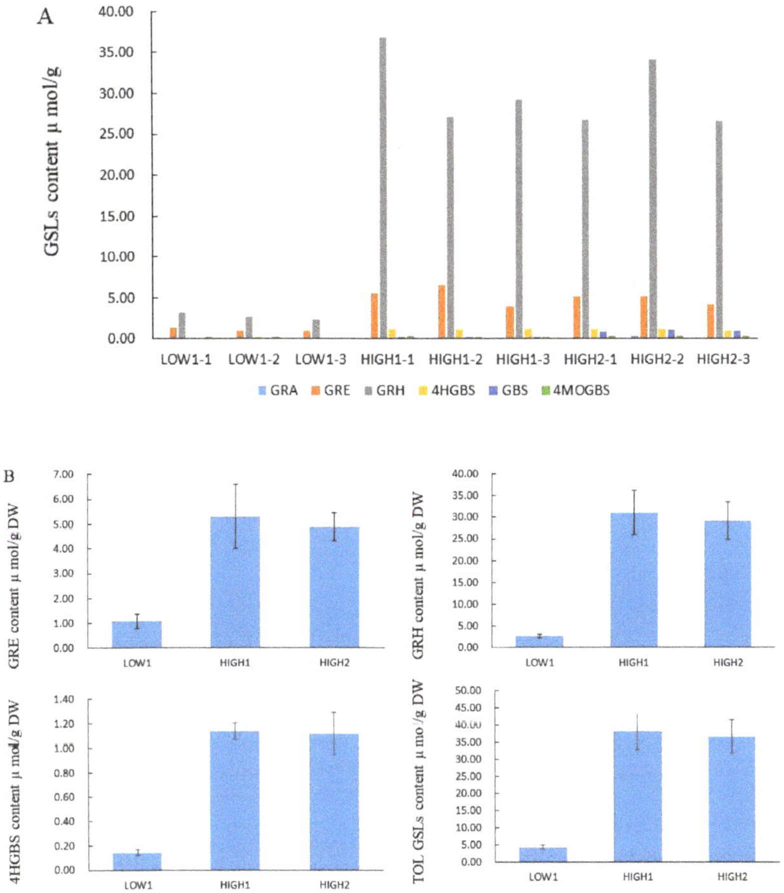

Figure 1. GSL profiles and content of various GSLs in the fleshy taproots of LOW1, HIGH1, and HIGH2 genotypes. (**A**): The x axis shows the GSL profiles of different biological replicates of LOW1, HIGH1, and HIGH2 fleshy taproots. The different color bars indicate GRA, GRE, GRH, 4-hydroxyglucobrassicin (4HGBS), glucobrassicin (GBS), and 4MeO-glucobrassicin (4MOGBS). The y axis shows the content of individual GSLs in LOW1, HIGH1, and HIGH2 fleshy taproots. (**B**): The main GSLs content in radish taproot.

Figure 2A and Table S4 show that more than 80% of the mapped reads were in exons. The number of transcripts identified in each sample is expressed in FPKMs and is shown in Figure 2B and Table S5. Genes with normalized read numbers lower than 0.5 FPKM were removed from the analysis. In total, 23,903, 23,522 and 24,077 transcripts were found to be expressed in HIGH1, HIGH2, and LOW1, respectively. Approximately 29% of the expressed genes were in the range of 0.5 to 5 FPKM, and 62% of the expressed genes were in the range of 5 to 100 FPKM (Figure 2B, Table S5). The comparison of HIGH1 and HIGH2 with LOW1 identified 3936 (2105 upregulated and 1831 downregulated) and

5505 (2651 upregulated and 2854 downregulated) DEGs, respectively. These results indicate that the majority of the genes were differentially expressed in fleshy taproots of different varieties at the mature stage.

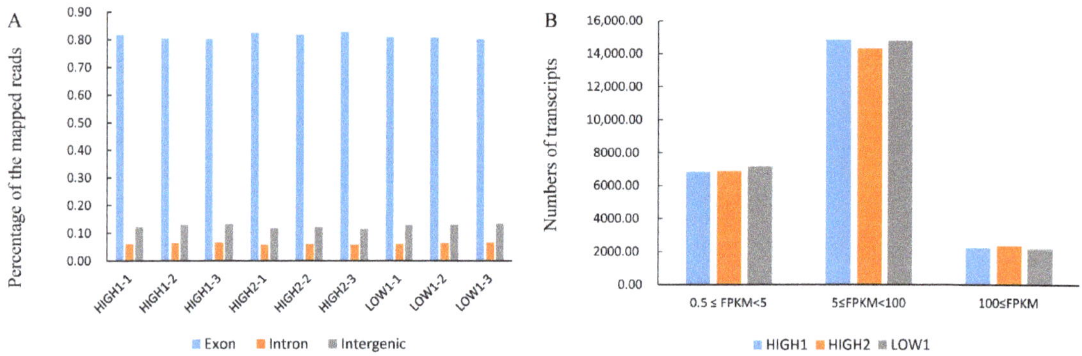

Figure 2. Transcriptome analysis of fleshy taproots of three radish genotypes. (**A**) Percentages of the mapped reads distributed in different sites (intron, intergenic, and exon) of genes in HIGH1, HIGH2, and LOW1. The x axis represents three replicates of the sequenced varieties—namely, HIGH1, HIGH2, and LOW1. The y axis shows the percentage of mapped reads distributed in different sites of genes. (**B**) Numbers of transcripts with different FPKM values in the roots of the three cultivars. The x axis represents the sequenced genotypes—namely, LOW1, HIGH1 and HIGH2. The y axis shows the numbers of transcripts. The different colors represent the different FPKM values.

2.3. *Expression of Genes Related to GSLs Biosynthesis and Degradation in Fleshy Taproots of Three Radish Varieties*

Firstly, we identified the radish whole genome genes involved in the GSLs metabolic pathway by referring to the GSLs biosynthetic pathway in Arabidopsis. The method and criteria referred to reference [29]. Secondly, based on the transcriptome data, we selected out the DEGs and made the heatmap by comparing the expression of all the identified genes based on the FC and FDR. According to the expression of these genes being involved in the three independent stages of GSLs biosynthesis in three varieties, we identified four DEGs related to side chain elongation, including branched-chain aminotransferase 4 (*BCAT4*), isoropylmalate dehydrogenase (*IPMDH1*), and methylthioalkylmalate synthase (*MAM1*). The expression levels of *RsBCAT4*, *RsIPMDH1*, *RsMAM1a*, and *RsMAM1b* were significantly higher in HIGH1 and HIGH2 than in LOW1 (Figures 3B and 4A).

Totally, we identified sixteen genes functioning in GSL core structure formation in three radish taproot samples (Figure 3B). Among these genes, there were seven DEGs (*RsCYP79F1*, *RsGSTF9*, one copy of *RsGGP1* (*Rsa10019051*), *RsSUR1*, *RsUGT74C1*, one copy of *RsST5b* (*Rsa10034275*), and *RsAPK1*) that showed extremely markedly and repeatedly higher expression in HIGH1 and HIGH2 compared with LOW1 (Figures 3B and 4A); however, *CYP83A1* was not detected in the three cultivars. Among the genes involved in side chain modification, only two copies of *RsGSL-OH* were significantly differentially expressed among three radish genotypes. The *GSL-OH* locus is responsible for biosynthesis of the hydroxylated alkenyl GSLs 2-hydroxybut-3-enyl GSLs, whose production is dependent on the presence of both *AOP2* and *MAM1* [56].

In Arabidopsis, *TGG1*, *TGG2*, *PEN2*, and *PEN3* are four myrosinase genes involved in the degradation of GSLs, and *TGG1* and *TGG2* are well known as important myrosinases [57]. Among these, *PEN2* and *PEN3* cleave indole GSLs. Aliphatic GSLs cannot be degraded in the double mutant, whereas indole GSLs can be reduced slightly in tgg1tgg2, which indicates that *TGG1* and *TGG2* mainly degrade aliphatic GSLs and exert only slight effects on indole GSLs. However, none of these genes were identified as DEGs between the high- and low-GSLs genotypes.

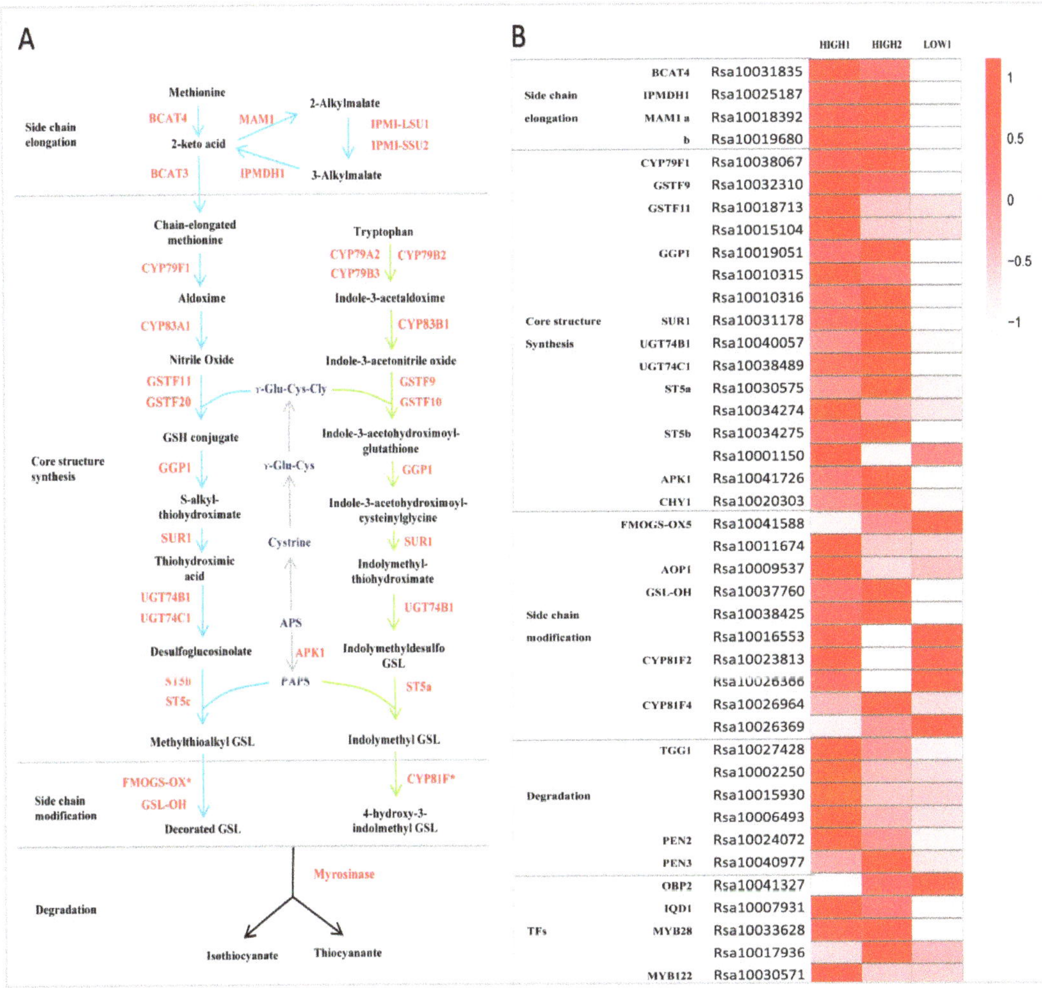

Figure 3. GSLs pathway and its expression pattern in radish fleshy taproots. (**A**) GSLs pathway map. (**B**) Heatmap showing the expression level of genes encoding enzymes and regulators in the GSLs pathway. The color code indicates the differential expression values, and the scale on the right represents the gene expression levels based on the log2 values of reads per kilobase of exon model per million mapped reads (FPKM). BCAT, branched-chain amino acid aminotransferase; MAM, methylthioalkylmalate synthase; IPMI-LSU1, isopropylmalate isomerase large subunit 1; IPMI-SSU2, isopropylmalate isomerase small subunit 2; IPMDH, isopropylmalate dehydrogenase; CYP, cytochrome P450; GSTF, glutathione S-transferase; GGP, gamma-glutamyl peptidase; SUR, C-S lyase SUPERROOT; UGT, UDP-glucosyltransferase; ST, desulfoglucosinolate sulfotransferase; FMO-GSOX, flavin-monooxygenase glucosinolate S-oxygenase; AOP, alkenyl hydroxalkyl producing; GSL-OH, Fe (II)-dependent oxygenase superfamily protein; ESP, epithiospecifier protein; APK, adenylyl sulfate kinase; GSH, gamma-glutamylcysteine synthetase; APS, adenosine-5′-phosphosulfate; PAPS, 3′-phosphoadenosine-5′-phosphosulfate; OBP and IQD, DOF zinc finger proteins; and MYB, MYB domain protein.

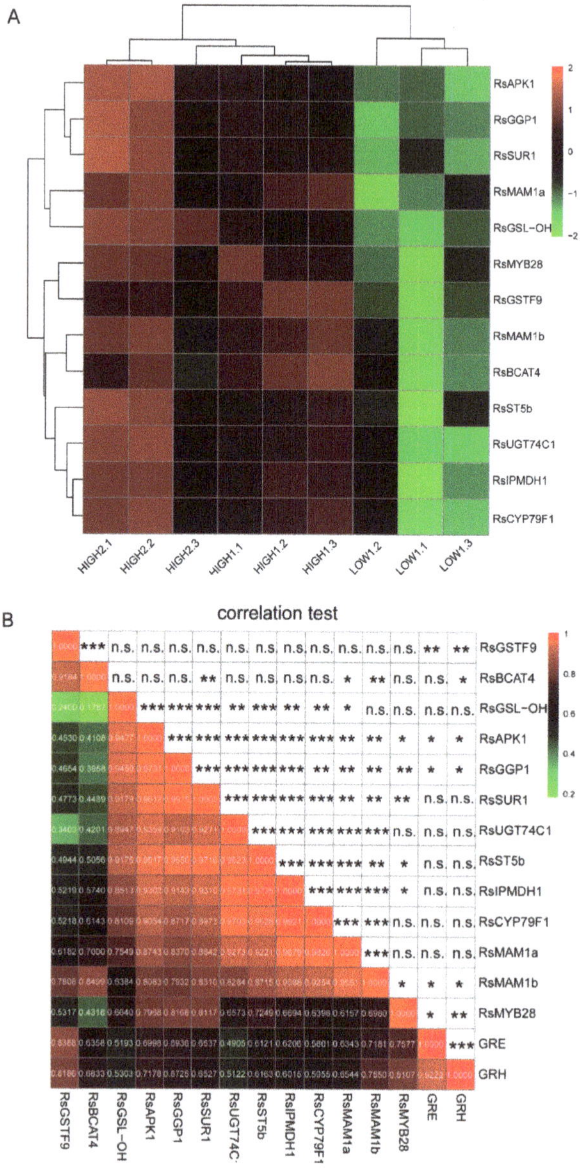

Figure 4. Differential expression of key GSLs pathway-related genes and their correlations in different radish samples. (**A**) Heatmap visualization of the relative gene expression levels in different radish taproots. The legend shows the up- and downregulation expression pattern that was maintained in each sample. (**B**) The Pearson's coefficient (r) and correlation significance (asterisks) are elucidated in a symmetric matrix prepared using the same variables as in A, with the addition of the GRH and GRE content. The location of the variables in the heatmap are based on the results of hierarchical clustering; the negative and positive correlations are shown in green and red squares, respectively, and the r values are shown in the squares. *, **, and *** indicate significance at $p \leq 0.05$, $p \leq 0.01$, and $p \leq 0.001$, respectively; n.s., not significant.

2.4. Performance of Transcription Factors Related to GSLs Biosynthesis in the Fleshy Taproots of Three Radish Varieties

In *A. thaliana*, *MYB28*, *MYB29*, and *MYB76* are commonly defined as TFs in the biosynthesis of aliphatic GSLs, and these TFs can specifically activate aliphatic GSLs biosynthesis-related genes such as *MAM3*, *CYP79F1*, and *CYP83A1* [30,58]. In contrast, *MYB34*, *MYB51*, and *MYB122* exclusively transactivate the promoters of *TSB1*, *CYP79B2*, and *CYP79B3*, which are involved in the indole GSLs biosynthetic pathway [59]. In this study, the expression of five genes (*Rsa10041327*, *Rsa10007931*, *Rsa10033628*, *Rsa10017936*, and *Rsa10030571*) were detected and classified as TFs related to GSLs biosynthesis in the expression profiles of radish taproots (Figure 3B). Notably, we noticed that *MYB28* had three copies in radish, which were *Rsa10033628*, *Rsa10019636*, and *Rsa10017936*. Among these three genes, the expression of one copy of *MYB28* homologs (*Rsa10033628*) was significantly higher in HIGH1 and HIGH2 than in LOW1. Furthermore, IQD1 expression was significantly higher in HIGH1 and HIGH2 than in LOW1, whereas other TFs exhibited no significant differences among the three radish varieties.

2.5. Establishment of a Co-Expression Network of the Key Genes for GSLs Biosynthesis in Radish Fleshy Taproots

Based on the differential expression analysis of genes related to the synthesis of GSLs, 13 significant DEGs were found among the three varieties (Figure 4A). After conducting correlation analyses among these key genes and the principal GSLs, we found that the *RsMYB28* transcription was extremely significantly positively correlated with the GRH content. Furthermore, *RsMYB28* was significantly positively associated with *RsSUR1*, which can convert S-alkylthiohydroximates to thiohydroximates during formation of the GSLs core structure (Figure 4B). The 12 key DEGs on the circle (Figure 5) showing distinct expression based correlation strengths are involved in the formation of two types of GSLs, respectively: aliphatic GSLs, which can be formed by the co-expression of 11 key unigenes in GSLs biosynthesis, and indole GSLs, which are affected by two unigenes, *RsBCAT4* and *RsGSTF9*. No significant correlations were found between other GSLs biosynthesis genes and the corresponding GSLs components. The analyses allowed depiction of a putative gene/metabolite network (Figure 5), which encompassed TFs, *RsMYB28*, 12 GSLs biosynthesis genes, GRH, and GRE. This model will aid further study of the relationship of the candidate genes and GSL accumulation in radish taproots.

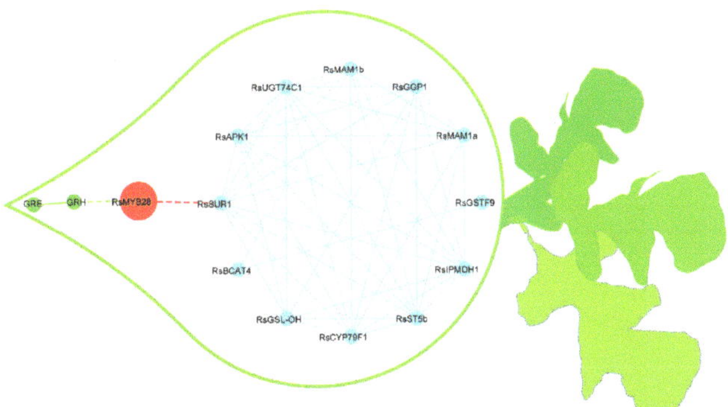

Figure 5. Putative metabolic network for the synthesis of GSLs in radish fleshy taproots. The blue lines represent the relationships among GSLs biosynthesis genes; the red lines represent the relationships between *RsMYB28* and GSLs biosynthesis genes; the green lines represent the relationships between genes and GSLs. The correlation strength according to the r coefficient ranges was used to determine whether the lines are shown as solid or dashed (dashed: $0.7 \leq r \leq 0.85$; solid: $r > 0.85$). The red nodes represent transcription factors, the blue nodes indicate GSLs biosynthesis genes, and the green nodes are GSLs.

2.6. qRT–PCR Analysis of DEGs in Three Different Radish Cultivars

In order to verify the reliability of the transcriptome analysis, we validated the sequencing data for 13 unigenes by qRT–PCR, and the results show a similar expression pattern, with the genes exhibiting higher expression levels in HIGH1 and HIGH2 than in LOW1 (Figure 6).

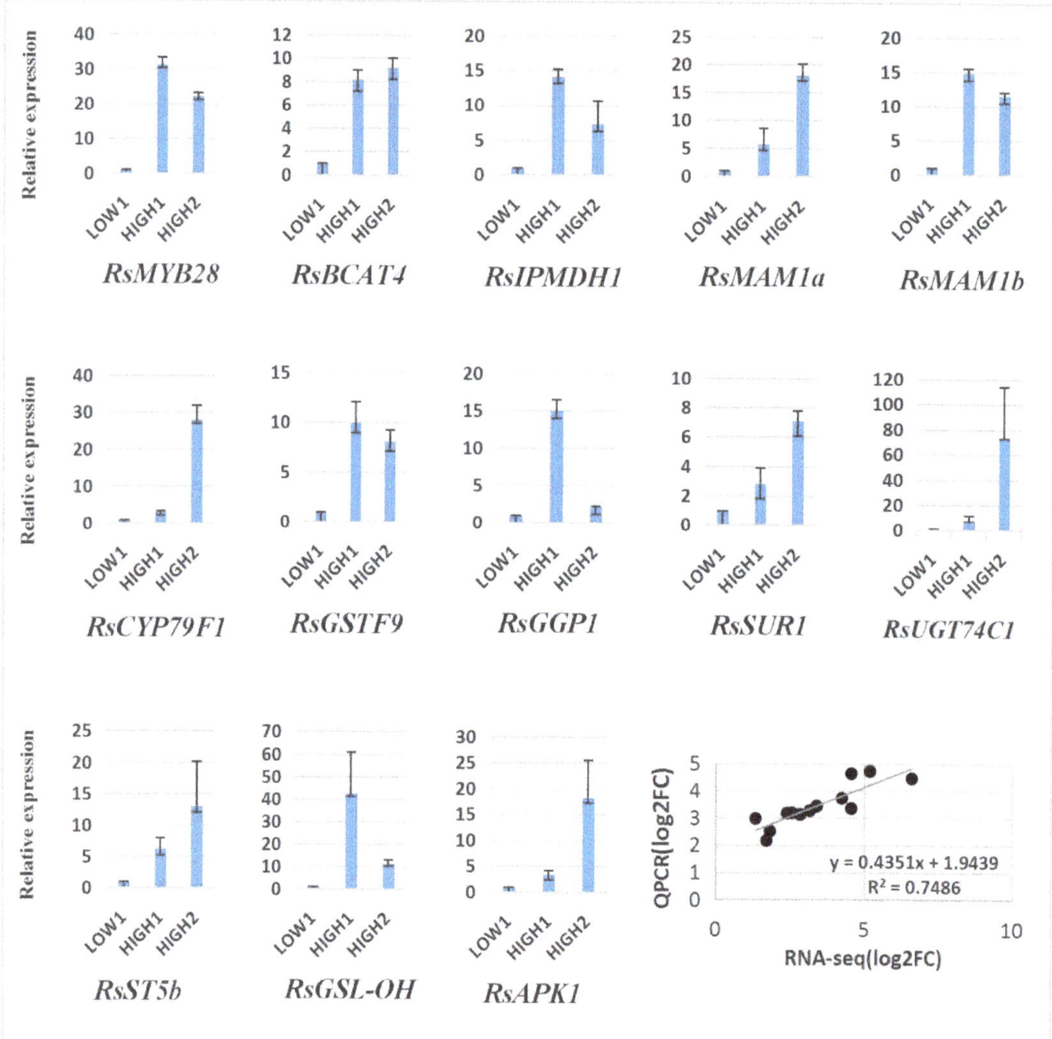

Figure 6. Validation of the expression of 13 differentially expressed genes in LOW1, HIGH1, and HIGH2 fleshy taproots by qRT–PCR. The x axis represents the three cultivar taproots. The y axis shows the relative expression of one specific gene to the reference gene *RsGAPDH*, and the expression of genes in LOW1 was set to "1" as a control. The last panel shows the results from the RNA-seq/qRT–PCR correlation analysis (bottom right) expressed as log2-fold changes in the low-GSLs vs. high-GSLs genotypes; a significant positive correlation was found between the fold changes in expression, as determined by Pearson's correlation ($R^2 = 0.75$; $p < 0.001$).

3. Discussion

3.1. The Content of GRH as the Main GSLs Component in Radish Taproot Is Significantly Different among Different Genotypes

Through HPLC analysis, we found that the GSLs profiles were similar and that GRH was the main component among the six types of GSLs in the radish taproot, followed by GRE. The total GSLs content and components content were significantly different among the three radish genotypes. Additionally, the GRH values in genotype HIGH1 were almost 13 times as high as those in genotype LOW1. Kang et al. [53] found six aliphatic and four indole GSLs in 59 radish accessions and significant difference in total GSLs content and GRH content among different genotypes, although it was puzzling that GRH content was higher than the total GSLs content in RA157-74 and in IT119238-8. Kim et.al [46] detected 13 types of GSLs in radish fleshy taproot skin and flesh and found that the highest total GSLs was 168.44 μmol/g DW in radish skin and 71.09 μmol/g DW in radish flesh. Maldini et al. [42] identified and quantified GSLs in different parts of wild radish and showed that GRH (56 mg 100 g^{-1} FW) was followed by GRE. Ishida et al. [42] analyzed GSLs content in 28 cultivated radishes and found that the GSLs content ranged from 8.6 μmol/g to 137.7 μmol/g and that the predominant GSLs were 4-methylthio-3-butenyl GSLs. Yi et al. [45] analyzed 71 radish accessions and also proved that GRH was dominant (84.5% of total GSLs) in radish root. The types and content of glucosinolates detected in different studies were somewhat different, which may be due to different materials, the precision of analytical equipment, and experimental conditions. In general condition, the types and content of glucosinolates detected in a batch of materials under the controlled condition of a same study were credible and comparable. In general, GRE accounts for 70–95% of total GSLs in seeds, while GRH accounts for 90% of the total GSLs in radish taproots [29,42], differing from what is found in *Brassica* species [60]. Therefore, the radish genotypes with great differences in GSLs content as the model for studying the GRH biosynthesis and metabolic mechanism are valuable.

3.2. RsMYB28 Regulatory Model Responsible for GSLs Accumulation in Radish Fleshy Taproots

The expression of *RsMYB28* (*Rsa10033628*) was significantly correlated with the GRH content in radish fleshy taproots (Figures 4B and 5). Meanwhile, *Rsa10033628* was co-expressed with 12 important DEGs related to GSL biosynthesis. It could be inferred that *RsMYB28* was a key regulatory factor responsible for GSLs accumulation in radish fleshy taproots. Kang et al. deduced that it is *MYB29* not *MYB28* that is regulated in GRH biosynthesis in radish root only based on its co-expression with 13 aliphatic GSL metabolic DEGs [53], which are not totally identical to the 12 aliphatic GSL biosynthesis DEGs we identified in the present study. It was a pity that they did not actually analyze the relationship among GRH content, *MYB29*, and GSL biosynthesis genes. The differences in results may be due to different genotypes in these two different studies, which needs to be confirmed by further experiments. However, in *A. thaliana*, *MYB28* is commonly defined as the main transcriptional regulator in the biosynthesis of aliphatic GSLs [30,33,61]. In *B. juncea*, four homologous genes encoding *MYB28* were found to participate in the regulation of aliphatic GSLs biosynthesis [36], and *BoaMYB28* has the potential to alter the aliphatic GSLs content in *Brassica oleracea* at the genetic level [34]. Our investigation confirmed that *RsMYB28* may also be the key gene regulating the content of aliphatic GRH, which is the most abundant GSL in radish fleshy taproots in its own way of cooperating with other genes in GSLs biosynthesis.

Most of the genes related to aliphatic GSLs are regulated by *MYB28* and *MYB29* [62]. A previous study showed that *BrAOP2* and *BrGSL-OH* are negatively and positively regulated in *BrMYB28.1*-overexpressing lines by all three *BrMYB28s*, which indicates that the regulatory mechanism of GSLs biosynthesis in *B. rapa* differs from that in *A. thaliana* [63]. Our results show that *RsMYB28* exhibited a higher significance of expression correlation and co-expression trend with RsSUR1, which converts S-alkyl-thiohydroximates to thiohydroximates (Figures 4B and 5). Therefore, we speculated that *RsMYB28* may promote the accumulation of GSLs in radish fleshy taproots by regulating the expression of *RsSUR1* and another eleven co-expressing DEGs that may be the key genes involved in GSLs biosynthesis.

3.3. Key Gene of GSLs Biosynthesis in Radish Fleshy Taproots

Similar key DEGs related to GSL biosynthesis were detected in our study and Kang's study [53], such as *RsBCAT4*, *RsMAMa*, *RsMAMb RsIPMDH1*, *RsCPY79F1*, *RsSUR1*, *RsUTG74C1*, and *RsGSL-OH*. However, *RsGSTF9*, *RsGGP1*, *RsST5b*, and *RsAPK1* related to GSL biosynthesis had different expression levels and expression patterns between our study and Kang's result. This may result from the difference in the genetic background of different radish genotypes. We found that several key genes (*RsBCAT4*, *RsMAM1(2)*, *RsMYB28*, and *RsGSL-OH*) of GSLs biosynthesis were closely co-expressed. The expression levels of *BCAT4* and *MAM1*, which belong to gene families encoding proteins involved in Leu biosynthesis, exhibited an extraordinarily strong correlation, which indicates that a close evolutionary relationship may exist between the pathways of GSLs chain elongation and Leu formation [19]. In addition, *RsIPMDH1* is a key gene of side chain extension in GSLs synthesis and is closely co-expressed with *RsMAM1(2)* (Figure 5). *MAM1* catalyzed the condensations in the first three elongation cycles, which were responsible for the C3/C4 chain length variation in Met-derived glucosinolates in Arabidopsis [21]. In the cabbage transgenic line M1-1, the accumulation of both GNA (C4 GSLs) and GBN (C5 GSLs) was 3.5- and 2-fold higher than that in the control, respectively [64]. The two most important GSLs in radish fleshy taproots are both 4C GSLs; therefore, we speculated that side chain extension in GSLs synthesis could be catalyzed by *RsMAM1(2)* and that *RsBCAT4* is probably the key mechanism that controls the GSLs content.

Our results also show that *RsCYP79F1*, *RsGSTF9*, *RsGGP1*, *RsSUR1*, *RsUGT74C1*, *RsST5b*, and *RsAPK1* are the key genes in the biosynthesis and accumulation of GSLs in radish fleshy taproots (Figure 3B). Genetic and transgenic approaches have been used to confirm that the biosynthesis of 3-carbon (3C) GSLs can be regulated by *CYP79F1* in *B. juncea* [65]. However, *GSL-PRO* is a probable candidate gene responsible for 3C GSLs biosynthesis in *A. thaliana*, *Brassica napus*, and *B. oleracea* [66,67]. These core-structure genes were significantly co-expressed with *RsMAM1* (Figure 5). These genes may play indispensable roles in the accumulation of 4C GSLs in radish fleshy taproots. The synergistic expression of these key genes also indicates their important roles in the GSLs synthesis process.

3.4. Key Genes of Indole GSLs Biosynthesis in Radish Fleshy Taproots

A very close co-expression relationship exists between *RsBCAT4* and *RsGSTF9* (Figure 5), which are responsible for the key steps of transferring sulfur during the synthesis of indole-GSLs [68]. Their expression and possible interaction may regulate the content of indole GSLs in radish fleshy taproots.

In conclusion, a comparative metabolome and transcriptome combined analysis effectively assessed the differences in GSLs profiles and gene expression between different radish cultivars with high and low GSLs content and characterized the key genes involved in radish GSLs biosynthesis and accumulation. Specifically, the key role of *RsMYB28* in the accumulation of aliphatic GSLs in radish fleshy taproots was clarified by identification of DEGs and gene co-expression analysis, and the possible regulatory relationship among *RsMYB28*, *RsSUR1*, and other 11 DEGs related to GSLs biosynthesis was also predicted. The possible molecular mechanism of 4C GSLs accumulation in radish fleshy taproots was also explored. These provide new insights into a comprehensive understanding of the mechanism of GSLs biosynthesis and for possible genetic improvement in GSLs accumulation in radish fleshy taproots. Of course, in order to confirm and apply our results, further studies are needed, such as fine mapping, cloning, and transgenic function verification of crucial genes, especially *RsMYB28*, for constructing a usable and effective regulatory network of GRH synthesis in radish genome design breeding and also in synthetic biology research to produce more beneficial GSLs for promoting human health in the future.

4. Materials and Methods

4.1. Plant Material, Cultivation, and Sampling

Based on our preliminary identification of the GSLs content in the radish taproots, three inbred lines—P16Q-17-1 (HIGH1), P16Q-23-1 (HIGH2), and P16Q-24-1 (LOW1)—with significant differences in GSLs content were selected for this study. 'P16Q-17-1' was a radish advanced inbred line coming from a local variety 'Dahongpao' in Beijing, China, and had a larger elliptical fleshy taproot with red peel and white flesh. The advanced inbred line 'P16Q-24-1' was bred from a local variety 'Qiaotouqing' in Jilin province, China, and had a long conical fleshy taproot with green peel and light green flesh. The advanced inbred line 'P16Q-23-1' was developed from the Japanese radish variety 'Xiaqiumeinong'. It had a long cylindrical fleshy taproot with white peel and white flesh. A randomized block design was performed for three varieties with three replicates and fifteen plants for each replicate for each variety on the farm at the Institute of Vegetables and Flowers, Chinese Academy of Agricultural Sciences, Beijing, China. The seeds were sown in the middle of August, and fleshy taproots were harvested at the end of October 2017. The agronomic techniques were the same as those used in radish field production. The highest and lowest month average temperature was about 20 °C and 30 °C in August, about 15 °C and 28 °C in September, and 10 °C and 15 °C in October on the farm. The sandy loam with deep soil layer of water content in 65–80% and well drainage was adopted. In terms of fertilizers, we created a good balance in the application of three nutrient elements (15% N, 15% p, 15% K) [69]. During the fleshy taproot ripening periods, the whole taproots of five normally growing plants were collected and treated to form a replicate. Three replicates of each variety were rapidly generated, frozen in liquid nitrogen, and stored at −80 °C for GSLs analysis and RNA extraction.

4.2. GSLs Extraction and Analysis

GSLs were extracted according to the modified method described by He et al. [70]. Two hundred milligrams of frozen-dried taproot powder were weighed and placed into a 15 mL tube. Five milliliters of 100% methanol were added to the tube, and 100 µL of glucotropaeolin (benzylGS) was then added as an internal standard. The samples were then incubated at 83 °C for 20 min in a water bath and vortexed at 4 min to 5 min intervals The mixture was centrifuged at 3000× g and 4 °C for 10 min, and the supernatant was decanted into another tube and stored in an ice bath. The extraction was repeated twice from residues using the same procedure with 70% methanol, and the supernatant solution was incorporated to obtain the sample solution. The three supernatants were combined, and 2 mL of each GSLs extract was added to a mini-column filled with diethylaminoethanol (DEAE) activated with 0.02 M NaAc and desulfated by sulfatase (Helix Pomatia Type H-1, Sigma Company, New York, NY, USA). After reaction at room temperature overnight (16 h), the desulfated GSLs were eluted with 2 mL of deionized water and stored at −20 °C for high-performance liquid chromatography (HPLC) analysis.

The HPLC system included a UV detector (SPD-20A, SHIMADZU, Tsu, Japan) with a wavelength of 230 nm, a gradient elution pump (LC-20AD, SHIMADZU, Tsu, Japan), an automatic sampler (Gina 160, Gynkotek, Germany), and a LiChrospher RP-18 analysis column (2504 mm, Merck, Germany). Mobile phase A consisted of 0.1 mol/L ammonium acetate, and mobile phase B consisted of 0.1 mol/L ammonium acetate + 300 mL of methanol; gradient elution was performed for 34 min, and all GSLs were separated. The retention times of the reference substances obtained with the newly established ESI/MS method were compared to identify GSLs. Because no phenylthioside was found in the sample, this compound was used as the internal standard. The response factors are reported with reference to Moller et al. [71].

4.3. RNA Extraction, Library Construction, and RNA-Seq

RNA was extracted from each biological replicate of the radish fleshy taproots using the RNAprep Pure Plant Kit (TIANGEN). After characterization of the RNA purity using a Nanodrop 1000 spectrophotometer (Thermo Fisher Scientific, Waltham, MA, USA) and measurement of the RNA concentration with a Qubit R 2.0 Fluorometer (Thermo Fisher Scientific, Waltham, MA, USA), the RNA integrity was assessed with an Agilent Bioanalyzer 2100 system (Agilent Technologies Inc. Palo Alto, CA, USA). RNA samples with an integrity higher than 7.0 were selected to construct libraries. Illumina sequencing was performed at Mega Genomics Technologies Corporation (Beijing, China). After enrichment and purification with oligo(dT)-rich magnetic beads, the mRNAs were broken into short fragments, and these fragments were converted to first- and second-strand cDNA. The cDNA was purified using AMPure XP beads, and the 3′ ends of the cDNA fragments were repaired. Poly (A) was then added and ligated to adapters for selection of a template size range. The nine cDNA libraries were then enriched by PCR amplification and sequenced using an Illumina HiSeq™ 2500.

After removing the reads with only adaptors and unknown nucleotides larger than 5% or those with low quality, the clean reads were filtered from the raw reads. The clean RNA-seq reads were then mapped to the radish reference genome [55] using TopHat 2.1.1 [72]. The subsequent transcripts were then assembled, and the transcript abundance was measured based on the fragments per kilobase of exon per million fragments (FPKM) values using Cufflinks 2.2.1 [73] with the default parameters.

4.4. Functional Annotation of Expressed Genes and Differential Expression Analysis

Gene function was annotated based on the following databases: Nr (NCBI nonredundant protein sequences), Swiss-Prot (a manually annotated and reviewed protein sequence database), GO (Gene Ontology), KOG (EuKaryotic Orthologous Groups), COG (Clusters of Orthologous Groups of proteins), Pfam (Protein family), and KEGG (Kyoto Encyclopedia of Genes and Genomes).

A differential expression analysis among each pair of varieties was performed using the FPKM values by means of DESeq [74]. Genes with significant differential expression, denoted as DEGs, were identified based on the co-occurrence of the following two criteria: absolute value of fold change (FC) ≥ 2 and false discovery rate (FDR) value ≤ 0.01 [75]. Heatmaps were generated with the TBtoools software with log2 transformed FPKM values [76].

4.5. Co-Expression Analysis of the Key Genes for GSLs Biosynthesis and Their Relationship with GSLs Content in Radish Fleshy Taproots

Gene co-expression analysis and gene expression–metabolite production correlation analyses were then performed using R 3.5.0. The correlations between the expression of key genes and the GSLs content were determined by Pearson's correlation analysis based on the criteria $r \geq |0.7|$ and $p \leq 0.01$. Based on the expression levels of the selected important DEGs and the contents of the predominate GSLs components, we calculated their Pearson's coefficients (r) and their correlation significance. Then we determined the putative key genes and components in the network according to the r values at an extremely significant level and qPCR verification results. Further, we searched the existing relationships among these key genes and GSL components discovered by previous researchers in the STRING database [77] and validated their interaction relationship again. In this way, we decided what were sources and what were targets. After that, we prepared the files (Tables S6 and S7) that included the source, target, node, Pearson's coefficient, and the types and colors of lines designated in the network. Finally, we introduced these data into Cytoscape 3.3.0 [78] software for the interaction relationship visualization.

4.6. Gene Expression Validation and Analysis

Total RNA was extracted using the above-described method. A total of 800 ng of total RNA was reverse transcribed to synthesize first-strand cDNA using oligo dT primers and

EasyScript® One-Step gDNA Removal and cDNA Synthesis SuperMix (TransGen Biotech, Beijing, China) and diluted 20-fold to serve as templates for qPCR.

Quantitative real-time PCR was performed with a StepOne™ Real-Time PCR System (Applied Biosystems, Waltham, MA, USA) and TransStart® Green qPCR SuperMix (TransGen Biotech) following the manufacturer's instructions. Primers were designed using Primer-BLAST Web (https://www.ncbi.nlm.nih.gov/tools/primer-blast/index.cgi?LINK_LOC=BlastHome (accessed on 9 September 2019)). The length of primer is generally 15–30 bp, the product size is 100–200 bp, GC content is 40–60%, and temperature (Tm) is in the range of 57–63 °C. The list of genes and primers is shown in Table S1. Three independent biological and four technical replicates were performed. The data were analyzed using StepOne™ Software v.2.0 (Applied Biosystems, Waltham, MA, USA). Tubulin was used as an internal control. The relative expression levels were estimated using the $2^{-\Delta\Delta C_T}$ method [79].

Supplementary Materials: The following supporting information can be downloaded at: https://www.mdpi.com/article/10.3390/ijms23062953/s1.

Author Contributions: Conceived the study: X.L. (Xixiang Li); supervised the implementation: X.L. (Xixiang Li) and L.L.; designed the experiments: X.L. (Xixiang Li), X.L. (Xiaoman Li), J.W. and P.W.; performed the experiments: P.W., X.L. (Xiaoman Li), J.W., T.L., C.W. and H.Y.; prepared materials: J.W., X.L. (Xiaoman Li), P.W., J.S. and W.Y.; analyzed the data: P.W., X.L. (Xiaoman Li), J.W. and X.Z.; provided daily management: H.W. and J.W.; wrote the paper: X.L. (Xiaoman Li) and P.W.; revised the paper: X.L. (Xixiang Li) and L.L. All authors have read and agreed to the published version of the manuscript.

Funding: This work was supported by grants from the National Key Technology R & D Program of the Ministry of Science and Technology of China (2016YFD0100204-02), the National Natural Science Foundation of China (31772303), the National Horticulture Germplasm Resources Center Project (NHGRC2021-NH01), and the Science and Technology Innovation Program of the Chinese Academy of Agricultural Sciences (CAAS-XTCX2016016-4-4, CAAS-XTCX2016001-5-2).

Institutional Review Board Statement: Not applicable.

Informed Consent Statement: Not applicable.

Data Availability Statement: Not applicable.

Conflicts of Interest: The authors declare no conflict of interest.

References

1. Halkier, B.A.; Gershenzon, J. Biology and biochemistry of glucosinolates. *Annu. Rev. Plant Biol.* **2006**, *57*, 303–333. [CrossRef] [PubMed]
2. Sønderby, I.E.; Geu-Flores, F.; Halkier, B. Biosynthesis of glucosinolates–gene discovery and beyond. *Trends Plant Sci.* **2010**, *15*, 283–290. [CrossRef]
3. Scholl, C.; Eshelman, B.D.; Barnes, D.M.; Hanlon, P.R. Raphasatin is a more potent inducer of the detoxification enzymes than its degradation products. *J. Food Sci.* **2011**, *76*, C504–C511. [CrossRef] [PubMed]
4. Clarke, D.B. Glucosinolates, structures and analysis in food. *Anal. Methods* **2010**, *2*, 310–325. [CrossRef]
5. Petersen, B.L.; Chen, S.; Hansen, C.H.; Olsen, C.E.; Halkier, B. Composition and content of glucosinolates in developing *Arabidopsis thaliana*. *Planta* **2002**, *214*, 562–571. [CrossRef] [PubMed]
6. Yang, B.; ·Quiros, C.F. Survey of glucosinolate variation in leaves of *Brassica rapa* crops. *Genet Resour. Crop. Evol.* **2010**, *57*, 1079–1089. [CrossRef]
7. Sun, B.; Liu, N.; Zhao, Y.; Yan, H.; Wang, Q. Variation of glucosinolates in three edible parts of Chinese kale (*Brassica alboglabra* Bailey) varieties. *Food Chem.* **2011**, *124*, 941–947. [CrossRef]
8. Farnham, M.W.; Stephenson, K.K.; Fahey, J.W. Glucoraphanin level in broccoli seed is largely determined by genotype. *HortScience* **2005**, *40*, 50–53. [CrossRef]
9. Kakizaki, T.; Kitashiba, H.; Zou, Z.; Li, F.; Fukino, N.; Ohara, T.; Nishio, T.; Ishida, M. A 2-Oxoglutarate-Dependent Dioxygenase Mediates the Biosynthesis of Glucoraphasatin in Radish. *Plant Physiol.* **2017**, *173*, 1583–1593. [CrossRef]
10. Zhang, Y.; Kensler, T.W.; Cho, C.G.; Posner, G.H.; Talalay, P. Anticarcinogenic activities of sulforaphane and structurally related synthetic norbornyl isothiocyanates. *Proc. Natl. Acad. Sci. USA* **1994**, *91*, 3147–3150. [CrossRef]

11. Razis, A.F.A.; Bagatta, M.; De Nicola, G.R.; Iori, R.; Ioannides, C. Up-regulation of cytochrome P450 and phase II enzyme systems in rat precision-cut rat lung slices by the intact glucosinolates, glucoraphanin and glucoerucin. *Lung Cancer* **2011**, *71*, 298–305. [CrossRef] [PubMed]
12. Brader, G.; Mikkelsen, M.D.; Halkier, B.A.; Tapio Palva, E. Altering glucosinolate profiles modulates disease resistance in plants. *Plant J.* **2006**, *46*, 758–767. [CrossRef] [PubMed]
13. Lopez-Rodriguez, M.F.; Cymbaluk, N.; Epp, T.; Laarveld, B.; Recale, E.; Simoko, E.; Card, C. Effects of the glucosinolate sinigrin in combination with a non-iodine supplemented diet on serum iodine and thyroid hormone concentrations in non-pregnant mares. *J. Equine Vet. Sci.* **2020**, *91*, 103110. [CrossRef] [PubMed]
14. Dalmazi, G.D.; Giuliani, C. Plant constituents and thyroid: A revision of the main phytochemicals that interfere with thyroid function. *Food Chem. Toxicol.* **2021**, *152*, 112158. [CrossRef]
15. Matusheski, N.V.; Jeffery, E.H. Comparison of the bioactivity of two glucoraphanin hydrolysis products found in broccoli, sulforaphane and sulforaphane nitrile. *J. Agric. Food Chem.* **2001**, *49*, 5743–5749. [CrossRef]
16. Hanschen, F.S.; Herz, C.; Schlotz, N.; Kupke, F.; Rodríguez, M.M.B.; Schreiner, M.; Rohn, S.; Lamy, E. The Brassicaepithionitrile 1-cyano-2,3-epithiopropane triggers cell death in human liver cancer cellsin vitro. *Mol. Nutr. Food Res.* **2015**, *59*, 2178–2189. [CrossRef]
17. Grubb, C.D.; Abel, S. Glucosinolate metabolism and its control. *Trends Plant Sci.* **2006**, *11*, 89–100. [CrossRef]
18. Diebold, R.; Schuster, J.; Däschner, K.; Binder, S. The branched-chain amino acid transaminase gene family in arabidopsis encodes plastid and mitochondrial proteins. *Plant Physiol.* **2002**, *129*, 540–550. [CrossRef]
19. Schuster, J.; Knill, T.; Reichelt, M.; Gershenzon, J.; Binder, S. Branched-chain aminotransferase4 is part of the chain elongation pathway in the biosynthesis of methionine-derived glucosinolates in Arabidopsis. *Plant Cell* **2006**, *18*, 2664–2679. [CrossRef]
20. Knill, T.; Schuster, J.; Reichelt, M.; Gershenzon, J.; Binder, S. Arabidopsis branched-chain aminotransferase 3 functions in both amino acid and glucosinolate biosynthesis. *Plant Physiol.* **2008**, *146*, 1028–1039. [CrossRef]
21. Kroymann, J.; Textor, S.; Tokuhisa, J.G.; Falk, K.L.; Bartram, S.; Gershenzon, J.; Mitchell-Olds, T. A gene controlling variation in Arabidopsis glucosinolate composition is part of the methionine chain elongation pathway. *Plant Physiol.* **2001**, *127*, 1077–1088. [CrossRef] [PubMed]
22. He, Y.; Mawhinney, T.P.; Preuss, M.L.; Schroeder, A.C.; Chen, B.; Abraham, L.; Jez, J.M.; Chen, S. A redox-active isopropylmalate dehydrogenase functions in the biosynthesis of glucosinolates and leucine in Arabidopsis. *Plant J.* **2009**, *60*, 679–690. [CrossRef] [PubMed]
23. Hansen, C.H.; Wittstock, U.; Olsen, C.E.; Hick, A.J.; Pickett, J.A.; Halkier, B. Cytochrome P450 CYP79F1 from Arabidopsis catalyzes the conversion of dihomomethionine and trihomomethionine to the corresponding aldoximes in the biosynthesis of aliphatic glucosinolates. *J. Biol. Chem.* **2001**, *276*, 11078–11085. [CrossRef] [PubMed]
24. Chen, S.; Glawischnig, E.; Jørgensen, K.; Naur, P.; Jørgensen, B.; Olsen, C.E.; Hansen, C.H.; Rasmussen, H.; Pickett, J.A.; Halkier, B.A. CYP79F1 and CYP79F2 have distinct functions in the biosynthesis of aliphatic glucosinolates in Arabidopsis. *Plant J.* **2003**, *33*, 923–937. [CrossRef]
25. Hemm, M.R.; Ruegger, M.O.; Chapple, C. The Arabidopsis ref2 mutant is defective in the gene encoding CYP83A1 and shows both phenylpropanoid and glucosinolate phenotypes. *Plant Cell* **2003**, *15*, 179–194. [CrossRef]
26. Mikkelsen, M.D.; Naur, P.; Halkier, B.A. Arabidopsis mutants in the C-S lyase of glucosinolate biosynthesis establish a critical role for indole-3-acetaldoxime in auxin homeostasis. *Plant J.* **2004**, *37*, 770–777. [CrossRef]
27. Gachon, C.; Langlois-Meurinne, M.; Henry, Y.; Saindrenan, P. Transcriptional co-regulation of secondary metabolism enzymes in Arabidopsis: Functional and evolutionary implications. *Plant Mol. Biol.* **2005**, *58*, 229–245. [CrossRef]
28. Piotrowski, M.; Schemenewitz, A.; Lopukhina, A.; Müller, A.; Janowitz, T.; Weiler, E.W.; Oecking, C. Desulfoglucosinolate sulfotransferases from *Arabidopsis thaliana* catalyze the final step in the biosynthesis of the glucosinolate core structure. *J. Biol. Chem.* **2004**, *279*, 50717–50725. [CrossRef]
29. Wang, J.; Qiu, Y.; Wang, X.; Yue, Z.; Yang, X.; Chen, X.; Zhang, X.; Shen, D.; Wang, H.; Song, J.; et al. Insights into the species-specific metabolic engineering of glucosinolates in radish (*Raphanus sativus* L.) based on comparative genomic analysis. *Sci. Rep.* **2017**, *7*, 16040. [CrossRef]
30. Gigolashvili, T.; Yatusevich, R.; Berger, B.; Müller, C.; Flügge, U.-I. The R2R3-MYB transcription factor HAG1/MYB28 is a regulator of methionine-derived glucosinolate biosynthesis in *Arabidopsis thaliana*. *Plant J.* **2007**, *51*, 247–261. [CrossRef]
31. Celenza, J.L.; Quiel, J.A.; Smolen, G.A.; Merrikh, H.; Silvestro, A.R.; Normanly, J.; Bender, J. The Arabidopsis ATR1 Myb transcription factor controls indolic glucosinolate homeostasis. *Plant Physiol.* **2005**, *137*, 253–262. [CrossRef] [PubMed]
32. Gigolashvili, T.; Yatusevich, R.; Rollwitz, I.; Humphry, M.; Gershenzon, J.; Flügge, U.-I. The plastidic bile acid transporter 5 is required for the biosynthesis of methionine-derived glucosinolates in *Arabidopsis thaliana*. *Plant Cell* **2009**, *21*, 1813–1829. [CrossRef] [PubMed]
33. Sønderby, I.E.; Burow, M.; Rowe, H.C.; Kliebenstein, D.; Halkier, B.A. A complex interplay of three R2R3 MYB transcription factors determines the profile of aliphatic glucosinolates in Arabidopsis. *Plant Physiol.* **2010**, *153*, 348–363. [CrossRef] [PubMed]
34. Yin, L.; Chen, H.; Cao, B.; Lei, J.; Chen, G. Molecular characterization of MYB28 involved in Aliphatic glucosinolate biosynthesis in Chinese kale (*Brassica oleracea* var. *alboglabra Bailey*). *Front. Plant Sci.* **2017**, *8*, 1083. [CrossRef]
35. Augustine, R.; Mukhopadhyay, A.; Bisht, N.C. Targeted silencing of BjMYB28 transcription factor gene directs development of low glucosinolate lines in oilseed Brassica juncea. *Plant Biotechnol. J.* **2013**, *11*, 855–866. [CrossRef]

36. Augustine, R.; Majee, M.; Gershenzon, J.; Bisht, N.C. Four genes encoding MYB28, a major transcriptional regulator of the aliphatic glucosinolate pathway, are differentially expressed in the allopolyploid *Brassica juncea*. *J. Exp. Bot.* **2013**, *64*, 4907–4921. [CrossRef] [PubMed]
37. Araki, R.; Hasumi, A.; Nishizawa, O.I.; Sasaki, K.; Kuwahara, A.; Sawada, Y.; Totoki, Y.; Toyoda, A.; Sakaki, Y.; Li, Y. Novel bioresources for studies of B rassica oleracea: Identification of a kale MYB transcription factor responsible for glucosinolate production. *Plant Biotechnol. J.* **2013**, *11*, 1017–1027. [CrossRef]
38. Wang, B.; Wu, Z.; Li, Z.; Zhang, Q.; Hu, J.; Xiao, Y.; Cai, D.; Wu, J.; King, G.J.; Li, H.; et al. Dissection of the genetic architecture of three seed-quality traits and consequences for breeding in Brassica napus. *Plant Biotechnol. J.* **2018**, *16*, 1336–1348. [CrossRef]
39. Levy, M.; Wang, Q.; Kaspi, R.; Parrella, M.P.; Abel, S. Arabidopsis IQD1, a novel calmodulin-binding nuclear protein, stimulates glucosinolate accumulation and plant defense. *Plant J.* **2005**, *43*, 79–96. [CrossRef]
40. Skirycz, A.; Reichelt, M.; Burow, M.; Birkemeyer, C.; Rolcik, J.; Kopka, J.; Zanor, M.I.; Gershenzon, J.; Strnad, M.; Szopa, J.; et al. DOF transcription factor AtDof1.1 (OBP2) is part of a regulatory network controlling glucosinolate biosynthesis in Arabidopsis. *Plant J.* **2006**, *47*, 10–24. [CrossRef]
41. Crisp, P. Radish, *Raphanus sativus* (Cruciferae). In *Evolution of Crop Plants, 2nd ed*; Smartt, J., Simmonds, N.W., Eds.; Longman Scientific & Technical: Harlow, UK, 1995; pp. 86–89.
42. Maldini, M.; Foddai, M.; Natella, F.; Petretto, G.L.; Rourke, J.P.; Chessa, M.; Pintore, G. Identification and quantification of glucosinolates in different tissues of Raphanus raphanistrum by liquid chromatography tandem-mass spectrometry. *J. Food Compos. Anal.* **2017**, *61*, 20–27. [CrossRef]
43. Barillari, J.; Cervellati, R.; Paolini, M.; Tatibout, A.; Rollin, P.; Iori, R. Isolation of 4-Methylthio-3-butenyl Glucosinolate from *Raphanus sativus* Sprouts (Kaiware Daikon) and Its Redox Properties. *J. Agric. Food Chem.* **2005**, *53*, 9890–9896. [CrossRef] [PubMed]
44. Ishida, M.; Nagata, M.; Ohara, T.; Kakizaki, T.; Hatakeyama, K.; Nishio, T. Small variation of glucosinolate composition in Japanese cultivars of radish (*Raphanus sativus* L.) requires simple quantitative analysis for breeding of glucosinolate component. *Breed. Sci.* **2012**, *62*, 63–70. [CrossRef] [PubMed]
45. Yi, G.; Lim, S.; Chae, W.B.; Park, J.E.; Park, H.R.; Lee, E.J.; Huh, J.H. Root Glucosinolate Profiles for screening of radish (*Raphanus sativus* L.) genetic resources. *J. Agric. Food Chem.* **2016**, *64*, 61–70. [CrossRef]
46. Kim, S.-J.; Uddin, M.R.; Park, S.U. Glucosinolate accumulation in three important radish (*Raphanus sativus*) cultivars. *Australian J. Crop Sci.* **2013**, *7*, 1843–1847.
47. Carlson, D.G.; Daxenbichler, M.E.; Vanetten, C.H.; Hill, C.B.; Williams, P.H. Glucosinolates in radish cultivars. *J. Am. Soc. Hortic. Sci.* **1985**, *110*, 634–638.
48. Rampal, G.; Thind, T.S.; Vig, A.P.; Arora, S. Antimutagenic Potential of Glucosinolate-Rich Seed Extracts of Broccoli (*Brassica oleracea* L. var *Italica plenck*). *Int. J. Toxicol.* **2010**, *29*, 616–624. [CrossRef]
49. Voa, Q.V.; Trenerryb, C.; Rochfortc, S.; Wadesonc, J.; Hughes, A.B. Synthesis and anti-inflammatory activity of aromatic glucosinolates. *Bioorganic Med. Chem.* **2013**, *21*, 5945–5954. [CrossRef]
50. Dinkova-Kostova, A.T.; Kostov, R.V. Glucosinolates and isothiocyanates in health and disease. *Trends Mol. Med.* **2012**, *18*, 337–347. [CrossRef]
51. Zou, Z.; Ishida, M.; Li, F.; Kakizaki, T., Suzuki, S., Kitashiba, H., Nishio, T. QTL analysis using SNP markers developed by next-generation sequencing for identification of candidate genes controlling 4-Methylthio-3-Butenyl glucosinolate contents in roots of radish, *Raphanus sativus* L. *PLoS ONE* **2013**, *8*, e53541. [CrossRef]
52. Kuang, P.; Song, D.; Yuan, Q.; Lv, X.; Zhao, D.; Liang, H. Preparative separation and purification of sulforaphene from radish seeds by high-speed countercurrent chromatography. *Food Chem.* **2013**, *136*, 309–315. [CrossRef] [PubMed]
53. Kang, J.-N.; Won, S.Y.; Seo, M.-S.; Lee, J.; Lee, S.M.; Kwon, S.-J.; Kim, J.S. Induction of glucoraphasatin biosynthesis genes by MYB29 in radish (*Raphanus sativus* L.) roots. *Int. J. Mol. Sci.* **2020**, *21*, 5721. [CrossRef] [PubMed]
54. Hua, B. *Analysis and Evaluation on Glucosinolate Composition and Sulforaphane Content of Radish (Raphanus sativus* L.*) Germplasm Resources*; Chinese Academy of Agricultural Science: Beijing, China, 2013.
55. Zhang, X.; Yue, Z.; Mei, S.; Qiu, Y.; Yang, X.; Chen, X.; Cheng, F.; Wu, Z.; Sun, Y.; Jing, Y. A de novo genome of a Chinese radish cultivar. *Hortic. Plant J.* **2015**, *1*, 155–164.
56. Mithen, R.; Clarke, J.; Lister, C.; Dean, C. Genetics of aliphatic glucosinolates. III. Side chain structure of aliphatic glucosinolates in *Arabidopsis thaliana*. *Heredity* **1995**, *74*, 210–215. [CrossRef]
57. Barth, C.; Jander, G. Arabidopsis myrosinases TGG1 and TGG2 have redundant function in glucosinolate breakdown and insect defense. *Plant J.* **2006**, *46*, 549–562. [CrossRef]
58. Gigolashvili, T.; Engqvist, M.; Yatusevich, R.; Müller, C.; Flügge, U.I. HAG2/MYB76 and HAG3/MYB29 exert a specific and coordinated control on the regulation of aliphatic glucosinolate biosynthesis in *Arabidopsis thaliana*. *New Phytol.* **2008**, *177*, 627–642. [CrossRef]
59. Gigolashvili, T.; Berger, B.; Mock, H.P.; Müller, C.; Weisshaar, B.; Flügge, U.I. The transcription factor HIG1/MYB51 regulates indolic glucosinolate biosynthesis in *Arabidopsis thaliana*. *Plant J.* **2007**, *50*, 886–901. [CrossRef]
60. Farnham, M. A comprehensive program to enhance glucoraphanin content of broccoli heads and seed. In Proceedings of the International Symposium on Human Health Effects of Fruits and Vegetables, Québec City, QC, Canada, 17–20 August 2005; p. 30.

61. Hirai, M.; Sugiyama, K.; Sawada, Y.; Tohge, T.; Obayashi, T.; Suzuki, A.; Araki, R.; Sakurai, N.; Suzuki, H.; Aoki, K.; et al. Omics-based identification of Arabidopsis Myb transcription factors regulating aliphatic glucosinolate biosynthesis. *Proc. Natl. Acad. Sci. USA* **2007**, *104*, 6478–6483. [CrossRef]
62. Frerigmann, H. Glucosinolate regulation in a complex relationship–MYC and MYB–no one can act without each other. *Adv. Bot. Res.* **2016**, *80*, 57–97.
63. Seo, M.-S.; Jin, M.; Chun, J.-H.; Kim, S.-J.; Park, B.-S.; Shon, S.-H.; Kim, J.S. Functional analysis of three BrMYB28 transcription factors controlling the biosynthesis of glucosinolates in *Brassica rapa*. *Plant Mol. Biol.* **2016**, *90*, 503–516. [CrossRef]
64. Zang, Y.-X.; Kim, J.-H.; Park, Y.-D.; Kim, D.-H.; Hong, S.-B. Metabolic engineering of aliphatic glucosinolates in Chinese cabbage plants expressing Arabidopsis MAM1, CYP79F1, and CYP83A1. *BMB Rep.* **2008**, *41*, 472–478. [CrossRef] [PubMed]
65. Sharma, M.; Mukhopadhyay, A.; Gupta, V.; Pental, D.; Pradhan, A.K. BjuB.CYP79F1 regulates synthesis of propyl fraction of aliphatic glucosinolates in oilseed mustard *Brassica juncea*: Functional validation through genetic and transgenic approaches. *PLoS ONE* **2016**, *11*, e0150060. [CrossRef]
66. Magrath, R.; Bano, F.; Morgner, M.; Parkin, I.; Sharpe, A.; Lister, C.; Dean, C.; Turner, J.; Lydiate, D.; Mithen, R. Genetics of aliphatic glucosinolates. I. Side chain elongation in Brassica napus and *Arabidopsis thaliana*. *Heredity* **1994**, *72*, 290–299. [CrossRef]
67. Li, G.; Gao, M.; Yang, B.; Quiros, C.F. Gene for gene alignment between the Brassica and Arabidopsis genomes by direct transcriptome mapping. *Theor. Appl. Genet.* **2003**, *107*, 168–180. [CrossRef] [PubMed]
68. Dixon, D.P.; Skipsey, M.; Edwards, R. Roles for glutathione transferases in plant secondary metabolism. *Phytochemistry* **2010**, *71*, 338–350. [CrossRef]
69. Institute of Vegetables and Flowers, Chinese Academy of Agricultural Sciences. *Chinese Vegetable Cultivation*; China Agriculture Press: Beijing, China, 2009; pp. 276–291.
70. He, H.; Chen, H.; Schnitzler, W.H. Glucosinolate composition and contents in Brassica vegetables. *Sci. Agric. Sin.* **2002**, *35*, 192–197.
71. Møller, P.; Olsen, O.; Plöger, A.; Rasmussen, K.; Sørensen, H. Quantitative analysis of individual glucosinolates. In *Advances in the Production and Utilization of Cruciferous Crops*; Springer: Dordrecht, The Netherlands, 1985; pp. 111–126.
72. Kim, D.; Pertea, G.; Trapnell, C.; Pimentel, H.; Kelley, R.; Salzberg, S.L. TopHat2: Accurate alignment of transcriptomes in the presence of insertions, deletions and gene fusions. *Genome Biol.* **2013**, *14*, R36. [CrossRef]
73. Trapnell, C.; Williams, B.A.; Pertea, G.; Mortazavi, A.; Kwan, G.; Van Baren, M.J.; Salzberg, S.L.; Wold, B.J.; Pachter, L. Transcript assembly, differential expression, and differential regulation for RNA-Seq reveals unannotated transcripts and isoform switching during cell differentiation. *Nat. Biotechnol.* **2010**, *28*, 511–515. [CrossRef]
74. Anders, S.; Huber, W. Differential expression analysis for sequence count data. *Nat. Preced.* **2010**, *11*, R106.
75. Benjamini, Y.; Yekutieli, D. The control of the false discovery rate in multiple testing under dependency. *Ann. Stat.* **2001**, *29*, 1165–1188. [CrossRef]
76. Chen, C.; Chen, H.; Zhang, Y.; Thomas, H.R.; Frank, M.H.; He, Y.; Xia, R. TBtools: An integrative toolkit developed for interactive analyses of big biological data. *Mol. Plant* **2020**, *13*, 1194–1202. [CrossRef] [PubMed]
77. Mering, V.C. STRING: A database of predicted functional associations between proteins. *Nucleic Acids Res.* **2003**, *31*, 258. [CrossRef] [PubMed]
78. Shannon, P.; Markiel, A.; Ozier, O.; Baliga, N.S.; Wang, J.T.; Ramage, D.; Amin, N.; Schwikowski, B.; Ideker, T. Cytoscape: A software environment for integrated models of Biomolecular Interaction Networks. *Genome Res.* **2003**, *13*, 2498–2504. [CrossRef] [PubMed]
79. Livak, K.J.; Schmittgen, T.D. Analysis of relative gene expression data using real-time quantitative PCR and the $2^{-\Delta\Delta C_T}$ method. *Methods* **2001**, *25*, 402–408. [CrossRef]

International Journal of
Molecular Sciences

MDPI

Review

Mechanism and Utilization of Ogura Cytoplasmic Male Sterility in Cruciferae Crops

Wenjing Ren [1,2,†], Jinchao Si [1,†], Li Chen [1,2], Zhiyuan Fang [1], Mu Zhuang [1], Honghao Lv [1], Yong Wang [1], Jialei Ji [1], Hailong Yu [1,*] and Yangyong Zhang [1,*]

1 Institute of Vegetables and Flowers, Chinese Academy of Agricultural Sciences, Key Laboratory of Biology and Genetic Improvement of Horticultural Crops, Ministry of Agriculture, Beijing 100081, China
2 State Key Laboratory of Crop Genetics and Germplasm Enhancement, College of Horticulture, Nanjing Agricultural University, Nanjing 210095, China
* Correspondence: yuhailong@caas.cn (H.Y.); zhangyangyong@caas.cn (Y.Z.)
† These authors contributed equally to this work.

Abstract: Hybrid production using lines with cytoplasmic male sterility (CMS) has become an important way to utilize heterosis in vegetables. Ogura CMS, with the advantages of complete pollen abortion, ease of transfer and a progeny sterility rate reaching 100%, is widely used in cruciferous crop breeding. The mapping, cloning, mechanism and application of Ogura CMS and fertility restorer genes in *Brassica napus*, *Brassica rapa*, *Brassica oleracea* and other cruciferous crops are reviewed herein, and the existing problems and future research directions in the application of Ogura CMS are discussed.

Keywords: cruciferous vegetables; Ogura CMS; sterility gene; fertility restorer gene (FR gene); cytoplasmic male sterility; distant hybridization

1. Introduction

In 1973, the German botanist Joseph Gottlieb Kolreuter first observed the phenomenon of male sterility (MS). Researchers defined MS as plant failure to produce dehiscent anthers, functional pollen or energetic male gametes [1,2]. To date, researchers have observed MS in 43 families, 162 genera and approximately 617 species, including *Zea mays*, *Oryza sativa*, *Raphanus sativus*, *Gossypium hirsutum*, *Allium cepa*, *Sorghum bicolor* and *Glycine max* [3–6]. According to different inheritance characteristics, MS can be divided into genic MS (GMS) and cytoplasmic MS (CMS). GMS is controlled by one or more nuclear genes, while CMS is usually controlled by mitochondrial genes and a few nuclear genes. CMS has been found in more than 150 higher plants [7,8]. Because female organ function is not affected by CMS and CMS lines can be leveraged to maximize cost-effectiveness, the CMS/FR system has become the most widely used MS system in cruciferous crops [9,10]. To date, numerous different CMS systems have been reported in cruciferous crops [11], including Ogura CMS (Ogu CMS), Nap CMS, Polima CMS (Pol CMS), Kosena CMS (Kos CMS), Tour CMS, Moricandia arvensis CMS, Nsa CMS, Nca CMS, Hau CMS and inap CMS (Figure 1). In addition, several CMS-related genes have been identified, including *orf138* of CMS-Ogu in *Raphanus sativus* [12–21], *orf263* of CMS-Tour in *Brassica tournefortii* [22], *orf224* of CMS-Polima, *orf222* of CMS-Nap in *Brassica napus* [23,24], *orf220* of a new type of CMS line and *orf288* of CMS-hau in *Brassica juncea* [25–27] (Figure 1). Among all types of CMS, Ogura CMS is the most widely applied in cruciferous crops due to its significant advantages of stable sterility and complete abortion.

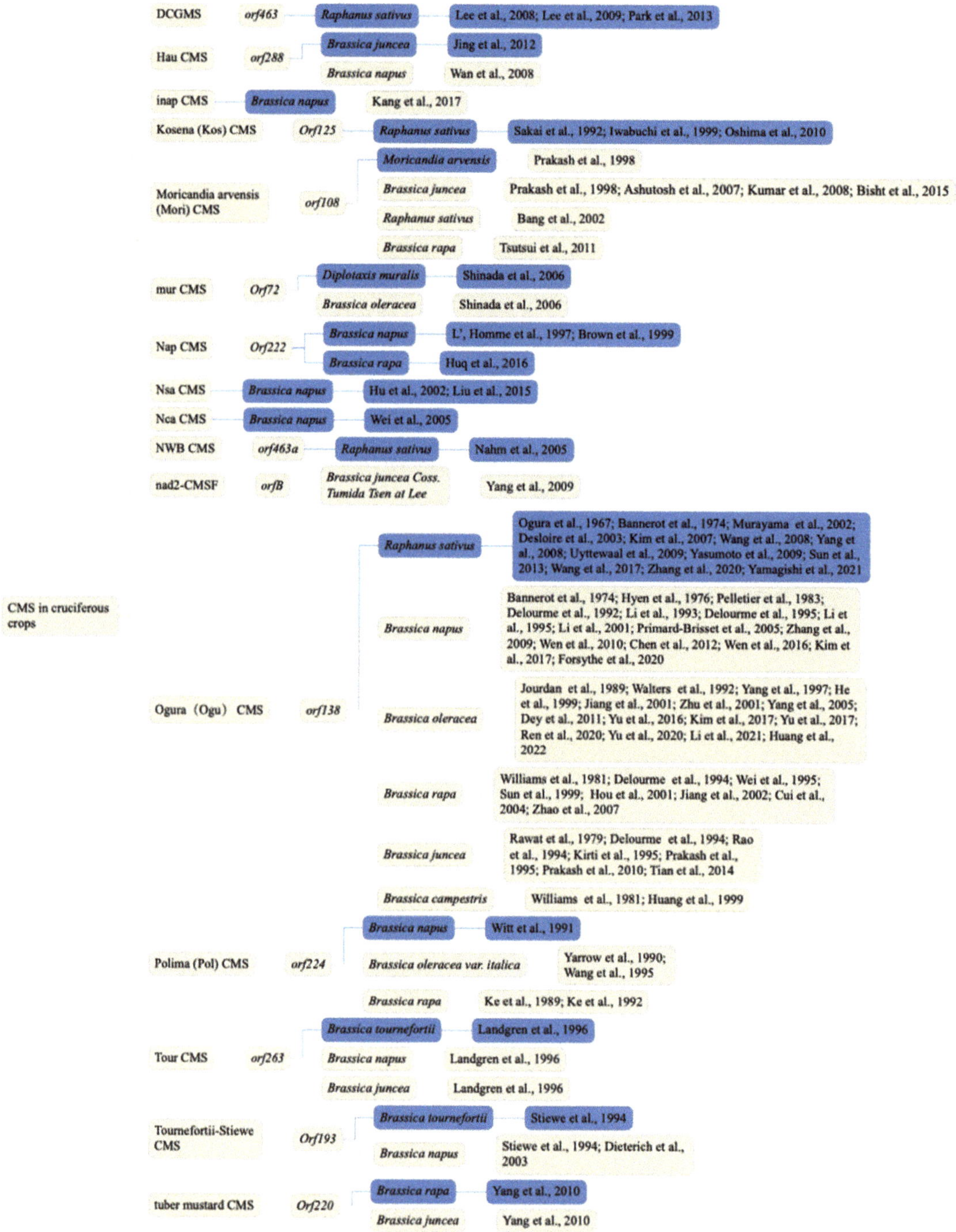

Figure 1. CMS systems reported in cruciferous crops. Blue indicates the origin of a cruciferous vegetable crop [22,23,26–112].

2. Mapping and Cloning of the Ogura CMS Gene

2.1. Identification of the Ogura Sterility Gene *orf138*

The determination of the sterility gene of the CMS system is a critical step in mechanistic research. Through the transcription pattern profiling approach, glutathione S-transferase (GST) fusion protein experiments and mitochondrial genome sequencing, the *orf138* gene was identified as the sterility gene of Ogura CMS. *orf138* was first identified because the transcription patterns of three mitochondrial structural genes (*atp6*, *atpA* and *coxI*) changed in Ogura CMS radish, but these changes were not strictly correlated with Ogura CMS or Ogura nuclear restorer genes [12,13]. An Ogura-specific mitochondrial DNA fragment containing two ORFs, *orf138* and *orf158*, was subsequently reported, which was consistent with the sterility phenotype in the 13 somatic hybrids of Ogura CMS *B. napus* obtained from protoplast fusions [14,15]. Sequence analysis and transcript analysis of this region showed that there were also transcripts of *orf158* in fertile plants but no transcripts of the *orf138* gene in fertile revertant plants. Therefore, it was speculated that the gene controlling Ogura CMS is the mitochondrial gene *orf138* [16]. In addition, a polypeptide confirmed to be the product of the *orf138* gene using antibodies against a glutathione S-transferase-ORF138 fusion protein was only found in sterile-plant mitochondria isolated from *B. napus* with Ogura CMS, which further confirmed that *orf138* was the sterility gene in Ogura CMS [17,18]. The mitochondrial genome sequencing of the Ogura CMS line and maintainer line promoted the identification of the Ogura CMS gene *orf138*. The unique presence of the *orf138* gene in Ogura CMS lines was confirmed by comparing the mitochondrial genomes of Cruciferae crops, and the key role of the MS-causing gene *orf138* was identified. The Ogura CMS mitochondrial genomes were generally larger and showed higher recombination and rearrangement than the mitochondrial genomes of the maintainer line, leading to the loss or generation of some genes (Table 1). For example, the mitochondrial genome sequencing of Ogura CMS radish showed that the mitochondrial genome had four special regions compared with that of normal fertile plants and that the *orf138* gene was located at the edge of the largest region (15,255 bp), which was occupied by the *cox1* gene in normal fertile radish [18–20]. Meanwhile, Ogura CMS radish possesses an additional copy of the genes *atp9* and *trnfM*.

Table 1. Sequenced mitochondrial genome information for Ogura CMS sterile lines and corresponding maintainers in cruciferous crops.

Cruciferae Crop	Material	Genome (bp)	Accession No.	Genes	Unique Regions	Reference
Raphanus sativus	Ogu CMS maintainer	244,036	AB694743	33 protein-coding genes, 3 rRNA genes and 17 tRNA sequences	/	[20]
Raphanus sativus	Ogu CMS	258,426	AB694744	Same genes with its maintainer line and additional *atp9* and tRNA gene *trnfM*	Four	[20]
Brassica oleracea	Ogu CMS maintainer	219,962	MW423604	31 known genes, 67 ORFs, 3 rRNA genes and 25 tRNA genes	/	[113]
Brassica oleracea	Ogu CMS	236,648	MW423605	30 known genes (lost *rps7* gene compared with the maintainer line), 69 ORFs, 3 rRNA genes and 25 tRNA genes	Five	[113]
Brassica oleracea	Ogu CMS maintainer	219,969	ON758774	33 protein-coding genes, 3 rRNA genes and 18 tRNA genes	/	[114]
Brassica oleracea	Ogu CMS	71,998, 185,431	ON758776, ON758777	Same genes with its maintainer line and additional *orf138* and three tRNA genes (*trnfM-CAU*, *trnN-GUU* and *trnY-GUA*)	Nine	[114]
Brassica juncea	Ogu CMS maintainer (RLM198)	219,776	MT675103	/	/	[115]
Brassica juncea	Ogu CMS (OgRLM)	258,462	MT675104	/	/	[115]
Brassica juncea	Ogu CMS (Og1)	250,999, 96,185	MT675105, MT675106	34 protein-coding genes, 3 rRNA genes and 15 tRNA genes	/	[115]
Brassica juncea	Ogu CMS revertant line (Og1-rt)	250,999	MT675105	33 protein-coding genes (lost the *orf138* gene compared with the sterile line Og1), 3 rRNA genes and 15 tRNA genes	/	[115]
Brassica napus	Ogu CMS	258,473	Unknown	33 protein-coding genes, 3 rRNA sequences and 23 tRNA sequences	Nine	[116]

2.2. Functional Verification of the Sterility Gene orf138

The function of *orf138* was verified through transient gene overexpression in *Arabidopsis thaliana* and knockdown approaches. The method of using mitochondrion-targeting peptides to transform sterility genes into the mitochondrial genome has been successfully applied in the functional verification of CMS genes in *G. max*, *O. sativa*, *Beta vulgaris*, *B. napus*, *B. juncea*, *Capsicum annuum* and other crops [50,117,118]. Zhang et al. (2015) cloned the *orf138* gene from the mitochondria of the Ogura CMS line in cabbage, constructed a mitochondrion-targeting expression vector driven by the CaMV35S promoter, transformed it into *A. thaliana* to induce MS, and verified the sterility-related function of the *orf138* gene [119]. Evidence suggests that stable mitochondrial transformation is currently lacking, while recent advances in gene editing have made it possible to easily knock out a gene. The 3′ end of the *orf138* gene has three consecutive 39-bp repeats. Apart from a 39-bp deletion and two single-nucleotide polymorphisms (SNPs), the *orf125* and *orf138* genes share 100% sequence identity with each other, which means that these two genes are highly homologous. Kos CMS caused by *orf125* and Ogura CMS caused by *orf138* can both be fertility restored by the fertility restorer gene *Rfo*/*Rfk* [114]. Kazama et al. (2019) knocked out the mitochondrial gene *orf125* by mitoTALENs in rapeseed, and the hypothesis that *orf125* is the driver of Kosena-type CMS was verified, which strongly suggested that the *orf138* gene is the sterility gene of Ogura CMS [120].

3. Mapping and Cloning of Ogura CMS Fertility Restorer Genes in Cruciferous Crops

3.1. Ogura CMS Fertility Restorer Genes in Raphanus Sativus

Parallel with the discovery of sterile cytoplasms, FR genes were identified as the genetic means for the recovery of the lost harmony between the mitochondrial and nuclear genomes [121]. To date, all the Ogura CMS FR genes (*Rfo*, *Rfk*, *RF1*, *RF2*, *RF3*, *Rfob*, *Rfoc*, *RsRf3-1*/*RsRf3-2*, *RsRf3-4*~*RsRf3-7*, *Rft* and *Rfs*) have been reported in radish and cloned from European radish, Japanese radish and Chinese radish (Table 2). The Ogura CMS fertility restorer gene *Rfo* was first cloned using map cloning and comparative genomic analysis in radish [51,122]. A fertility restorer gene of Ogura CMS named *Rfk* was cloned and identified as the same gene as *Rfo* [123]. Three populations were used to locate *RF1*, *RF2* and *RF3* in the upper region of the Rsl linkage group, the middle region of the Rs2 linkage group and the upper region of the Rs7 linkage group, respectively, and all three genes showed dominance and were mutually epistatic [124]. The *Rfob* and *Rfoc* genes can also restore the fertility of Ogura CMS radish [52]. *Rfob* is closely linked with the *Rfo* gene and has two SNPs compared with *Rfo*. *Rfoc* is generated by recombination between *Rfo* (*PPR-B*) and *PPR-C*. In addition, several new fertility restorer loci, namely *RsRf3-1*/*RsRf3-2* and *RsRf3-4*~*RsRf3-7*, in the Chinese radish materials 2007H, 9802H and 9606H were also cloned and sequenced [53,125,126]. A new restorer gene of Ogura CMS, *Rft*, which was found in the Tomioka population of Japanese wild radish, was also a homolog of *Rfo* and had 19 SNPs compared with *Rfo* [50]. Yamagishi et al. (2021) confirmed that Ogura CMS had a new fertility restorer gene, *Rfs*, which could restore the fertility of individuals by processing *orf138* mRNA and was different from *Rft* [54].

Several intragenic markers or closely linked markers have been developed, such as AFLP markers (R3, R15, R5 and AFLP190), RFLP-PCR markers (2F/1R), RAPD markers (OPA-14_{600} and OPK-17_{440}) and STS markers (STS-A14 and STS-K17) [51,122,127]. A high-throughput molecular marker suitable for radish population screening was also designed. Wang et al. (2017) developed a high-throughput SNP marker named Rfo-SNP1 based on the nucleotide sequences of *Rfo* in radish using KASP technology, which has been successfully applied in the high-throughput screening of radish restorer lines [128] (Table 2).

Table 2. Comparative analysis of Ogura CMS systems in different Cruciferae crops.

Cruciferae Crops	Sterility Gene Markers	FR Genes	FR Gene Loci	Linked FR Gene or Intragenic Molecular Markers	Advantages	Limitations	References
Raphanus sativus	Primer A/B, CMSF/R, orf138-F/R	*Rfo* (*Rfk*), *RF1*, *RF2*, *RF3*, *Rfob*, *Rfoc*, *RsRf3-1*/*RsRf3-2*, *RsRf3-4~RsRf3-7*, *Rft*, *Rfs*	C3	R3, R15, R5, AFLP190, Rfo-SNP1, RFLP-PCR2F/1R with the Ssp1 restriction enzyme site, OPA-14600,OPK-17440, STS-A14, STS-K17	(1) The most studied male sterility type in radish (2) Complete and stable sterility (3) Many FR genes in European radish, Japanese radish and Chinese radish varieties	(1) Low distribution frequency of the rfrf genotype (2) Maintainer line selection requires a large number of backcross offspring	[51–54,59–61]
Brassica napus	P51/P52	*Rfo*, *Rfob*, *Rft* (introduced from *R. sativus*)	N19, C3, A genome	PGI-2, $OPC02_{1150}$, $OPD02_{1000}$, $OPF06_{1200}$, $OPG02_{700}$	(1) Long history of creation and thorough study in Ogura CMS lines and restorer lines (2) The safest and most effective way to utilize heterosis (3) Three-line breeding and commercial production have been achieved	(1) Large radish alien genome segments around the FR gene and poor agronomic traits of restorer lines (2) Restorer lines and breeding methods with low glucosinolate content have been submitted for long-term patent protection, which greatly limits the application	[20,72,73,129–133]
Brassica oleracea	Bo138300BF/R; P1/P2; P12F/R; P13F/R; OKB-F/R	*Rfo* (introduced from *R. sativus*)	/	BnRFO-AS2F/BnRFO-NEW-R, Rfo-6F/R, Rfo-6eF/R, Rfo-8F/R, Rfo-11F/R, Rfo-page-4eF/R	(1) Complete and stable sterility (2) Fertility-restored materials have been created with 18 normal chromosomes, a normal seed setting rate and a closer genetic background to the parent cabbage	(1) Varied fertility restoration levels (pollen vitality) of fertility restorer lines (2) Low and abnormal *Rfo* gene transmission rate (3) Large proportion of radish and rapeseed genomic components	[80–83,114,134–136]
Brassica rapa	F/R; P11/P12; orf138 primer F/R	*Rfk1* (introduced from radish and homolog of *Rfo*)	Additional radish chromosome	F/R	(1) Fertility of Ogura CMS can be restored by the *Rfk1* gene	(1) Restorer gene was unstable in the turnip rape genome (2) Fertility restoration trait was unstable in subsequent generations	[137–139]
Brassica juncea	/	*Rfo* (introduced from *R. sativus*)	A09	KASP-RFO-1814	(1) Good seed set and agronomic performance in restorer lines	(1) A large radish introgression alien segment carrying the *Rfo* gene and linkage drag exist (2) Pollen contamination and impaired transmission frequency of the *Rfo* gene	[99,140]

3.2. Ogura CMS Fertility Restorer Genes in Other Cruciferous Crops

With the wide application of Ogura CMS, fertility restorer genes have been successfully introduced into different cruciferous crops and integrated at different genomic locations. In *B. napus*, three main fertility restorer genes in Ogura CMS have been localized and cloned, namely *Rfo*, *Rfob* and *Rft* [91,118,129,130]. They were integrated into the C genome (C09, C03) or A genome in different *B. napus* materials. The *Rfo* restorer gene in the restorer line R2000 created by INRA and the restorer line 46H02 created by Pioneer Hibred were both located in the N19 linkage group, and a large number of gene-specific markers and linked markers were developed [129,141]. In CLR650, a new fertility-restored material of Ogura CMS *B. napus*, the restorer gene was located in the N19 linkage group between the SSR markers BnGMS35 and BoGMSl97 [137]. In the Ogura CMS restorer line N1717 of *B. napus*, the *Rfo* gene was confirmed to be located in the C genome by bacterial artificial chromosome–fluorescence in situ hybridization (BAC-FISH), and it was inferred to be located on chromosome C03 according to the size of the chromosome and the location of the centromere [129]. In a new version of the Ogura CMS restorer line CLR6430, Wang et al. (2020) detected the restorer gene by FISH using the C genome and BAC as a probe and verified the presence of *Rfo* in the A genome of *B. napus* [130]. In *B. rapa*, the Ogura CMS fertility restorer gene was integrated into an additional radish chromosome. *Rfk1* of Kosena CMS is a homolog of the fertility restorer gene *Rfo* of Ogura CMS. It was transferred from *B. napus* to Ogura CMS *B. rapa*, and its location on the chromosome was identified by BAC-FISH technology [138]. In *B. juncea*, the *Rfo* gene was located in an ~108 kb radish chromosome fragment, which was positioned amidst a large C-genome translocation at the distal part of chromosome A09, and an intragenic KASPar marker (KASP-RFO-1814) for the marker-aided transfer of *Rfo* was developed [99,140].

4. Mechanisms of Ogura CMS in Major Cruciferous Crops

The factor that most directly causes MS in plants is abnormal pollen development [142]. Although the detailed process by which sterility genes affect floral organ and pollen development is still unclear, it is generally believed that mitochondrial dysfunction is mainly responsible for the disruption of pollen development [6,143]. Furthermore, it was shown that during the development of pollen grains, the mitochondrial content of tapetum cells was 40 times that of somatic cells, and the mitochondrial content of microspore cells was 20 times that of somatic cells. Therefore, anthers require more ATP and are extremely sensitive to changes in mitochondrial ATP synthesis [144]. The *orf138* gene encodes a mitochondrial transmembrane protein, similar to many other CMS genes [145]. In sterile plants, the ORF138 protein accumulates largely on the mitochondrial membrane, which may interfere with the expression of key genes in the electron transport chain, such as *ATP6*, *ATP8* and *COXI*, and inhibit anther synthesis (Figure 2). In addition, the occurrence of CMS is mostly related to some chimeric ORFs generated by mitochondrial genome rearrangement [146]. Tanaka et al. (2012) found that the Ogura CMS cytoplasm may be generated by mitochondrial genome rearrangement, and this rearrangement is likely to destroy mitochondrial structures, thereby resulting in Ogura CMS [20].

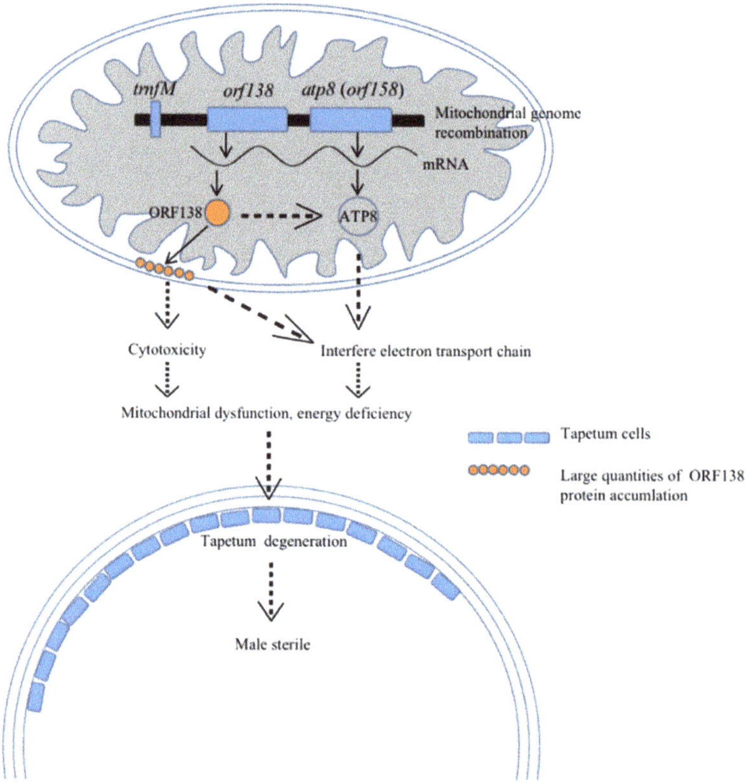

Figure 2. Model of the mechanisms of the Ogura CMS fertility restorer gene *Rfo* (*PPR-B*).

The abnormal development of tapetal cells was confirmed as an important cause of abnormal pollen development. Normal microsporogenesis requires the appropriate timing of tapetum degeneration [147]. Studies revealed that pollen abortion in Ogura CMS occurred after the tetrad stage [147]. Lin et al. (2019) found that aberrant anther development occurs during the transition from microspore mother cells to tetrads, and defective microspore development and the early clearing of tapetal cytoplasm led to shrunken anthers with collapsed locules. Some studies indicated that the autolysis process, rather than the normal programmed cell death (PCD) process, led to the premature death of tapetal cells, which hindered the development of pollen in the microspore stage of large vacuoles and finally led to MS [55,148]. In addition, the abnormal proliferation of tapetal cells was confirmed as an immediate cause of Ogura CMS in cabbage [113]. Combining cDNA-AFLP with microarray analysis, only one of twenty-nine AT hook nuclear localized (AHL) family genes, *BoMF2*, was differentially expressed in the anthers of Ogura CMS cabbage and might regulate tapetum proliferation during anther development [149].

Flavonoids play an important role in plant pollen development. Many studies in recent years have shown that the chalcone synthase (CHS) gene is related to MS. Related studies have mainly focused on petunia, maize, rice and other plants [147,150]. In radish, the expression of CHS in the flavonoid biosynthesis pathway of Ogura CMS anthers was reduced, and the decrease in CHS expression led to a decrease in flavonoid content, which may ultimately have a negative impact on the production of pollen grains [55].

The possible toxicity of ORF138 remains unclear, and direct evidence for a cytotoxicity model is still lacking. The ORF138 protein was expressed in almost all tissues, including hypocotyls, leaves, roots and buds, but its expression levels varied among the tissues. In vegetative tissues, no morphological or respiratory defects were detected, confirming that

ORF138 can cause complete pollen abortion without affecting vegetative development and female gametogenesis in Ogura CMS plants [151]. However, the expression of ORF138 in *Escherichia coli* can inhibit normal growth, indicating that ORF138 has a toxic effect on cells [152]. In addition, the expression of *orf138* in the nucleus was realized by fusing and transforming the CMS gene, green fluorescent protein (GFP) and a mitochondrion-targeting peptide into *A. thaliana* and yeast. The *orf138* gene changed the morphology of mitochondria in yeast and *Arabidopsis* transgenic lines but did not inhibit the respiration and growth of yeast and did not cause sterility in *Arabidopsis* [153].

Great progress in high-throughput sequencing, transcriptomics, proteomics and degradomics has provided a large amount of genetic data and unprecedented opportunities to identify differentially expressed genes (DEGs) that play important roles in CMS [6,154–158]. By comparing the Ogura CMS sterile line and the corresponding maintainer line, a large number of DEGs were identified (Table 3). Substantial numbers of DEGs have been shown to be involved in energy metabolism, which underpins the idea that insufficient energy supply may lead to MS [113,159–166]. Microsporocytes give rise to pollen via meiosis, and somatic cells, particularly tapetal cells, are essential for the normal development and release of pollen [167]. The failure to produce functional pollen in Ogura CMS lines is found to be accompanied by DEGs involved in microspore formation, anther development, pollen wall formation, exine formation (the formation and dissolution of the callose wall, fatty acid metabolism pathway and biosynthesis of sporopollenin precursors in the tapetum), the accumulation of the pollen coat and pectinesterase activity [159–161,163–168]. CMS always induces the expression of stress-related genes, especially HSP genes [160,162,164,168,169]. ORF138 may affect redox status in Ogura CMS plants, although only DEG evidence exists [159,160,164]. Caspase-like and metacaspase activity genes involved in cell apoptosis were also discovered among DEGs, suggesting that pollen abortion in Ogura CMS was related to PCD [164,165]. In addition, the plant hormone auxin plays a central role in plant growth and development [170]. Studies showed that the delayed expression of most auxin-related genes may have caused short filaments and reduced plant growth in Ogura CMS plants [160,171]. DEGs were also enriched in the mitochondrial retrograde signaling pathway [159]. Two novel miRNA/target cascades (novel-miR-335/H^+-ATPase and novel-miR-448/SUC1) may participate in the communication between the mitochondria and nucleus [172]. Genes that play important roles in the plant defense response were significantly upregulated in Ogu CMS lines, suggesting their likely involvement in Ogu CMS pollen abortion [163,164,171]. Although these -omics technology methods have been extensively applied in DEG exploration in cruciferous crops, the molecular mechanisms underlying Ogura CMS remain elusive. Furthermore, most impacts of *orf138*/ORF138 were primarily based on speculations, lacking direct experimental evidence to support such conclusions.

Table 3. Various -omics studies on Ogura CMS in cruciferous crops.

Cruciferae Crops	Materials	Omics Technologies	Functions of DEGs	Reference
Raphanus sativus	Sterile line, maintainer line	Transcriptomics	Fatty acid metabolism, pollen development and tapetum development	[56]
	Sterile line, maintainer line	Transcriptomics	Binding, catalytic activity, metabolic process, cellular process and response to stimulus	[162]
Brassica napus	Sterile line	Proteomics	Carbohydrate and energy metabolism, aldehyde dehydrogenase (ALDH), photosynthesis and flavonoid synthesis	[161]
	Sterile line, maintainer line	Transcriptomics	Plant hormone signal transduction, plant–pathogen interaction, peroxisome, pentose–glucuronate interconversions and starch–sucrose metabolism	[171]
Brassica oleracea	Sterile line, maintainer line	Transcriptomics	Biosynthesis of secondary metabolites, starch and sucrose metabolism, plant–pathogen interaction and glycerophospholipid metabolism	[163]
	Sterile line, maintainer line	Transcriptomics	Polygalacturonase metabolism, glycosyl hydrolases, oxidation reduction process, phenylalanine metabolism and phenylpropanoid biosynthesis	[164]
	Sterile line, fertile line	Transcriptomics and proteomics	Gibberellin-mediated signaling pathways regulating tapetum programmed cell death and sporopollenin synthesis	[165]
	Sterile line, maintainer line	Transcriptomics and proteomics	Sporopollenin synthesis, callose wall degeneration and oxidative phosphorylation	[166]
	Sterile line, maintainer line	Transcriptomics	Energy metabolic pathways	[113]
Brassica rapa	Sterile line, maintainer line	Transcriptomics	Mitochondrial retrograde signaling pathway, auxin response, ATP synthesis, pollen development and stress response	[160]
	Sterile line, maintainer line	Degradome analysis, miRNA analysis and transcriptomics	Pollen-development-related genes	[172]
	Sterile line	Transcriptomics	Stress-response genes, mitochondrial protein genes, ascorbic acid biosynthesis and thylakoid protein gene	[169]
	Sterile line, maintainer line	Transcriptomics	Anther development and microspore formation	[159]
	Sterile line, maintainer line	Transcriptomics	Pollen development, carbon metabolism, lipase activity, lipid binding, citrate cycle and oxidative phosphorylation	[173]
	Sterile line, maintainer line	Transcriptomics	Phenylpropane synthesis pathway and glutathione oxidation–reduction	[174]

5. Mechanisms of Ogura CMS fertility Restorer Genes in Major Cruciferous Crops

The *PPR-B* gene, but not the *PPR-A* and *PPR-C* genes, restored the fertility of the Ogura CMS materials. Ogura CMS is regulated by the *orf138* gene, and fertility can be restored when the *Rfo* gene is present [51,122]. The *Rfo* gene is present as three highly similar genes: *PPR-A*, *PPR-B* and *PPR-C*. Three homologous copies are arranged in a series, encoding highly similar proteins [51,57,122,123,175]. *PPR-A* and *PPR-C* do not have fertility restoration ability and have no effect on the synthesis of sterility proteins, while the *PPR-B* gene plays a role in the fertility restoration of Ogura CMS materials [57].

Several important functional sites and binding sites of the fertility restorer gene *Rfo* and the sterility gene *orf138* have been identified. Most restorer genes encode pentatricopeptide repeat (PPR) proteins, and all identified Rf-PPR genes evolved from a unique subset of the PPR gene family called the Rf-like or RFL gene family [176,177]. PPR proteins are characterized by tandem repeats of 35-amino-acid motifs, most of which are considered sequence-specific RNA-binding proteins, and they regulate mitochondrial and chloroplast gene expression through editing, splicing and cleavage [178–181]. In the fertility restoration process of Ogura CMS, all PPR-B repeats are indispensable for complete fertility restoration [182]. An allele sequence analysis of the fertility restorer gene revealed four substituted amino acids (the 118th, 153rd, 170th and 171st amino acids) in the second and third repeats of the PPR, suggesting an essential role in the fertility restoration of these domains formed by these repeats in the *Rfo* gene-encoding protein ORF687 [123]. Yamagishi et al. (2021) and Imai et al. (2002) also considered that the 118th and 153rd amino acids in ORF687 play a crucial role in fertility restoration [54,183]. By comparing the transcript sequences of *rfo*/*rfo* and *Rfo*/*Rfo* homozygote plants, it was found that four amino acids (the 176th to 179th amino acids) of the *Rfo* gene were essential, and the deletion of these four amino acids in the central region of the *Rfo* gene-encoding protein reduced fertility restoration ability [182]. In the interaction of a nuclear fertility restorer gene and a mitochondrial CMS gene, Uyttewaal et al. [57] proposed that ORF687 could associate with the 5′ untranslated region (5′ UTR) of *orf138* mRNA. Yamagishi et al. (2021) found that direct ORF687 binding to the coding region of *orf138*, not the 5′ UTR inferred previously, was essential for fertility restoration by *Rfo* [54]. It was also found that either a single nucleotide substitution (the 61st nucleotide) in *orf138* or two amino acid substitutions (the 118th and 153rd amino acids) in ORF687 will prevent the *Rfo* gene from restoring the fertility of Ogura CMS materials with the *orf138* gene [54]. Wang et al. (2021) demonstrated that *PPR-B* bound within the coding sequence of *orf138* and found that the GTAAAGTTAGTG-TAATA sequence of the *orf138* transcript was the *PPR-B* binding site [184]. Interestingly, the 61st nucleotide was in the binding site, which confirmed the validity and reliability of this conclusion.

The Ogura CMS restorer gene acted at the translation level and restored fertility by blocking the translation elongation of *orf138* mRNA. The mechanisms of different restorer genes are diverse among crops. Fertility restoration in CMS plants is mainly regulated at the genomic level, post-transcriptional level, translational level, post-translational level and metabolic level [6,63]. The function of Ogura CMS restorer genes is generally considered to involve the processing and editing of sterility-related gene transcripts at the post-transcriptional or translational level and then the inhibition of the accumulation of ORF138 and the elimination of the negative effects of sterility proteins on pollen or tapetum development [6,184]. The ORF138 protein showed a significant decrease in the mitochondria of flowers and leaves after the fertility of Ogura CMS *R. sativus* was restored. However, the presence or absence of fertility restorer genes did not affect the size, abundance or RNA editing pattern of *orf138* transcripts, which suggested that the Ogura CMS fertility restorer gene affects the translational level of the *orf138* gene [6,18,57]. Immunolocalization experiments showed that fertility restoration was related to the complete elimination of the ORF138 protein in the tapetum. Therefore, *PPR-B* restores fertility mainly by inhibiting the synthesis of the ORF138 protein in the anther tapetum. Previous research suggested that the precursor RNA of the *orf138* gene in sterile plants was stable because of the formation of a 'neck-loop' structure during the splicing process, and it was degraded during the splicing process because of the formation of an unstable 3′ end in fertile plants [185,186]. Therefore, it was concluded that the restorer gene *Rfo* acts post-translationally on the stability of the Ogura CMS-associated protein ORF138 in the reproductive tissues of rapeseed cybrids. By using a restored rapeseed transgenic line containing four copies of *PPR-B*, Wang et al. (2021) confirmed that ORF138 disappeared from mitochondria in the presence of *PPR-B* and rejected the hypothesis regarding a fertility restoration mechanism for the increased instability of ORF138. Wang et al. (2021) demonstrated that this specific translational inhibition of *orf138* mRNA and ORF687 acted as a ribosome blocker to specifically impede translation elongation along the *orf138* mRNA by in organello synthesis, polysome sedimentation and Ribo-Seq analyses [184]. Based on the results of Wang et al. (2021) and other studies, a schematic diagram of the mechanisms of Ogura CMS fertility restorer genes is shown in Figure 3. In addition, another Ogura CMS fertility restorer gene (*Rft*) found in most Japanese wild radishes can affect the expression of *orf138* mRNA [184]. This finding showed that the *Rf* gene, with different molecular mechanisms, evolved in radish and inhibited the expression of the sterility gene *orf138* in Ogura CMS lines.

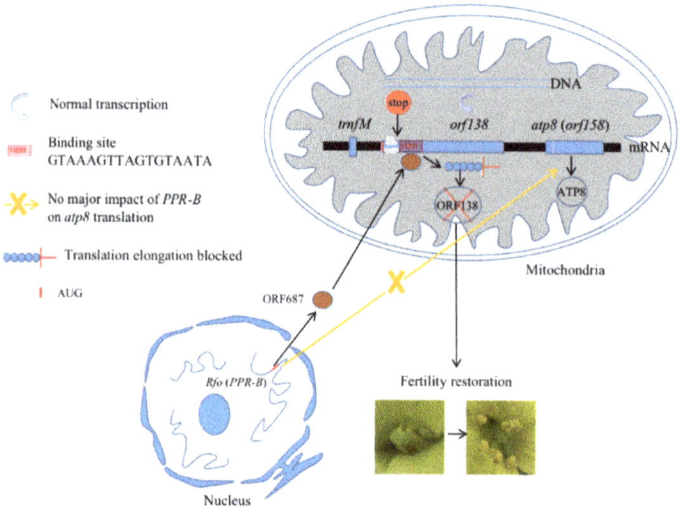

Figure 3. Model of the mechanisms of the Ogura CMS fertility restorer gene *Rfo* (*PPR-B*) (Wang et al., 2021).

6. Application of Ogura CMS and Creation of Ogura CMS Fertility Restorer Lines

The Ogura cytoplasm has been introduced into different *Brassica* crops, including *B. napus* (AACC), *B. oleracea* (CC), *B. rapa* (AA) and *B. juncea* (AABB), by intergeneric hybridization, somatic hybridization and repeated backcrossing. In rapeseed, it is essential for the *Rfo* restorer gene. In vegetables, because the leafy organ is used as a product for consumers, it is not necessary for fertility restoration to be possible in the vegetative growth stage. However, the wide usage of Ogura CMS has led to a new problem: all offspring not carrying the fertility-restored gene exhibit MS, which has inhibited germplasm innovation and breeding. Therefore, there is an urgent need to create a fertility restorer to promote the innovation and reutilization of germplasm resources in all Ogura CMS *Brassica* vegetables. Corresponding restorer lines have also been successfully created in crops such as *B. napus* (AACC), *B. oleracea* (CC) and *B. juncea* (AABB).

6.1. Brassica Napus

The Ogura CMS cytoplasm was successfully transferred to *B. napus*, and the whole process consisted of two stages. (1) The French scholar Bannerot introduced the nucleus of European oilseed rape into Japanese radish sterile cytoplasm and transferred the Ogura

CMS cytoplasm into *B. napus* through continuous backcrossing [63]. However, due to cytonuclear discordance, which is a hallmark of many introgression events, many undesirable traits, such as young leaf chlorosis under low temperature and poor nectary development, affect normal growth and development [58,64]. (2) In 1983, Pelletier successfully transferred Ogura CMS to *B. napus* by protoplast fusion and solved existing problems in the previously created Ogura CMS *B. napus*. The male-sterile line exhibited normal green leaves at 12 °C, the normal development of nectaries and the stable inheritance of sterility [65]. Ogura CMS lines of *B. napus* were also introduced in China, and massive restorer line screening was performed [66–68]. Li et al. (1995, 2001) introduced the FR gene into *B. napus* by hybridization with the radish variety 'Makino' and finally gained the addition line Ad-6, which unfortunately could not be applied in hybrid seed production because of unstable fertility and poor agronomic traits. Chen et al. (2012) used grafting technology to overcome the reproductive obstacles of distant hybridization. Using an interspecific doubling hybrid of oilseed rape and radish (AACCRR), the restorer line CLR650 with high glucosinolate content was selected after multigeneration backcrossing and screening, but it still needs to be improved [69]. Wen et al. (2010,2016) obtained the restorer line R2008 by multigeneration backcrossing, test-crossing and microspore culture and further improved it to obtain six restorer lines with better agronomic traits [68,70].

The creation and improvement of Ogura CMS *B. napus* restorer lines have undergone 30 years of development, and the resulting lines have already been applied in commercial production. Through a large number of screenings, scientists found that there are no natural fertility restorer lines of Ogura CMS in *Brassica* species, and the restorer genes in radish must be transferred to *Brassica* species by intergeneric hybridization. The creation of a fertility restorer line for Ogura CMS in *B. napus* can be divided into three stages. (1) Fertility-restored interspecific materials were first obtained by hybridization and/or protoplast fusion between *R. sativus* with a restorer gene and *B. napus* [65,71]. However, due to the influence of large redundant radish fragments, *B. napus* fertility restorer lines have poor agronomic traits, low seed setting rates and even high glucosinolate content under the influence of close linkage between glucosinolate genes and the restorer gene *Rfo* [127]. (2) With the realization of marker-assisted selection (MAS) in many crops and the development of *Rfo*-linked markers, the improvement of restoration materials has been greatly accelerated [72]. MAS combined with backcross breeding was used to obtain low-glucosinolate-content restorer lines, but some redundant genes closely linked to *Rfo* were still retained in the created materials, resulting in poor agronomic traits [73]. (3) INRA finally provided the restorer line R2000 with both low glucosinolate and low erucic acid levels by combining gamma-ray irradiation with multigeneration backcross breeding [74]. Gamma-ray irradiation greatly increased the genome recombination rate, and the improved fertility restorer line had good agronomic traits, a normal seed setting rate and good *Rfo* transmission efficiency, which have been successfully applied to seed production in Ogura CMS *B. napus*. Pioneer Hi-Bred International Inc. screened a series of *Brassica* mutants with reduced radish fragments from tens of thousands of plants obtained by gamma-ray irradiation and finally bred a new Ogura CMS restorer line (SRF line) of *B. napus* with fewer radish fragments and a lower glucosinolate content than R2000 [75].

6.2. *Brassica Oleracea*

Ogura CMS was transferred into cabbage by distant hybridization between radish and cabbage combined with embryo rescue, and OguraR_1 CMS cabbage was created [58]. However, because the obtained Ogura CMS cabbage has a radish cytoplasm, problems frequently occur, such as nuclear–cytoplasmic incompatibility, chlorosis at low temperature (15 °C), nectary dysplasia and poor seed set. OguraR_2 CMS materials were successfully obtained by replacing the radish chloroplast in the OguraR_1 CMS system with the cauliflower chloroplast by protoplast fusion [77]. Most likely because of the large proportion of the radish cytoplasm in protoplast fusion plants, the growth vigor of two sterile lines derived from OguraR_2 CMS materials decreased, while the numbers of abnormal flowers and pods increased after 5–6 generations of transformation breeding [78]. The Asgrow seed company applied the asymmetric protoplast fusion method to reduce the proportion of radish mitochondria and obtained Ogura $CMSR_3$ with normal sterility and pistil structure. With the Ogura $CMSR_3$ material as the sterile source, several Ogura CMS lines of cabbage with stable sterility and normal flowering and seed set have been bred and widely used in the production and breeding of cabbage hybrids. A number of new varieties have been approved or identified, such as 'Zhonggan 22', 'Zhonggan 192', 'Zhonggan 96' and 'Zhonggan 101' in China [79].

The *Rfo* restorer gene was introduced from *B. napus* to *B. oleracea* by the bridge material *Brassica alboglabra* Bailey [80–82]. Using triploid and hexaploid methods, several Ogura CMS fertility-restored plants of *B. alboglabra* Bailey were successfully created. A series of fertility-restored cabbage plants in the BC_5 generation were obtained by crossing the BC_2 generation of *B. alboglabra* Bailey fertility-restored plants with cabbage materials, and these materials were successfully used for the fertility restoration of Ogura CMS clubroot resistance resources in *B. oleracea* [83]. Li et al. (2021) produced the Ogura CMS *Rfo* restorer by transforming the modified Rfo^B restorer gene into the Ogura CMS line 'CMS2016' of *B. oleracea* var. *capitata* and developed 18 different morphological Ogura CMS restorers [84]. The polyploid materials, introgressive lines and interspecific hybrid materials created from the above distant hybridization approaches will also provide important materials for several essential research topics, such as polyploidization, distant hybridization, the influence of alien fragments, the innovation of gene function and the formation of important traits.

6.3. Brassica Rapa

Ogura CMS was successfully transferred into Chinese cabbage [91,92]. Williams and Heyn (1981) introduced the sterility gene from Japanese radish into Chinese cabbage and bred many male-sterile lines of Chinese cabbage [92]. However, there were different defects in these germplasm resources after the screening of Ogura sterile lines. Delourme et al. (1994) also reported the transformation breeding of the Ogura cytoplasm from *B. napus* to *B. rapa* [91]. In China, sterile resources were collected from the United States, Germany and other countries. Some Ogura CMS Chinese cabbage lines were introduced from abroad, and a number of sterile lines that basically overcame the defects of seedling yellowing, nectary degeneration, bud abortion and a low seed setting rate of the original Ogura sterile lines of *B. rapa* were developed and obtained after continuous multigeneration backcross breeding [93–95]. Hou et al. (2001) transferred Ogura CMS into Chinese cabbage by an asymmetric fusion technique [96]. Cui et al. (2004) used Ogura CMS *B. napus* as the female parent and inbred lines as the recurrent parent to conduct hybridization and continuous backcross breeding and finally obtained three CMS lines of *B. rapa* with stable sterility and nonchlorosis [97]. Zhao et al. (2007) introduced the sterility of RC97-1 into Chinese cabbage by using Ogura CMS *B. napus* RC97-1 as a sterile source and then bred a new Ogura CMS line, RC7, which overcame many shortcomings of the original Ogura CMS material through continuous backcrossing and strict economic trait selection [98]. Although Ogura CMS Chinese cabbage has been developed over a long period, it is generally believed that the lack of excellent Ogura CMS Chinese cabbage varieties with good economic traits and large-scale production is due to the problems of disease resistance, poor combination ability, low yield, negative cytoplasmic effects and the severe degradation of backcross generations despite some progress in addressing the problems of yellowing and nectary development. In addition, restorer lines of *B. rapa* have not been created or reported.

6.4. Brassica Juncea

CMS lines of *B. juncea* are obtained in three main ways: (1) spontaneous CMS; (2) CMS created by introducing the *B. juncea* nucleus into a heterologous cytoplasm, with the resulting CMS given a name that reflects the species, such as Moricandia CMS [100], Trachystoma CMS [101] and Siifolia CMS [102]; and (3) CMS, first widely used in *B. napus*, transferred into *B. juncea* through protoplast fusion, somatic hybridization and conventional backcrosses, such as Tournefortii CMS [103], Oxyrrhina CMS [104] and Ogura CMS [62]. Delourme et al. (1994) used protoplast fusion to transfer Ogura CMS from *B. napus* to *B. juncea*. Kirti et al. (1995) also reported the transfer of Ogura CMS to the *B. juncea* variety RLM198 by using continuous backcrossing combined with variety screening. However, the obtained sterile line exhibited serious defects, such as seedling yellowing, late flowering time, poor seed set and small or abnormally shaped pods. Therefore, Kirti et al. (1995) further fused the protoplast of the Ogura sterile line and normal cytoplasmic variety RLM198 and obtained an improved Ogura CMS line with great improvement in the above defects by continuous backcrossing. INRA also created the Ogura sterile line and restorer line of *B. juncea*, and Tian et al. (2014) introduced it in Canada [99]. The sterile line was proven to be stable, while the introduced Ogura CMS restorer line in *B. juncea* had a low seed setting rate and many unfavorable agronomic traits, preventing the restorer line from being used. Using MAS combined with backcross breeding, Tian et al. (2014) successfully developed a homozygous Ogura CMS restorer line with a high seed set and good agronomic traits in *B. juncea*, and the linkage redundancy of the improved restorer line was greatly reduced [99].

6.5. Other Cruciferous Crops

In *Brassica campestris*, an Ogura CMS line was obtained by crossing the Ogura CMS line of *B. napus* with *B. campestris* and backcrossing with the original male parent [92].

In *B. campestris* L. var. *purpurea* Bailey, a corresponding CMS line was obtained by using Ogura CMS material in *B. napus* after three consecutive years of backcrossing [105]. Since this variety is grown for its consumable vegetative organs rather than its seeds, there is no need for restorer lines.

In *B. oleracea* L. var. *botrytis* L., an Ogura CMS line with resistance to atrazine herbicide was obtained by protoplast fusion technology [85]. Jiang et al. (2001) used Ogura CMS cauliflower introduced from the United States as the female parent to improve Ogura CMS cauliflower in China [86]. The sterile line bred by Jiang et al. exhibited normal growth and development, better disease resistance, good agronomic traits and a good combination ability. Two excellent cross combinations were developed for demonstration and promotion in production. Dey et al. (2011) developed and characterized an Ogura-based improved CMS line of cauliflower [87]. Shu et al. (2019) evaluated the effects of different CMS sources on the morphology and size of floral organs, the attractiveness to pollinators and the yield of hybrid seeds and selected four CMS sources (including Ogura CMS) suitable for future use in broccoli breeding [187]. Singh et al. (2021) characterized a large genetic stock of cauliflower encompassing 76 CMS (consisting of 67 Ogura cytoplasm) and DH lines and laid a good foundation of F_1 hybrids creation through the effective utilization of CMS and doubled-haploid (DH) lines in cauliflowers [188].

In *B. oleracea* var. *italica* Plenck, He et al. (1999) overcame the defects of pistil dysfunction, bud closure, bud abortion, stigma exsertion and nectary deletion in Ogura CMS donors when the donors were backcrossed with broccoli [88]. Zhu et al. (2001) used the Ogura CMS line 92-08 as the female parent and different broccoli types as the male parent to breed the sterile line BC7-19, which basically overcame the defect of yellowing at low temperatures in the seedling stage [89]. The flower organ structure was normal, and the seed setting ability was strong. Using this broccoli sterile line, the first-generation hybrid 'Luqing No. 1' of broccoli was prepared and had the advantages of strong growth potential, early maturity, excellent comprehensive traits and significant yield benefits. Using Chinese kale 16Q2-11 containing the fertility restorer gene *Rfo* as an intermediate material, the *Rfo* gene was introduced into six Ogura CMS broccoli lines by hybridization and subsequent backcrossing [90].

7. Discussions

7.1. How Do CMS Sterility Genes and Fertility Restorer Genes Come about?

Plant mitochondrial genomes are rich in repeated sequences, with estimates of up to 38% of the mitochondrial genome occupied by tandem repeats, short repeats and large repeats [189,190]. Repetitive sequences are the most common and important sequences in the mitochondrial intergenic region. They are related to the mutation and recombination of the mitochondrial genome and play an important role in the evolution of plant mitochondrial genomes. The comparison of the mitochondrial genomes of CMS lines and maintainer lines showed that there were almost no deleterious mutations in CMS line genomes except that of *Beta vulgaris* ssp. *maritima* with G-type CMS, which means that deleterious mutations are not the root cause of CMS [191,192]. Together with the reported mitochondrial genome sequences, comparative analysis showed that most CMS types are closely related to repeating sequences. Heng et al. (2014) found that repeats in the CMS line were typically twice the size as those in the maintainer line and three large repeats were present downstream of the hau CMS-associated gene *orf288*, which implicated the strong ties between repeat sequences and the formation of this chimera [193,194]. Comparative analysis between the mitogenomes of CMS and male-fertile lines of pepper (*Capsicum annuum* L.) showed that the CMS candidate genes *orf507* and *ψatp6-2* were proximal to the edges of highly rearranged CMS-specific DNA regions, whose evolution may be the result of nearby intermediate or large-sized repeats [195]. Tanaka et al. (2014) reported that the Ogura-type mitochondrial genome was highly rearranged compared with the normal-type genome by recombination through one large repeat and multiple short repeats. A high rearrangement rate is a common feature of CMS mitochondrial genomes. Recombination via repeat sequences is believed to be responsible for extensive structural changes [20]. The structure of plant mitochondrial genomes evolves rapidly, driven by rearrangements that result from high rates of recombination [196]. Thus, CMS leading to mitochondrial genome rearrangement may be associated with DNA recombination mediated by repeating sequences and probably arose in progenitor species, which was then revealed when hybridization removed the corresponding nuclear restorer gene. Mutation and selection are two primary forces that drive evolutionary processes [197]. Any nuclear allele able to restore the function of the male gametophyte in plants with sterility genes will be favored and subsequently fixed in a population. As the frequency of the CMS gene increases in the population, selection will more strongly favor the restorer alleles, which should subsequently spread in the population.

7.2. What Are the Future Development Directions of Ogura CMS?

Ogura CMS is the most extensively studied and widely used CMS system in Brassicaceae crops [54]. Nevertheless, many open questions remain, and this situation has changed in recent years with the development of new techniques. According to the limitations and problems during the application of Ogura CMS in existing research, future directions are discussed and then summarized into two major aspects.

More direct evidence for the functional gene validation of the sterility gene *orf138* is still needed. Although the overexpression of the *orf138* gene in transgenic *Arabidopsis* and the knockdown of the *orf125* gene, which is the homologous gene of *orf138*, have been achieved, functional genomics research needs to employ diverse experimental approaches to investigate gene functions [198]. Gene editing of *orf125* was implemented by using transcription activator-like effector nucleases (TALENs) with mitochondrial localization signals (mitoTALENs). This approach has also been applied to validate the mitochondrial gene *orf312* of Tadukan-type CMS and the mitochondrial gene *orf79* of CMS-BT in rice [120,199]. CRISPR-based gene editing has revolutionized targeted gene editing in plants in recent years [200]. CRISPR-mediated genome editing was realized in the mitochondria of yeast and the chloroplasts of Chlamydomonas [201]. Kang et al. (2021) achieved DNA-free editing in chloroplasts and generated lettuce calli and plantlets resistant to streptomycin or spectinomycin [202]. The Northwest Institute of Plateau Biology, Chinese Academy of Sciences, also unveiled a method to develop male-sterile lines in plants using a mitochondrial gene editing system [203]. One of the most direct pieces of evidence would be to verify the function of the *orf138* gene by editing ORFs related to CMS with mitochondrial gene editing to silence the *orf138* gene and directly realize fertility conversion. In addition, the transcription-specific targeting of mitochondrial gene expression is difficult, and it is also necessary to realize the mitochondrially targeted expression of the *orf138* gene in Cruciferae crops, except for *Arabidopsis*.

A great deal remains to be determined about sterility and fertility restoration mechanisms. (1) The mechanism by which sterility genes affect pollen abortion but not female gametes and other tissues remains unclear. Research on CMS-S in maize has led to a breakthrough in solving this problem. Xiao et al. (2020) found that the nuclear-encoded DREB transcription factor *ZmDREB1.7* was specifically expressed in anthers and promoted the expression of the CMS-S gene *orf355* [204]. The accumulation of ORF355 activated mitochondrial retrograde signaling and in turn induced *ZmDREB1.7* expression. *ZmDREB1.7* and *orf355* formed a positive-feedback transcriptional regulation loop in microspores of CMS-S maize, which eventually led to the accumulation of the ORF355 protein and abortion. Similar transcription factors may also exist in Ogura CMS and influence pollen abortion without affecting other tissues. Recent advances in single-cell sequencing technologies would be useful for addressing this issue. Zhang et al. (2021) isolated the anthers of CMS lines and restorer lines, analyzed their transcriptional expression profiles by single-cell RNA sequencing, and then expounded the mechanism of CMS-C sterility genes and nuclear restorer genes affecting meiosis and microspore development [205]. There are still many research gaps in the genetics, cytology and molecular biology fields related to male and female gametophyte development. Breakthroughs in future leading-edge technology are expected to provide insights into the regulatory mechanism of male and female gametophyte development at the single-cell level and into basic theoretical research on sterility and fertility restoration mechanisms. (2) Problems such as large differences in pollen amount and pollen viability and a low *Rfo* gene transmission rate still exist in fertility restorer lines [83]. Su et al. (2016) performed bulked segregant RNA-Seq and identified six potential associated genes in minor effect QTLs contributing to fertility instability, which inspired us [206]. Studies have shown that the introgression of exogenous fragments has an effect on genetic variation in chromosomes possessing chromatin fragments and may result in both whole-genome shock and local chromosomal shock [207,208]. The effect of exogenous fragments from *R. sativus* and *B. napus* will be an important direction for future research on the Ogura CMS/FR system.

Author Contributions: Conceptualization, W.R. and Y.Z.; funding acquisition, Y.Z.; writing—original draft, W.R.; writing—review and editing, W.R., J.S., L.C., Z.F., M.Z., H.L., Y.W., J.J., H.Y. and Y.Z. All authors have read and agreed to the published version of the manuscript.

Funding: This work was supported by grants from the National Natural Science Foundation of China (32172578), the earmarked fund for the Modern Agro-Industry Technology Research System, China (CARS-23) and the Science and Technology Innovation Program of the Chinese Academy of Agricultural Sciences (CAAS-ASTIP-IVFCAAS). The work was performed in the Key Laboratory of Biology and Genetic Improvement of Horticultural Crops, Ministry of Agriculture, Beijing 100081, China. No funders were involved in study design, data analysis or writing.

Institutional Review Board Statement: Not applicable.

Informed Consent Statement: Not applicable.

Data Availability Statement: Not applicable.

Acknowledgments: The authors are grateful to all lab members for their useful suggestions, support and encouragement.

Conflicts of Interest: The authors declare no conflict of interest.

References

1. Xu, H.; Kong, X.; Chen, F.; Huang, J.; Lou, X.; Zhao, J. Transcriptome analysis of *Brassica napus* pod using RNA-Seq and identification of lipid-related candidate genes. *BMC Genom.* **2015**, *16*, 858. [CrossRef]
2. Fan, Y.; Zhang, Q. Genetic and molecular characterization of photoperiod and thermo-sensitive male sterility in rice. *Plant Reprod.* **2017**, *31*, 3–14. [CrossRef]
3. Mayr, E. Joseph gottlieb kolreuter's contributions to biology. *Osiris* **1986**, *2*, 135–176. [CrossRef]
4. Kaul, M.L.H. *Male Sterility in Higher Plants*; Springer: New York, NY, USA, 1988.
5. Bhatnagar-Mathur, P.; Gupta, R.; Reddy, P.S.; Reddy, B.P.; Reddy, D.S.; Sameerkumar, C.V.; Saxena, R.K.; Sharma, K.K. A novel mitochondrial *orf147* causes cytoplasmic male sterility in pigeonpea by modulating aberrant anther dehiscence. *Plant Mol. Biol.* **2018**, *97*, 131–147. [CrossRef] [PubMed]
6. Chen, L.; Liu, Y. Male sterility and fertility restoration in crops. *Annu. Rev. Plant Biol.* **2014**, *65*, 579–606. [CrossRef] [PubMed]
7. Carlsson, J.; Leino, M.; Sohlberg, J.; Sundström, J.; Glimelius, K. Mitochondrial regulation of flower development. *Mitochondrion* **2008**, *8*, 74–86. [CrossRef]
8. Tang, H.; Zheng, X.; Li, C.; Xie, X.; Chen, Y.; Chen, L.; Zhao, X.; Zheng, H.; Zhou, J.; Ye, S.; et al. Multi-step formation, evolution, and functionalization of new cytoplasmic male sterility genes in the plant mitochondrial genomes. *Cell Res.* **2017**, *27*, 130–146. [CrossRef]
9. Kaminski, P.; Dyki, B.; Stepowska, A.A. Improvement of cauliflower male sterile lines with *Brassica nigra* cytoplasm, phenotypic expression and possibility of practical application. *J. Agric. Sci.* **2012**, *4*, 190. [CrossRef]
10. Yamagishi, H.; Bhat, S.R. Cytoplasmic male sterility in Brassicaceae crops. *Breed. Sci.* **2014**, *64*, 38–47. [CrossRef]
11. Heng, S.; Gao, J.; Wei, C.; Chen, F.; Li, X.; Wen, J.; Yi, B.; Ma, C.; Tu, J.; Fu, T.; et al. Transcript levels of *orf288* are associated with the hau cytoplasmic male sterility system and altered nuclear gene expression in *Brassica juncea*. *J. Exp. Bot.* **2018**, *69*, 455–466. [CrossRef]
12. Makaroff, C.A.; Palmer, J.D. Mitochondrial DNA rearrangements and transcriptional alterations in the male-sterile cytoplasm of Ogura radish. *Mol. Cell. Biol.* **1988**, *8*, 1474–1480.
13. Makaroff, C.A.; Apel, I.J.; Palmer, J.D. The *atp6* coding region has been disrupted and a novel reading frame generated in the mitochondrial genome of cytoplasmic male-sterile radish. *J. Biol. Chem.* **1989**, *264*, 11706–11713. [CrossRef]
14. Bonhomme, S.; Budar, F.; Férault, M.; Pelletier, G.A. 2.5 kb *NcoI* fragment of Ogura radish mitochondrial DNA is correlated with cytoplasmic male-sterility in Brassica cybrids. *Curr. Genet.* **1991**, *19*, 121–127. [CrossRef]
15. Bonhomme, S.; Budar, F.; Lancelin, D.; Small, I.; Defrance, M.C.; Pelletier, G. Sequence and transcript analysis of the *Nco2.5* Ogura-specific fragment correlated with cytoplasmic male sterility in Brassica cybrids. *Mol. Gen. Genet.* **1992**, *235*, 340–348. [CrossRef]
16. Singh, S.; Dey, S.S.; Bhatia, R.; Kumar, R.; Behera, T.K. Current understanding of male sterility systems in vegetable Brassicas and their exploitation in hybrid breeding. *Plant Reprod.* **2019**, *32*, 231–256. [CrossRef]
17. Grelon, M.; Budar, F.; Bonhomme, S.; Pelletier, G. Ogura cytoplasmic male-sterility (CMS)-associated *orf138* is translated into a mitochondrial membrane polypeptide in male-sterile Brassica cybrids. *Mol. Gen. Genet.* **1994**, *243*, 540–547. [CrossRef]
18. Krishnasamy, S.; Makaroff, C.A. Organ-specific reduction in the abundance of a mitochondrial protein accompanies fertility restoration in cytoplasmic male-sterile radish. *Plant Mol. Biol.* **1994**, *26*, 935–946. [CrossRef]
19. Terachi, T.; Yamaguchi, K.; Yamagishi, H. Sequence analysis on the mitochondrial *orfB* locus in normal and Ogura male-sterile cytoplasms from wild and cultivated radishes. *Curr. Genet.* **2001**, *40*, 276–281. [CrossRef]
20. Tanaka, Y.; Tsuda, M.; Yasumoto, K.; Yamagishi, H.; Terachi, T. A complete mitochondrial genome sequence of Ogura-type male-sterile cytoplasm and its comparative analysis with that of normal cytoplasm in radish (*Raphanus sativus* L.). *BMC Genom.* **2012**, *13*, 352. [CrossRef]
21. Pelletier, G.; Budar, F. Brassica Ogu-INRA cytoplasmic male sterility: An example of successful plant somatic fusion for hybrid seed production. In *Somatic Genome Manipulation*; Li, X.Q., Donnelly, D., Jensen, T., Eds.; Springer: New York, NY, USA, 2015.
22. Landgren, M.; Zetterstrand, M.; Sundberg, E.; Glimelius, K. Alloplasmic male-sterile Brassica lines containing *B. tournefortii* mitochondria express an ORF 3′ of the *atp6* gene and a 32 kDa protein. *Plant Mol. Biol.* **1996**, *32*, 879–890. [CrossRef]
23. Brown, G.G. Unique aspects of cytoplasmic male sterility and fertility restoration in *Brassica napus*. *J. Hered.* **1999**, *90*, 351–356. [CrossRef]
24. Menassa, R.; Homme, Y.L.; Brown, G.G. Post-transcriptional and developmental regulation of a CMS-associated mitochondrial gene region by a nuclear restorer gene. *Plant J.* **1999**, *17*, 491–499. [CrossRef]
25. Brown, G.G.; Domaj, M.; DuPauw, M.; Jean, M.; Li, X.; Landry, B.S. Molecular analysis of Brassica CMS and its application to hybrid seed production. In Proceedings of the International Symposium Brassica 97, Xth Crucifer Genetics Workshop, Rennes, France, 1 April 1998.
26. Yang, J.; Liu, X.; Yang, X.; Zhang, M. Mitochondrially-targeted expression of a cytoplasmic male sterility-associated *orf220* gene causes male sterility in *Brassica juncea*. *BMC Plant Biol.* **2010**, *10*, 231. [CrossRef]
27. Jing, B.; Heng, S.; Tong, D.; Wan, Z.; Fu, T.; Tu, J.; Ma, C.; Yi, B.; Wen, J.; Shen, J. A male sterility-associated cytotoxic protein ORF288 in *Brassica juncea* causes aborted pollen development. *J. Exp. Bot.* **2012**, *63*, 1285–1295. [CrossRef]
28. Lee, Y.; Park, S.; Lim, C.; Kim, H.; Lim, H.; Ahn, Y.; Sung, S.; Yoon, M.; Kim, S. Discovery of a novel cytoplasmic male-sterility and its restorer lines in radish (*Raphanus sativus* L.). *Theor. Appl. Genet.* **2008**, *117*, 905–913. [CrossRef]

29. Lee, Y.; Kim, S.; Lim, H.; Ahn, Y.; Sung, S. Identification of mitochondrial genome rearrangements unique to novel cytoplasmic male-sterility in radish (*Raphanus sativus* L.). *Theor. Appl. Genet.* **2009**, *118*, 719–728. [CrossRef]
30. Park, J.Y.; Lee, Y.P.; Lee, J.; Choi, B.S.; Kim, S.; Yang, T.J. Complete mitochondrial genome sequence and identification of a candidate gene responsible for cytoplasmic male sterility in radish (*Raphanus sativus* L.) containing DCGMS cytoplasm. *Theor. Appl. Genet.* **2013**, *126*, 1763–1774. [CrossRef]
31. Wan, Z.; Jing, B.; Tu, J.; Ma, C.; Shen, J.; Yi, B.; Wen, J.; Huang, T.; Wang, X.; Fu, T. Genetic characterization of a new cytoplasmic male sterility system (hau) in *Brassica juncea* and its transfer to *B. napus*. *Theor. Appl. Genet.* **2008**, *116*, 355–362. [CrossRef]
32. Kang, L.; Li, P.; Wang, A.; Ge, X.; Li, Z. A novel cytoplasmic male sterility in *Brassica napus* (inap CMS) with carpelloid stamens via protoplast fusion with Chinese woad. *Front. Plant Sci.* **2017**, *8*, 529. [CrossRef]
33. Sakai, T.; Imamura, J. Alteration of mitochondrial genomes containing *atpA* genes in the sexual progeny of cybrids between *Raphanus sativus* CMS line and *Brassica napus* cv. *Westar*. *Theor. Appl. Genet.* **1992**, *84*, 923–929. [CrossRef]
34. Iwabuchi, M.; Koizuka, N.; Fujimoto, H.; Sakai, T.; Imamura, J. Identification and expression of the kosena radish (*Raphanus sativus* cv. *Kosena*) homologue of the Ogura radish CMS-associated gene, *orf138*. *Plant Mol. Biol.* **1999**, *39*, 183–188. [CrossRef] [PubMed]
35. Oshima, M.; Kikuchi, R.; Imamura, J.; Handa, H. Origin of the CMS gene locus in rapeseed cybrid mitochondria: Active and inactive recombination produces the complex CMS gene region in the mitochondrial genomes of *Brassicaceae*. *Genes Genet. Syst.* **2010**, *85*, 311–318. [CrossRef] [PubMed]
36. Prakash, S.; Kirti, P.B.; Bhat, S.R.; Gaikwad, K.; Kumar, V.D.; Chopra, V.L. A Moricandia arvensis-based cytoplasmic male sterility and fertility restoration system in *Brassica juncea*. *Theor. Appl. Genet.* **1998**, *97*, 488–492. [CrossRef]
37. Kumar, P.; Kumar, D.V.; Sharma, P.C.; Prakash, S.; Bhat, S.R. A novel *orf108* co-transcribed with the *atpA* gene is associated with cytoplasmic male sterility in *Brassica juncea* carrying Moricandia arvensis cytoplasm. *Plant Cell Physiol.* **2008**, *49*, 284–289.
38. Ashutosh, P.; Sharma, P.C.; Prakash, S.; Bhat, S.R. Identification of AFLP markers linked to the male fertility restorer gene of CMS (Moricandia arvensis) *Brassica juncea* and conversion to SCAR marker. *Theor. Appl. Genet.* **2007**, *114*, 385–392. [CrossRef]
39. Bisht, D.S.; Chamola, R.; Nath, M.; Bhat, S.R. Molecular mapping of fertility restorer gene of an alloplasmic CMS system in *Brassica juncea* containing Moricandia arvensis cytoplasm. *Mol. Breed.* **2015**, *35*, 14. [CrossRef]
40. Bang, S.W.; Kaneko, Y.; Matsuzawa, Y.; Bang, K.S. Breeding of *Moricandia arvensis* monosomic chromosome addition lines (2n = 19) of alloplasmic (*M. arvensis*) *Raphanus sativus*. *Breed. Sci.* **2002**, *52*, 193–199. [CrossRef]
41. Tsutsui, K.; Jeong, B.H.; Ito, Y.; Bang, S.W.; Kaneko, Y. Production and characterization of an alloplasmic and monosomic addition line of *Brassica rapa* carrying the cytoplasm and one chromosome of *Moricandia arvensis*. *Breed. Sci.* **2011**, *61*, 373–379. [CrossRef]
42. Shinada, T.; Kikucho, Y.; Fujimoto, R.; Kishitani, S. An alloplasmic male-sterile line of *Brassica oleracea* harboring the mitochondria from *Diplotaxis muralis* expresses a novel chimeric open reading frame, *orf72*. *Plant Cell Physiol.* **2006**, *47*, 549–555. [CrossRef]
43. L'Homme, Y.; Stahl, R.J.; Li, X.; Hameed, A.; Brown, G.G. Brassica nap cytoplasmic male sterility is associated with expression of a mtDNA region containing a chimeric gene similar to the pol CMS associated *orf224* gene. *Curr. Genet.* **1997**, *31*, 325–335.
44. Huq, M.A.; Akter, S.; Jung, Y.J.; Nou, I.S.; Cho, Y.G.; Kang, K.K. Genome sequencing, a milestone for genomic research and plant breeding. *Plant Breed. Biotechnol.* **2016**, *4*, 29–39. [CrossRef]
45. Liu, J.; Xiang, R.; Wang, W.; Mei, D.; Li, Y.; Mason, A.S.; Fu, L.; Hu, Q. Cytological and molecular analysis of Nsa CMS in *Brassica napus* L. *Euphytica* **2015**, *206*, 279–286. [CrossRef]
46. Hu, Q.; Andersen, S.B.; Dixelius, C.; Hansen, L.N. Production of fertile intergeneric somatic hybrids between *B. napus* and *S. arvensis* for the enrichment of rapeseed gene pool. *Plant Cell Rep.* **2002**, *21*, 147–152.
47. Wei, W.L.; Wang, H.Z.; Liu, G.H. Anatomical observations of anther development of NCa, a cytoplasmic male sterile line in rapeseed (*Brassica napus* L.). *Sci. Agric. Sin.* **2005**, *38*, 1232–1237.
48. Nahm, S.H.; Lee, H.J.; Lee, S.W.; Joo, G.Y.; Harn, C.H.; Yang, S.G.; Min, B.W. Development of a molecular marker specific to a novel CMS line in radish (*Raphanus sativus* L.). *Theor. Appl. Genet.* **2005**, *111*, 1191–1200. [CrossRef]
49. Yang, J.; Zhang, M.; Yu, J. Mitochondrial *nad2* gene is cotranscripted with CMS-associated *orfB* gene in cytoplasmic male-sterile stem mustard (*Brassica juncea*). *Mol. Biol. Rep.* **2009**, *36*, 345–351. [CrossRef]
50. Yasumoto, K.; Terachi, T.; Yamagishi, H. A novel *Rf* gene controlling fertility restoration of Ogura male sterility by RNA processing of *orf138* found in Japanese wild radish and its STS markers. *Genome* **2009**, *52*, 495–504. [CrossRef]
51. Desloire, S.; Gherbi, H.; Laloui, W.; Marhadour, S.; Clouet, V.; Cattolico, L.; Falentin, C.; Giancola, S.; Renard, M.; Budar, F. Identification of the fertility restoration locus, *Rfo*, in radish, as a member of the pentatricopeptide-repeat protein family. *EMBO Rep.* **2003**, *4*, 588–594. [CrossRef]
52. Wang, Z.; Yu, J.; Chang, P.; Shi, Y.; Yuan, Z.; Guo, P.; Wang, T. A new fertility restorer locus linked closely to the *Rfo* locus for cytoplasmic male sterility in radish. *Theor. Appl. Genet.* **2008**, *117*, 313–320. [CrossRef]
53. Wang, Z.; Wang, C.; Cai, Q.; Mei, S.; Gao, L.; Zhou, Y.; Wang, T. Identification of promoter exchange at a male fertility restorer locus for cytoplasmic male sterility in radish (*Raphanus sativus* L.). *Mol. Breed.* **2017**, *37*, 82. [CrossRef]
54. Yamagishi, H.; Jikuya, M.; Okushiro, K.; Hashimoto, A.; Fukunaga, A.; Takenaka, M.; Terachi, T. A single nucleotide substitution in the coding region of Ogura male sterile gene, *orf138*, determines effectiveness of a fertility restorer gene, *Rfo*, in radish. *Mol. Genet. Genom.* **2021**, *296*, 705–717. [CrossRef]
55. Yang, S.; Terachi, T.; Yamagishi, H. Inhibition of chalcone synthase expression in anthers of *Raphanus sativus* with Ogura male sterile cytoplasm. *Ann. Bot.* **2008**, *102*, 483–489. [CrossRef]

56. Sun, X.; Wang, Y.; Chen, Y.; Xu, L.; Jiang, L.; Gong, Y.; Wang, L.; Liu, L. Differential gene expression profiling of Ogura CMS line and its maintainer in radish (*Raphanus sativus* L.). *Acta Physiol. Plant.* **2013**, *35*, 3413–3425. [CrossRef]
57. Uyttewaal, M.; Arnal, N.; Quadrado, M.; Martin-Canadell, A.; Vrielynck, N.; Hiard, S.; Gherbi, H.; Bendahmane, A.; Budar, F.; Mireau, H. Characterization of *Raphanus sativus* pentatricopeptide repeat proteins encoded by the fertility restorer locus for Ogura cytoplasmic male sterility. *Plant Cell* **2009**, *20*, 3331–3345. [CrossRef]
58. Bannerot, H.; Boulidard, L.; Cauderon, Y.; Tempe, J. Transfer of cytoplasmic male sterility from *Raphanus sativus* to *Brassica oleracea*. *Proc. Eucarpia Meet. Crucif.* **1974**, *25*, 52–54.
59. Murayama, S.; Habuchi, T.; Yamagishi, H.; Terachi, T. Identification of a sequence-tagged site (STS) marker linked to a restorer gene for Ogura cytoplasmic male sterility in radish (*Raphanus sativus* L.) by non-radioactive AFLP analysis. *Euphytica* **2002**, *129*, 61–68. [CrossRef]
60. Kim, S.; Lim, H.; Park, S.; Cho, K.H.; Sung, S.K.; Oh, D.G.; Kim, K.T. Identification of a novel mitochondrial genome type and development of molecular markers for cytoplasm classification in radish (*Raphanus sativus* L.). *Theor. Appl. Genet.* **2007**, *115*, 1137–1145. [CrossRef] [PubMed]
61. Zhang, L.; Wang, Q.; Wang, Y. Molecular marker assisted selective breeding of radish male sterile line. *China Veg.* **2020**, *2*, 40–44.
62. Ogura, H. Studies on the new male sterility in Japanese radish, with special references to the utilization of this sterility towards the practical raising of hybrid seeds. *Mem. Fac. Agric. Kagoshima Univ.* **1967**, *6*, 39–78.
63. Kim, Y.J.; Zhang, D. Molecular control of male fertility for crop hybrid breeding. *Trends Plant Sci.* **2017**, *23*, 53–65. [CrossRef] [PubMed]
64. Forsythe, E.S.; Nelson, A.; Beilstein, M.A. Biased gene retention in the face of introgression obscures species relationships. *Genome Biol. Evol.* **2020**, *12*, 1646–1663. [CrossRef]
65. Pelletier, G.; Primard, C.; Vedel, F.; Chetrit, P.; Remy, R.; Rousselle; Renard, M. Intergeneric cytoplasmic hybridization in Cruciferae by protoplast fusion. *Mol. Gen. Genet.* **1983**, *191*, 244–250. [CrossRef]
66. Li, Y.; Cai, M.; Zheng, Y.; Hu, Q.; Liu, G. Systematic study on Ogu CMS and test for restorer-maintenancer relationship in *B. napus*. *Chin. J. Oil Crop Sci.* **1993**, *3*, 8–11.
67. Li, X.; Li, L.; Qin, J.; Wu, S.; Yuan, X.; Wang, J.; Yang, F. Cytogenetic study on the restored material F_2 (TC_1) from test cross progenies of *B. napus* with Ogu CMS. *Sci. Agric. Sin.* **2001**, *34*, 108–110.
68. Wen, Y.; Zhang, S.; Wang, J.; Zhu, J.; Zhao, L.; Cao, J. Screening and preliminary studies of restorers of radish cytoplasmic male sterile lines in *Brassica napus* L. *Acta Agric. Boreali Sin.* **2010**, *25*, 102–106.
69. Chen, W.; Li, M.; Wang, T.; Hui, R.; Tu, J.; Fu, T. Creative process of a new restorer material with Ogu CMS in *Brassica nupus*. *Sci. Agric. Sin.* **2012**, *45*, 1465–1474.
70. Wen, Y.; Zhang, S.; Wang, J.; Zhu, J.; He, J.; Cai, D.; Cao, J.; Zhao, L.; Wang, D. Studies on rapid improvement techniques of radish cytoplasmic male sterile restorers in *Brassica napus* L. *Acta Agric. Boreali Sin.* **2016**, *31*, 102–109.
71. Heyn, F.W. Transfer of restorer genes from *Raphanus* to cytoplasmic male sterile *Brassica napus*. *Eucarpia Crucif. Newsl.* **1976**, *1*, 15–16.
72. Delourme, R.; Eber, F. Linkage between an isozyme marker and a restorer gene in radish cytoplasmic male sterility of rapeseed (*Brassica napus* L.). *Theor. Appl. Genet.* **1992**, *85*, 222–228. [CrossRef]
73. Delourme, R.; Eber, F.; Renard, M. Breeding double low restorer lines in radish cytoplasmic male sterility of rapeseed (*Brassica napus* L.). In Proceedings of the 9 International Rapeseed Congress, Cambridge, UK, 4 July 1995.
74. Primard-Brisset, C.; Poupard, J.P.; Horvais, R.; Eber, F.; Pelletier, G.; Renard, M.; Delourme, R. A new recombined double low restorer line for the Ogu-INRA CMS in rapeseed (*Brassica napus* L.). *Theor. Appl. Genet.* **2005**, *111*, 736–746. [CrossRef]
75. Zhang, Y.; Patel, J.; Tulsieram, L. Brassica Ogura Restorer Lines with Shortened Raphanus Fragment (SRF). U.S. Patent 8466347B2, 13 August 2009.
76. Li, X.; Yang, Y.; Wang, Y.; Luo, P. Rapeseed (*Brassica napus* L) chromosome enineering—transferring characters of *Raphanus sativus* L. var. *raphanistroides makino* to *B. napus*. *J. Sichuan Agric. Univ.* **1995**, *32*, 599–604.
77. Walters, T.W.; Mutschler, M.A.; Earle, E.D. Protoplast fusion-derived Ogura male sterile cauliflower with cold tolerance. *Plant Cell Rep.* **1992**, *10*, 624–628. [CrossRef]
78. Yang, L.; Liu, Y.; Wang, X.; Sun, P.; Zhuang, M.; Fang, Z. Preliminary observation on main botanical characters of cytoplasmic male sterility (CMS) materials in Cabbage. *China Veg.* **1997**, *6*, 24–25.
79. Yang, L.; Fang, Z.; Liu, Y.; Zhuang, M.; Wang, X.; Zhang, Y.; Sun, P. Several new F_1 hybrid varieties and hybrid crossing lines bred with male sterile lines and hybrid lines. *China Veg.* **2005**, *1*, 23–25.
80. Yu, H.; Fang, Z.; Liu, Y.; Yang, L.; Zhuang, M.; Lv, H.; Li, Z.; Han, F.; Liu, X.; Zhang, Y. Development of a novel allele-specific *Rfo* marker and creation of Ogura CMS fertility-restored interspecific hybrids in *Brassica oleracea*. *Theor. Appl. Genet.* **2016**, *129*, 1625–1637. [CrossRef]
81. Yu, H.; Li, Z.; Ren, W.; Han, F.; Yang, L.; Zhuang, M.; Lv, H.; Liu, Y.; Fang, Z.; Zhang, Y. Creation of fertility-restored materials for Ogura CMS in *Brassica oleracea* by introducing *Rfo* gene from *Brassica napus* via an allotriploid strategy. *Theor. Appl. Genet.* **2020**, *133*, 2825–2837. [CrossRef]
82. Yu, H.; Li, Z.; Yang, L.; Liu, Y.; Zhuang, M.; Zhang, L.; Lv, H.; Li, Z.; Han, F.; Liu, X.; et al. Morphological and molecular characterization of the second backcross progenies of Ogu-CMS Chinese kale and rapeseed. *Euphytica* **2017**, *213*, 55. [CrossRef]

83. Ren, W.; Li, Z.; Han, F.; Zhang, B.; Li, X.; Fang, Z.; Yang, L.; Zhuang, M.; Lv, H.; Liu, Y.; et al. Utilization of Ogura CMS germplasm with the clubroot resistance gene by fertility restoration and cytoplasm replacement in *Brassica oleracea* L. *Hortic. Res.* **2020**, *7*, 61. [CrossRef]
84. Li, Q.F.; Xu, B.B.; Du, Y.M.; Peng, A.; Ren, X.S.; Si, J.; Song, H.Y. Development of Ogura CMS restorers in *Brassica oleracea* subspecies via direct Rfo^B gene transformation. *Theor. Appl. Genet.* **2021**, *134*, 1123–1132. [CrossRef] [PubMed]
85. Jourdan, P.S.; Earle, E.D.; Mutschler, M.A. Synthesis of male sterile, triazine-resistant *Brassica napus* by somatic hybridization between cytoplasmic male sterile *B. oleracea* and atrazine-resistant *B. campestris*. *Theor. Appl. Genet.* **1989**, *78*, 445–455. [CrossRef]
86. Jiang, P.; Zhu, Z.; Zheng, Y.; Chen, J.; Chen, W. Selection of alloplasmic cytoplasmic male sterility line C50-2 in cauliflower. *Fujian J. Agric. Sci.* **2001**, *16*, 39–41.
87. Dey, S.S.; Sharma, S.R.; Bhatia, R.; Kumar, P.R.; Parkash, C. Development and characterization of "Ogura" based improved CMS lines of cauliflower (*Brassica oleracea* var. *botrytis* L.). *Indian J. Genet.* **2011**, *71*, 37–42.
88. He, C.; Guo, S.; Zhang, Z.; Li, W.; Qi, J. Preliminary studies on the alloplasmic male sterile materials in cauliflower. *Acta Hortic.* **1999**, *26*, 125–127.
89. Zhu, Y.; Wu, X.; Gong, J.; Jin, H.; Yao, W.; Li, S.; Wu, Q. Breeding and utilization of a male sterile line with radish cytoplasm in broccoli. *J. Nanjing Agric. Univ.* **2001**, *24*, 15–18.
90. Huang, J.; Liu, W.; Liu, Y.; Han, F.; Fang, Z.; Yang, L.; Zhuang, M.; Zhang, Y.; Lv, H.; Wang, Y.; et al. Creation of fertility restorer materials for Ogura CMS broccoli and the study of its genetic background. *Acta Hortic.* **2022**, *49*, 533–547.
91. Delourme, R.; Eber, F.; Renard, M. Transfer of radish cytoplasmic male sterility from *Brassica napus* to *B. juncea* and *B. rapa*. *Crucif. Newsl.* **1994**, *16*, 79.
92. Williams, P.H.; Heyn, F.W.; Talekar, N.S.; Griggs, T.D. *The Origin and Development of Cytoplasmic Male Sterile Chinese Cabbage*; AVRDC: Tainan City, China, 1981.
93. Wei, B.; Wei, Y. Selection and breeding of Ogura male sterile lines of Chinese cabbage. *China Veg.* **1995**, *1*, 18–21.
94. Sun, R. Studies on Male Sterility of Radish Cytoplasmic Chinese Cabbage. Ph.D. Thesis, Chinese Academy of Agricultural Sciences, Beijing, China, 1999.
95. Jiang, S.; Chen, H.; Yang, X.; Yin, Y.; Dong, J. Study on the introgression of cytoplasmic male sterlity in *Brassica campestris* spp. *chinensis*. *China Veg.* **2002**, *1*, 28–29.
96. Hou, X.; Cao, S.; She, J.; Lu, W. Synthesis of cytoplasm hybrid of non-heading Chinese cabbage through asymmetric electric fusion of protoplast cell. *Acta Hortic.* **2001**, *28*, 532–537.
97. Cui, H.; Cao, J.; Zhang, M.; Yao, X.; Xiang, X. Production of the Ogura cytoplasmic male sterile (CMS) lines of Chinese cabbage-pak-choi (*Brassica campestris* L. ssp. *chinensis* var. *communis*) and turnip (*B. campestris* L. ssp. *rapifera*) and cytological observation of their sterile organs. *Acta Hortic.* **2004**, *31*, 467–471.
98. Zhao, L.; Ke, G. Breeding of a radish CMS in Chinese cabbage (RC_7) and the research of its traits. *Acta Bot. Boreali Occident. Sin.* **2007**, *27*, 2404–2410.
99. Tian, E.; Roslinsky, V.; Cheng, B. Molecular marker-assisted breeding for improved Ogura CMS restorer line (*RfoRfo*) and mapping of the restorer gene (*Rfo*) in *Brassica juncea*. *Mol. Breed.* **2014**, *34*, 1361–1371. [CrossRef]
100. Prakash, S.; Kirti, P.; Chopra, V. Cytoplasmic male sterility (CMS) systems other than Ogu and Polima in Brassica: Current status. In Proceedings of the International Rapeseed Congress, Cambridge, UK, 4 July 1995.
101. Kirti, P.B.; Banga, S.S.; Prakash, S.; Chopra, V.L. Transfer of Ogu cytoplasmic male sterility to *Brassica juncea* and improvement of the male sterile line through somatic cell fusion. *Theor. Appl. Genet.* **1995**, *91*, 517–521. [CrossRef] [PubMed]
102. Rao, G.U.; Batra-Sarup, V.; Prakash, S.; Shivanna, K.R. Development of a new cytoplasmic male-sterility system in *Brassica juncea* through wide hybridization. *Plant Breed.* **1994**, *112*, 171–174. [CrossRef]
103. Rawat, D.S.; Anand, I.J. Male sterility in Indian mustard. *Indian J. Genet. Plant Breed.* **1979**, *39*, 412–414.
104. Prakash, S.; Chopra, V.L. Synthesis of alloplasmic *Brassica campestris* as a new source of cytoplasmic male sterility. *Plant Breed.* **2010**, *101*, 253. [CrossRef]
105. Huang, B.; Li, W.; Ju, C.; Zhou, Y. Introduction of Ogura male sterile cytoplasm into *Brassica rapa* L. ssp. *chinensis* var. *purpurea* Baile. and heterosis utilization report. *Seed* **1999**, *3*, 57.
106. Witt, U.; Hansen, S.; Albaum, M.; Abel, W.O. Molecular analyses of the CMS-inducing 'Polima' cytoplasm in *Brassica napus* L. *Curr. Genet.* **1991**, *19*, 323–327. [CrossRef]
107. Yarrow, S.A.; Burnett, L.A.; Wildeman, R.P.; Kemble, R.J. The transfer of 'Polima' cytoplasmic male sterility from oilseed rape (*Brassica napus*) to broccoli (*B. oleracea*) by protoplast fusion. *Plant Cell Rep.* **1990**, *9*, 185–188. [CrossRef]
108. Wang, H.M.; Ketela, T.; Keller, W.A.; Gleddie, S.C.; Brown, G.G. Genetic correlation of the *orf224/atp6* gene region with Polima CMS in Brassica somatic hybrids. *Plant Mol. Biol.* **1995**, *27*, 801–807. [CrossRef]
109. Ke, G.; Song, Y. Breeding of alloplasmic male sterile line in Chinese cabbage (*Brassica campestris* L. ssp. *pekinensis* (Lour) olsson) and its application. *Shaanxi J. Agric. Sci.* **1989**, *3*, 9–11.
110. Ke, G.; Zhao, Z.; Song, Y.; Zhang, L.; Zhao, L. Breeding of alloplasmic male sterile line CMS3411-7 in Chinese cabbage (*Brassica campestris* L. ssp. *pekinensis* (Lour) olsson) and its application. *Acta Hortic. Sin.* **1992**, *19*, 333–340.
111. Stiewe, G.; Robbelen, G. Establishing cytoplasmic male sterility in *Brassica napus* by mitochondrial recombination with *Brassica tournefortii*. *Plant Breed.* **1994**, *113*, 294–304. [CrossRef]

112. Dieterich, J.H.; Braun, H.P.; Schmitz, U.K. Alloplasmic male sterility in *Brassica napus* (CMS 'Tournefortii-Stiewe') is associated with a special gene arrangement around a novel *atp9* gene. *Mol. Genet. Genom.* **2003**, *269*, 723–731. [CrossRef]
113. Zhong, X.H.; Chen, D.H.; Cui, J.; Li, H.L.; Huang, Y.X.; Kang, J.G. Comparative analysis of the complete mitochondrial genome sequences and anther development cytology between maintainer and Ogura-type cytoplasm male-sterile cabbage (*B. oleracea* Var. *capitata*). *BMC Genom.* **2021**, *22*, 646. [CrossRef]
114. Chen, L.; Ren, W.; Zhang, B.; Chen, W.; Fang, Z.; Yang, L.; Zhuang, M.; Lv, H.; Wang, Y.; Ji, J.; et al. Organelle comparative genome analysis reveals novel alloplasmic male sterility with *orf112* in *Brassica oleracea* L. *Int. J. Mol. Sci.* **2021**, *22*, 13230. [CrossRef]
115. Vasupalli, N.; Kumar, V.; Bhattacharya, R.; Bhat, S.R. Analysis of mitochondrial recombination in the male sterile *Brassica juncea* cybrid Og1 and identification of the molecular basis of fertility reversion. *Plant Mol. Biol.* **2021**, *106*, 109–122. [CrossRef]
116. Wang, J.; Jiang, J.; Li, X.; Li, A.; Zhang, Y.; Guan, R.; Wang, Y. Complete sequence of heterogenous-composition mitochondrial genome (*Brassica napus*) and its exogenous source. *BMC Genom.* **2012**, *13*, 675. [CrossRef]
117. He, S.; Abad, A.R.; Gelvin, S.B.; Mackenzie, S.A. A cytoplasmic male sterility-associated mitochondrial protein causes pollen disruption in transgenic tobacco. *Proc. Natl. Acad. Sci. USA* **1996**, *93*, 11763–11768. [CrossRef]
118. Hu, J.; Wang, K.; Huang, W.; Liu, G.; Gao, Y.; Wang, J.; Huang, Q.; Ji, Y.; Qin, X.; Wan, L.; et al. The rice pentatricopeptide repeat protein RF5 restores fertility in Hong-Lian cytoplasmic male-sterile lines via a complex with the glycine-rich protein GRP162. *Plant Cell* **2012**, *24*, 109–122. [CrossRef]
119. Zhang, Q.; Zhang, X.W.; Yuan, Y.X.; Yao, Q.J.; Zhao, Y.Y.; Jiang, W.S.; Wei, X.C.; Wang, Z.Y. Study on molecular mechanism of homeosis in cabbage Ogura CMS. *J. Henan Agric. Sci.* **2015**, *44*, 111–116.
120. Kazama, T.; Okuno, M.; Watari, Y.; Yanase, S.; Koizuka, C.; Tsuruta, Y.; Sugaya, H.; Toyoda, A.; Itoh, T.; Tsutsumi, N.; et al. Curing cytoplasmic male sterility via TALEN-mediated mitochondrial genome editing. *Nat. Plants* **2019**, *5*, 722–730. [CrossRef]
121. Bohra, A.; Jha, U.C.; Adhimoolam, P.; Bisht, D.; Singh, N.P. Cytoplasmic male sterility (CMS) in hybrid breeding in field crops. *Plant Cell Rep.* **2016**, *35*, 967–993. [CrossRef]
122. Brown, G.G.; Formanová, N.; Jin, H.; Wargachuk, R.; Dendy, C.; Patil, P.; Laforest, M.; Zhang, J.; Cheung, W.Y.; Landry, B.S. The radish *Rfo* restorer gene of Ogura cytoplasmic male sterility encodes a protein with multiple pentatricopeptide repeats. *Plant J.* **2003**, *35*, 262–272. [CrossRef]
123. Koizuka, N.; Imai, R.; Fujimoto, H.; Hayakawa, T.; Kimura, Y.; KohnoMurase, J.; Sakai, T.; Imamura, J. Genetic characterization of a pentatricopeptide repeat protein gene, *orf687*, that restores fertility in the cytoplasmic male-sterile Kosena radish. *Plant J.* **2003**, *34*, 407–415. [CrossRef]
124. Bett, K.E.; Lydiate, D.J. Mapping and genetic characterization of loci controlling the restoration of male fertility in Ogura CMS radish. *Mol. Breed.* **2004**, *13*, 125–133. [CrossRef]
125. Wang, Z.; Wang, C.; Gao, L.; Mei, S.; Zhou, Y.; Xiang, C.; Wang, T. Heterozygous alleles restore male fertility to cytoplasmic male-sterile radish (*Raphanus sativus* L.): A case of overdominance. *J. Exp. Bot.* **2013**, *64*, 2041–2048. [CrossRef]
126. Wang, Z.; Wang, C.; Mei, S.; Gao, L.; Zhou, Y.; Wang, T. An insertion-deletion at a pentatricopeptide repeat locus linked to fertility transition to cytoplasmic male sterility in radish (*Raphanus sativus* L.). *Mol. Breed.* **2015**, *35*, 108. [CrossRef]
127. Giancola, S.; Marhadour, S.; Desloire, S.; Clouet, V.; Falentin-Guyomarc'h, H.; Laloui, W.; Falentin, C.; Pelletier, G.; Renard, M.; Bendahmane, A. Characterization of a radish introgression carrying the Ogura fertility restorer gene *Rfo* in rapeseed, using the Arabidopsis genome sequence and radish genetic mapping. *Theor. Appl. Genet.* **2003**, *107*, 1442–1451. [CrossRef] [PubMed]
128. Wang, Q.; Zhang, L.; Wen, C.; Yang, J. Development and application of high-throughput SNP markers for Ogura-CMS fertility restorer gene in radish. *Acta Hortic. Sin.* **2017**, *44*, 1309–1318.
129. Feng, J.; Primomo, V.; Li, Z.; Zhang, Y.; Jan, C.C.; Tulsieram, L.; Xu, S.S. Physical localization and genetic mapping of the fertility restoration gene *Rfo* in canola (*Brassica napus* L.). *Genome* **2009**, *52*, 401–407. [CrossRef]
130. Wang, T.; Guo, Y.; Wu, Z.; Xia, S.; Hua, S.; Tu, J.; Li, M.; Chen, W. Genetic characterization of a new radish introgression line carrying the restorer gene for Ogura CMS in *Brassica napus*. *PLoS ONE* **2020**, *15*, e0236273. [CrossRef]
131. Wei, W.; Wang, H.; Liu, G. Molecular identification of the sterile cytoplasm of NCa of a cytoplasmic male sterile line in rapeseed (*Brassica napus* L.). *Sci. Agric. Sin.* **2005**, *38*, 1965–1972.
132. Zhang, S.H. Identification of New Restored Lines in Ogura CMS and Analysis of Transfer of the Gene *Rfo* in *Brassica napus*. Master's Thesis, Huazhong Agricultural University, Wuhan, China, 2013.
133. Szała, L.; Sosnowska, K.; Popławska, W.; Liersch, A.; Olejnik, A.; Kozłowska, K.; Bocianowski, J.; Cegielska-Taras, T. Development of new restorer lines for CMS ogura system with the use of resynthesized oilseed rape (*Brassica napus* L.). *Breed. Sci.* **2016**, *66*, 516–521. [CrossRef]
134. Zhu, Q.; Kang, Z.L.; Jian, Y.; Ding, Y.; Kang, J. The molecular characteristics of Ogura cytoplasmic male sterility related gene *orf138* in cabbage (*Brassica oleracea* var. capitata). *Sci. Bull.* **2012**, *28*, 104–109.
135. Zhang, Y.; Wang, X.; Li, C.; Song, H.; Ren, X.; Si, J. Molecular identification of *Brassica oleracea* CMS and the morphology response of flower to nuclear background. *Acta Hortic. Sin.* **2010**, *37*, 915–922.
136. Shu, J.; Liu, Y.; Li, Z.; Zhang, L.; Fang, Z.; Yang, L.; Zhuang, M.; Zhang, Y.; Lv, H. Detection of the diversity of cytoplasmic male sterility sources in broccoli (*Brassica oleracea* var. *italica*) using mitochondrial markers. *Front. Plant Sci.* **2016**, *7*, 927. [CrossRef]
137. Zhang, R.; Hu, S.; Yan, J.; Sun, G. Cytoplasmic diversity in *Brassica rapa* L. investigated by mitochondrial markers. *Genet. Resour. Crop Evol.* **2013**, *60*, 967–974. [CrossRef]

138. Niemelä, T.; Seppänen, M.; Badakshi, F.; Rokka, V.M.; Heslop-Harrison, J.P. Size and location of radish chromosome regions carrying the fertility restorer *Rfk1* gene in spring turnip rape. *Chromosome Res.* **2012**, *20*, 353–361. [CrossRef]
139. Niemelä, T.; Seppänen, M.; Jauhiainen, L.; Tulisalo, U. Transfer of the Kosena *Rfk1* gene, required in hybrid seed production, from oilseed rape to turnip rape. *Euphytica* **2010**, *175*, 1–12. [CrossRef]
140. Gudi, S.; Atri, C.; Goyal, A.; Kaur, N.; Akhtar, J.; Mittal, M.; Kaur, K.; Kaur, G.; Banga, S.S. Physical mapping of introgressed chromosome fragment carrying the fertility restoring (*Rfo*) gene for Ogura CMS in *Brassica juncea* L. Czern & Coss. *Theor. Appl. Genet.* **2020**, *133*, 2949–2959.
141. Hu, X.; Sullivan-Gilbert, M.; Kubik, T.; Danielson, J.; Hnatiuk, N.; Marchione, W.; Greene, T.; Thompson, S.A. Mapping of the Ogura fertility restorer gene *Rfo* and development of *Rfo* allele-specific markers in canola (*Brassica napus* L.). *Mol. Breed.* **2008**, *22*, 663–674. [CrossRef]
142. Liu, Z.H.; Shi, X.Y.; Li, S.; Hu, G.; Zhang, L.L.; Song, X.Y. Tapetal-delayed programmed cell death (PCD) and oxidative stress-induced male sterility of *Aegilops uniaristata* cytoplasm in wheat. *Int. J. Mol. Sci.* **2018**, *19*, 1708. [CrossRef]
143. Hanson, M.R.; Bentolila, S. Interactions of mitochondrial and nuclear genes that affect male gametophyte development. *Plant Cell* **2004**, *16*, 154–169. [CrossRef]
144. Warmke, H.E.; Lee, S.J. Pollen abortion in T cytoplasmic male sterility corn (*Zea mays*): A suggested machanism. *Science* **1978**, *200*, 561–563. [CrossRef]
145. Duroc, Y.; Hiard, S.; Vrielynck, N.; Ragu, S.; Budar, F. The Ogura sterility-inducing protein forms a large complex without interfering with the oxidative phosphorylation components in rapeseed mitochondria. *Plant Mol. Biol.* **2009**, *70*, 123–137. [CrossRef]
146. Melonek, J.; Duarte, J.; Martin, J.; Beuf, L.; Murigneux, A.; Varenne, P.; Comadran, J.; Specel, S.; Levadoux, S.; Bernath-Levin, K.; et al. The genetic basis of cytoplasmic male sterility and fertility restoration in wheat. *Nat. Commun.* **2021**, *12*, 1036. [CrossRef] [PubMed]
147. Kong, X.J.; Khan, A.; Li, Z.L.; You, J.Y.; Munsif, F.; Kang, H.D.; Zhou, R.Y. Identification of chalcone synthase genes and their expression patterns reveal pollen abortion in cotton. *Saudi J. Biol. Sci.* **2020**, *27*, 3691–3699. [CrossRef] [PubMed]
148. González-Melendi, P.; Uyttewaal, M.; Morcillo, C.N.; Hernández Mora, J.R.; Fajardo, S.; Budar, F.; Lucas, M.M. A light and electron microscopy analysis of the events leading to male sterility in Ogu-INRA CMS of rapeseed (*Brassica napus*). *J. Exp. Bot.* **2008**, *59*, 827–838. [CrossRef] [PubMed]
149. Kang, J.; Guo, Y.; Chen, Y., Li, H.; Zhang, L.; Liu, H. Upregulation of the AT-hook DNA binding gene *BoMF2* in OguCMS anthers of *Brassica oleracea* suggests that it encodes a transcriptional regulatory factor for anther development. *Mol. Biol. Rep.* **2014**, *41*, 2005–2014. [CrossRef]
150. Xie, Y.; Zhang, W.; Wang, Y.; Xu, L.; Zhu, X.; Muleke, E.; Liu, L. Comprehensive transcriptome-based characterization of differentially expressed genes involved in microsporogenesis of radish CMS line and its maintainer. *Funct. Integr. Genom.* **2016**, *16*, 529–543. [CrossRef]
151. Yamagishi, H. Distribution and allelism of restorer genes for Ogura cytoplasmic male sterility in wild and cultivated radishes. *Genes Genet. Syst.* **1998**, *73*, 79–83. [CrossRef]
152. Duroc, Y.; Gaillard, C.; Hiard, S.; Defrance, M.C.; Pelletier, G.; Budar, F. Biochemical and functional characterization of *orf138*, a mitochondrial protein responsible for Ogura cytoplasmic male sterility in Brassiceae. *Biochimie* **2005**, *87*, 1089–1100. [CrossRef]
153. Duroc, Y.; Gaillard, C.; Hiard, S.; Tinchant, C.; Berthomé, R.; Pelletier, R.; Budar, F. Nuclear expression of a cytoplasmic male sterility gene modifies mitochondrial morphology in yeast and plant cells. *Plant Sci.* **2006**, *170*, 755–767. [CrossRef]
154. Sara, Z.; Alberto, F.; Enrico, G.; Luciano, X.; Marianna, F.; Giovanni, M.; Diana, B.; Mario, P.; Massimo, D. Characterization of transcriptional complexity during berry development in *Vitis vinifera* using RNA-Seq. *Plant Physiol.* **2010**, *152*, 1787–1795.
155. Guo, J.; Wang, P.; Cheng, Q.; Sun, L.; Wang, H.; Wang, Y.; Kao, L.; Li, Y.; Qiu, T.; Yang, W.; et al. Proteomic analysis reveals strong mitochondrial involvement in cytoplasmic male sterility of pepper (*Capsicum annuum* L.). *J. Proteom.* **2017**, *168*, 15–27. [CrossRef]
156. Wang, P.; Lu, Q.; Ai, Y.; Wang, Y.; Li, T.; Wu, L.; Liu, J.; Cheng, Q.; Sun, L.; Shen, H. Candidate gene selection for cytoplasmic male sterility in pepper (*Capsicum annuum* L.) through whole mitochondrial genome sequencing. *Int. J. Mol. Sci.* **2019**, *20*, 578. [CrossRef]
157. Yang, Z.; Li, G.; Tieman, D.; Zhu, G. Genomics approaches to domestication studies of horticultural crops. *Hortic. Plant J.* **2019**, *5*, 240–246. [CrossRef]
158. Lv, J.; Liu, Z.; Liu, Y.; Ou, L.; Deng, M.; Wang, J.; Song, J.; Ma, Y.; Chen, W.; Zhang, Z.; et al. Comparative transcriptome analysis between cytoplasmic male-sterile line and its maintainer during the floral bud development of pepper. *Hortic. Plant J.* **2020**, *6*, 89–98. [CrossRef]
159. Lin, S.; Miao, Y.; Su, S.; Xu, J.; Jin, L.; Sun, D.; Peng, R.; Huang, L.; Cao, J. Comprehensive analysis of Ogura cytoplasmic male sterility-related genes in turnip (*Brassica rapa* ssp. *rapifera*) using RNA sequencing analysis and bioinformatics. *PLoS ONE* **2019**, *14*, e0218029. [CrossRef]
160. Dong, X.; Kim, W.; Lim, Y.; Kim, Y.; Hur, Y. Ogura-CMS in Chinese cabbage (*Brassica rapa* ssp. *pekinensis*) causes delayed expression of many nuclear genes. *Plant Sci.* **2013**, *199–200*, 7–17. [CrossRef]
161. Sheoran, I.S.; Sawhney, V.K. Proteome analysis of the normal and Ogura (ogu) CMS anthers of *Brassica napus* to identify proteins associated with male sterility. *Botany* **2010**, *88*, 217–230. [CrossRef]

162. Mei, S.; Liu, T.; Wang, Z. Comparative transcriptome profile of the cytoplasmic male sterile and fertile floral buds of radish (*Raphanus sativus* L.). *Int. J. Mol. Sci.* **2016**, *17*, 42. [CrossRef]
163. Wang, S.; Wang, C.; Zhang, X.; Chen, X.; Liu, J.; Jia, X.; Jia, S. Transcriptome de novo assembly and analysis of differentially expressed genes related to cytoplasmic male sterility in cabbage. *Plant Physiol. Biochem.* **2016**, *105*, 224–232. [CrossRef]
164. Shu, J.; Zhang, L.; Liu, Y.; Li, Z.; Fang, Z.; Yang, L.; Zhuang, M.; Zhang, Y.; Lv, H. Normal and abortive buds transcriptomic profiling of broccoli ogu cytoplasmic male sterile line and its maintainer. *Int. J. Mol. Sci.* **2018**, *19*, 2501. [CrossRef]
165. Xing, M.; Sun, C.; Li, H.; Hu, S.; Lei, L.; Kang, J. Integrated analysis of transcriptome and proteome changes related to the Ogura cytoplasmic male sterility in cabbage. *PLoS ONE* **2018**, *13*, e0193462. [CrossRef] [PubMed]
166. Han, F.; Zhang, X.; Yang, L.; Zhuang, M.; Zhang, Y.; Li, Z.; Fang, Z.; Lv, H. iTRAQ-based proteomic analysis of Ogura-CMS cabbage and its maintainer line. *Int. J. Mol. Sci.* **2018**, *19*, 3180. [CrossRef] [PubMed]
167. Zhang, C.; Qi, L.; Hou, X.; Shi, G.; Zhang, J. Differential gene expression analysis of a new Ogura CMS line and its maintainer in non-heading Chinese cabbage by cDNA-AFLP. *Acta Physiol. Plant.* **2010**, *32*, 781–787. [CrossRef]
168. Jeong, S.W.; Yi, H.; Song, H.; Lee, S.S.; Hur, Y. Chlorosis of Ogura-CMS *Brassica rapa* is due to down-regulation of genes for chloroplast proteins. *J. Plant Biotechnol.* **2017**, *44*, 115–124. [CrossRef]
169. Fujii, S.; Yamada, M.; Fujita, M.; Itabashi, E.; Hamada, K.; Yano, K.; Kurata, N.; Toriyama, K. Cytoplasmic-nuclear genomic barriers in rice pollen development revealed by comparison of global gene expression profiles among five independent cytoplasmic male sterile lines. *Plant Cell Physiol.* **2010**, *51*, 610–620. [CrossRef] [PubMed]
170. Lavy, M.; Prigge, M.J.; Tao, S.; Shain, S.; Kuo, A.; Kirchsteiger, K.; Estelle, M. Constitutive auxin response in Physcomitrella reveals complex interactions between Aux/IAA and ARF proteins. *eLife* **2016**, *5*, e13325. [CrossRef]
171. Ding, B.; Hao, M.; Mei, D.; Zaman, Q.; Sang, S.; Wang, H.; Wang, W.; Fu, L.; Cheng, H.; Hu, Q. Transcriptome and hormone comparison of three cytoplasmic male sterile systems in *Brassica napus*. *Int. J. Mol. Sci.* **2018**, *19*, 4022. [CrossRef]
172. Wei, X.; Zhang, X.; Yao, Q.; Yuan, Y.; Li, X.; Wei, F.; Zhao, Y.; Zhang, Q.; Wang, Z.; Jiang, W.; et al. The miRNAs and their regulatory networks responsible for pollen abortion in Ogura-CMS Chinese cabbage revealed by high-throughput sequencing of miRNAs, degradomes, and transcriptomes. *Front. Plant Sci.* **2015**, *6*, 894. [CrossRef]
173. Wei, X.; Lv, Y.; Zhao, Y.; Nath, U.; Yuan, Y.; Wang, Z.; Yang, S.; Jia, H.; Wei, F.; Zhang, X. Comparative transcriptome analysis in Chinese cabbage (*Brassica rapa* ssp. *pekinesis*) for DEGs of Ogura-, Polima-CMS and their shared maintainer. *Mol. Biol. Plants* **2020**, *26*, 719–731. [CrossRef]
174. Hu, L.; Zhang, X.; Yuan, Y.; Wang, Z.; Yang, S.; Li, R.; Nath, U.; Zhao, Y.; Tian, B.; Shi, G.; et al. Comparative transcriptome identifies gene expression networks regulating developmental pollen abortion in Ogura cytoplasmic male sterility in Chinese cabbage (*Brassica rapa* ssp. *pekinensis*). *Horticulturae* **2021**, *7*, 157. [CrossRef]
175. Imai, R.; Koizuka, N.; Fujimoto, H.; Hayakawa, T.; Sakai, T.; Imamura, J. Delimitation of the fertility restorer locus *Rfk1* to a 43-kb contig in Kosena radish (*Raphanus sativus* L.). *Mol. Genet. Genom.* **2003**, *269*, 388–394. [CrossRef]
176. Fujii, S.; Bond, C.S.; Small, I.D. Selection patterns on restorer-like genes reveal a conflict between nuclear and mitochondrial genomes throughout angiosperm evolution. *Proc. Natl. Acad. Sci. USA* **2011**, *108*, 1723–1728. [CrossRef]
177. Ning, L.; Wang, H.; Li, D.; Li, Y.; Chen, K.; Chao, H.; Li, H.; He, J.; Li, M. Genome-wide identification of the restorer-of-fertility-like (*RFL*) gene family in *Brassica napus* and expression analysis in Shaan2a cytoplasmic male sterility. *BMC Genom.* **2020**, *21*, 765. [CrossRef]
178. Zhang, Z.; Cui, X.; Wang, Y.; Wu, J.; Gu, X.; Lu, T. The RNA editing factor *WSP1* is essential for chloroplast development in rice. *Mol. Plant* **2017**, *10*, 86–98. [CrossRef]
179. Ito, A.; Sugita, C.; Ichinose, M.; Kato, Y.; Yamamoto, H.; Shikanai, T.; Sugita, M. An evolutionarily conserved P-subfamily pentatricopeptide repeat protein is required to splice the plastid *ndhA* transcript in the moss *Physcomitrella patens* and *Arabidopsis thaliana*. *Plant J.* **2018**, *94*, 638–648. [CrossRef]
180. McCormick, S. Binding sites for pentatricopeptide repeat proteins differentially activate chloroplast transgenes. *Plant J.* **2018**, *94*, 6–7. [CrossRef]
181. Sun, Y.; Huang, J.; Zhong, S.; Gu, H.; He, S.; Qu, L. Novel DYW-type pentatricopeptide repeat (PPR) protein BLX controls mitochondrial RNA editing and splicing essential for early seed development of *Arabidopsis*. *J. Genet. Genom.* **2018**, *45*, 155–168. [CrossRef]
182. Qin, X.; Warguchuk, R.; Arnal, N.; Gaborieau, L.; Mireau, H.; Brown, G.G. In vivo functional analysis of a nuclear restorer PPR protein. *BMC Plant Biol.* **2014**, *14*, 313. [CrossRef]
183. Imai, R.; Koizuka, N.; Fujimoto, H.; Hayakawa, T.; Kimura, Y.; KohnoMurase, J.; Sakai, T.; Imamura, J. Identification and characterization of a fertility restorer gene, *rfk1*, for Kosena cms. (3) Analysis of allelic polymorphism relevant to *rfk1* gene function. *Breed. Res.* **2002**, *4*, 184.
184. Wang, C.; Lezhneva, L.; Arnal, N.; Quadrado, M.; Mireau, H. The radish Ogura fertility restorer impedes translation elongation along its cognate CMS-causing mRNA. *Proc. Natl. Acad. Sci. USA* **2021**, *118*, e2105274118. [CrossRef]
185. Bellaoui, M.; Pelletier, G.; Budar, F. The steady-state level of mRNA from the Ogura cytoplasmic male sterility locus in Brassica cybrids is determined post-transcriptionally by its 3′ region. *EMBO J.* **1997**, *16*, 5057–5068. [CrossRef] [PubMed]
186. Bellaoui, M.; Grelon, M.; Pelletier, G.; Budar, F. The restorer *Rfo* gene acts post-translationally on the stability of the ORF138 Ogura CMS-associated protein in reproductive tissues of rapeseed cybrids. *Plant Mol. Biol.* **1999**, *40*, 893–902. [CrossRef] [PubMed]

187. Shu, J.S.; Liu, Y.M.; Zhang, L.L.; Li, Z.S.; Fang, Z.Y.; Yang, L.M.; Zhuang, M.; Zhang, Y.Y.; Lv, H.H. Evaluation and selection of sources of cytoplasmic male sterility in broccoli. *Euphytica* **2019**, *215*, 125. [CrossRef]
188. Singh, S.; Bhatia, R.; Kumar, R.; Das, A.; Ghemeray, H.; Behera, T.K.; Dey, S.S. Characterization and genetic analysis of OguCMS and doubled haploid based large genetic arsenal of Indian cauliflowers (*Brassica oleracea* var. *botrytis* L.) for morphological, reproductive and seed yield traits revealed their breeding potential. *Genet. Resour. Crop Evol.* **2021**, *68*, 1603–1623. [CrossRef]
189. Mower, J.P.; Sloan, D.; Alverson, A. Plant mitochondrial genome diversity: The genomics revolution. In *Plant Genome Diversity*; Springer: Vienna, Austria, 2012; Volume 1, pp. 123–144.
190. Bi, C.W.; Paterson, A.H.; Wang, X.L.; Xu, Y.Q.; Wu, D.Y.; Qu, Y.S.; Jiang, A.N.; Ye, Q.L.; Ye, N. Analysis of the complete mitochondrial genome sequence of the diploid cotton gossypium raimondii by comparative genomics approaches. *BioMed Res. Int.* **2016**, *2016*, 5040598. [CrossRef]
191. Ducos, E.; Touzet, P.; Boutry, M. The male sterility G cytoplasm of wild beet displays modified mitochondrial respiratory complexes. *Plant J.* **2001**, *26*, 171–180. [CrossRef]
192. Kubo, T.; Kitazaki, K.; Matsunaga, M.; Kagami, H.; Mikami, T. Male sterility-inducing mitochondrial genomes: How do they differ? *Crit. Rev. Plant Sci.* **2011**, *30*, 378–400. [CrossRef]
193. Heng, S.P.; Wei, C.; Jing, B.; Wan, Z.J.; Wen, J.; Yi, B.; Ma, C.Z.; Tu, J.X.; Fu, T.D.; Shen, J.X. Comparative analysis of mitochondrial genomes between the hau cytoplasmic male sterility (CMS) line and its iso-nuclear maintainer line in *Brassica juncea* to reveal the origin of the CMS-associated gene *orf288*. *BMC Genom.* **2014**, *15*, 322. [CrossRef]
194. Chen, Z.W.; Zhao, N.; Li, S.S.; Grover, C.E.; Nie, H.; Wendel, J.F.; Hua, J.P. Plant mitochondrial genome evolution and cytoplasmic male sterility. *Crit. Rev. Plant Sci.* **2017**, *36*, 55–69. [CrossRef]
195. Jo, Y.D.; Choi, Y.; Kim, D.H.; Kim, B.D.; Kang, B.C. Extensive structural variations between mitochondrial genomes of CMS and normal peppers (*Capsicum annuum* L.) revealed by complete nucleotide sequencing. *BMC Genom.* **2014**, *15*, 561. [CrossRef]
196. Wallet, C.; Le Ret, M.; Bergdoll, M.; Bichara, M.; Dietrich, A.; Gualberto, J.M. The RECG1 DNA translocase is a key factor in recombination surveillance, repair, and segregation of the mitochondrial DNA in Arabidopsis. *Plant Cell* **2015**, *27*, 2907–2925. [CrossRef]
197. Tkadlec, J.; Pavlogiannis, A.; Chatterjee, K.; Nowak, M.A. Population structure determines the tradeoff between fixation probability and fixation time. *Commun. Biol.* **2019**, *2*, 138. [CrossRef]
198. Rubanova, N.; Pinna, G.; Kropp, J.; Campalans, A.; Radicella, J.P.; Polesskaya, A., Harel-Bellan, A.; Morozova, N. MasterPATH: Network analysis of functional genomics screening data. *BMC Genom.* **2020**, *21*, 632. [CrossRef]
199. Takatsuka, A.; Kazama, T.; Arimura, S.I.; Toriyama, K. TALEN-mediated depletion of the mitochondrial gene *orf312* proves that it is a Tadukan-type cytoplasmic male sterility-causative gene in rice. *Plant J.* **2022**, *110*, 994–1004. [CrossRef]
200. Badhan, S.; Ball, A.S.; Mantri, N. First report of CRISPR/Cas9 mediated DNA-free editing of *4CL* and *RVE7* genes in Chickpea Protoplasts. *Int. J. Mol. Sci.* **2021**, *22*, 396. [CrossRef]
201. Yoo, B.C.; Yadav, N.S.; Orozco, E.M., Jr.; Sakai, H. Cas9/gRNA-mediated genome editing of yeast mitochondria and Chlamydomonas chloroplasts. *PeerJ* **2020**, *8*, e8362. [CrossRef]
202. Kang, B.C.; Bae, S.J.; Lee, S.; Lee, J.S.; Kim, A.; Lee, H.; Baek, G.; Seo, H.; Kim, J.; Kim, J.S. Chloroplast and mitochondrial DNA editing in plants. *Nat. Plants* **2021**, *7*, 899–905. [CrossRef]
203. Li, B.; Li, Y.; Cao, D.; Zong, Y.; Fan, S.; Jiang, Y. Methods for Breeding Male Sterile Lines of Plants Using Mitochondrial Gene Editing System. China Patent 25217475, 28 May 2021.
204. Xiao, S.; Zang, J.; Pei, Y.; Liu, J.; Liu, J.; Song, W.; Shi, Z.; Su, A.; Zhao, J.R.; Chen, H. Activation of mitochondrial *orf355* gene expression by a nuclear-encoded DREB transcription factor causes cytoplasmic nale sterility in maize. *Mol. Plant* **2020**, *13*, 1270–1283. [CrossRef]
205. Zhang, H.; Yang, H.; Hu, D.; Li, B.; Lin, Y.; Yao, W.; Guo, Z.; Li, H.; Ding, D.; Zhang, Z.; et al. Single-cell RNA sequencing of meiocytes and microspores reveals the involvement of the *Rf4* gene in redox homeostasis of CMS-C maize. *Crop J.* **2021**, *9*, 1237–1247. [CrossRef]
206. Su, A.; Song, W.; Xing, J.; Zhao, Y.; Zhang, R.; Li, C.; Duan, M.; Luo, M.; Shi, Z.; Zhao, J. Identification of genes potentially associated with the fertility instability of S-type cytoplasmic male sterility in maize via bulked segregant RNA-Seq. *PLoS ONE* **2016**, *11*, e0163489. [CrossRef] [PubMed]
207. Wang, M.; Ji, Y.; Feng, S.; Liu, C.; Xiao, Z.; Wang, X.; Wang, Y.; Xia, G. The non-random patterns of genetic variation induced by asymmetric somatic hybridization in wheat. *BMC Plant Biol.* **2018**, *18*, 244. [CrossRef] [PubMed]
208. Xu, W.; Li, Y.; Li, Y.; Liu, C.; Wang, Y.; Xia, G.; Wang, M.C. Asymmetric somatic hybridization affects synonymous codon usage bias in wheat. *Front. Genet.* **2021**, *12*, 682324. [CrossRef] [PubMed]

MDPI AG
Grosspeteranlage 5
4052 Basel
Switzerland
Tel.: +41 61 683 77 34

International Journal of Molecular Sciences Editorial Office
E-mail: ijms@mdpi.com
www.mdpi.com/journal/ijms

www.ingramcontent.com/pod-product-compliance
Lightning Source LLC
Chambersburg PA
CBHW061838210926
51850CB00006B/918

* 9 7 8 3 7 2 5 8 5 3 7 5 5 *